Y0-CCU-407

The Dow Jones-Irwin Handbook of Telecommunications

The Dow Jones-Irwin Handbook of Telecommunications

James Harry Green

DOW JONES-IRWIN
Homewood, Illinois 60430

ISBN 0-87094-661-7

Library of Congress Catalog Card No. 85-52445

Printed in the United States of America

1 2 3 4 5 6 7 8 9 0 K 3 2 1 0 9 8 7 6

To my family:
Pat, Jim, Becky, Sally, Peggy, Kristy

PREFACE

The telephone industry is a little like Abe Lincoln's ax—three new handles and two new heads, but still the same old ax. In the hundred plus years since Alexander Graham Bell uttered his famous words, "Watson, come here; I want you," the telephone industry has undergone changes unlike any other industry in our society, and the transformation continues.

Three powerful forces are converging at this moment to wrest the telephone industry from its first century of steady but unspectacular change into the next century of radical reconstruction. These forces—technology, deregulation, and the merger of telecommunications and the computer—are the driving forces that will dramatically reconfigure the nation's telecommunications network from a single entity owned and operated by the Bell System and connecting independent telephone companies to multiple networks owned by companies that will compete vigorously for a growing amount of business.

Changes as dramatic as the ones we are currently experiencing are not always welcome because they force us to reexamine our fundamental assumptions and to modify the methods of operation we have become comfortable with.

Moreover, these changes force us to learn more about the technology to use it effectively.

Some technologies can be used effectively with little or no understanding of how they function. For example, one can gain the benefits of the aviation industry with no understanding of aerodynamics, air traffic control, or airline tariffs. From 1876 when Bell introduced the telephone, and from 1945 when the Electronic Numerical Integrator and Calculator (ENIAC) first went into service in Los Alamos, we could safely ignore telecommunications and computer technologies unless we were one of the experts. But that changed in about 1980. If you are in a business or government organization, you are forced to learn to use the computer and communications. To be sure, experts will handle the details of making them work, but you as a manager will manage the experts on your behalf, and that requires understanding their technology.

The jury is still out on the question of whether these changes are beneficial for the public as a whole, but for large organizations, the limited choices of the past have been replaced by a minefield of options. To complicate the problem, the companies that owned the telecommunications equipment and furnished the services in the past also controlled a substantial portion of the talent. Before the historic 1968 Carterfone decision, in which the FCC ruled that the telephone companies could not prohibit the interconnection of customer-owned equipment, the bulk of the people with telecommunications knowledge worked for the Bell System or one of the large manufacturers who deal with independent telephone companies and overseas markets. Telecommunications technology was largely taught by the companies themselves to students with backgrounds in engineering or with experience gained on the job. Because the industry was largely inbred, telecommunications knowledge was more limited than computer knowledge, which is learned by most university graduates today.

Changes in the regulatory outlook were not the only forces driving the change in telecommunications to a more open technology—microelectronics also played a significant role. Older analog equipment used specialized components such as filters, repeat coils, terminal strips, and patch panels that had little or no commercial application outside the telecommunications industry. Because commercial components were not available, telecommunications equipment was manufactured by a few companies capable of designing and fabricating their own ingredients. The advent of microelectronics and of digital telephony changed that. Components and circuits that comprised computers became the ingredients of modern telecommunications equipment. Encouraged by the loosening of regulatory restraints, hundreds of companies turned their electronic skills toward the enormous telecommunications market.

The third factor driving the change has been the advent of distributed data processing. Telecommunications networks are becoming extensions of computer resources that are, in turn, being removed from their hermetically sealed environments and placed among the people they are designed to support. Telecommunications, which is defined as the movement of information, becomes the transport mechanism for the data produced by distributed computer systems.

While these changes have brought options that were previously unavailable, they have also brought the hazards of the trade down to the working level. There are the inevitable hucksters that clothe their products in a cloak of jargon that is confusing and often intimidating. And there are the inevitable compatibility problems. Telecommunications equipment has a lot in common with computers in that respect. It is not purchased as a unit but is supplied as a series of separate assemblies that either act harmoniously or not at all. These changes have also brought the problems of preannounced or underdeveloped products to the market, and telecommunications managers are discovering, to their chagrin, that software bugs are no longer the exclusive province of their brethren in the computer room.

The days are gone when an organization merely turns its telecommunications problems over to the telephone company and has a single point of contact when things go wrong. Although many people can get by knowing how to work the system but not how the system works, every large organization is rapidly approaching the time when someone must have the expertise to deal with telecommunications affairs.

The alternative, of course, is to turn these affairs over to an outside concern, but even that requires knowledge of how to choose a company and evaluate its work.

This book is addressed to people who must know how the U.S. telecommunications network functions and how its parts interact. It is a book for users, not for engineers, although engineers who understand in depth how the equipment of their specialty functions may find this book a useful explanation of the rest of the network. Network technology is explained in nonmathematical terms. To deal effectively in the world of telecommunications it is essential to speak the language, if only as a means of self-preservation. This volume explains the meaning of technical terms, both in context and in the glossary found at the end of each chapter.

Each chapter is designed to stand largely on its own; however, the telecommunications network is a system, and a system cannot be understood by merely inspecting its components. Thus, a certain amount of cross-referencing is inevitable to avoid repetition. Each chapter follows a similar format. After a short introduction, the essentials of the technology and the necessary terminology are explained. Next is a section on standards that explains the source of the criteria against which that portion of the network is designed. This is followed by an applications section that is in two parts. Where appropriate, a case history is presented of an organization that has applied the technology to a private network. The second portion of the applications section is a summary of the most important criteria in selecting equipment or evaluating the services described in the chapter. The final three sections are a glossary, a bibliography, and a listing of some manufacturers of products covered in the chapter.

Three appendixes are included to aid readers who may be taking their first plunge into the world of telecommunications. Appendix A is a brief overview of the principles of electricity as they are applied to telecommunications. Appendix B lists the sources of technical information published by standards agencies and manufacturers who are referenced in the applications and standards sections. Appendix C lists the addresses and telephone numbers of the manufacturers who are referenced in the product listings in each chapter.

Readers should not consider these product listings as complete. Space does not permit an exhaustive listing of all products. Instead, the chapters list the dominant manufacturers and some who are less dominant, but who fill a particular niche in the market.

Although this volume is addressed primarily to telecommunications managers, it will be equally valuable to anyone who wants an overview of the elements of telecommunications and who wants to understand how the nation's networks function. It should therefore be useful as an introductory text or as a reference for people who must sell, acquire, or apply telecommunications equipment. Readers should be cautioned that each of the 25 chapters in this book contains subjects that are complex enough to warrant a volume in themselves. In fact, volumes have been written on each of these topics, but most are written for engineers and scientists, and not for users.

This book emphasizes breadth rather than depth. Readers with a high level of knowledge about a subject will object that the technology is more complex than I have portrayed it here. In this they will be correct because I have attempted to favor clarity over precision to give readers a background in what they need to know to apply the technology without burdening them with a discussion of exceptions to the general case.

With the rate of change in the telecommunications industry on an exponential curve, it is likely that many changes will occur that cannot be foreseen at the time of this writing. The final chapter in this book presents trends that are apparent in 1985, but some may fail to materialize, or, more likely, they will materialize in a different direction than they are foreseen today. Whatever occurs, we are today in the midst of exciting changes that have their foundation in the concepts that are presented in this volume.

James Harry Green

ACKNOWLEDGMENTS

It would be impossible to name all the friends and associates in the Bell System and its suppliers who have, over the years, taught me something of this technology. In the preparation of this book I am particularly indebted to several active and retired employees of Pacific Northwest Bell who consented to review portions of this manuscript, and whose advice clarified concepts and rectified errors that had crept into the writing. Dean Doel, Arthur P. Lanier, Larry Zanella, Richard A. Greulich, and George J. Williams struggled through early drafts of this book and offered comments that are deeply appreciated.

I am also indebted for contributions made by others to portions of this book. Dr. Joseph G. Robertson, Director of Telecommunications at Rockwell International, was generous with his time and writing talents in describing his company's extensive private network. Don Jacobs, Manager of the Advanced Communications Technology Group at Martin Marietta Data Systems, described their application of fiber optics in a private network. Carolyn Miller of Hercules Corporation and Fritz Witti of Satellite Business Systems provided information that helped me to understand the Hercules Integrated Telecommunications System. Stan Distel and

George Shadduck of the AT&T Communications Consultant Liaison Program were most generous with time and assistance to explain AT&T's products and services. I also appreciate the assistance of Robert W. Swanson, Technical Support Manager at San Diego State University, in furnishing information about their local area network.

Finally, thanks also go to Dominick Abel, my agent, for his assistance and encouragement in this project, and to Kristy Green who spent long hours at the word processor in preparing the manuscript for the publisher.

J.H.G.

CONTENTS

List of Figures xxii
List of Tables xxviii

1 Introduction to Telecommunications 1

The Major Telecommunications Systems: *Customer Premises Equipment. Subscriber Loop Plant. Local Switching Systems. Interoffice Trunks. Tandem Switching Offices. Interexchange Trunks. Transmission Equipment.* Fundamentals of Multiplexing: *Frequency Division Multiplexing. Time Division Multiplexing. Higher Order Multiplexing. Data Multiplexing.* Analog and Digital Transmission Concepts. Switching Systems: *Early Switching Systems. Common Control Switching Offices. Computer Controlled Switching Systems. Digital Central Offices.* Numbering Systems: *The Switching Hierarchy. The Nationwide Numbering Plan. The Worldwide Numbering Plan.* Interexchange Carrier Access to Local Networks. Private Telephone Systems.

2 Transmission Concepts 22

Transmission Impairments: *Volume or Level. Noise. Bandwidth. Echo. Amplitude Distortion. Envelope De-*

lay Distortion. Elements of Transmission Design: *Variables in Transmission Quality. Subscriber Loop Transmission. Toll Connecting Trunks. Intertoll Trunks. Local Interoffice Trunks. Special-Purpose Trunks. Loss and Noise Grade of Service.* Transmission Measurements: *Loss Measurements. Noise Measurements. Return Loss Measurements. Envelope Delay Measurements.*

3 Data Communications Systems 56

Data Network Facilities: *Point-to-Point Circuit. Circuit Switching. Message Switching. Packet Switching.* Data Communications Fundamentals: *Coding. Data Communications Speeds. Modulation Methods. Full- and Half-Duplex Mode. Synchronizing Methods.* Evaluating Data Communication Services: *Modem Selection. Data Network Facilities. Reliability and Availability. Error Detection and Correction. Protocols. Throughput. Line Conditioning.*

4 Pulse Code Modulation and Digital Carrier 106

Digital Carrier Technology: *Digital Transmission Facilities. Digital Signal Timing.* The T–1 Carrier System: *Channel Banks. Digital Cross-Connect Panel. Distributing Frame. T Carrier Lines.* Digital Cross-Connect Systems. The Digital Signal Hierarchy. T–1 Data Multiplexers. T Carrier Data Compression Systems: *T Carrier Line Encoding. Data Compression Equipment. Delta Modulation.*

5 Frequency Division Multiplex 139

Analog Carrier Technology: *The L Multiplex System. Short-Haul Carrier Systems.* Analog to Digital Connectors. Time Assignment Speech Interpolation.

6 Station Equipment 156

Telephone Set Technology: *Elements of a Telephone Set. Protection. Coin Telephones. Cordless Telephones. Voice Recording Equipment. Multiple Line Equipment.*

Network Channel Terminating Equipment. Station Wiring Plans.

7 Outside Plant 189

Outside Plant Technology: *Supporting Structures. Cable Characteristics. Loop Resistance Design. Feeder and Distribution Cable. Subscriber Carrier. Range Extenders. Cable Pressurization.* Electrical Protection.

8 Signaling Systems 212

Signaling Technology: *Signaling System Overview. E and M Signaling. Signaling Irregularities.* Direct Current Signaling Systems. Trunk Signaling Systems: *Single-Frequency Signaling. Addressing Signals. Common Channel Signaling. Private Line Signaling. Coin Telephone Signaling.*

9 Circuit Switched Network Systems 230

Network Terminology. Network Architecture: *The Changing Network Environment. Switching System Architecture.* Switching System Control. Switching Networks. Direct Control Switching Systems. Common Control Central Offices. Stored Program Control: *Electronic Switching Networks. Other Switching Networks. Comparison of Digital and Analog Switching Networks.* Line, Trunk, and Service Circuits: *Line Circuit Functions. Trunk Circuits. Service Circuits.* Access to the Local Network: *Predivestiture Network Architecture. Beginnings of Competition in the Toll Network. Equal Access. Local Access Transport Areas.*

10 Local Switching Systems 265

Digital Central Office Technology: *Distributed Processing. SPC Central Office Memory Units. Redundancy. Maintenance and Administrative Features. Line Equipment Features. Trunk Equipment Features.* Local Central Office Equipment Features: *Alarm and Trouble Indicating Systems. Automatic Number Identification (ANI). Automatic Message Accounting (AMA). Coin*

Telephone Interface. Common Equipment. Local Measured Service (LMS). Traffic Measuring Equipment. Network Management. Local Central Office Service Features.

11 Private Branch Exchanges 290

PBX Technology: *Contrasts between PBXs and Local Central Offices. Voice/Data Integration in the PBX.* Interfaces: *Line Interfaces. Ping Pong. PBX Trunk Interfaces. Tie Trunks. Special Trunks. Gateways and External Interfaces. Modem Pooling. Code, Speed, and Protocol Conversion.* Principal PBX Features: *Least-Cost Routing (LCR). Station Message Detail Recording (SMDR). Voice Store and Forward (Voice Mail). PBX Voice Features. Attendant Features. Automatic Call Distributors. Office Automation Features.*

12 Tandem Switching Systems 319

Tandem Trunking Facilities. Tandem Switch Technology: *Wideband Switching Capability.* Tandem Switch Features: *Public Tandem Switch Features. Private Tandem Switching Machine Features.*

13 Network Design Concepts 353

The Network Design Problem: *Queuing Theory. Traffic Load. Busy Hour Determination. Grade of Service. Traffic Measurements. Alternate Routing. Simulation. Network Topology.* Data Network Design. Network Management: *Network Administration Tools. Automatic Network Controls.* Network Management: *Network Administration Tools. Automatic Network Controls.*

14 Power, Distributing Frames, and Other Common Equipment 378

Cable Racking and Interbay Wiring. Distributing Frames: *Protector Frames. Combined and Miscellaneous Distributing Frames.* Ringing and Tone Supplies. Alarm and Control Equipment. Power Equipment.

15 Microwave Radio **391**

Microwave Characteristics. Microwave Technology: *Modulation Methods. Data under Voice (DUV). Bit Error Rate. Diversity. Microwave Impairments. Heterodyning versus Baseband Repeaters. Multiplex Interface. Microwave Antennas, Waveguides, and Towers. Entrance Links. Digital Termination System. Short-Haul Digital Microwave.*

16 Lightwave Communications **416**

Lightwave Technology: *Lightguide Cables. Fiber Optic Terminal Equipment. Wavelength Division Multiplexing. Star Couplers.* Lightwave System Design Criteria: *Information Transfer Rate. System Attenuation and Losses. Cutoff Wavelength.* Free Space Lightwave Transmission.

17 Satellite Communications **440**

Satellite Technology: *Physical Structure. Transponders. Attitude Control Apparatus. Power Supply. Telemetry Equipment. Station Keeping Equipment.* Earth Station Technology: *Radio Relay Equipment. Satellite Communications Control. Satellite Transmission. Rain Absorption. Sun Transit Outage. Interference. Carrier to Noise Ratio.* Representative Satellite Services: *Satellite Business Services. International Maritime Satellite Service (INMARSAT). Direct Broadcast Satellite (DBS).*

18 Data Communications Networks **471**

Point-to-Point Network Technology: *Addressing. Polling. Point-to-Point Digital Facilities.* Packet Network Technology: *Packet Switching Nodes. Packet Assembly and Disassembly. Virtual Circuits. Network Access Methods. CCITT X.25 Protocol.* Value Added Networks: *Public Packet Switched Data Networks.* Local Area Data Transport. Switched 56 KB/S Service. Network Control Centers.

19 Local Area Networks **502**

Network Topology. Access Method: *Contention Access. Noncontention Access.* Modulation Methods: *Baseband. Broadband.* Transmission Media: *Twisted Pair Wire or Ribbon Cable. Coaxial Cable. Fiber Optics. Microwave Radio. Light.* Throughput. Size of the Network. Interconnecting LANs with Other Networks. Local Area Network Standards: *The IEEE 802 Committee. The CSMA/CD LAN IEEE 802.3. The Token Bus LAN IEEE 802.4. The Token Ring LAN IEEE 802.5.*

20 Video Systems **535**

Video Technology: *Cable Television Systems. Two-Way CATV Systems. Video Compression. Freeze-Frame Video. High-Definition Television.* Video Services and Applications: *Cable Television Services. Video Conferencing.*

21 Facsimile Transmission **561**

Facsimile Technology: *Facsimile Machine Characteristics. Scanners. Printing. Special Telecommunications Features. Group 4 Facsimile.*

22 Mobile Radio **578**

Conventional Mobile Telephone Technology. Cellular Mobile Radio Technology: *Cell-Site Operation. Mobile Telephone Serving Office. Mobile Units. Mobile Telephone Features. Cellular Radio Services.* Radio Paging. Mobile Data Transmission.

23 Testing Principles **601**

Test Access Methods. Analog Circuit Testing Principles: *Loss Measurements. Noise Measurements. Envelope Delay. Return Loss. Phase Jitter. Peak-to-Average Ratio. Harmonic Distortion.* Subscriber Loop Measurements: *Local Test Desks. Automatic Testing. Manual Loop Tests. Network Interface Devices.* Trunk Transmission Measurements: *Manual Switched Circuit Test*

Systems. Automatic Switched Circuit Test Systems. Tests on Special Service Trunks. Data Circuit Testing: *Interface Tests. Loopback Tests. End-to-End Tests. Protocol Analyzers. PCM Test Sets. Technical Control.*

24 Network Management **630**

System Records: *Equipment Identification. Equipment Documentation. Location Records. Interconnection Records.* System Usage Management: *Line Usage Measurements. Traffic Usage Measurements.* Trouble Handling. Telecommunications Costs. Service Monitoring. Network Reconfiguration. Preventing Failures.

25 Future Developments in Telecommunications **644**

Development of Network Intelligence. The Integrated Services Digital Network (ISDN): *Typical ISDN Services.* The Information Revolution. An Explosion in the Use of Personal Computers. Digital Communications. Changes in Local Loops. Voice/Data Integration: *Voice/Data Integration in the Switching System. Voice/Data Integration in Transmission Facilities.* Further Deregulation: *The Advent of Equal Access. Effect on Resellers. Coin Telephone Deregulation. Competition in the Local Network.*

Appendix A Principles of Electricity Applied to Telecommunications **663**

Appendix B Sources of Technical Information **684**

Appendix C List of Selected Manufacturers and Vendors of Telecommunications Products and Services **688**

Index **701**

LIST OF FIGURES

1.1 Major Classes of Telecommunications Equipment 3
1.2 Phantom Telephone Circuit 7
1.3 Frequency Division Multiplexing System 8
1.4 Time Division Multiplexing System 8
1.5 Switching Hierarchy 14
2.1 Typical Telecommunications Circuit 23
2.2 Typical Long-Haul Connection 27
2.3 Digital Echo Canceler 29
2.4 Voice Channel Attenuation 30
2.5 Transmission Levels 32
2.6 C Message Weighting Response Curve 33
2.7 Loss of 10,000 Feet of 26-Gauge Cable 36
2.8 Noise Current Flow in Cable 37
2.9 Effects of Thermal and Intermodulation Noise on Circuit Noise 40
3.1 Polled Multidrop Network 58
3.2 Circuit Switched Network 59
3.3 Packet Switched Network 60
3.4 Parallel and Serial Data Conversion 61
3.5 Data Modulation Methods 65
3.6 Asynchronous Data Transmission 66
3.7 Synchronous Data Transmission (IBM SDLC Frame) 67
3.8 Time Division Multiplexing 74
3.9 Statistical Multiplexing 75
3.10 Network Configuration 76
3.11 Data Network Topologies 77
3.12 Character Parity 80
3.13 Longitudinal Redundancy Checking 81
3.14 International Standards Organization Open Systems Interconnection (OSI) Protocol Model 84
3.15 Effect of Line Error Rate on Throughput 89
3.16 IBM Systems Network Architecture 95
4.1 Wescom 360 D4 Channel Bank 108
4.2 Voice Sampling 109
4.3 Companding in a PCM Channel Bank 109
4.4 PCM Frame 110

4.5	Wescom 3440–01 T–1 Line Repeater	111
4.6	Block Diagram of T–1 Carrier System	112
4.7	Block Diagram of a PCM Channel Bank	113
4.8	Block Diagram of Dataport Application	116
4.9	T Carrier Line Signals and Faults	118
4.10	Block Diagram of T Carrier Regenerator	119
4.11	U.S. Digital Signal Hierarchy	122
4.12	M1–3 Multiplexer	123
4.13	General Datacom Megamux Plus T–1 Multiplexer	124
4.14	T–1 Multiplexer Layout Showing Drop-and-Insert Capability and Submultiplexing Low-Speed Channels	125
4.15	Comparison of Pulse Code Modulation and Adaptive Differential Pulse Code Modulation	127
4.16	Adaptive Differential Pulse Code Modulated Transcoder	128
5.1	Single Sideband Suppressed Carrier (SSBSC) Channel Block Diagram	143
5.2	Basic L–600 Mastergroup	144
5.3	Direct-to-Line Multiplex Unit	145
5.4	N–1 Carrier Modulation Plan	148
5.5	120-Channel Transmultiplexer	150
6.1	Functional Diagram of a Telephone Set	158
6.2	DTMF Dialing Frequency Combinations	159
6.3	Multiparty Signaling in a Telephone System	161
6.4	Diagram of a Station Protector	163
6.5	Code-A-Phone Model 2570 Answering Set	165
6.6	Diagram of 1A Key Telephone System	168
6.7	Electronic Key Telephone Instrument	169
6.8	NCTE on User's Premises	173
6.9	Station Wiring Plan	175
7.1	Diagram of Outside Plant	190
7.2	Multipair Cable	193
7.3	Aerial Splice Cases	194
7.4	Feeder and Distribution Service Areas	197
7.5	Terminals	198
7.6	Bridged Tap in Cable Pairs	199
7.7	Toroidal Load Coil	199
7.8	Block Diagram of Remote Line Concentrator	201

7.9	Station and Central Office Protection Equipment	204
7.10	Multipair Protected Building Entrance Terminal	206
8.1	Signaling on an Interoffice Connection	214
8.2	Single Frequency Signaling Simplified Block Diagram	219
8.3	Common Channel Signaling System	222
8.4	End-to-End versus Link-by-Link Signaling	226
9.1	Direct and Tandem Trunks in Single-Level and Hierarchical Networks	233
9.2	Switching Hierarchy	234
9.3	Block Diagram of a Switching System	235
9.4	Switching Network Diagram	238
9.5	Direct Control Switching System (Step-by-Step)	239
9.6	Block Diagram of Crossbar Switching Machine	242
9.7	Crossbar Switch	243
9.8	Crossbar Switch Schematic Diagram	244
9.9	Block Diagram of Stored Program Switching Machine	247
9.10	Time Multiplexed Bus System	249
9.11	Time Slot Interchange	250
9.12	Digital Switching Network	251
9.13	Feature Group A and D Access to Local Telephone Network	257
9.14	Equal Access through Telephone Company Provided Tandem	260
10.1	Major Components of a Digital Central Office	268
10.2	Digital Central Office Line Circuit Architecture	274
10.3	Digital Remote Line Equipment	276
11.1	Conceptual Diagram of a Fourth Generation PBX	293
11.2	Comparison of Call Attempts and Holding Time for Voice and Data in a PBX	296
11.3	PBX to LAN Interface	300
12.1	Use of a Digital Tandem Switch for Alternate Voice and Video Service	322

12.2 Automatic Remote Trunk Testing System 325
12.3 Traffic Operator Service Position Access 326
12.4 Sources of Transmission Problems with Remote Access to a Tandem Switch 329
12.5 Rockwell Integrated Digital Network 345
13.1 Hourly Variation in Calls for a Typical Local Central Office 355
13.2 Waiting Time as a Function of Circuit Occupancy 357
13.3 Poisson Distribution of Call Arrivals 359
13.4 Exponential Distribution of Service Times 360
13.5 Load/Service Curve of Typical Common Control Equipment 361
13.6 Circuit Capacity as a Function of Size of Circuit Group 365
13.7 Sources and Dispositions of Network Load 366
13.8 High-Usage and Final Routes in a Three-Node Network 366
13.9 Distribution of Holding Time Probability in a Data Network 368
13.10 Multinode Concentrator Data Network 369
13.11 Multidrop Polled Network 371
13.12 Multidrop Polled Network Generalized Response Time Model 372
14.1 Central Office Distributing Frame 380
14.2 Protector Module 382
14.3 Central Office Power Plant 385
15.1 Microwave Antennas 393
15.2 Direct and Reflected Microwave Paths between Antennas 394
15.3 Microwave Protection System 399
15.4 Microwave Path Profile 401
15.5 Channel Dropping at Main Microwave Repeater Stations 403
15.6 General Model of an Analog Microwave System 404
15.7 Antenna Radiation Patterns 405
16.1 Block Diagram of a Typical Fiber Optic System 419
16.2 Spectral Loss for a Typical Optical Fiber 420

16.3	Light Ray Paths through a Step Index Optical Fiber	421
16.4	Wave Propagation through Different Types of Optical Fiber	422
16.5	Multiple Strand Fiber Optic Cable	423
16.6	Splicing Fiber Optic Cable	424
16.7	Fiber Optic Terminal	426
16.8	Wave Division Multiplexer	427
16.9	Martin Marietta Fiber Optic Network	433
17.1	Communication Satellites Are Spaced in an Equatorial Orbit 22,300 Miles above the Earth's Surface	442
17.2	Satellite Circuit	445
17.3	Satellite Transponder	446
17.4	K Band Satellite Earth Station Mounted on Rooftop	449
17.5	Data Transmission through Satellite Delay Compensators	452
17.6	Satellite Business Systems Network Access Centers	456
17.7	Direct Broadcast Television	458
17.8	The Hercules Network Includes Nine Earth Stations in the United States	461
17.9	Block Diagram of Hercules Earth Station	463
17.10	Hercules Telecommunications Network Diagram	465
18.1	Polled Multidrop Data Network	476
18.2	Packet Network Showing Access Options	479
18.3	Level 2 Frame Enclosing X.25 Packet	480
18.4	Alternatives for Access to Packet Switched Networks	483
18.5	Local Area Data Transport	488
18.6	Generalized Data Network Response Time Model	496
19.1	LAN Topologies	505
19.2	Collisions in a Contention Network	506
19.3	Token Ring Network	509
19.4	Token Bus Network	510
19.5	CATV Frequency Allocations	511
19.6	IEEE 802 Standard	520

19.7 Maximum Configuration of IEEE 802.3 Network 521
20.1 Interlaced Scanning 538
20.2 Synchronizing and Blanking in a Television Signal 540
20.3 Block Diagram of a CATV System 541
20.4 Head End Equipment 542
20.5 Cable Television Frequency Bands 543
20.6 Two-Way Cable Frequency Splitting 546
21.1 Desktop Facsimile Transceiver 564
21.2 Block Diagram of a Facsimile System 565
21.3 Encoding a Character with Facsimile 568
21.4 Analog and Digital Facsimile Scanning Process 569
22.1 Diagram of Conventional Mobile Telephone Service 581
22.2 Frequency Reuse in a Cellular Mobile Serving Area 583
22.3 Cellular Radio Serving Plan 584
22.4 Cell-Site Controller 587
22.5 Sectored Cell 588
22.6 Increasing Capacity by Splitting Cells 588
22.7 DMS Mobile Telephone Exchange 590
22.8 Block Diagram of a Cellular Mobile Radio Transceiver 591
22.9 Metropolitan Radio Data Network 595
23.1 Transmission Measurement on a Circuit between TLPs 606
23.2 Phase Jitter 608
23.3 Common Cable Faults 610
23.4 Cable Testing Circuit 610
23.5 4TEL Central Office Line Testers 612
23.6 Automatic Remote Trunk Testing System 616
23.7 Switched Access Remote Testing 617
23.8 Modem Loopback Paths 618
23.9 Protocol Analyzer 620
24.1 Stromberg Carlson Centralized Maintenance and Administration Center 639
25.1 Intelligent Network Concept 646
25.2 Software Defined Network 647
25.3 Integrated Services Digital Network Concepts 650

25.4 The Ratio between the Computing Power of Desktop and Mainframe Computers is Narrowing 653
A.1 Schematic Diagram of a Relay 665
A.2 Current Flow in a Resistive Circuit 667
A.3 Chart of Power and Voltage Ratios versus Decibels 668
A.4 Series and Parallel Resistance 669
A.5 Alternating Current Sine Wave 670
A.6 Phase Relationship between Two Sine Waves 671
A.7 Derivation of Square Waves 672
A.8 Frequency Response of Loaded and Nonloaded Voice Cable 673
A.9 Frequency Response of a Typical Telephone Channel 673
A.10 Oscillator Circuit 676
A.11 Transformer 676
A.12 Band Pass Filter 678
A.13 Block Diagram of a Radio Transmitter 679
A.14 Amplitude Modulation 681
A.15 Mixer 682
A.16 Block Diagram of a Superheterodyne Radio Receiver 683

LIST OF TABLES

3.1 American Standard Code for Information Interchange (ASCII) 62
3.2 Extended Binary Coded Decimal Interchange Code (EBCDIC) 63
3.3 Data Transmission Speeds and Applications 64
3.4 Split-Channel Modem Frequencies (Bell 103) 69
4.1 Digital Carrier Typical Channel Unit Types 114
6.1 Key Telephone Systems Typical Features 171
8.1 Single-Frequency Signaling Tone Frequencies 221
8.2 Voice Frequency Terminating and Signaling Units 225
9.1 Dialing Plan under Equal Access 261
12.1 Trunk Maintenance Test Lines 324
13.1 Partial Poisson Traffic Table 364

15.1	Common Carrier and Operational Fixed (Industrial) Microwave Frequency Allocations in the United States	392
17.1	Communications Satellite Frequency Bands	441
21.1	Proposed CCITT Group 4 Facsimile Terminal Characteristics	572
A.1	Common Units of Electric Measurements	664
A.2	Frequencies and Wavelengths of Radio Frequencies	670

CHAPTER ONE

Introduction to Telecommunications

One of the greatest impediments to learning any discipline is the terminology. As a reminder to those of us who often confuse words with reality, Alfred Korzybski, the founder of general semantics, was fond of declaring, "The map is not the territory." Korzybski was right, of course, but a map is an indispensable aid to understanding the territory. When one is surrounded by mountain ranges, a map conserves the energy it would otherwise take to climb a mountain just for a look around. In this chapter we are going to draw a map. We will describe telecommunications systems so that when we cover them in more detail in later chapters it will be clear how the parts fit to form a whole.

When we use terminology, we will define it in context to demonstrate its meaning. It seems curious that a discipline based so heavily on the physical sciences would lack precision in the meaning of its terms, but as with other disciplines, telecommunications is fraught with ambiguity, and terms can be translated into concepts only in context.

The nation's telecommunications network has accurately been described as the most complex time-shared computer on earth, one that can best be understood by analyzing its component systems. Just as the human body can be viewed either as a unit or as an assembly of systems such as the digestive, respiratory, and circulatory, the telecommunications network can be understood on the basis of its systems. We will first discuss the major systems in broad terms, and in subsequent chapters we will examine them in greater detail.

THE MAJOR TELECOMMUNICATIONS SYSTEMS

Figure 1.1 shows the major classes of telecommunications equipment and how they fit together to form a communications network or *one large system.* In the telecommunications industry, as with the computer industry, standards have been set primarily by the manufacturers, and compliance with equipment standards set by others has been voluntary. Unlike the systems of the human body, the systems in the large telecommunications system or network are not tightly bound. Each of the elements in Figure 1.1 is largely autonomous. The telecommunications network is created as its systems exchange signals across the interfaces.

Customer Premises Equipment

Located on the user's premises, *station* or *terminal* equipment is the only part of the large telecommunications system that the users normally contact directly. This station equipment includes the telephone instrument itself and the wiring in the user's building that connects to the telephone company's equipment. For our purposes, station equipment also includes other apparatus such as *private branch exchanges* (sometimes also called computer branch exchanges or private automatic branch exchanges) and *local area networks* (LANs) that are normally dedicated to a single organization. It also includes multiple line equipment used to select, hold, and conference calls. This equipment is generally known as *key telephone equipment.* Auxiliary equipment such as speaker phones, automatic dialers, answering sets, and the like also fall into the station equipment classification.

Customer premises equipment has two primary functions. It is used for intraorganizational communication within a

FIGURE 1.1

Major Classes of Telecommunications Equipment

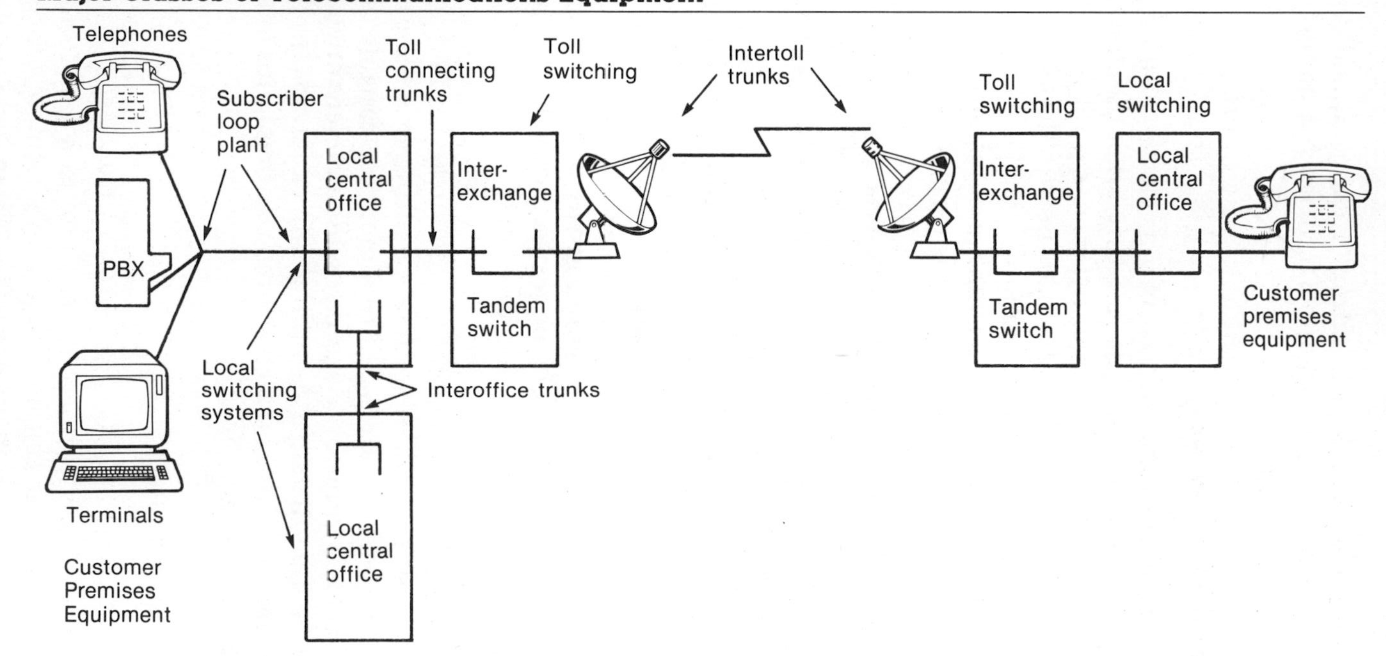

narrow range, usually a building or campus, and it connects to private or common carrier facilities for communication over a wider range.

Subscriber Loop Plant

The *subscriber loop,* also known as the *local loop,* consists of the wire, poles, terminals, conduit, and other outside plant items that connect customer premise equipment to the telephone company's central office. Before deregulation, telephone companies had a monopoly on the local loop. Now, however, alternatives are becoming available. For some services it is possible to transport information over cable television, and in other cases to use privately owned facilities such as microwave radio to bypass telephone company equipment.

Local Switching Systems

The objective of the large telecommunications system is to interconnect users, whether they are people communicating over a telephone line or machines communicating over specially designed data circuits. These connections are either dialed by the user or are wired in the telephone company's central office and remain connected until the service is discontinued. This latter kind of circuit is called *private line* or *dedicated.*

The local central switching office (often called an end office) is the point where local loops terminate. Loops used for switched services are wired to a machine capable of switching to other loops or to *trunks,* which are channels to other local or long-distance switching offices. Loops used for private line services are directly wired to other loops or to trunks to distant central offices.

Interoffice Trunks

Because of the huge concentrations of wire that converge in local central switching offices, there is a practical limit to how many users can be served from a single office or *wire center.* Therefore, in major metropolitan areas, multiple central offices are strategically placed according to population density and are interconnected by interoffice trunks. Central offices exchange signals and establish talking connections over these trunks to set up paths to addresses corresponding to telephone numbers dialed by the users.

Tandem Switching Offices

As the number of central offices in a region increases, it becomes impractical to connect every office to every other office with trunks. For one thing, the number of groups of trunks would be unmanageable. Furthermore, some central offices have too little traffic demand between them to justify the cost of directly connected trunks. To solve these problems, the telephone network is equipped with *tandem switches* to interconnect trunks. *Local tandem* switches connect local trunks together, and *toll tandem,* also called *interexchange tandem,* switches connect central offices to interexchange trunks leading outside the free calling area. A special type of tandem switch known as a *gateway* is used to interconnect the telephone networks of different countries when their networks are incompatible.

Tandem switches may also be privately owned to interconnect circuits under the control of a single organization. Before the breakup of the Bell System, toll tandem switches were owned exclusively by the telephone companies. Now, with multiple long-distance carriers offering service, each carrier connects tandem switches to the end office or to a telephone company-owned *exchange access tandem.*

Interexchange Trunks

Telephone companies divide their serving areas into classifications known as *exchanges.* Most exchanges correspond roughly to the boundaries of cities and their surrounding area. Interexchange trunks that connect offices within the free calling area are known as *extended area service* (EAS) trunks, while those that connect outside the free calling area are called *toll* trunks.

Transmission Equipment

The process of transporting information in any form including voice, video, and data between users is called *trans-*

mission in the telecommunications industry. The earliest transmission took place over open wire suspended on poles equipped with crossarms and insulators, but that technique has now largely been displaced. For short ranges, trunks are carried on pairs of copper wire that are twisted into cables of as many as 2,400 pairs. For longer ranges, interoffice and interexchange trunks are transported over twisted pair wire, terrestrial microwave radio, fiber optic light guides, or satellites. These backbone transmission facilities are divided into voice channels by *multiplexing* equipment.

FUNDAMENTALS OF MULTIPLEXING

The basic building block of the telephone network is the *voice grade* communications channel occupying 300 to 3,200 Hz of bandwidth. This bandwidth is a far cry from the bandwidth of high-fidelity systems that typically reproduce 30 to 20,000 Hz faithfully, but for ordinary voice transmission, a voice grade circuit is entirely satisfactory. For the first six or seven decades of telephony, open wire or multiple-pair cables were the primary transmission medium. At first, each pair of wires carried one voice channel. However, these media (known as *facilities* in telecommunications vernacular) have enough bandwidth to carry several channels. For example, most stereo systems use twisted pair wire between the amplifier and speaker, and they have bandwidths six times that of a voice grade channel.

The telecommunications industry has always invested heavily in research and development. Much of it has been directed toward how to superimpose an increasing amount of information on a single transmission medium. The process for placing multiple voice channels over one facility is known as *multiplexing.*

The term *multiplexing* is as broad in meaning as the capacity of the transmission medium it seeks to expand. In its earliest and simplest form, multiplexing did not even require electronic equipment. Telecommunications engineers found that if two identical circuits were used to carry a third circuit as shown in Figure 1.2, the resulting circuit, known as a *phantom,* was of even higher quality than the two original

FIGURE 1.2
Phantom Telephone Circuit

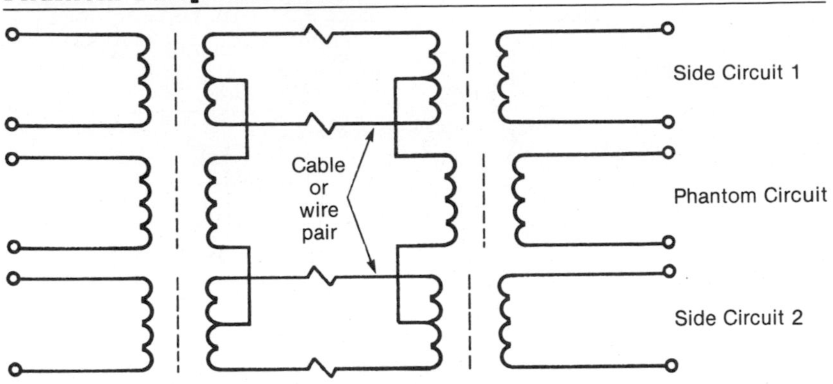

side circuits. For the price of a simple transformer (telephone people call it a *repeat coil*), the carrying capacity of a pair of wires could be increased by 50 percent.

Frequency Division Multiplexing

The phantom circuit still, however, left a great deal of bandwidth unused. With the advent of the vacuum tube in the 1920s, another form of multiplexing known as *carrier* became feasible. The earliest carrier systems increased the capacity of a pair of wires by means of a technique known as *frequency division multiplexing* (FDM). Broadcast radio is an example of FDM in action (except that it does not allow two-way communications). As shown in Figure 1.3, in an FDM carrier system, each channel is assigned a transmitter-receiver pair or *modem* (a term derived from the words *modulator* and *demodulator*). These units operate at low power levels, are connected to cable rather than an antenna, and therefore do not radiate as a radio does. Otherwise, the concepts of radio and carrier are identical.

FDM carrier systems in use today range in capacity from one additional voice or data channel over a pair of wires to as many as 13,200 voice channels on a broadband coaxial cable.

Time Division Multiplexing

Although FDM is an efficient way of increasing the capacity of a transmission medium, the techniques used for its man-

FIGURE 1.3
Frequency Division Multiplexing System

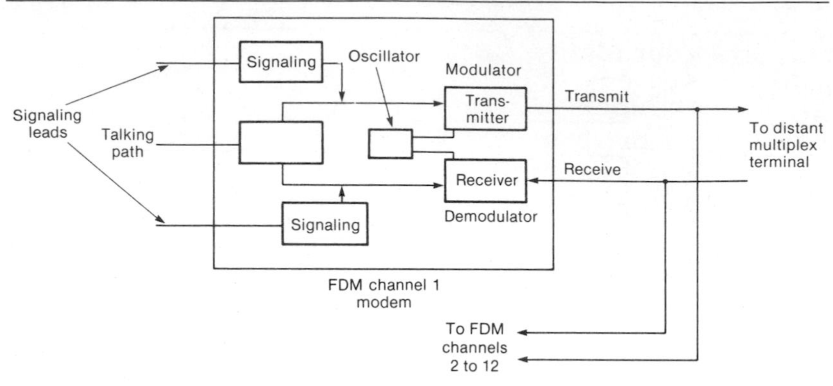

FIGURE 1.4
Time Division Multiplexing System

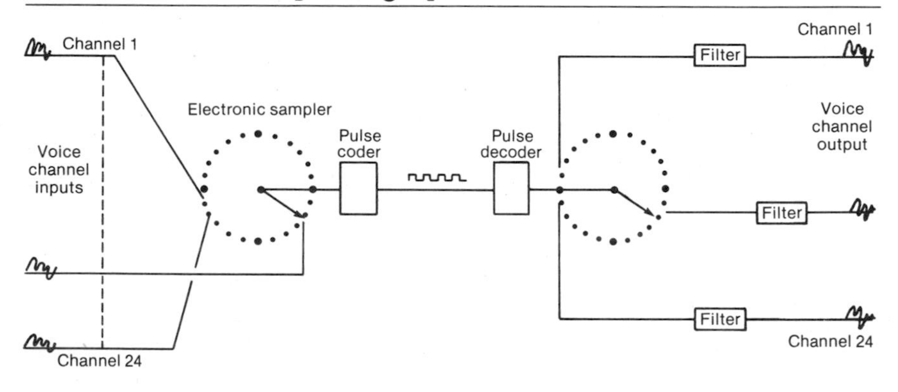

ufacture do not lend themselves to large scale integration. Therefore, FDM is gradually being replaced by *time division multiplexing* (TDM), which uses techniques similar to those used in computers. In TDM, shown conceptually in Figure 1.4, instead of the bandwidth of the transmission medium being divided into frequency segments, each user is given access to the full bandwidth of the system for a small amount of time. As shown in Figure 1.4, electronic equipment distributes a signal from each user to an appropriate time segment of the transmission medium. The capacity of the transmission medium is so great that users are not aware that they are sharing it. The concept of time division multiplexing is described in detail in Chapter 4.

Higher Order Multiplexing

The basic building block of both analog and digital multiplexing systems (explained in the next main section of this chapter) is the *group*. A group consists of 12 voice grade circuits combined into a band of analog frequencies, or into a stream of digital data. Groups are formed in equipment called a *channel bank*. Analog channel banks consist of 12 circuits; digital channel banks derive 24 circuits known as a *digroup*. The output of channel banks is combined by higher order multiplexing into hundreds and thousands of circuits that can be transmitted over broadband facilities such as coaxial cable, microwave radio, and fiber optics.

Data Multiplexing

When data communications people speak of multiplexing, they are usually using it in a different sense than we have just discussed. Data multiplex equipment subdivides a voice channel into several lower bandwidth data channels. In later chapters we will discuss this multiplexing in more detail. For now it is important to know that the term *multiplex* must be understood in context because it is a term that is employed with many different meanings.

ANALOG AND DIGITAL TRANSMISSION CONCEPTS

When people speak into a telephone instrument, the voice actuates a transmitter to cause current flowing in the line to vary proportionately or analogous to the changes in sound pressure. Because people speak and hear in analog, there was, until recent years, little reason to convert an analog signal to digital. Now, there are three primary reasons digital transmission is important. First, digital equipment is less expensive to manufacture than analog; second, an increasing amount of communication takes place between digital terminal equipment such as computers; and third, digital transmission has higher quality in many respects than analog.

The higher quality of digital signals results from the difference in the methods of amplifying the signal. In analog transmission, the signal, together with any noise on the line, is boosted by an audio amplifier. With digital transmission, regenerators detect the incoming bit stream and create an entirely new signal that is identical to the original signal. If a digital signal is regenerated before noise causes errors to occur, the result is a channel that is practically noise-free.

The system for transmitting telephone signals digitally is known as *pulse code modulation* (PCM). In a PCM channel bank, voice is sampled, converted to an eight-bit digital word, and then transmitted over a line interspersed with digital signals from 23 other channels. The 24-channel signal is regenerated by repeaters spaced at appropriate intervals.

The theory of PCM is not new—it was developed by an ITT scientist in England in 1938. However, even though the system was technically feasible then, it was not economical because of the high cost of the electronics needed to make the analog-to-digital conversion. With the invention of the transistor, the development of solid state electronics, and particularly the growth of large-scale integration, the economics shifted in favor of digital transmission. PCM is replacing analog techniques in all parts of the large telecommunications system except at the source, the telephone. However, telephone sets in which the voice is digitized in the instrument are beginning to appear in PBXs. Presently, digital telephone sets are not practical in public telephone networks for technical and economic reasons, but the drawbacks will disappear in time.

SWITCHING SYSTEMS

For many applications, fixed circuits between points, known as *point-to-point* or *dedicated* circuits, are desirable. In the majority of cases, however, the real value of a telecommunications system is in its ability to access a wide range of users wherever they are located. This is the role of telephone switching systems.

Early Switching Systems

The earliest switching systems were manually operated. Telephone lines and trunks were terminated on jacks, and operators interconnected lines by inserting plug-equipped cords into the jacks.

In 1891, a Kansas City undertaker named Almon B. Strowger patented an electromechanical switch that could be controlled by pulses from a rotary dial. The Strowger system, also known as the *step-by-step system,* is still in widespread use in the United States today. Although the step-by-step system is disappearing in large metropolitan areas, the majority of the small offices serving rural communities still use this system. The distinguishing feature of the step-by-step system is that electrical pulses created by a dial directly control the motion of the switches. This point will be important to remember when we discuss some of the limitations in providing completely uniform telecommunications services between users and competing long-distance common carriers.

Common Control Switching Offices

All switching offices have a common characteristic: they contain a limited amount of equipment that is shared by many users. With manual and step-by-step systems, all the equipment used to establish a talking path remains connected for the duration of the call. Keeping the equipment occupied for the duration of the call is not economical. More important, it is also inflexible. As a call progresses through the switching machine, if blockage is encountered, the machine is incapable of rerouting the call to a different path. It can only signal the user to hang up and try again. These drawbacks can be overcome by use of *common control* switching systems.

Under the common control concept, the talking path is established through a switching network by equipment that is released when the connection has been established. This common control equipment is not called on again until the connection is to be taken down. In this respect, common control equipment serves a function similar to the manual switchboard operator. Although common control equipment is more complex than directly controlled switching, it is also more efficient and much faster.

Contrasted to the step-by-step system where the user builds a connection gradually with pulls of the dial, under the common control system the user transmits dial pulses or tone pulses into a circuit that registers the digits. An advantage to this system is that when dialing is complete, logic circuits can inspect the digits, determine the destination of the call, and can choose an alternate route if all trunks in the preferred route are busy. This capability, known as *alternate routing,* is a characteristic shared by all modern switching machines.

The earliest common control equipment, introduced in the 1920s, was a system known as *panel,* followed about two decades later by the *crossbar* system. The systems take their names from the method of interconnecting lines and trunks. Panel equipment has all but disappeared, but crossbar equipment is still in widespread use in the United States.

Computer Controlled Switching Systems

Common control central offices use electromechanical relays in their logic circuits. Relays have an electronic counterpart, the logic gate. Both gates and relays are binary logic devices. That is, they are either on or off, and if a decision can be reduced to a series of "yes/no" responses to outside conditions, both logic gates and relays can perform the same functions.

It was natural, therefore, to develop the electronic equivalent of a common control central office. Early electronic switching systems used wired logic, that is, they were not programmable. In the early 1960s, the age of the stored program control (SPC) central office was born. The common control equipment consists of an electromechanical switching network driven by a software program.

Digital Central Offices

When large-scale integrated circuits were perfected in the 1970s, it became technically feasible to develop a digital switching network to replace the electronic network in SPC central offices. The state of the art of central office technology consists of a digital switching network controlled by a programmable central processor. Virtually all modern switching equipment, ranging from small PBXs to large toll tandem switches capable of handling thousands of trunks, use this technology. At the present time, further research is underway to develop even less costly switching machines capable of switching light streams rather than electrical pulses.

NUMBERING SYSTEMS

Routing between switching machines and selecting terminating stations within a switching machine is accomplished with addressing by an area code and telephone number. Without controlled assignment of telephone numbers, the telephone system could not function. In this section we will look briefly at how telephone numbering operates both in this country and worldwide.

The Switching Hierarchy

The public telephone network operated by AT&T and connecting local telephone companies is divided into five classes of switching machines as illustrated in Figure 1.5. Class 5 central offices, the lowest class in the hierarchy, are the machines that directly serve the end users. Higher class offices are tandem offices used for connecting toll trunks.

Trunks between switching machines are classified as either *high-usage* or *final* trunks. Within certain limits, high-usage trunks can be established between any two offices if the traffic volume justifies it. Common control switching machines contain the intelligence to enable them to decide what group of trunks to choose to route calls to the destination, always attempting first to connect to a distant central office over a high-usage trunk. Calls progress from machine to machine until the terminating office is reached. Machines route calls by exchanging signals over either the trunk that provides the talking path or over a separate data network.

From Figure 1.5 it can be seen that if all high-usage trunks are busy, a call from a Class 5 office could be routed over as

FIGURE 1.5
Switching Hierarchy

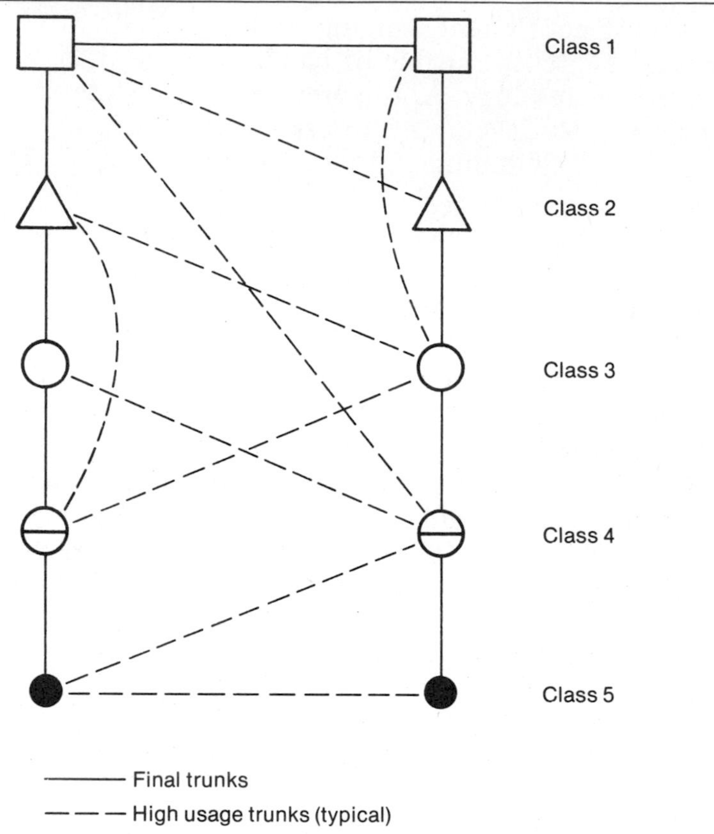

many as nine final trunks in tandem before it reaches the terminating Class 5 office. In practice, this rarely, if ever, happens. Traffic engineers monitor the amount of traffic flowing between terminating points and order the proper quantity of trunks to cause calls to complete over the most economical route.

The Nationwide Numbering Plan

Every telephone line in the nation has a unique 10 digit address. Obviously, if the addresses were duplicated, switching would be impossible because machines would be unable to determine which address was the intended destination.

The first division is a three-digit area code. Within the area, each central office is assigned a three-digit central office code. Within the central office, each customer is assigned a line number between 0000 and 9999. Each central office code is, therefore, capable of handling 10,000 lines. It is common for a single switching machine to serve multiple central office codes.

The Worldwide Numbering Plan

Not all nations in the world use the same numbering plan as the United States. This is of little consequence if the gateway offices that interconnect the countries are able to translate the dialed digits to their destination. Each nation in the worldwide plan is assigned a two- or three-digit country number. Worldwide dialing is accomplished without operators by dialing an international direct distance dialing (IDDD) access code, a country code, and the terminating telephone number.

INTEREXCHANGE CARRIER ACCESS TO LOCAL NETWORKS

The 10-digit telephone addresses we have been discussing allow completing a call to any telephone in the country after you have obtained access to the network administered by the interexchange carrier (IC) you select to handle your long distance calls. Until recently, the only public switched network that spanned the country was owned and administered by AT&T through its Long Lines Division (now AT&T Communications) and through the Bell Operating Companies. The operating companies generally handled interexchange communications within state boundaries, and in a few instances between states. The bulk of the interstate calling, however, was handled by Long Lines.

The toll network was accessed by dialing "1." In the early years of user-dialed long distance calls, operators identified the calling telephone number by bridging on the call momentarily and asking the calling party for the billing number. Gradually, automatic number identification equipment was added to identify the billing number.

In 1976, the Federal Communications Commission (FCC), which regulates interstate telephone service, opened long distance telephone service to competition from other ICs. These ICs gained access to the local telephone network through an ordinary seven-digit telephone number. This form of access has technical drawbacks that result in poorer quality transmission for reasons that are explained in Chapter 2. An equally important drawback is the necessity of dialing a seven-digit telephone number for access to the long-distance network instead of the digit "1" used for access to AT&T's network. Furthermore, automatic number identification is not feasible over this method of access, so users are required to dial a personal identification number (PIN). With a five- or six-digit PIN, this form of access requires dialing 22 or 23 digits to reach a terminating telephone number in another area code.

In 1982, AT&T and the Department of Justice signed a consent decree that resulted in AT&T's divesting its operating telephone companies. An element of the decree requires the telephone companies to give equal access to all ICs. Under equal access, all carriers have connections that are technically equivalent to AT&T's connection to Class 5 central office. When a user originates a call, the switching equipment must be able to determine which IC the user wants to handle the call. In central offices equipped with stored program controllers that have been programmed for equal access, each user presubscribes to a preferred IC. That IC is selected by the central office when the user dials "1." Other ICs that serve the central office are accessed by dialing a five-digit code 10XXX, where XXX is a number assigned to the IC. Automatic number identification is a standard equal access feature, so PIN dialing is not required.

Equal access is being programmed into most Bell SPC offices through 1986. Because central offices lacking SPC equipment are unable to select multiple ICs, equal access in these offices will not be possible unless they are replaced or unless additional equipment is developed. In many locations, equal access tandem offices are being installed to allow access to multiple ICs.

Another significant feature of the agreement between AT&T and the Department of Justice is the subdivision of the country into serving areas called *local access transport areas* (LATAs). Under the terms of the agreement, the local Bell Operating Companies can offer toll telephone service only within the LATA. Between LATAs, AT&T and any other authorized IC can furnish long-distance services.

PRIVATE TELEPHONE SYSTEMS

Many organizations operate private telecommunications systems. These systems range in size from the Federal Telephone System (FTS), which is larger than the telecommunications systems in many countries, to small PBXs.

Private networks normally must be connected to the public network so users can place calls to points that are not covered by the private network, and so people outside the private network can reach them. This is accomplished by connecting to a Class 5 switching machine in a manner similar to the pre-equal access connection described for ICs. These connections are usually switched through a PBX or through private tandem switching machines.

Private networks are designed to criteria similar to networks operated by the telephone companies. There is no reason that private systems must conform to the nationwide numbering plan. If they do not, however, the result is often a dual numbering system, one for calls placed on the private network, and another for calls placed on the public network.

Transmission quality is also an important consideration in private networks. When private networks connect to the public telephone network, it can add another link to the switching hierarchy. Because each link degrades transmission somewhat, careful design is required to prevent unsatisfactory transmission quality, a topic covered in more detail in Chapter 2.

STANDARDS

The close interrelatedness of the telecommunications network requires standards for devices to communicate successfully with one another. Unfortunately, the need for standards and the need for technical progress often conflict because standards cannot be set until the technology has been proven in practice, and the only way to prove the technology is through extensive use. Therefore, when it comes time to set a standard, a large base of installed equipment designed to proprietary standards is already in place. The policies of many standards-setting organizations preclude their adopting proprietary standards, even if the manufacturer is willing to make them public. Also, competing manufacturers are represented on the standards-setting bodies of the United States, which often militates against their accepting proprietary standards. As a result, even after standards are adopted, a considerable amount of equipment exists that does not conform to the standard.

The International Telecommunications Union (ITU) was formed in 1865 to promote mutual compatibility of the communications systems that were then emerging. The ITU, now a United Nations–sponsored organization to which 160 countries belong, disseminates international standards through its two consultative committees. The International Radio Consultative Committee (CCIR) is responsible for frequency allocations and sponsors international agreements to promote the use of radio and to prevent interference.

The International Telephone and Telegraph Consultative Committee (CCITT) is the primary international standards body for telecommunications. CCITT does its work through 15 study groups that work in four-year time increments. At the end of a four-year session, the study groups' work is presented to a plenary assembly for approval. For example, the 1984 plenary assembly dealt with emerging standards on the Integrated Services Digital Network (ISDN) and on high speed facsimile.

In some countries where telecommunications is operated as a state-owned agency, CCITT recommendations bear the force of law. In the United States, compliance is voluntary. Large companies such as IBM and AT&T have enough market power to set their own standards, to which others must adhere to be compatible. IBM's Systems Network Architecture (SNA) is the most widely used data communications architecture in the world, yet SNA is not an international standard. Instead, the International Standards Organization

(ISO) has designed its Open Systems Interconnection (OSI) model as a set of building blocks in a standard data network architecture. However, OSI is not a working model but rather is a theoretical architecture that must be translated into practice by interface standards pertaining to each of its seven layers. More is said about data communications architectures and standards in Chapter 3.

The voice networks in the United States are largely designed to AT&T's proprietary standards. Before the divestiture of AT&T, these standards were released to other manufacturers through the U.S. Independent Telephone Association (now U.S. Telephone Association). In some cases, equipment was manufactured by others through cross-licensing agreements with AT&T, but in most cases, compatibility information was unavailable until several years after the technology was successfully introduced by AT&T. Following the breakup of the Bell System, the Bell and independent telephone companies formed the Exchange Carrier Standards Association (ECSA) to deal with telephone standards. The ECSA is currently working on T-1 carrier standards.

In some cases, AT&T standards have been adopted by CCITT. For example, the standards for the U. S. version of digital multiplex is a CCITT recommendation, but it is not compatible with the European version, which is also a CCITT standard. In other cases, CCITT standards and AT&T proprietary standards conflict. For example, signaling between AT&T's long distance switching offices uses a protocol known as Common Channel Interoffice Signaling (CCIS). The international standard is CCITT Signaling System No. 7, which is incompatible with CCIS.

The standards body that represents the United States in its dealings with ISO and ITU is the American National Standards Institute (ANSI). ANSI is a nongovernmental, nonprofit organization comprising 300 standards committees. Both consumers and manufacturers are represented on ANSI committees, and much of its work is done by cooperating trade groups that follow ANSI procedures. The Institute of Electrical and Electronic Engineers (IEEE) and the Electronic Industries Association (EIA) are two prominent organizations that promulgate standards through ANSI.

EIA has produced many standards that are important to the telecommunications industry. For example, the EIA RS-232-C interface standard is employed by most data terminal devices in their interconnection with circuit equipment. The IEEE is a professional association that has had an important effect on standards activities such as the local area network standards developed by its "802" committee. These standards, discussed in more detail in Chapter 19, use the framework of ISO's Open Systems Interconnect model and CCITT protocols to develop three local network alternatives.

In the years before the FCC and the courts opened AT&T's network to interconnection and competition, users could avoid compatibility problems (except in data communications where incompatible protocols were a frequent problem) by turning the responsibilities over to the telephone company. Now that station equipment is no longer owned by the telephone company and long distance networks are a complex combination of common carrier and private facilities, compatibility is a concern of virtually every user. The need for compatibility thrusts the issue of standards to the forefront because users' options are limited if the manufacturer's interfaces are proprietary and not made publicly available.

Each chapter of this book lists some of the standards relevant to the technology. Appendix B tells where to obtain information from the organization that published the standard. For the most part these are the standards of public bodies such as CCITT and EIA. Some information that is published by Bell Communications Research (the research arm of the seven Bell Operating regions) is also listed because of its widespread acceptance as a virtual standard.

SUMMARY

One cannot help being awed by the intricacy of the nation's large telecommunications system. The complexity is evident from this brief overview, but becomes even more impressive as the details emerge. The wonder is that the system can cover such a vast geographical area, can be administered by hundreds of thousands of workers, contain countless pieces of electrical apparatus, and still function as reliably as it does. As we discuss these elements in greater detail, the techniques that create this high-quality service will become more understandable.

CHAPTER TWO

Transmission Concepts

In common carrier and private telecommunications systems, quality is assured by systematic maintenance, careful circuit design, and high quality equipment. Transmission quality is an entirely different concept than switching system quality. Switching systems, which are discussed in Chapters 9 through 13, are go/no-go devices. The connection is either established or it isn't. With transmission systems, quality is a statistical measure. The telecommunications system always introduces some impairments into the talking path between users. Transmission design is a compromise between quality and cost. Although it is possible to build a telecommunications system that will reproduce voices with near-perfect fidelity and clarity, such quality is neither necessary nor economical.

Most transmission standards and design objectives used in the United States have been developed by AT&T and the Bell Telephone companies. The primary references to telephone transmission in this chapter relate to transmission over public telephone networks. However, the same design considerations also apply to private networks, whether they are composed of privately owned facilities or facilities obtained from common carriers.

FIGURE 2.1
Typical Telecommunications Circuit

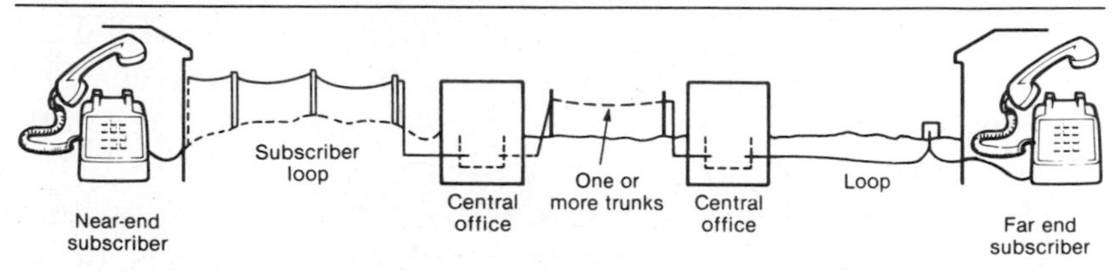

TRANSMISSION IMPAIRMENTS

For voice communication, four variables are most important for adequacy of communications: level or volume, noise, bandwidth, and echo. For data communication, which comprises an increasingly large portion of telecommunications traffic, envelope delay and amplitude distortion are also important.

Volume or Level

Consider the simple telecommunications circuit illustrated in Figure 2.1. The telephone instrument converts the changes in sound pressure from a talker's voice to a varying electrical current that is an analog of the acoustic signal. (See Chapter 6 for an explanation of how the telephone instrument functions.) The electrical characteristics of the circuit between the sending and receiving telephone modify the electrical signal in such a way as to reduce its volume (or increase its *loss*), change the bandwidth of the signal, and introduce extraneous signals such as noise, crosstalk, and distortion.

The unit of loss is the *decibel* (dB). An increase of signal volume that doubles the power of the signal is defined as a 3 dB increase. Similarly, a drop in signal power that halves the signal is a 3 dB reduction. The smallest change that the human ear can detect is about 1 dB; a 3 dB change is apparent to a listener concentrating on hearing the change. The concept of the decibel is essential for understanding transmission theory and is explained in Appendix A for those readers not yet familiar with this method of measuring differences in electrical quantities. It is essential to understand that the dB is not an absolute unit of measurement as are the volt, ampere, and watt. The dB measures only the ratio of two quantities of the same unit. A signal power of 1 *milliwatt* (mw), or 0.001 watt, is an almost universally used standard power against which other power levels are compared. The dB is often used to express power levels compared to 1 milliwatt; for example, 0 dBm = 1 mw.

The amount of loss that can be tolerated in a circuit depends on the tolerance of the listener for weak signals, noise impairments, and other distortions that alter the character of the received signal. This, of course, depends greatly on individual preference and varies widely among users, ranging from those with hearing impairment to those with acute hearing sensitivity. Because of the wide differences in preference, transmission objectives are statistically based with signal and noise standards designed to satisfy the majority of users.

Loss is easily overcome by amplification in telecommunications circuits. Amplifiers or *repeaters,* however, cause undesired side effects as well as the desired effect of reducing loss. Not only do they add cost to the circuit, but they also add distortion in the form of limited bandwidth, additional noise, and other undesirable changes to the signal they amplify. Designers attempt to minimize the use of repeaters, particularly in local subscriber loops, to the greatest extent possible. With more than 150 million subscriber loops in the United States, the cost of amplifying more than a small fraction of them would be substantial. Instead, amplification is provided for trunks, which are shared by all users, and for a limited number of long loops. The design of all the elements of the telecommunications network is driven by the characteristics and economics of the ordinary telephone and the subscriber loop, which are discussed in Chapters 6 and 7.

Noise

Noise is defined as any unwanted signal in a circuit. Hum, crackling or "frying," and crosstalk from adjacent circuits are all examples of unwanted signals that are controlled by careful design and maintenance of a circuit. There are definite trade-offs among the various noise impairments and the quality of the signal as perceived by the listener. A uniform level of hiss, for example, may be tolerable if no other impairments exist and the signal level is high. The most important measurement of noise is the *signal-to-noise ratio* (expressed in dB).

Data signals exhibit an entirely different tolerance to noise than do humans. A data signal will be satisfactory in the presence of uniform steady hissing noise (white noise) that would be bothersome to humans. On the other hand, impulse noise (clicks, pops, or sometimes a frying noise) will destroy a data signal on a circuit that might be acceptable for speech communication.

Circuit noise originates from three primary sources. The first is interference from external sources. Electric power lines, lightning, industrial apparatus such as electric motor commutators, crosstalk from adjacent circuits, and other such sources can cause circuit noise.

The second source is thermal noise developed within the telecommunications apparatus itself. Any conductor carrying current at a temperature higher than absolute zero (−273 degrees C.) generates noise from internal electron movement. Some types of circuit elements such as vacuum tubes generate more thermal noise than others, but it is present in all circuit elements, even including passive elements such as wire.

The third source of noise is distortion generated by nonlinearities in circuit elements, primarily amplifiers. Amplifiers do not precisely reproduce the signal that is introduced into their inputs. The small imperfections in an amplifier's transfer characteristic (the output of the amplifier compared to the input) distort the amplified signal so that extra signal components appear in the output signal. The effect is aggravated by operating the amplifier beyond its design capability. This effect is called intermodulation distortion.

Bandwidth

Bandwidth is the circuit attribute that, along with frequency response, controls the naturalness of transmitted speech. As with level, this is a subjective evaluation. The human ear can detect tones in the range of 20 to 16,000 Hz, but because the voice has few frequency components below about 300 Hz or above 3,500 or 4,000 Hz, a telephone circuit that transmits a band of frequencies in this range is quite adequate for voice communication. Channels for voice transmission are usually designed to pass a nominal bandwidth of 300 to 4,000 Hz. Telephone receivers have been designed to be most sensitive to the frequency spectrum between 500 and 2,500 Hz because research has shown that most of the frequency components of ordinary speech fall into this range.

Because of the technical difficulty of constructing filters and amplifiers with uniform transmission at all frequencies within the pass band, the high- and low-frequency ends of the transmitted spectrum suffer more loss or attenuation than frequencies in the center of the band. Circuits that are designed for program audio or high-speed data transmission are made with much wider bandwidths than those for voice transmission.

Echo

When telephone signals traverse a transmission facility, they move with a finite, although very high, speed. Electrical signals propagated in free space, a radio broadcast signal for example, advance at the speed of light (300 million meters per second). Signal propagation on a physical transmission circuit, on the other hand, advances at about 50 to 80 percent of the speed of light, depending on the type of transmission medium and the amount of amplification or filtering applied to the circuit. Because the propagation time of these circuit elements is less than the speed of light, a signal is delayed as it transits a network.

If in traversing the circuit the electrical signal encounters an impedance irregularity, a reflection will occur just as a reflection occurs to a sound propagated in a large, empty room. The reflected signal returns to the sending end of the circuit and sounds to the talker like the echo from a long, hollow pipe. The greater the distance from the talker to the irregularity, the greater will be the time delay in the reflected signal.

Echo is detrimental to transmission in proportion to the amount of delay suffered by the signal and the amplitude of

FIGURE 2.2
Typical Long-Haul Connection

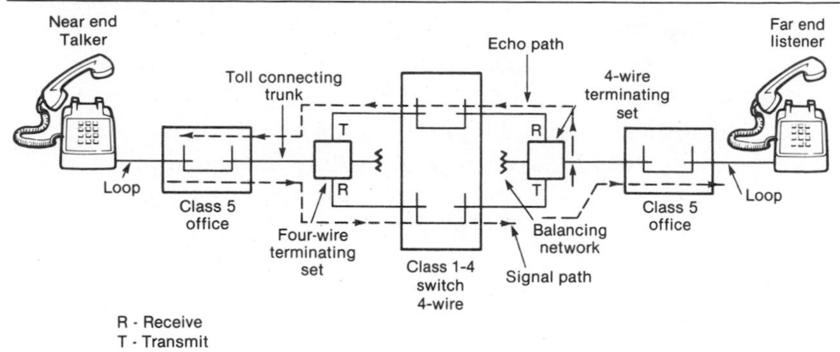

the echoed signal. A communication circuit that displays even a small amount of echo is unfit for service if the delay is too long. On the other hand, a small amount of echo occurs in the ordinary telephone where it is heard as *sidetone,* which is the sound of the speaker's voice in his or her own receiver. Because sidetone is not delayed, it does not interfere with communication. In fact, a certain amount of sidetone is deliberately introduced in a telephone to regulate the talker's voice level. A lack of sidetone gives the talker the perception that the instrument is dead, while a sufficient amount causes talkers to lower their voices somewhat.

The most serious form of echo in communications circuits arises from *four-wire terminating sets* or *hybrids.* These are devices that convert the transmission circuit from *four wires* to *two wires* as shown in Figure 2.2. Economics impels the designer to use two wires in as much of a communications network as possible to minimize costs. Carrier and radio systems inherently transmit in only one direction so that a separate path or channel is required in each direction to obtain two-way transmission. The two directions of transmission must be combined into a single two-way, two-wire circuit at each end by means of a four-wire terminating set, or hybrid, for extension through two-wire switching systems and to two-wire local loops.

In a hybrid, the two-wire portion of a circuit is balanced against a network that approximates the electrical characteristics of the two-wire transmission line. When the balancing network is identical to the transmission line, the hybrid is in balance and energy received over the four-wire transmission path is coupled to the two-wire path. To the degree that the balancing network fails to match the two-wire line, a signal is fed back to the talker at the distant end in the form of an echo. The farther away from the talker the echo occurs, the greater the time delay introduced into the return signal and the greater the impediment to good transmission. Figure 2.2 shows the configuration of a typical long-distance circuit showing four-wire terminating sets, four-wire transmission facilities, and the echo path.

The loss of a signal traversing a circuit, through the hybrid and back to the sending end, is called *return loss.* If the return loss of the circuit does not equal or exceed the amplification in both paths of the four-wire circuit, oscillation or "singing" occurs. This is the same effect observed when the volume is advanced on a public-address system until the system squeals, or "sings." The hybrid's balancing network must be adjusted to at least 10 dB more loss than is needed to prevent the circuit from singing. Adjustments to the hybrid balance network become more important as the circuit becomes longer. At about 2,000 miles, the round trip delay on a circuit becomes excessive and any appreciable echo becomes disturbing to the talkers. Echo is controlled in long circuits with *echo suppressors.*

Echo suppressors are devices that automatically insert loss in the return path of a four-wire circuit. The echo suppressor switches back and forth between the two transmission paths, following first one talker then the other. Properly adjusted, an echo suppressor inserts only enough loss in the circuit so that a listener can interrupt a talker, but otherwise the reflected signal is attenuated by approximately 15 dB. With long circuits such as satellite circuits with round trip delays of about 0.5 second, a more effective method of eliminating echo is required. On such circuits, *echo cancelers* are used. Echo cancelers perform the same function as echo suppressors but operate by creating a replica of the near-end signal and subtracting it from the echo to cancel out the echo. A photograph of a digital echo canceler is shown in Figure 2.3.

Echo is controlled in shorter circuits by adjusting the loss of the circuit. The whole subject of loss and echo control is

FIGURE 2.3
Digital Echo Canceler

Courtesy Granger Associates.

embodied in what is known as the *via net loss* plan, a set of design rules that optimize transmission performance and economics in the national telecommunications network. The advent of modern digital transmission and switching facilities is causing the via net loss plan to evolve toward a fixed loss plan that integrates zero loss digital transmission facilities and digital echo cancelers into the network. The via net loss plan is discussed in a later section.

Amplitude Distortion

Telecommunications channels rarely have a perfectly flat response across the voice frequency band. Figure 2.4a shows an equalized channel; Figure 2.4b shows the frequency response characteristic of a channel before equalization. Although equipment used in voice frequency channels is manufactured to close tolerances, the accumulated small deviations inherent in the manufacturing process add to produce the irregular response illustrated. The channels can be brought into close tolerance by the addition of equalizers

FIGURE 2.4
Voice Channel Attenuation

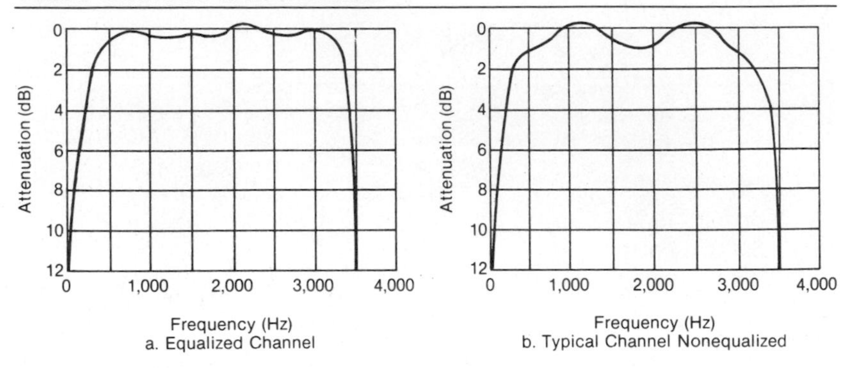

where the cost of this treatment is justified by the demands of the service.

Envelope Delay Distortion

The design of electronic amplifiers and multiplexers requires components that introduce varying amounts of delay to frequencies within the voice frequency passband. This varying delay is known as *envelope delay.* Envelope delay has no discernible effect on voice frequency signals. However, for data signals, which are composed of complex voice frequency tones, envelope delay results in tones arriving at the receiver slightly out of phase with one another. Envelope delay distortion can be compensated by the addition of delay equalizers either in the telecommunications circuit or in the data terminal equipment.

ELEMENTS OF TRANSMISSION DESIGN

Transmission design is the process of balancing loss, noise, and echo against circuit costs. The fourth variable, bandwidth, is not adjustable; it is inherent in the design of amplifiers and multiplex equipment. The remainder of this chapter discusses transmission from the point of view of the quality requirements of voice frequency circuits. Data circuits traveling over the switched telephone network must accept the characteristics of the circuits as they are. On private line data circuits, transmission can be controlled more closely by *conditioning,* which will be discussed in Chapter 3.

Telecommunications circuits are either switched or dedicated; that is, the connection is dialed by the user and released on call termination, or it is permanently connected between two or more points. The characteristics of dedicated or private line circuits can be specifically designed for the application, but because of the random originations and terminations and the variation of the characteristics of switched circuits, design control is less exact. Switched circuit design is a compromise based on probability of the user's receiving a connection that is of satisfactory quality a high percentage of the time.

Each switched connection is composed of three types of circuits: a subscriber loop on each end, a toll-to-local office connecting trunk (*toll connecting trunk*) on each end, and one or more interoffice trunks (*intertoll trunks*) in between. These types of circuits are designed using different rules. Each circuit type is allocated a share of the end-to-end impairment, with the objective of providing an overall connection of satisfactory quality to the user.

Variables in Transmission Quality

In the preceding sections, we introduced several terms that are essential to understanding transmission quality. In this section, additional terminology is introduced, and the method by which transmission is measured and controlled is discussed.

It is important to remember the difference between loss and level in a circuit. Level is a measurement of signal power at a specified point in the circuit known as a *transmission level point* (TLP). Loss is the difference in level between TLPs. A voice frequency signal is a complex amalgam of tones that vary widely in frequency and amplitude. Such a complex signal is often measured by broadcast engineers as *volume units* (VU) using a meter with a highly damped movement to smooth out the peaks and valleys of the voice signal. Other than the dynamic characteristics, a VU meter is the same as a dB meter.

FIGURE 2.5
Transmission Levels

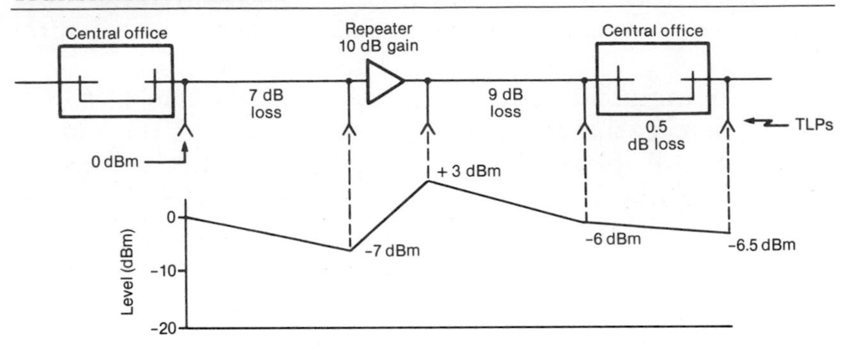

A voice frequency signal is impractical for measuring level and making adjustments to circuits in which level is adjusted in 0.1 dB increments. Level is measured by applying a 1,004 Hz single-frequency tone to the circuit at a standard TLP and then measuring the test tone at some other TLP. The measured tone is then compared to a standard 1 milliwatt test source to establish a test tone level. Most transmission testing equipment is calibrated against a 1 milliwatt source so that measurements may be made directly with the test set. The test tone frequency is set at 1,004 Hz because a frequency of 1,000 Hz is an exact submultiple of many of the carrier frequencies used in multichannel carrier systems. These tones may interact with these carriers to produce confusing effects on the measurements made at the distant end of the circuit. Figure 2.5 illustrates the concept of TLPs, loss, and gain.

Transmission Level Points

Several measurement points in circuits are traditionally set at a specified level. For example, the output of a switch is considered a 0 TLP. This does not, however, mean that signals leave the switch at 0 dBm; usually they are lower. For example, if a 0 dBm signal is inserted at the end of a subscriber loop that has 5 dB of loss, and if the switch inserts an additional 0.5 dB of loss, the output measurement at the 0 TLP will be −5.5 dBm.

Carrier channels are normally designed with TLPs of −16 dBm into the transmitting port and +7 dBm out of the re-

FIGURE 2.6
C Message Weighting Response Curve

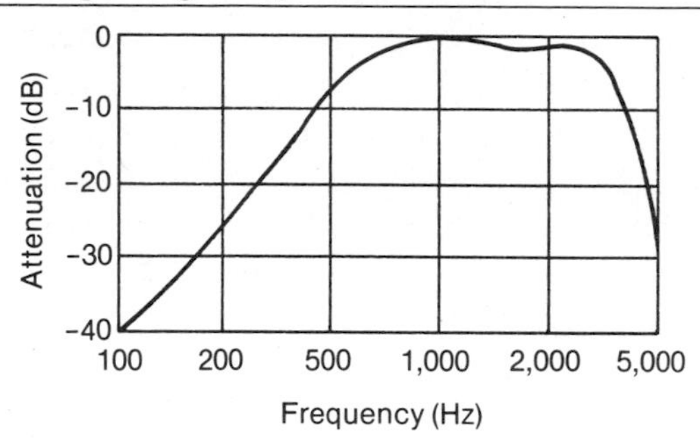

ceiver. Thus, a carrier system introduces into the circuit 23 dB of gain which can be used to overcome other circuit losses. If this is more gain than needed, it is adjusted with fixed loss pads.

Insertion Loss

Circuit design is simplified by treating certain elements as "black boxes" with identifiable loss characteristics. For example, when a connection traverses a switching machine, it is generally assumed to have a 0.5 dB loss. Many PBXs have a loss of 4 or 6 dB. Circuit designers refer to this as *insertion loss* or *inserted connection loss* (ICL). Insertion losses are additive. If two trunks each with 5 dB of loss are connected into a circuit, they will introduce 10 dB of loss into the connection.

Reference Noise

Noise is measured with respect to an arbitrary reference noise (rn) level of −90 dBm. This level, defined as 0 dBrn, is at the threshold of audibility. As mentioned earlier, not only the level but also the frequency of noise determines its interfering effects. If noise is evenly distributed across a voice frequency band (called *white noise*), the noise in the 500–2,500 Hz range will be more annoying to the listener than low- and high-frequency noise because both the ear and the telephone are more sensitive to these middle frequencies. Therefore, noise is measured through a filter known as "C message weighting." This weighting, shown in Figure 2.6, passes noise in roughly the same proportion as the sensitivity of the human ear. The interfering effect of noise on voice communication is usually expressed as dBrnc, with the "C" indicating the use of a C message weighting filter. When noise is measured at a zero level TLP or mathematically adjusted to a 0 TLP, it is expressed as dBrnc0. Special service circuits such as data and broadcast audio do not have the same tolerance for high- and low-frequency noise. Therefore, they are measured without the C message weighting filter.

A noise measurement is a measure of the noise power in a connection and is not directly additive in the same way as loss. Doubling the noise power in a circuit increases the noise by 3 dB. Thus, if two circuits that each have a noise level of 20 dBrn are connected in tandem, the result will be a noise level of 23 dBrn.

Echo Return Loss

As mentioned previously, listeners are most sensitive to interference in the frequency spectrum between about 250 Hz and 2,500 Hz. Measuring a circuit's return loss in that band produces a value called *echo return loss*. The measurement is made by transmitting a band of white noise that is limited to 250 to 2,500 Hz and measuring the returned noise energy. A useful companion measurement is the *singing point* in which an amplifier is inserted into the circuit to increase the gain to the point that the circuit just begins to sing. The circuit tends to sing at the frequency where it has the poorest return loss. Measurements of echo return loss, singing loss, and singing frequency are important indicators of circuit performance.

Subscriber Loop Transmission

Of the circuit elements, subscriber loop transmission is the most difficult to control because of the varying distance of users from the telephone central office and the varying composition of the circuits that serve them. Subscriber loop losses vary from 0 dB (some are inside the telephone central office) to as much as 8 or 10 dB. It is technically possible to design all subscriber loops to some target figure, say, 5 dB. This could be done by inserting resistance networks (pads) into shorter loops and amplifiers into loops with more than 5 dB of loss. However, with so many subscriber loops, the cost of designing them all to a fixed loss would be prohibitive.

Loop Loss

Loops are designed with a primary objective of allowing at least 23 milliamps of current to flow in the combination of the telephone set, cable circuit, and central office battery feed circuitry. Of these elements, the loss of the cable is the most variable because its resistance varies with the wire gauge and the length of the loop. Cables are built to design objectives that typically result in circuits of 8 dB or less of loss. Except for those few that are derived on multiplex equipment, subscriber loops are connected to the central office with cable. Loop design balances cost against transmission quality. To the greatest degree possible, fine-gauge cable is used because its cost is lower and its smaller diameter fits into conduit more readily. Most loops leave the central office in cables composed of 26-gauge wire. This provides adequate transmission to about 15,000 feet from the central office. The effects of loss in longer loops are overcome by using coarser gauge cable; 24-, 22-, and, rarely, 19-gauge wire is used.

The frequency response of cable is reasonably linear. Cable loss is a composite of resistance and capacitance loss. Cable pairs, acting like large capacitors, attenuate the high more than the low frequencies. The effects of high-frequency loss are overcome in loops more than 18,000 feet long by using inductance coils in series with the conductors. This technique, known as *loading,* improves the loss of the loop at the expense of frequencies above 4 kHz, which are attenuated by the load coils. Figure 2.7 shows the frequency response of loaded and nonloaded cable pairs. While loading improves voice frequency characteristics of cable, it blocks data communications above 9,600 b/s (bits/second) without the use of sophisticated and expensive modems.

Shielding

Properly constructed loops are enclosed in shielded cables and are not greatly affected by external noise. A large percentage of the loops in the United States are built jointly

FIGURE 2.7
Loss of 10,000 Feet of 26-Gauge Cable

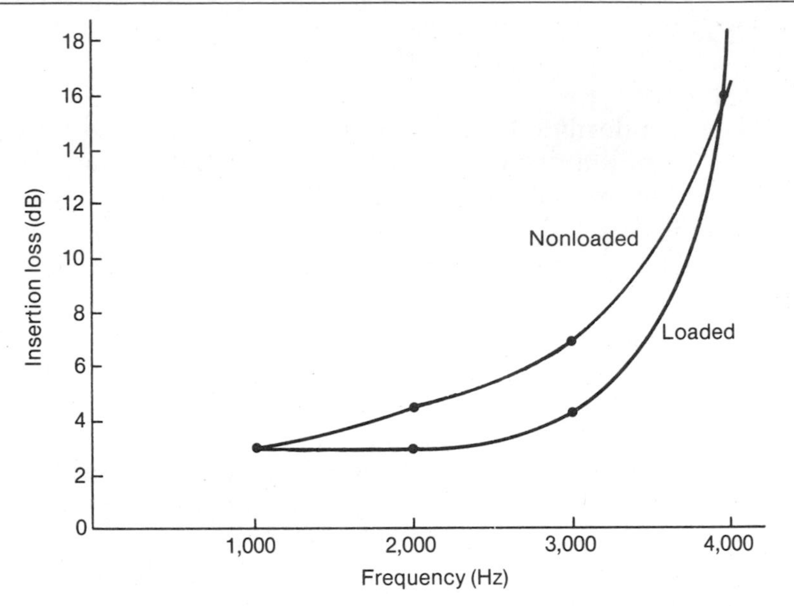

with electric power transmission lines, which are a potential source of noise. Continuity of the metallic shield that surrounds cable pairs must be strictly maintained if noise influence is to be minimized. Failure to attend to shielding and balance, described in the next section, accounts for the majority of noise problems in telephone circuits.

Balance

Induction, by itself, is not necessarily detrimental to telephone circuits. In rural areas, unshielded open wires often share pole lines with electric power circuits without interference. The degree of interference depends on the *balance* of the telephone circuit. Balance means that each wire in the cable has exactly the same amount of exposure to interference. In a balanced environment, interference induced in one of the wires of the pair will be canceled out by the induced interference in the second wire. As shown in Figure 2.8a, when the two wires of a cable pair are identically balanced in resistance and isolation from ground, induced voltages are equal when measured from each side of the pair to ground.

FIGURE 2.8
Noise Current Flow in Cable

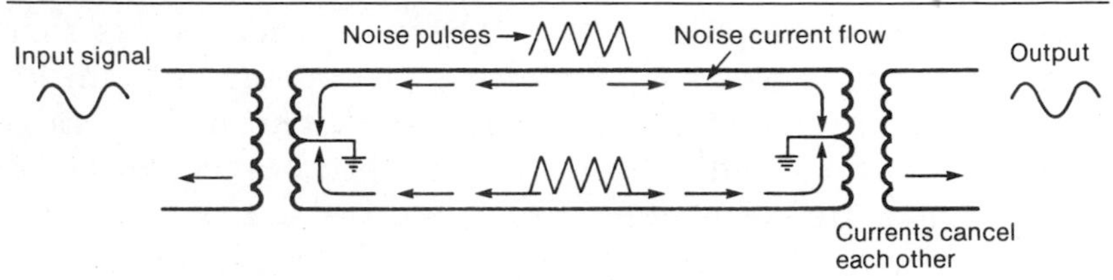

a. Balanced Cable Pair

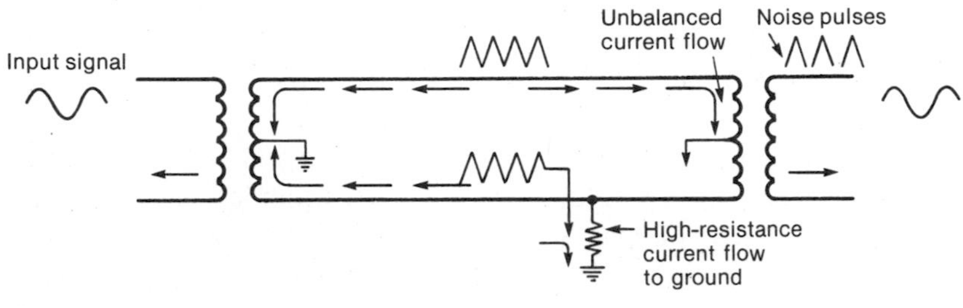

b. Unbalanced Cable Pair

Telephone sets and amplifiers, which detect voltages across the two wires of a pair, are insensitive to these balanced voltages, but any imbalance results in noise in the output circuit as shown in Figure 2.8b. Chapter 7 describes how cable balance is controlled. For now, it is necessary only to understand its effects. Noise on most subscriber loops should be in the order of 5 or 10 dBrnc0. When the noise exceeds 20 dBrnc0, it is generally an indication that corrective action is called for.

Loops exist in a hostile environment. They are exposed to weather, flooded manholes, ice, winds, and disruptions by excavations; the hazards are numerous and difficult to avoid. Therefore, this portion of a connection exhibits the greatest variability in transmission performance. Bandwidth, envelope delay, and amplitude distortion are less affected by the hazards that confront subscriber loops, but loss and noise are greatly affected by variability in subscriber loops.

Toll Connecting Trunks

Trunks connecting an interexchange carrier's toll office to a local Class 5 switching machine are called *toll connecting trunks.* These trunks are directly connected from the toll office to the local office, or they are connected through an *access tandem* switching office. These are the circuits that connect the user to an operator or to recording and billing equipment for directly dialed calls, or that terminate calls from a distant toll office to the local central office. Some of these trunks are connected over voice frequency cable circuits, but the trend is toward carrier. In any event, these trunks almost invariably have adjustable gain, and the loss, therefore, can be controlled within close limits. The design objectives for toll connecting trunks prescribed by the Bell System's *Notes on the Network* range from 2.0 to 5.4 dB of loss.

Noise on toll connecting trunks is also controllable. Trunks connected over carrier facilities are subject to thermal and intermodulation noise. Voice frequency trunks are subject to the same external interfering effects as subscriber loops, although their cables are usually installed in a less hostile environment and are more carefully balanced.

All such trunks are subject to interference from within the telephone central office. In electromechanical central offices, relays and switches are a source of interference that affect all types of trunks. This interference usually takes the form of impulse noise—short spikes of noise that have little effect on voice communication but are a serious source of errors in data transmission.

Intertoll Trunks

Intertoll trunks are circuits that interconnect Class 4 and higher switching offices. These trunks are, with rare exception, deployed over carrier facilities. Thus, the loss is controllable, and the causes of noise found in cables are of small concern.

Intertoll circuits are designed with loss according to via net loss design rules to aid in controlling the interfering effects of echo. Analog intertoll trunks operate with a variable amount of loss determined by via net loss rules and are tested with 2 dB pads in each end to optimize the signal level with respect to the TLPs. Similarly, digital trunks are operated at zero loss and are tested with 3 dB pads in the receiving path at each end to provide consistency with the analog environment. When analog circuits are connected in tandem to build up a long circuit, the overall connection of intertoll trunks operates at via net loss. Similarly, digital circuits operate at zero loss when connected in tandem.

Via Net Loss

The variable component in trunk design is determined by a factor known as the *via net loss* (VNL) factor of the circuit. A VNL factor of 0.0015 dB per mile is used for analog terrestrial carrier circuits; for example, a circuit 1,000 miles long would require 1.5 dB of loss. In practice, somewhat more loss than this is added for administrative reasons. The amount is unimportant for this discussion.

Intertoll circuit loss is administered by the interexchange carrier. Private network users who obtain circuits from a carrier will be provided with circuits with a *net loss* that is designed and controlled by the carrier. On dialed-up connections, considerable variability in end-to-end loss is possible because as many as seven intertoll trunks can be connected in tandem in a five-level hierarchy such as that used by AT&T Communications. Each of these circuits may introduce from 0 to 2.9 dB of loss, assuming loss is maintained precisely according to design criteria.

The net loss of circuits is subject to variation because of maintenance actions, component aging, and equipment troubles. Statistically, the variations between circuits should compensate for one another. However, because of the random nature of connections through the switching network, it is possible for a small percentage of the connections to be established with either so much loss that it is difficult to hear or so little that the circuit sounds hollow or, in the worst case, may oscillate or sing.

Noise Sources

Intertoll circuits are susceptible to the same noise sources as toll connecting trunks. As intertoll circuits rarely terminate in electromechanical Class 5 offices, these are not an important noise source. However, the worldwide span of intertoll circuits subjects them to noise sources that are not

FIGURE 2.9
Effects of Thermal and Intermodulation Noise on Circuit Noise

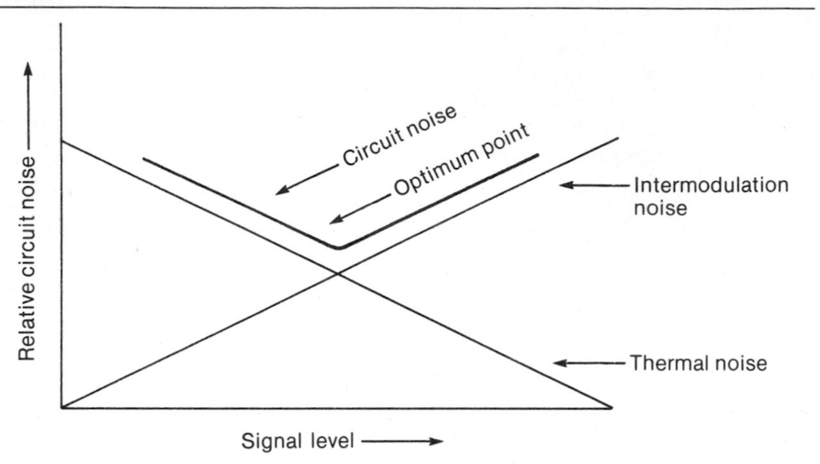

normally encountered in subscriber loops and toll connecting trunks.

Most trunks over 500 miles long, as discussed in Chapters 5 and 15, are carried over coaxial cable or microwave radio. Lightwave communications systems, which provide superior transmission performance, are being implemented over these longer distances, but the percentage of intertoll circuits carried on lightwave will not be significant for several years.

Both microwave and the multiplex equipment used on microwave and coaxial cable are subject to intermodulation noise. As mentioned earlier, both thermal noise and intermodulation noise are present in electronic communications systems. The overall noise effect on a circuit is the sum of both of these types of noise. The effect of noise is optimized by adjusting the operating level of the signal applied to the circuit. When the signal level is very low, the thermal noise becomes controlling, and when the signal level is high, it causes a high level of intermodulation noise. As Figure 2.9 shows, there is an optimum operating level for the signals where the noise from both sources is at a minimum.

Microwave radio systems are also susceptible to other noise sources, particularly noise in the first receiver amplifier stages. This noise becomes perceptible when the received radio signal fades. As the received signal falls, it approaches the noise level generated in the front end of the receiver. This noise is then amplified along with the desired signal, and appears as a deteriorated signal-to-noise ratio. Microwave systems are usually equipped with standby transmitters and receivers that can take over when the regular channel fades or fails. This protection or *diversity* is discussed more fully in Chapter 15.

Coaxial cable systems become noisy as amplifiers age and their signal-to-noise ratio degenerates. Fiber optic systems present the best noise performance of any transmission medium in use today, but fiber optic facilities are frequently unavailable in common carrier networks and are expensive for many private network applications.

Echo Control

Echo is an important variable in intertoll network design. As previously mentioned, short delay time echo is controlled by introducing a small amount of loss proportional to the delay of the circuit. Beyond a delay in the order of 10 to 20 milliseconds, however, the loss begins to be too great for satisfactory transmission, and echo suppressors or cancelers must be inserted in the circuit.

Local Interoffice Trunks

In a metropolitan network composed of more than one central office, the end or local central offices are linked by local interoffice trunks. These trunks are normally designed to 3 dB of loss, but the loss may be as high as 5 dB. They are used for calls within a local calling area, and thus are not used for access to interexchange facilities. Therefore, loss and noise are less critical than in toll connecting trunks, and echo is of no consequence. Balance is still important because a poor return loss will result in hollowness or singing. The facilities and equipment used in interoffice trunks are identical to equipment used in toll connecting trunks and the same loss and noise considerations apply.

Special-Purpose Trunks

Telephone companies use a variety of trunks for special applications. Examples are directory assistance, intercept (used for recorded announcements and operator announcement of disconnected numbers), repair service, and verification (used for operators to verify busy/idle status of lines). The design of these trunks is wholly within the purview of the telephone company, and is of little concern to the user.

Private networks may employ special-purpose trunks for a variety of applications. Examples are PBX tie lines, WATS lines, 800 (INWATS) lines, point-to-point voice and data, broadcast audio, wired music, and telemetering. These circuits are deployed over private circuits within the user's premises, over private or common carrier facilities in a metropolitan area, or over leased or common carrier facilities worldwide. Satisfactory transmission quality will be obtained only by careful design of these circuits. Common carriers usually offer design assistance. However, the common carrier often lacks knowledge of the total makeup of a circuit and cannot control transmission variables. It is, therefore, essential that all users of special services understand the effects of transmission design on the systems they are purchasing.

Loss and Noise Grade of Service

Transmission measurements can be made with precision, and with modern equipment, a high degree of level stability and noise performance are achievable. This does not answer the question of how good a circuit must be to satisfy its users. This evaluation is subjective. A hard-of-hearing person finds a connection perfectly satisfactory that others complain is too loud. It is clear that transmission quality must be evaluated against a widely varying base of opinion.

Transmission objectives are based on the results of numerous opinion samples measured by Bell Laboratories. In these tests, varying amounts of loss and noise are introduced into connections, and users are asked to rate the quality. The result of these samples is a family of loss/noise grade-of-service curves. These opinion curves are translated into design and maintenance objectives for both loss and noise.

Transmission quality is assured by controlling three elements:

- Circuit design.

- Circuit maintenance.
- Overall connection evaluation.

In circuit design, loss is the only adjustable characteristic. The other variables, principally noise and echo, are functions of how well telecommunications equipment is designed and maintained. It is unrealistic to expect equipment design and maintenance to overcome the effects of poor circuit design. Circuits with excessive loss will produce connections with excessive noise. Circuits with too little loss will result in singing, increases in intermodulation noise, and complaints from users of excessive level or hollowness.

Conversely, properly designed circuits cannot be expected to provide satisfactory transmission service when equipment is not maintained to high standards. The nation's telecommunications network is composed of equipment ranging from new to more than 40 years old. Although most new equipment is capable of excellent transmission performance with minimum maintenance, regular testing is required to ensure satisfactory performance.

Equipment and circuit maintenance are not enough to ensure that virtually all connections will be satisfactory to the user. In both public and private networks, a regular sampling program of end-to-end connection quality is an important quality assurance technique. By making repeated loss and noise measurements to terminating locations resembling the calling patterns of the users, irregularities can be detected before they result in user complaints.

TRANSMISSION MEASUREMENTS

Transmission quality assurance requires systematic measurement with accurately calibrated test equipment. This section describes transmission measurement techniques and equipment.

Loss Measurements

Loss is measured with a transmission measuring set (TMS) that consists of a tone generator or oscillator and a level detector with a meter or digital readout. Many TMSs also include noise measuring apparatus. A TMS is used to measure level at specified TLPs and at impedances that match the impedance of the TLP. Loss is nominally specified at 1,000 Hz. In practice, measurements are made at 1,004 Hz to prevent interference with digital carrier equipment.

Measurements are made by sending a single tone into the circuit at a transmitting TLP and measuring the level at a receiving TLP. Levels are adjusted by changing pads so the design loss is achieved within specified limits. A TMS is also used to measure frequency distortion of a circuit by sending and receiving a band of frequencies rather than a single tone.

Most common carrier and many private networks are equipped to make automatic transmission measurements on circuits. Both the transmitting and receiving ends are equipped with testing systems that automatically send and measure test tones in both directions. Equipment at the far end, known as a *responder,* records measurements from the near end and reports the results over a data circuit. Most automatic test equipment is also designed to measure noise and to test signaling. Refer to Chapters 12 and 23 for more information on automatic circuit testing.

Noise Measurements

Noise is measured with a noise measuring set that can either be a separate instrument or part of a TMS. Noise measurements are made at a TLP with the far end terminated in its characteristic impedance. If the circuit is used for voice communication, measurements are made through a C message filter. If the TLP is not a 0 level point, noise measurements are adjusted to 0 and expressed as dBrnc0. For example, if noise measures 27 dBrnc0 at a +7 TLP, the 7 dB gain would be subtracted and the circuit noise expressed as 20 dBrnc0.

Return Loss Measurements

Return loss measurements are made by sending a signal into the input port of a four-wire terminating set and measuring the signal returned to the output port. If a 1 kHz tone is used for the test signal, the resulting measurement may not reveal the worst-case return loss because the hybrid balance is invariably not uniform across the voice frequency spectrum. To determine the degree of hybrid balance in a working circuit, it is necessary to measure return loss at enough frequencies across the voice band to permit plotting a return loss curve that shows the worst-case frequency.

To simplify testing, a white noise source in the band of 250 to 2,500 Hz is used as a test signal. Typical return loss measuring instruments are equipped with a white noise source, band limiting filters, and high and low pass filters so return loss measurements can be made over the entire voice frequency band.

Envelope Delay Measurements

Envelope delay measurements are made with test sets that send a pair of closely spaced frequencies from the sending end of the circuit to a synchronized test set at the receiving end. The combination measures directly the relative delay of signal frequencies at various points within the circuit pass band. The ideal circuit would display linear delay across the pass band so that all components would be transmitted in a perfect phase relationship.

STANDARDS

Transmission standards are derived from a variety of experiments performed by AT&T and CCITT. In the United States, suppliers of telecommunications circuits and services are free to design their networks to whatever criteria they choose. Users should be alert to differences in transmission quality and may wish use CCITT recommendations as references in comparing alternate sources. The following are the principal transmission recommendations of the standards agencies:

ANSI

ANSI/IEEE 455 Test Procedure for Measuring Longitudinal Balance of Telephone Equipment Operating in the Voice Band.

ANSI/IEEE 661 Method for Determining Objective Loudness Ratings of Telephone Connections.

ANSI/IEEE 269 Method of Measuring Transmission Performance of Telephone Sets.

ANSI/IEEE 820 Telephone Loop Performance Characteristics.

Bell Communications Research

TR NPL-000002 Item 362 Estimated Transmission Performance of Switched Access Service Feature Group-D.

CCITT

G.101 The transmission plan.

G.102 Transmission performance objectives and recommendations.

G.113 Transmission impairments.

G.114 Mean one-way propagation time.

G.117 Transmission aspects of unbalance about earth.

G.120 Transmission characteristics of national networks.

G.122 Influence of national networks on stability and echo losses in national systems.

G.123 Circuit noise in national networks.

G.125 Characteristics of national circuits on carrier systems.

G.131 Stability and echo.

G.132 Attenuation distortion.

G.133 Group-delay distortion.

G.134 Linear crosstalk.

G.141 Transmission characteristics of exchanges.

G.143 Circuit noise and the use of compandors.

G.152 Characteristics appropriate to long-distance circuits of a length not exceeding 2,500 km.

G.153 Characteristics appropriate to international circuits more than 2,500 km in length.

G.161 Echo suppressors suitable for circuits having either short or long propagation times.

G.162 Characteristics of compandors for telephony.

G.163 Call concentrating systems.

G.164 Echo suppressors.

G.165 Echo cancelers.

G.171 Transmission characteristics of leased circuits forming part of a private telephone network.

G.221 Overall recommendations relating to carrier-transmission system.

G.222 Noise objectives for design of carrier-transmission systems of 2,500 km.

G.299 Unwanted modulation and phase jitter.

G.445 Noise objectives for communication-satellite system design.

G.441 Permissible circuit noise on frequency-division multiplex radio-relay systems.

G.611 Characteristics of symmetric cable pairs for analog transmission.

M.717 Testing point (transmission).

M.733 Transmission routine maintenance measurements on automatic and semiautomatic circuits.

P.10 Vocabulary of terms on telephone transmission quality and telephone sets.

P.11 Effect of transmission impairments.

P.16 Subjective effects of direct crosstalk: thresholds of audibility and intelligibility.

P.63 Methods for the evaluation of transmission quality on the basis of objective measurements.

P.64 Determination of sensitivity/frequency characteristics of local telephone systems to permit calculation of their loudness ratings.

P.71 Measurement of speech volume.

P.74 Methods for subjective determination of transmission quality.

P.77 Method for evaluation of service from the standpoint of speech transmission quality.

Q.29 Causes of noise and ways of reducing noise in telephone exchanges.

Q.40 The transmission plan.

Q.43 Transmission losses, relative levels.

Q.44 Attenuation distortion.

V.2 Power levels for data transmission over telephone lines.

V.50 Standard limits for transmission quality of data transmissions.

APPLICATIONS

A satisfactory transmission grade of service must be designed into every common carrier and private network. This chapter has discussed transmission in only the broadest terms to give the reader an understanding of the importance and principles of transmission quality control. In public networks, transmission design is under control of the company supplying the service. When only one network existed in the United States, transmission design was controlled by AT&T and its subsidiaries. The numerous independent companies connected to the network designed their circuits to the same objectives so the network could be considered a single entity. Private networks were largely composed of circuits obtained from telephone companies and were also designed to AT&T-specified standards and objectives.

Today, the nation's telecommunications system is composed of a multitude of networks designed and controlled by the seven Bell operating regions, the independent telephone companies within their territories, and numerous interexchange carriers of which AT&T Communications is one. Although the networks of AT&T and its ex-affiliates are not significantly changed from their predivestiture designs, the shape of the remaining network is changing in ways that are described in Chapter 9. Each interexchange carrier sets its own transmission objectives, and the quality is therefore not necessarily uniform.

Users of telecommunications services need to be aware that the facilities offered by the carrier of their choice may provide different grades of service. This is not to imply that the service will be unsatisfactory. The nature of today's network is to provide choices at different cost levels; and for some applications, lower transmission quality may be acceptable. It is important to be able to evaluate quality and to understand the implications of the different alternatives.

The most important implication of the multiple networks of the kind we have in the United States today is that the vendor may not assume responsibility for end-to-end circuit performance. If the user has telephone sets made by one vendor, cable by another, a PBX by a third vendor, connected by circuits provided by a local telephone company and one or more interexchange carriers, it may be difficult to determine which vendor is at fault when transmission quality is impaired. This makes it imperative that an overall network design be developed before the equipment and circuits are procured.

The design of a telecommunications network must include performance specifications for each of the elements. For example, in designing a PBX tie trunk to a particular net loss, it is important to know how much of each impairment to assign to the loops, the PBXs, the interexchange access lines, and the long-haul trunks provided by the interexchange carrier. If calculations are made in advance and each vendor agrees to provide equipment and services that meet these specifications, offending circuit elements can be readily identified.

Do not assume that shortcomings in one part of a circuit can be compensated for in another part. For example, if the interexchange trunk has excessive loss, it cannot be made up by amplifiers at the station. To do so would likely result in noise in the receiving direction and excessive output level in the transmitting direction. Excessive output level causes crosstalk and distortion.

Transmission Traps for the Unwary

This section presents examples of some common causes of unsatisfactory transmission in private networks. No attempt is made to describe all the traps that can occur; the purpose is to demonstrate that users must be alert to avoid obtaining telecommunications services that provide unsatisfactory results.

Add-On Conferencing

Loops are designed to a maximum loss of 8 to 10 dB. With these losses, most connections will be satisfactory, depending on the talker's volume, circuit noise, and room noise. Many PBXs and key telephone systems allow multiparty conferencing by directly connecting lines together. Each time two lines are tied together, the received signal power is divided equally between them. This introduces 3 dB of loss, turning a loop with 8 dB of loss into one with 11 dB of loss. Depending on the loop loss, the number of stations tied together, and the loss of the circuits connecting them, this form of conferencing may be satisfactory; however, the results are not dependable because of the variability in the end-to-end circuit loss.

The most reliable, although more expensive way to handle multiport conferencing is through special apparatus known as *conference bridges.* These devices are mounted in the telephone company central office where they can be dialed up, or on the user's premises where they can be connected by dialing or by a PBX operator.

High-Loss PBX Switching Networks

Modern digital PBXs use four-wire switching networks. In contrast to older two-wire electromechanical PBXs, these machines have hybrids in all line interface circuits. To reduce costs, the hybrids often insert 2 dB of loss in each line circuit. Thus, on line-to-line connections, the loss across the switching network is 4 dB. Trunk hybrids are usually zero loss, so the line-to-trunk loss is 2 dB. For connections within most offices, the amount of cross-PBX loss is inconsequential. Several special situations can, however, result in transmission problems.

The first case involves a PBX with numerous off-premise lines. If lines to these distant stations have, say, 5 dB of loss, users in that location will experience 14 dB of line-to-line loss (two 5 dB loops plus 4 dB switching loss). This loss is tolerable to most talkers, providing the room and circuit noise are not too high. With high noise, these connections are apt to be the source of complaints.

The contrast in volume is also a frequent source of complaint. Users at the off-premise location may find it difficult to understand why connections to someone in the main PBX location are so much better than with someone in the next room.

When the PBX includes WATS lines or tie trunks to a distant machine, the extra loss from the PBX may result in complaints. Such trunks are designed to a net loss of about 5 dB. The quality of the connection depends on noise and other losses in the circuit. When long loops are present in both ends of the connection, loss is likely to be excessive. When add-on conferencing is used with these trunk calls, transmission is almost certain to be poor.

The electronic hybrids used in some PBXs are also a source of noise. In a noisy environment, room noise feeds into the telephone transmitter and is amplified in the hybrid. The circuit noise is therefore increased by the room noise.

The solution to PBX transmission problems is careful evaluation of the system before purchase. The PBX should be evaluated in an environment similar to that in which it will be used. Transmission calculations should be made on all worst-case combinations of line and trunk connections. When services are procured from interexchange carriers and the local telephone company, the companies' transmission specifications should be obtained. Sometimes it may be necessary to purchase a higher grade of service to obtain satisfactory transmission. In other cases, it may be necessary to make regular end-to-end loss and noise measurements to ensure that circuit elements are meeting their specifications.

Satellite Services

Geostationary satellites orbit the earth at an altitude of 22,300 miles. At the speed of light, it takes a radio signal about one-quarter second to travel from earth to satellite and back. Although the cost of satellite circuits may often be less than the cost of terrestrial circuits, the 0.25 second path delay of a satellite circuit makes it unsatisfactory for certain types of communication. Voice communication users usually become accustomed to the delay and are not adversely affected by it, but many data protocols will not function over a satellite. These will be discussed in greater detail in Chapters 3 and 17.

Remote Office Locations

Companies choosing a new office location should always consider its distance from the telephone company's central office. Loop design standards call for loop losses up to 8 dB before amplification is applied. Customers requiring loops with lower loss than this may be required to pay a higher rate for special service lines. When long loops are clustered in a single location, many complaints of poor transmission can result. A large percentage of the calls will be satisfactory, but organizations at the outer fringes of a central office serving area can expect a greater than average number of complaints. PBXs are not so greatly affected by these types of transmission problems because the central office trunks are designed to more exacting standards than ordinary loops.

GLOSSARY

Amplitude distortion: Any variance in the level of frequencies within the passband of a communications channel.

Balance: The degree of electrical match between the two sides of a cable pair or between a two-wire circuit and the matching network in a four-wire terminating set.

Balancing network: A network used in a four-wire terminating set to match the impedance of the two-wire circuit.

Bandpass: The range of frequencies that a channel will pass without excessive attenuation.

C message weighting: A factor used in noise measurements to approximate the lesser annoying effect on the human ear of high- and low-frequency noise compared to mid-range noise.

Conditioning: The process of applying special treatment to a private line circuit to improve its data carrying ability.

Crosstalk: The unwanted coupling of a signal from one transmission path into another.

dB: An abbreviation for decibel, a measure of relative signal voltage or power.

DBm: A measure of power level relative to the power of 1 milliwatt.

DBrn: A measure of noise power relative to a reference noise of −90 dBm.

DBrnc: A measure of noise power through a C message weighting filter.

DBrnc0: A measure of C message noise referred to a zero test level point.

Echo: The reflection of a portion of a signal back to its source.

Echo canceler: An electronic device that processes the echo signal and cancels it out to prevent annoyance to the talker.

Echo return loss: The weighted return loss of a circuit across a band of frequencies from 500 to 2,500 Hz.

Echo suppressor: A device that opens the receive path of a circuit when talking power is present in the transmit path.

Envelope delay: The difference in propagation speed of different frequencies within the pass band of a telecommunications channel.

Facility: Any set of transmission paths that can be used to transport voice or data. Facilities can range from a cable to a carrier system or a microwave radio system.

Four-wire circuit: A circuit that uses separate paths for each direction of transmission.

Four-wire terminating set: A device that combines the separate transmit and receive paths of a four-wire circuit into a two-wire circuit.

Grade of service: A quality indicator used in transmission measurements to specify the quality of a circuit based on both noise and loss.

Hybrid: A multiwinding coil or electronic circuit in a four-wire terminating set or switching system line circuits to separate the four-wire and two-wire paths.

Hz: An abbreviation for hertz, a unit of frequency. One Hz is equal to one cycle per second.

Insertion loss: The amount of loss a circuit element introduces into a telecommunications circuit.

Intermodulation distortion: Distortion or noise generated in electronic circuits when the power carried is great enough to cause nonlinear operation.

KHz: Kilohertz or thousandths of cycles per second.

Level: The signal power at a given point in a circuit.

Loading: The process of inserting fixed inductors in series with both wires of a cable pair to reduce voice frequency loss.

Loss: The drop in signal level between points on a circuit.

Milliwatt: One one-thousandth of a watt. Used as a reference power for signal levels in telecommunications circuits.

Noise: Any unwanted signal in a transmission path.

Pad: A device inserted into a circuit to introduce loss.

Reference noise (rn): The threshold of audibility to which noise measurements are referred, equal to −90 dBm.

Repeater: An electronic device that adds gain or amplification to a circuit.

Return loss: The degree of isolation, expressed in dB, between the transmit and the receive ports of a four-wire terminating set.

Sidetone: The sound of a talker's voice audible in the handset of the telephone instrument.

Signal-to-noise ratio: The ratio between signal power and noise power in a circuit.

Singing: The tendency of a circuit to oscillate when the return loss is too low.

Singing return loss: The loss at which a circuit oscillates or sings at the extreme low and high ends of the voice band.

Thermal noise: Noise created in an electronic circuit by the movement and collisions of electrons.

Transmission: The process of transporting voice or data over a network or facility from one point to another.

Transmission level point (TLP): A designated measurement point in a circuit where the transmission level has been specified by the designer.

Via net loss (VNL): A design procedure for long-haul trunks that controls echo by adding loss to a circuit, based on propagation speed of the transmission medium.

Volume unit (VU): A unit of speech or music level determined by reading an audio signal on a meter.

White noise: Noise frequencies that are equally distributed across all frequencies of a passband.

BIBLIOGRAPHY

American Telephone & Telegraph Co. *Notes on the Network*, 1980.

———. *Telecommunications Transmission Engineering*. Bell System Center for Technical Education, vol. 1, 1974; vol. 2, 1977; and vol. 3, 1975.

Chorafas, Dimitris N. *The Handbook of Data Communications and Computer Networks*. Princeton, N.J.: Petrocelli Books, 1985.

Freeman, Roger L. *Telecommunication System Engineering*. New York: John Wiley & Sons, 1980.

———. *Telecommunication Transmission Handbook*. New York: John Wiley & Sons, 1975.

Martin, James. *Telecommunications and the Computer*. Englewood Cliffs, N.J.: Prentice Hall, Inc., 1976.

Stuck, B. W., and E. Arthurs. *A Computer and Communications Network Performance Analysis Primer*. Englewood Cliffs, N.J.: Prentice Hall, Inc., 1985.

CHAPTER THREE

Data Communications Systems

In one form or another, data communication has been around since Samuel F. B. Morse first demonstrated the telegraph in 1844. Since 1904 when the teletypewriter arrived and began to replace manual telegraphers with electromechanical equipment, the trend of increasing machine-to-machine communication has continued unabated. Data communication today comprises a significant portion of the traffic on the telephone system, and its share is increasing.

This explosion in data communication was brought about by several factors, the most important being the dramatic drop in computer costs. Applications that formerly required an expensive mainframe computer can now be processed on mini- or microcomputers, bringing computing power closer to people. As the white collar work force increasingly uses computers to aid their work, the demand for more computing resources has burgeoned and has in turn spawned a need for shared data bases that are accessed over telecommunications facilities.

Telecommunications and the computer are vital partners in a marriage that is changing the way people store, access, and use information. New applications such as automatic teller machines, airline reservation systems, and credit card verification networks are made possible by the merger of the computer and telecommunications. However, the telephone network was designed and constructed for voice, which means that the computer, a digital device, must adapt to a network that is largely analog.

DATA NETWORK FACILITIES

Private line networks can be constructed over all-digital facilities, but for the next decade or two, most switched data communications will be handled over voice facilities. For those applications requiring nearly full-time use of a channel between fixed points, dedicated or private line facilities are the most economical. Many applications, however, require switching because they transmit data between multiple points or send only a few short messages each day. Further, many applications do not use the full capacity of a channel, so the only economical way they can be implemented is by sharing the capacity of the facility. Generally, the facilities they share can be classed as one of four types: point-to-point circuit, circuit switching, message switching, or packet switching.

Point-to-Point Circuit

A *point-to-point* circuit is permanently wired between all stations on the network. The simplest point-to-point circuits are between two stations that share exclusive access to the circuit. More complex configurations are the multidrop circuits such as that shown in Figure 3.1, in which only one station is permitted to transmit at a time. The access is allocated by a master control station, usually a central processing unit (CPU), that *polls* the slave stations. In a polling network, the CPU is attached to the network through a communications controller called a *front-end processor.* The processor sends a polling message to each station in turn. If the station has traffic, it sends the traffic. Otherwise, the station responds negatively.

A polling system keeps circuits fully occupied. However, a great deal of the circuit time is consumed with polling messages and negative responses, and these overhead characters do not contribute to information flow.

FIGURE 3.1
Polled Multidrop Network

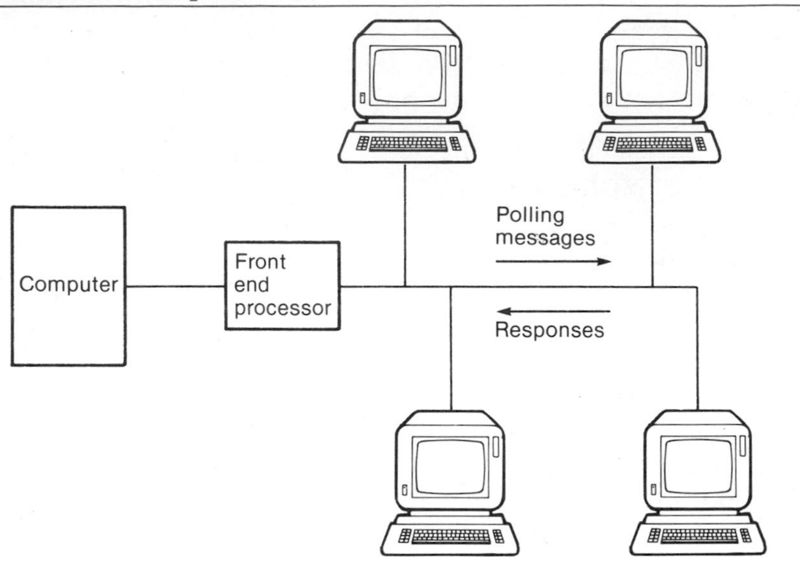

Circuit Switching

In a circuit switched network, a controller is connected to stations in a star configuration as shown in Figure 3.2. Communication is between the stations and the controller, or the controller establishes circuits between two or more stations. The connection is established by signaling from the stations to the controller, and when the stations have sent their traffic, they signal the controller to disconnect the circuit.

In circuit switched networks, the circuits to the stations (called loops) are not fully occupied. When the circuits are short, the cost of idle time is acceptable, but with long circuits, costs may be excessive if usage is light. Utilization can be improved by using a hierarchical network. In a hierarchical network, the controller is placed close to the stations so the circuits can be short and less costly. The more costly long circuits between the controllers (called intermachine trunks) are engineered for a higher occupancy rate.

Message Switching

Message switched networks are sometimes called *store and forward.* Stations home on a computer that accepts messages,

FIGURE 3.2
Circuit Switched Network

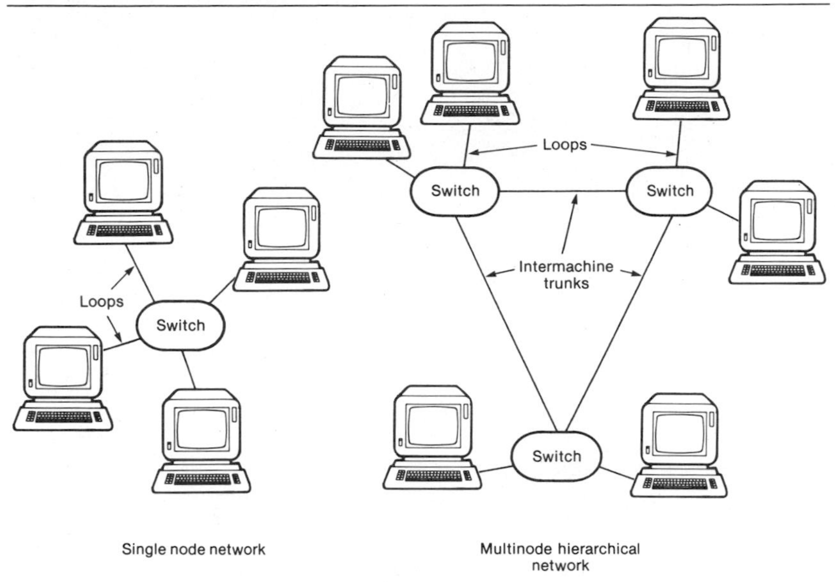

stores them, and delivers them to their destination. The storage turnaround time is either immediate for interactive applications or the message may be delayed for forwarding at a future time when circuits are idle, rates are lower, or a busy terminal is available.

Packet Switching

A packet switched network consists of control nodes that host the stations as shown in Figure 3.3. Nodes are interconnected by trunks sized to accommodate the traffic load. Data is sent from the station to the node in *packets,* which are blocks of data characters. The node moves the packet toward its destination by handing it off to the next node in the chain. Nodes are controlled by software with routing algorithms that determine the choice of the next station in the chain. In contrast to a circuit switched network where circuits are physically switched between stations, a packet switched network establishes *virtual circuits* between stations.

FIGURE 3.3
Packet Switched Network

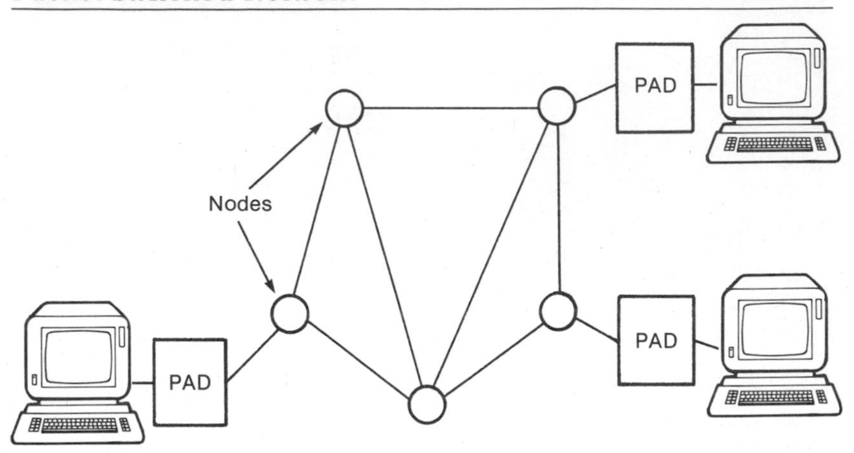

PAD = Packet Assembler/Disassembler

Virtual circuits are of two types. In a *permanent virtual circuit* mode ,the routing between stations is fixed and packets always take the same route. In a *switched virtual circuit* mode, the routing is determined with each packet.

DATA COMMUNICATIONS FUNDAMENTALS

The devices in a data network that originate and receive data are collectively called *data terminal equipment* (DTE). These can range from computers to simple receive-only terminals. DTE is coupled to the telecommunications network by *data circuit-terminating equipment* (DCE), a device that converts the DTE output to a signal suitable for the transmission medium. DCE ranges from line drivers to complex modulator/demodulators (*modems*).

The basic information element processed by a computer is the *binary digit* (*bit*). A bit is the smallest information element in the binary numbering system and is represented by the two digits, 0 and 1, corresponding to two different voltage states within the DTE. Processors manipulate data in groups of eight bits known as a *byte* or *octet.* Within the computer's circuits, bytes are transmitted over parallel paths

FIGURE 3.4
Parallel and Serial Data Conversion

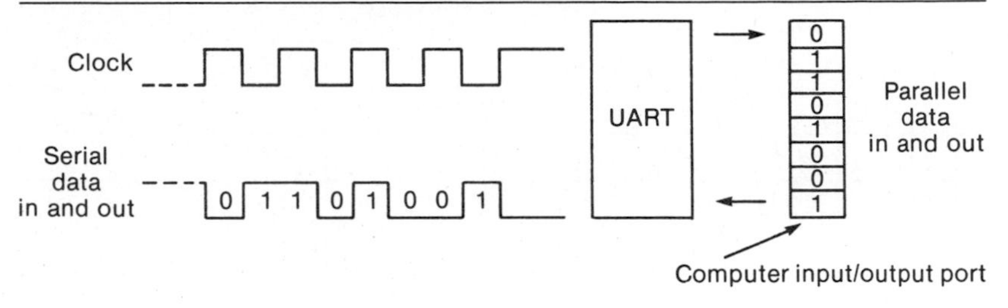

UART = Universal Asynchronous Receiver/Transmitter

that may be extended to output ports for connection to peripherals such as printers.

The range of parallel ports is limited to a few feet. Although the range can be extended with DCE, extending eight circuits in parallel over long distances is uneconomical. Therefore, most DTE is equipped with a *serial interface* to convert the eight parallel bits into a serial bit stream as shown in Figure 3.4. This serial bit stream can be coupled to telecommunications circuits through a modem or line driver.

Coding

The number of characters that can be encoded with binary numbers is a function of the number of bits in the code. Early teletypewriters used a five-level code called *baudot* that was capable of 2^5 or 32 characters. A five-level code limits communications because there are insufficient combinations to send a full range of upper- and lowercase characters plus special characters. To overcome this limitation, a seven-level code known as the American Standard Code for Information Interchange (ASCII) was introduced. This code, which is shown in Table 3.1, provides 2^7 or 128 combinations. In ASCII transmissions, eight bits are transmitted, with the eighth bit used for error detection as described later.

Several other codes are used for data communications. The most predominant is the Extended Binary Coded Decimal Interchange Code (EBCDIC) which is shown in Table 3.2. EBCDIC is an eight-bit code, allowing a full 256 characters to be encoded.

TABLE 3.1
American Standard Code for Information Interchange (ASCII)

$b_7\ b_6\ b_5$	0 0 0	0 0 1	0 1 0	0 1 1	1 0 0	1 0 1	1 1 0	1 1 1
$b_4\ b_3\ b_2\ b_1$								
0 0 0 0	NUL	DLE	SP	0	@	P	`	p
0 0 0 1	SOH	DC1	!	1	A	Q	a	q
0 0 1 0	STX	DC2	"	2	B	R	b	r
0 0 1 1	ETX	DC3	#	3	C	S	c	s
0 1 0 0	EOT	DC4	$	4	D	T	d	t
0 1 0 1	ENQ	NAK	%	5	E	U	e	u
0 1 1 0	ACK	SYN	&	6	F	V	f	v
0 1 1 1	BEL	ETB	'	7	G	W	g	w
1 0 0 0	BS	CAN	(	8	H	X	h	x
1 0 0 1	HT	EM	)	9	I	Y	i	y
1 0 1 0	LF	SUB	*	:	J	Z	j	z
1 0 1 1	VT	ESC	+	;	K	[	k	{
1 1 0 0	FF	FS	,	<	L	\	l	¦
1 1 0 1	CR	GS	—	=	M	]	m	}
1 1 1 0	SO	RS	.	>	N	^	n	~
1 1 1 1	SI	US	/	?	O	—	o	DEL

Code compatibility between machines is essential. Because EBCDIC and ASCII are both widely used, in some applications code conversion will be required. Most intelligent terminals can be programmed for code conversion, but with nonprogrammable terminals, external provisions are necessary. This can be a separate code converter, or can be accomplished as a value-added function of the network.

Data Communications Speeds

Table 3.3 shows the range of speeds and typical applications used for data communications. The speed a circuit can accommodate is a function of its bandwidth, which is limited to 300–3,300 kHz over voice frequency telephone channels for reasons explained in Chapter 2. Where wider bandwidths

TABLE 3.2

Extended Binary Coded Decimal Interchange Code (EBCDIC)

Bits			4	0	0	0	0	0	0	0	0	1	1	1	1	1	1	1	1
			3	0	0	0	0	1	1	1	1	0	0	0	0	1	1	1	1
			2	0	0	1	1	0	0	1	1	0	0	1	1	0	0	1	1
			1	0	1	0	1	0	1	0	1	0	1	0	1	0	1	0	1
8	7	6	5																
0	0	0	0	NUL	SOH	STX	ETX	PF	HT	LC	DEL			SMM	VT	FF	CR	SO	SI
0	0	0	1	DLE	DC_1	DC_2	DC_3	RES	NL	BS	IL	CAN	EM	CC		IFS	IGS	IRS	IUS
0	0	1	0	DS	SOS	FS		BYP	LF	EOB	PRE			SM			ENQ	ACK	BEL
0	0	1	1			SYN		PN	RS	UC	EOT					DC_4	NAK		SUB
0	1	0	0	SP										¢	.	<	(	+	|
0	1	0	1	&										!	$	*	)	;	¬
0	1	1	0	-	/										,	%	—	>	?
0	1	1	1										\	:	#	@	'	=	"
1	0	0	0		a	b	c	d	e	f	g	h	i						
1	0	0	1		j	k	l	m	n	o	p	q	r						
1	0	1	0			s	t	u	v	w	x	y	z						
1	0	1	1																
1	1	0	0		A	B	C	D	E	F	G	H	I						
1	1	0	1		J	K	L	M	N	O	P	Q	R						
1	1	1	0			S	T	U	V	W	X	Y	Z						
1	1	1	1	0	1	2	3	4	5	6	7	8	9						□

PF –Punch Off
HT –Horizontal Tab
LC –Lower Case
DEL –Delete
SP –Space

UC –Upper Case
RES –Restore
NL –New Line
BS –Backspace
IL –Idle

PN –Punch On
EOT –End of Transmission
BYP –Bypass
LF –Line Feed
EOB –End of Block

PRE –Prefix (ESC)
RS –Reader Stop
SM –Start Message
Others –Same as ASCII

TABLE 3.3

Data Transmission Speeds and Applications

	50 b/s	75 b/s	100 b/s	150 b/s	300 b/s	1,200 b/s	2.4 kb/s	4.8 kb/s	7.2 kb/s	9.6 kb/s	19.2 kb/s	56 kb/s	64 kb/s	1.5 mb/s	10 mb/s
Telemetry	←	→													
Telex	←	→													
Teleprinters			←	—	—	→									
Interactive Terminals					←	—	—	—	—	—	→				
Medium-Speed Data							←	—	—	—	→				
High-Speed Data											←	—	—	→	
Digital Video													←	—	→
Local Area Networks														←	→

FIGURE 3.5
Data Modulation Methods

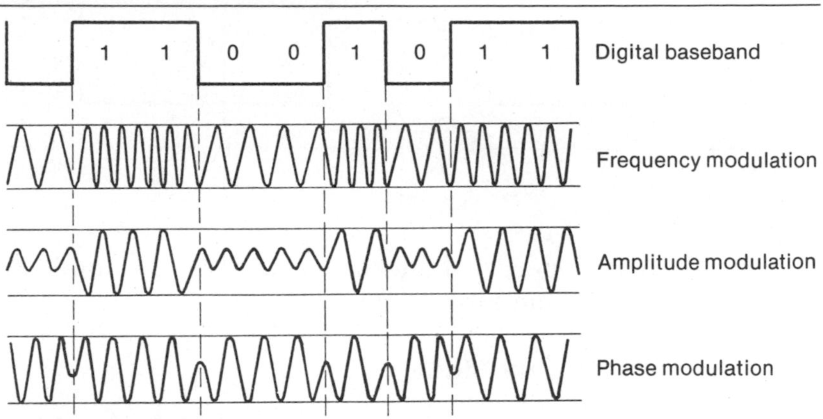

are required, special service circuits must be obtained over private facilities or through common carrier tariffs.

Two terms used to express the data-carrying capacity of a circuit are bit and *baud.* The baud rate of a channel describes the number of cycles or reversals the channel is capable of accommodating. The 3,000 Hz bandwidth of a voice channel can handle a 2,400 baud signal. If the data are coded at one bit per Hz, the channel is limited to 2,400 b/s. Higher bit rates are transmitted by encoding more than one bit per Hz.

Modulation Methods

A data signal leaves the serial interface of the DTE as a series of baseband voltage pulses as shown in Figure 3.5. Baseband pulses can be transmitted over limited distances by using a *limited distance modem* or a driver that matches the serial interface to the cable. For longer distances, the pulses are modulated by the modem into a combination of analog tones that fit within the passband of a voice channel.

The digital signal modulates either the frequency, amplitude, or phase of an audio signal as shown in Figure 3.5. Amplitude modulation is the least used method because it is susceptible to noise-generated errors. Frequency modulation is an inexpensive method used with low-speed modems. To reach higher speeds, phase shift modems are em-

FIGURE 3.6
Asynchronous Data Transmission

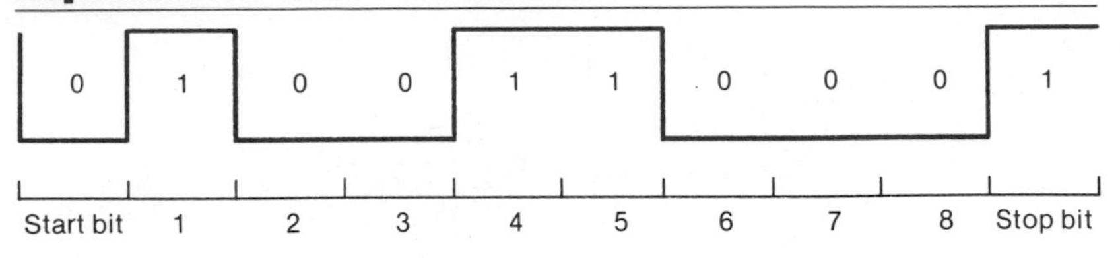

ployed. To reach speeds of more than 2,400 b/s, modems encode more than one bit per Hz.

Full- and Half-Duplex Mode

Full-duplex circuits transmit data in both directions simultaneously. *Half-duplex* circuits transmit in only one direction at a time; the channel is reversed for transmission in the other direction.

Full-duplex circuits are formed by using separate transmit and receive paths, or by using a *split channel* modem. Split channel modems divide the voice channel into two frequency segments, one for transmit and one for receive. The 2,400 baud bandwidth of a channel limits a full-duplex modem to 1,200 b/s in each direction. Modems with more sophisticated modulation are available at higher cost to provide 2,400 b/s full-duplex communication over two-wire circuits using the CCITT V.22 *bis* modulation method and 9,600 b/s using V.32 modulation.

Synchronizing Methods

Synchronization is required in all data communications channels to keep the sending and receiving ends in step with each other. The signal on a data communications channel is a series of rapid voltage reversals, and synchronization enables the receiving terminal to determine which pulse is the first bit in a character.

The simplest synchronizing method is *asynchronous*, sometimes called *stop-start* synchronization. Asynchronous signals, illustrated in Figure 3.6, begin with a start bit at the zero level followed by eight data bits, and a stop bit at the one level. Asynchronous signals are transmitted in a char-

FIGURE 3.7
Synchronous Data Transmission (IBM SDLC Frame)

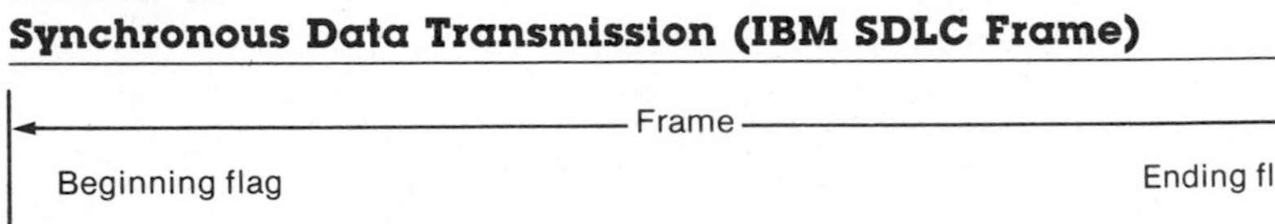
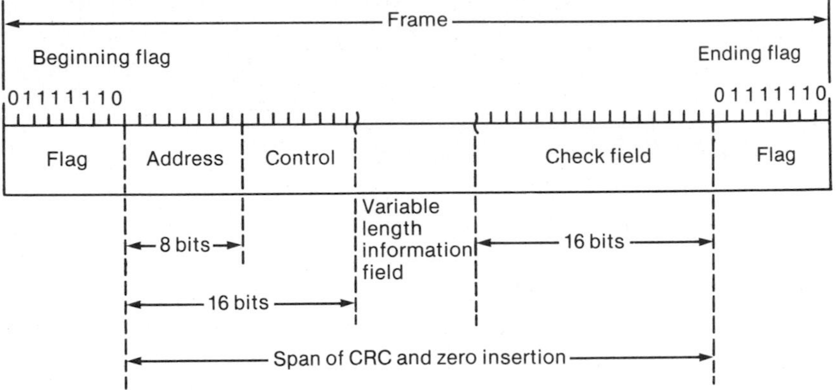

acter mode: that is, each character is individually synchronized. The chief drawback to asynchronous communication is the extra two bits per byte that have no information content. These noninformation bits are called *overhead bits.*

To reduce the amount of overhead, data can be transferred in a *synchronous* mode as illustrated in Figure 3.7. Synchronous data are sent in a block mode with information characters sandwiched between header and trailer records. The header and trailer contain the overhead bits; the information bits are transferred without start and stop bits. The two terminals are kept in synchronization by a clock signal that is derived by the modem from the incoming bit stream.

The drawbacks to synchronous signals are their complexity and lack of standardization. Variables in the data block, such as block length, error-checking routine, and structure of the header and trailer records, are functions of the protocol. Although there are some standard protocols such as High-level Data Link Control (HDLC) recommended by CCITT, many data manufacturers have their own protocols that are incompatible with one another.

Whereas asynchronous data terminals can communicate with each other if the speed, code, and error-checking conventions are identical, synchronous terminals require protocol compatibility and intelligence in the DTE. Synchronous data communication systems have offsetting advantages of greater throughput and the ability to use sophisticated error correction techniques that are not compatible with the character mode of transmission. Error detection and correction will be discussed in a later section.

EVALUATING DATA COMMUNICATION SERVICES

An effective data communication network is a compromise between many variables. The nature of data transmission varies so greatly with the application that designs are empirically determined. The network designer arrives at the most economical balance of performance and cost by evaluating alternatives as discussed in this section.

Modem Selection

One consideration in choosing a modem is whether it is directly or acoustically coupled to the telephone network. An acoustically coupled modem provides a cradle for the handset to rest in. Acoustic coupling is subject to errors and interference from room noise, defective telephone handsets, and poorly seated connections between the handset and the coupler. Direct coupling, which requires the modem to be registered with the FCC or applied through a registered coupler, is by far the most effective, and has replaced most acoustically coupled modems.

Compatibility

The primary issue in selecting modems is compatibility. The interface between DTE and the modem has been standardized in the United States, with the predominant interfaces the EIA RS-232-C, RS-422, and RS-449. These standards specify the functions of the interface circuits but do not specify the physical characteristics of the interface connector. Connectors have been adopted by convention; for example, the DB 25 connector used by AT&T Network Systems (Western Electric) has become a *de facto* standard for the RS-232-C interface. The physical connector is a minor problem to users because units can be easily interconnected with adapters if they are electrically compatible.

TABLE 3.4
Split-Channel Modem Frequencies (Bell 103)

	1	*0*
Low Frequency	1,270	1,070
High Frequency	2,225	2,025

Modem compatibility is often related to the Bell System modem numbering plan. Modems may be designated as "compatible" with a Bell modem, which means they are end-to-end compatible, or they may be designated as "equivalent," which means that they perform the same functions but are not necessarily end-to-end compatible with the Bell modem of that number.

Asynchronous modems are generally end-to-end compatible with one another. The accepted send and receive frequencies listed in Table 3.4 have been adopted by most manufacturers for split-channel operation at speeds of 300 b/s and below. Asynchronous communication is sometimes used at higher bit rates, but the bulk of the communications above 2,400 b/s is conducted in a synchronous mode.

The primary consideration in choosing an asynchronous modem is an evaluation of cost and features. The cost is closely related to the features that enhance the operation.

Features

A considerable share of asynchronous data communication is carried on over the switched telephone network. Therefore, many modem features are designed to emulate a telephone set. The most sophisticated modems, together with a software package in an intelligent terminal, are capable of fully unattended operation. Modems designed for unattended, and many designed for attended, operation are equipped with these features:

- Recognition of dial tone.
- Tone and dial pulse dialing.
- Monitoring call progress tones such as busy and reorder.
- Automatic answer.

- Call termination.

Most asynchronous 300 and 1,200 baud modems operate in a full-duplex split-channel mode. The modem answers an incoming call by sending a tone to indicate whether it will send on the high- or low-frequency segment of the channel. The other modem adjusts automatically, or the operator must manually set the modem to the receive frequency.

Synchronous modems used on half-duplex lines may be equipped with an optional reverse channel. This channel generates a signal that is either on or off, enabling it to interrupt the transmitting end. The reverse channel is usually transmitted over a narrow frequency band at the low end of the voice frequency channel.

When synchronous modems are used at speeds above 4,800 b/s, they may require channel conditioning from the common carrier. This precludes the use of these higher speeds on the switched voice telephone network. However, most manufacturers offer modems equipped with an *adaptive equalizer,* a device that automatically adjusts the modem to compensate for irregularities in the telephone channel. The adaptive equalizer substitutes for line conditioning and enables the use of 9,600 b/s on a voice channel.

Circuit throughput can be improved by using *data compression,* a system that replaces the original bit stream with another bit stream that has fewer bits. With data compression techniques and adaptive equalization, it is possible to operate at 9,600 b/s or higher over nonconditioned voice grade lines.

Other modems accomplish higher speeds by integrating *forward error correction* (FEC) modules with the modem. FEC processes a data bit stream through a complex algorithm before transmission. Additional bits are added to the data block to enable the receiving end to determine if the data block was received correctly. If it was not, the additional bits enable the receiver to reconstruct the original block.

Special Modem Types

A modem is not required for data transmission over some wire circuits. Line drivers are satisfactory over nonloaded facilities over ranges of about 10 miles and at speeds of up to 19.2 kb/s. Many telephone companies offer limited distance data services (LDDS), which are nonmultiplexed channels that are provided where nonloaded metallic facilities are available.

Special modems are available to transmit data over fiber optics or coaxial cable circuits. This equipment is used for point-to-point circuits over ranges limited only by the length of the transmission medium. Repeaters can extend the range within a metropolitan area at speeds of 10 megabits or more.

Simultaneous voice and data operation is desirable in many applications. A special type of modem called *data over voice* (DOV) multiplexes data and voice on the same point-to-point circuit. A DOV modem applies a full-duplex data channel to a carrier frequency above the voice channel. The carrier is shifted at speeds up to 19.2 kb/s with filters separating the data and voice signals. The range of DOV modems is limited to about 3 miles, which is the voice transmission range of a nonloaded cable pair.

Data Network Facilities

The data communication user has several key decisions to make in selecting network facilities. First is a question of whether the network should be owned or leased. Private ownership is the rule with local data networks, but in metropolitan networks the cost of right-of-way often precludes private ownership. Global networks are feasible for private ownership by only the largest companies. This section discusses the primary network choices the user must make in implementing a data communication network.

Common Carrier or Value Added Carrier

A common carrier delivers data communications circuits to an interface point on the user's premises. The carrier transports the signal but does not process the data.

A *value added* carrier not only transports data but also may process it or add other services such as store-and-forward switching, error correction, and authentication. Other message processing services such as electronic mail, filing, and message logging and receipting may also be provided. The value added carrier furnishes the user a dial-up or dedicated interface and provides the equivalent of a private network over shared facilities.

Switched or Dedicated Facilities

The nature of the application determines whether a switched or dedicated network service is required. With terrestrial services, the cost of a channel is directly related to length and duration of connection. If multiple users communicate over distances greater than a few miles and send a few short messages, switched services tend to be most effective. If messages are long and the number of points limited, dedicated or private line services tend to be most economical. Three types of switched services can be obtained.

Circuit switching is the connection of channels through a centralized switching machine. The telephone network is an example of a circuit switched network. Its primary advantage is its worldwide accessibility to any location with telephone service. Another advantage is that messages cannot arrive out of sequence because the circuit is intact from end to end. Its disadvantages are its cost, lack of error detection and correction capability, and limited bandwidth.

Message switching or *store-and-forward switching* is a service that accepts a message from a user, stores it, and forwards it to its destination at a later time. The storage time may be so short that it is virtually instantaneous, or messages may be stored for longer periods while a receiving terminal is temporarily unavailable, or while awaiting more favorable rates.

The primary advantage of store-and-forward switching is that the sender and receiver do not need to be on line simultaneously. If the receiving terminal is unavailable, the network can queue messages and release the originating terminal. If instantaneous delivery is not important, the store-and-forward technique can make maximum use of circuit capacity. Its primary disadvantages are the added cost of storage facilities in the switching device and the longer terminal response time compared to circuit or packet switching.

Packet switching, as described earlier, controls the flow of packets of information through a network by means of routing algorithms in each node. Stations interface the packet nodes on a dedicated or dial-up basis and deliver preaddressed packets to the node, which routes them through the network to the destination.

Packet switching is flexible. Service can be maintained during temporary overloads or circuit outages by the dynamic routing capability of the network. Its primary disadvantage for private ownership is its cost. However, value added carriers offer packet switched services at a cost competitive with other switching methods. Also, packets can arrive out of sequence, so the network must be capable of buffering received packets and reassembling them into finished messages.

Terrestrial or Satellite Circuits

Communications satellites offer a cost-effective alternative to terrestrial circuits for distances of about 1,000 miles or more. Because the tariff rates of terrestrial circuits are based on distance, terrestrial circuits are less expensive over shorter routes. Beyond a break-even point, satellite circuits are less expensive because their costs are independent of distance within the coverage field of a single satellite.

With all else equal, terrestrial circuits are more effective for data than satellite circuits. However, at longer distances, the greater cost of terrestrial circuits often offsets the disadvantages of a satellite. The primary disadvantage of a satellite circuit is the round trip propagation delay, which is approximately 0.5 seconds. Some protocols will not operate with this much delay. Other protocols are functional with throughput reduced because of error retransmission time. Satellite delay compensators, which are discussed in Chapter 17, can alleviate some of the protocol problems, but they introduce the additional disadvantage of loss of end-to-end error correction.

Another disadvantage of satellites is the lack of coverage by an earth station in some locations. If the earth station is owned by the common carrier, it may be located at such a distance that the cost of a terrestrial link may offset the cost savings. If the earth station is privately owned, its cost must be added to the cost of the data circuits.

Concentration

Many data applications, by their nature, are incapable of fully utilizing a data circuit. Rather than flowing in a steady

FIGURE 3.8
Time Division Multiplexing

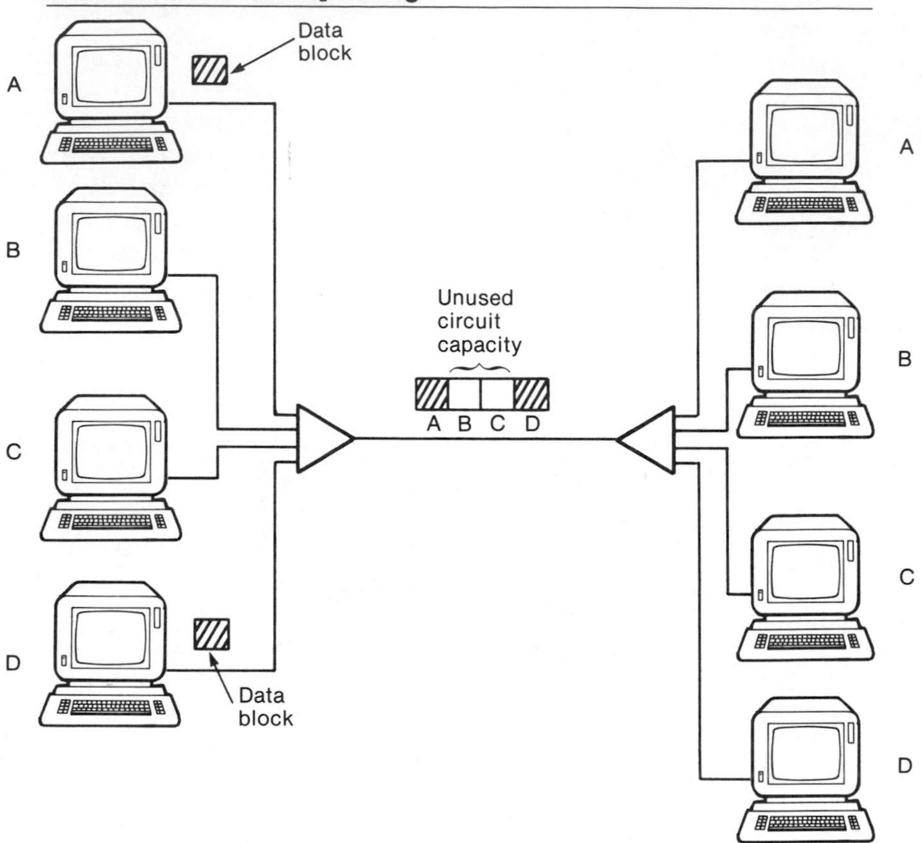

stream, data tends to flow in short bursts with long intervening idle periods. To make use of this idle capacity, *data multiplexers* are employed to collect data from multiple stations and combine it into a single high-speed bit stream.

Data multiplexers are of two types, *time division multiplexers* (TDM) and *statistical multiplexers* (statmux). In a TDM, each station is assigned a time slot and the multiplexer collects data from each station in turn. If the station has no data to send, its time slot goes unused. TDM operation is illustrated in Figure 3.8.

A statmux, illustrated in Figure 3.9, is able to make use of the idle time periods in a data circuit by assigning time slots to pairs of stations according to the amount of traffic

FIGURE 3.9
Statistical Multiplexing

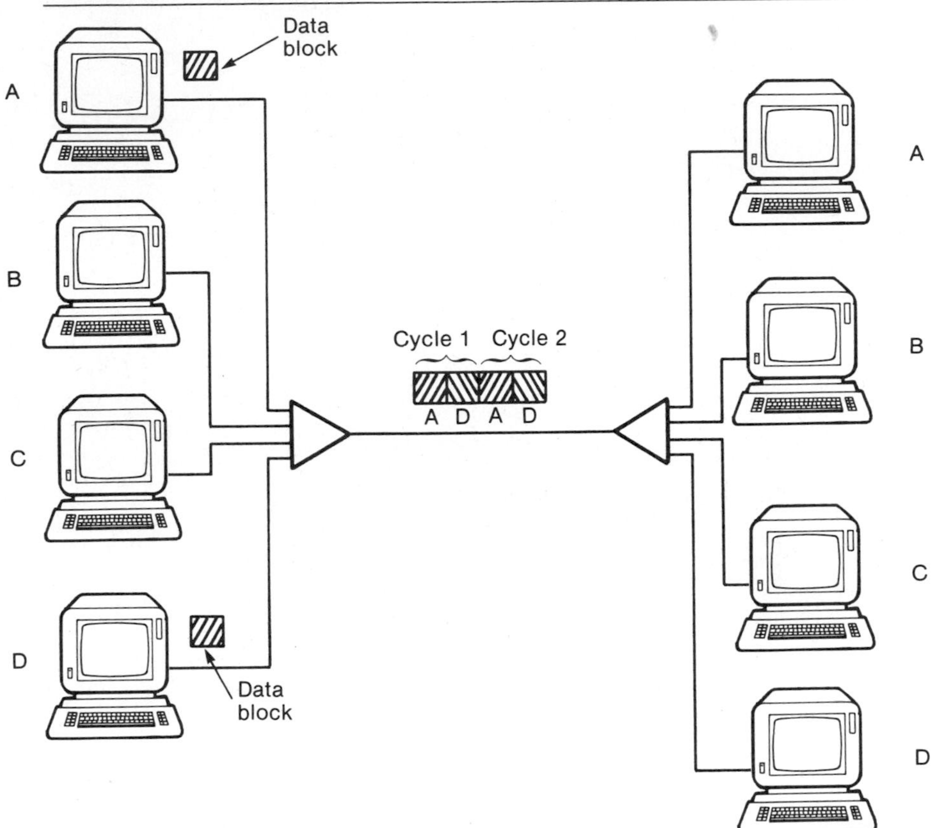

they have to send. The multiplexer collects data from the DTE and then sends the data to the distant end with the address of the receiving terminal.

Statistical multiplexers improve circuit utilization by minimizing idle time between transmissions. They are more costly than TDMs, however, and must be monitored to prevent overloads, with stations added or removed to adjust the load to the maximum the circuit will handle while meeting response time objectives.

Analog, or frequency division multiplexers, are also available to divide a voice channel into multiple segments for data transmission. These devices assign each data channel to a frequency and use frequency shift techniques similar to

FIGURE 3.10
Network Configuration

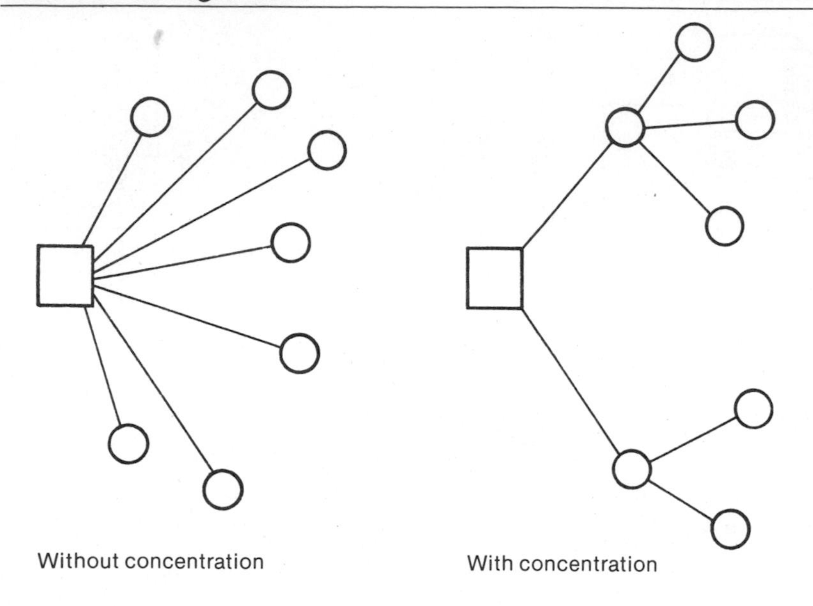

those used in a modem. Their primary use is to connect multiple slow-speed data terminals over voice channels.

Another important factor in data network design is the location of concentration points. Figure 3.10 shows two different ways of interconnecting multiple points on a circuit. The network design task is to minimize the cost of circuits, modems, and concentrators while keeping throughput at an acceptable level. In a large multipoint network, the number of possible configurations is substantial. Network designers use one of several empirical algorithms for reaching an optimum configuration.

Closely related to the location of concentration points is network *topology,* which refers to the configuration in connecting circuits. The topologies shown in Figure 3.11 are usually a function of the network access method. Circuit switched networks are usually deployed in a star or hierarchical configuration. Packet switched networks normally employ a mesh topology, while polled networks employ a tree or bus topology. The ring topology is applied in local area networks that use a token passing access method.

FIGURE 3.11
Data Network Topologies

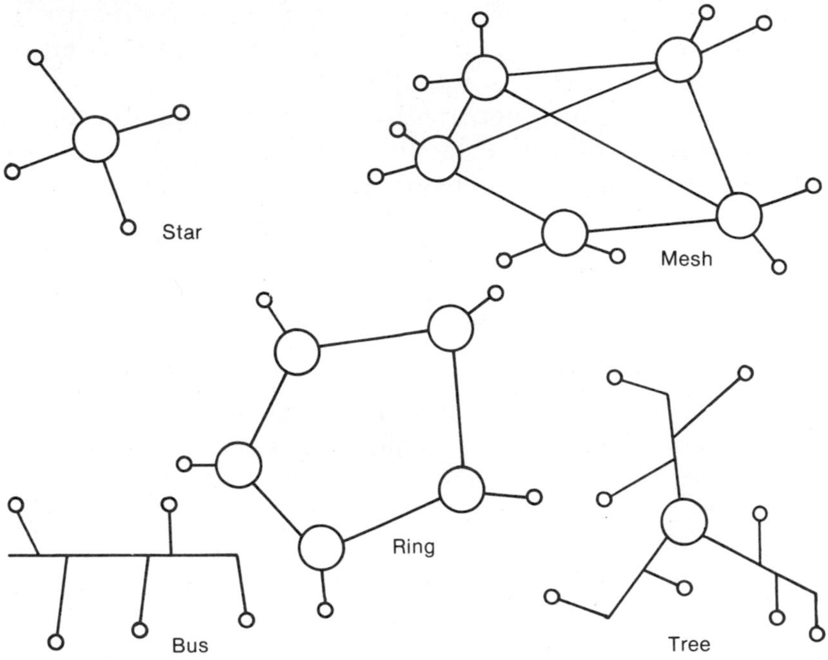

Digital or Analog Facilities

The bulk of global facilities today are deployed over analog microwave facilities, but digital circuits are rapidly becoming available over both terrestrial and satellite facilities. Digital circuits do not require a modem; instead, DTE is directly connected to the circuit or to a port of a multiplex. Common carrier network services such as AT&T's Dataphone® Digital Service (DDS) provide the equipment to multiplex circuits from the base speed of 2.4 to 56 kb/s up to the backbone transmission speed of 1.544 mb/s. At the distant end, the speed is demultiplexed to the base speed. This method of deriving digital circuits is discussed in Chapter 4. Bulk digital circuits operating at 1.544 mb/s (T-1) are also becoming widely available. To apply these services, the user supplies the multiplexers to subdivide the wideband channel into narrow data or voice channels.

Digital facilities have the additional advantage of providing improved performance. The carrier is able to monitor the bit error rate performance of backbone circuits and take corrective action when it exceeds the advertised limits. Also, as such circuits are inherently a full-duplex, four-wire circuit and have no modem, the time consumed in modem reversals and modem failures is eliminated.

The trend in both metropolitan and long haul circuits is toward the use of digital facilities. Although these circuits are more expensive than analog, several developments such as the construction of long-haul fiber optic facilities are reducing their costs.

Reliability and Availability

Data circuit reliability is the frequency of circuit failure, expressed as mean time between failures (MTBF). A related factor is circuit availability, which is the percentage of time the circuit is available to the user. Availability is a function of how frequently the circuit fails and how long it takes to repair it. The average length of time to repair a circuit is expressed as mean time to repair (MTTR). The formula for determining the percentage of availability is:

$$\text{Availability} = \frac{(\text{MTBF} - \text{MTTR})(100)}{\text{MTBF}}$$

For example, a circuit with an MTBF of 1,000 hours and an MTTR of 2 hours would have an availability of:

$$\frac{(1{,}000 - 2)(100)}{1{,}000} = 99.8 \text{ percent}$$

It is important to reach an understanding with the supplier of the conditions under which a circuit is considered failed. When the circuit is inoperative, it is clear that a failure condition exists, but it is less clear when a circuit is impaired by a high error rate. The bit error rate (BER) in a data circuit is usually expressed as a ratio of error bits to transmitted bits. For example, a data circuit with an error rate of 1×10^{-5} will have one bit in error for every hundred thousand bits transmitted. Reliability and error rate have a significant effect on throughput. Most error correction systems involve retransmitting a block when it is received in error, and retransmission of blocks reduces throughput.

Error Detection and Correction

Errors occur in all data communications circuits. Where the transmission is text that will be interpreted by people, a few errors can be tolerated because the meaning can be derived from context. In many applications, such as those involving transmission of bank balances, computer programs, and other numerical data, errors can have catastrophic effects. In these applications, nothing short of complete accuracy is acceptable. This section discusses causes, detection, and correction of data communication errors.

Causes of Data Errors

The type of transmission medium and the modulation method have the greatest effect on the error rate. Any transmission medium using analog modulation techniques is subject to external noise, which tends to affect the amplitude of the signal. Atmospheric conditions, such as lightning, that cause static bursts can induce noise into data-carrying analog radio and carrier systems. Relay and switch operations in electromechanical central offices, switching to standby channels in microwave, fiber optics, and carrier systems all cause momentary interruptions that result in data errors.

Any communication circuit is subject to errors during maintenance actions, or is subject to interruption by vandalism or external damage. Even local networks that are contained within a single building are subject to occasional interruptions from equipment failure or damage to the transmission medium.

Whatever the causes, errors are a fact of life in data circuits. The best error mitigation program is a design that reduces the susceptibility of the service to errors. Following that, the next most important consideration is to design the application to detect, and if possible, correct the errors.

Parity Checking

The simplest way of detecting errors is *parity checking* or *vertical redundancy checking* (VRC), a technique used on

FIGURE 3.12
Character Parity

	Bit 8	Bit 7	Bit 6	Bit 5	Bit 4	Bit 3	Bit 2	Bit 1
ASCII a		1	1	0	0	0	0	1
Odd Parity	0	1	1	0	0	0	0	1
Even Parity	1	1	1	0	0	0	0	1

asynchronous circuits. In the ASCII code set, the eighth bit is reserved for parity. Parity is set as odd or even, referring to the number of "one" bits in the character. As shown in Figure 3.12, DTE adds an extra bit, if necessary, to cause each character to match the parity of the network.

Most ASCII terminals can be set to send and receive odd, even, or no parity. When a parity error occurs, some form of alarm is registered at the terminal. Parity has two drawbacks: there is no way to tell what the original character should have been, and worse, if an even number of errors occur, parity checking will fail to detect the error. Therefore, parity is useful only for showing that an error occurred, but it is ineffective when transmission accuracy is required.

In terminals operating at 300 or 1,200 b/s, or more, characters are received with such speed that it is difficult to determine which character was in error when the parity alarm is registered. DTE can be programmed to flag an error character by substituting a special character such as an ampersand in its place. The error can be corrected by communication with the sending end.

Echo Checking

Over full-duplex circuits, errors can be detected by programming the receiving modem to retransmit the received characters to the sending end. This technique, called *echo checking,* is suitable for detecting errors in some forms of text. It is, however, subject to all the drawbacks of proofreading; it is far from infallible. Moreover, an error in an echoed character is as likely to have occurred on the return trip as in the original transmission. At 300 b/s a reader can keep up with echoed characters, but at 1,200 b/s it is im-

FIGURE 3.13
Longitudinal Redundancy Checking

Bit Position	Data Characters 1	2	3	4	5	6	Block Parity*
1	0	1	0	1	1	0	1
2	1	1	0	0	1	1	0
3	0	1	0	1	1	0	1
4	0	0	1	0	1	1	1
5	1	0	1	1	0	1	0
6	0	1	0	1	0	1	1
7	1	0	1	0	1	1	0
p*	1	0	1	0	1	1	0

Direction of transmission

*Even parity.

possible to read with any degree of reliability. The DTE can be programmed to make the echo check automatically, but correcting errors is just as difficult as with parity.

Although several systems have been used for error detection and correction in the asynchronous mode, none has been adapted as a standard, which is a drawback of asynchronous transmissions.

Longitudinal Redundancy Checking

Longitudinal redundancy checking (LRC) is a system used for error detection and correction in a block transmission mode. As shown in Figure 3.13, each character is checked for parity, and each bit position within the block is also checked. The last character in the block is a parity character created by the protocol to establish odd or even parity in each bit position in the block. If the block is received in error, the receiving end instructs the transmitting end to resend the block. The block is resent until it is received without error or until the protocol signals an alarm.

Although LRC is more reliable than simple parity checking and offers the additional advantage of error correction, it still suffers from the same inability to detect all parity errors. For example, if the same two bits were received in error in two characters in a block, the error would not be detected. The probability, however slight, is not acceptable for applications that need assurance of data integrity.

Cyclical Redundancy Checking

Cyclical redundancy checking (CRC) is used in most synchronous data networks. All the characters in a block are processed against a complex polynomial, and the mathematical result is entered in an error check block that is transmitted following the data block. Figure 3.7 illustrates a synchronous data block with a CRC field in the trailer record.

At the receiving end the data block is decoded with the same polynomial. If the two CRC fields fail to match, the protocol causes the block to be retransmitted. The probability of an uncorrected error with CRC is so slight that it can be considered virtually error-free.

Protocols

Protocol is the name given to the software functions that manage information transfer in a data communications network. The principal functions of a protocol are:

- To establish mutual compatibility of the DTE.
- To establish the session as half or full duplex.
- To establish the send and receive modem frequencies.
- To synchronize the transmitting and receiving terminals.
- To define the physical interface between DTE and the network.
- To convey the length of the data block to the receiving DTE.
- To acknowledge receipt of blocks.
- To detect errors.
- To arrange retransmission of error blocks.
- To verify the authenticity of parties to the session.
- To terminate the session.
- To determine charging and billing for the session.

Protocol compatibility and standardization are the most important issues in data communications. Most of the major computer manufacturers have developed proprietary protocols that are incompatible with those of other manufacturers. Even standard protocols developed by international agencies such as CCITT's HDLC and X.25 provide multiple options and are not always interchangeable between applications.

Incompatible protocols can communicate with each other through protocol converters. A protocol converter known as a *gateway* communicates with the connecting protocols using their own languages. A *bridge* processes the connecting protocols to enable them to communicate with each other as if they were compatible.

Certain value added networks also offer protocol conversion. They communicate with the DTE using that DTE's own protocol, convert it to the network protocol, and transport it to the distant DTE in its own language.

The incompatibility of protocols has been a stumbling block in the way of full interconnectibility of data networks. Although there are signs of development of standards, the problems of incompatibility are likely to remain for many years.

Layered Protocols

The most significant standardization efforts in protocols have revolved around developing layered protocols. A layer is a discrete set of functions that the protocol is designed to accomplish. The International Standards Organization (ISO) has published a seven-layer protocol model, the Open Systems Interconnect (OSI) model, which is illustrated in Figure 3.14.

Controlling communications in layers adds some extra overhead because each layer communicates with its counterpart through header records, but layered protocols are easier to administer than simpler protocols and provide opportunity for standardization. Although protocols are complex, functions in each layer can be modularized so the complexity can be dealt with separately by system designers. The seven OSI layers are defined below.

Layer 1—Physical. The first layer describes the method of physical interconnection over a circuit. The physical layer is concerned with the transmission of bits between machines and with the standardization of pin connections between

FIGURE 3.14
International Standards Organization Open Systems Interconnection (OSI) Protocol Model

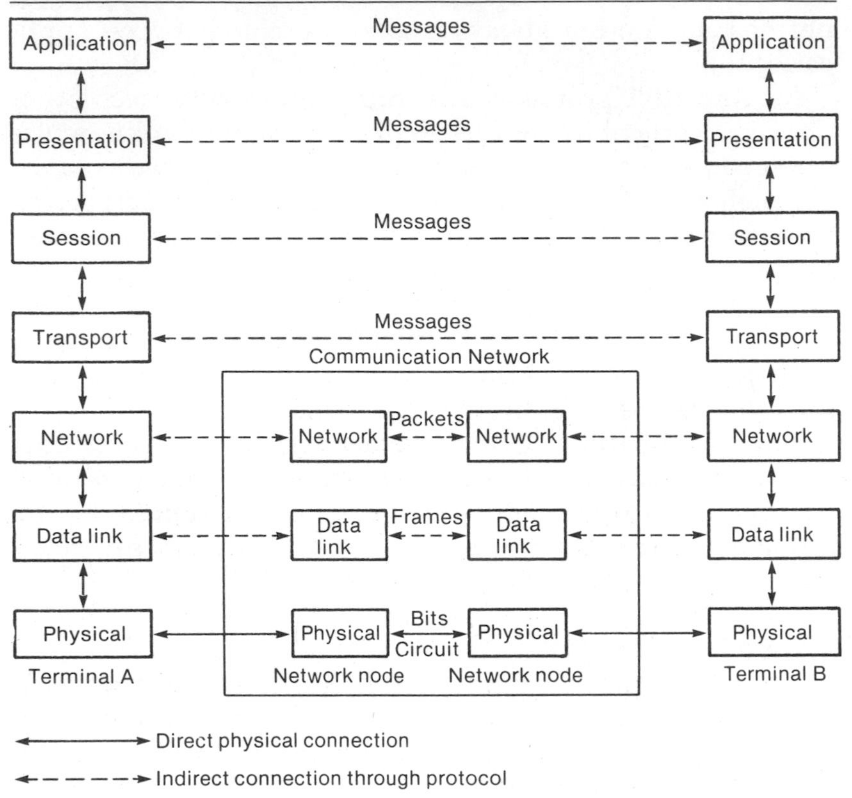

DCE and DTE. Typical standards are CCITT V.24 and X.21, and EIA RS-232-C and RS-449.

Layer 2—Data Link. Data link protocols are concerned with the transmission of frames of data between terminals. The protocol in the data link layer detects and corrects errors so that an error-free circuit is presented to the user. The data link layer takes raw data characters, creates frames of data from them, and processes acknowledgment messages from the receiver. When frames are lost or mutilated, the logic in this layer arranges retransmission. Protocols contain flags and headers so DTE can recognize the start and end of a frame. A frame of information, as shown in Figure 3.7, consists of flags to signal the beginning and ending of the frame, a header containing address and control information, an information field, and a trailer containing CRC bits for error correction. Typical link layer protocols are ISO's HDLC and IBM's Synchronous Data Link Control (SDLC).

Layer 3—Network. The network layer accepts messages from the higher layers, breaks them into packets, delivers them to the distant end through the link and physical layers, and reassembles them in the same form in which the sending end delivered them to the network. The network layer controls the flow of packets, controls congestion in the network, and routes between nodes to the destination. A typical packet switching protocol is CCITT X.25.

Layer 4—Transport. The transport layer controls end-to-end integrity between DTE devices, establishing and terminating the connection. Network management functions are provided by the transport layer because of the wide variety of types of network over which data may travel. The transport layer provides a standard service regardless of the characteristics of the network. It accomplishes its function by communicating with its corresponding layer in the distant machine through message headers and control messages. No standards have been developed for the transport and higher layers.

Layer 5—Session. The user communicates directly with the session layer, furnishing an address that the session layer converts to the address required by the transport layer. The conventions for the session are established in this layer. For example, the validity of the parties can be verified by passwords. The session can be established as a full- or a half-duplex session. The session layer determines whether machines can interrupt one another. It establishes how to begin and terminate a session and how to restore the connection in case the session is interrupted by failure.

Layer 6—Presentation. This layer interprets the character stream that flows between terminals during the session. For example, if encryption or bit compression are used, the presentation layer will provide it. Other machine control functions such as skipping, tabbing, form feed, and cursor movement are presentation layer functions.

Layer 7—Application. The application layer is specified by the user and defines the task being performed. For example, in facsimile communication, a task would be to transmit a page of information between two terminals. Applications are highly specialized and unique to the vendor and the user, and so can be standardized to only a limited degree in the future.

Layered control offers an opportunity for standardization and interconnection between the proprietary architectures of different manufacturers. Generally, the degree of standardization is greatest at the first layer and becomes increasingly disparate in the higher layers. Finally, the last layer, the application itself, is largely without standardization among vendors.

The objective of the OSI reference model is to establish a framework that will allow any conforming system or network to connect and exchange signals, messages, packets, and addresses. The model makes it possible for communications to become independent of the manufacturer that devised the technology. It should be understood that although the OSI model can be used to develop standards, it is not a standard itself.

X.25

An example of an important protocol that is used in the OSI model is CCITT's X.25. X.25 describes the interface between DTE and a packet switched network. It forms a network from the physical, link, and network layers of the OSI model. The physical layer interface is the X.21 standard, a new digital interface that is not yet widely available. Another version of the standard, X.21 *bis,* is nearly identical to RS-232-C, and is more commonly used.

The link layer uses a derivation of HDLC called Link Access Protocol-B (LAP-B) to control packet transfer, to control errors, and to establish the data link. The third, or network, layer provides two types of channels between equipment. A permanent virtual circuit offers the equivalent of a leased end-to-end channel by predefining a fixed routing through the network. The other option, a switched virtual circuit, is one in which the routing is established with each packet.

The primary use of X.25 is to connect between a host mainframe and a public packet switched data network. Presently there are few terminals that support X.25 interfaces.

For an individual terminal, the major packet carriers offer dial-up access to their networks, but the advantage of end-to-end error correction is lost. Large commercial users can interface public packet networks in one of three ways: over a carrier-certified software interface, a carrier-supplied interface processor that resides on the user's premises and connects to the user's computer, or through a user-supplied packet assembler/disassembler (PAD).

X.25 is far from static. It has undergone two major revisions since its provisional introduction by CCITT in 1974. Little commercial use was made of the first issue, but it was revised again in 1980 to add numerous options that were not included in earlier versions and has since gradually become accepted.

Throughput

The critical measure of a data communication circuit is its throughput, or the amount of information it is capable of transporting per unit of time. Although it would be theoretically possible for the throughput of a data channel to approach its maximum speed, in practice this can never be realized because of overhead bits and the retransmission of error blocks. The following are the primary factors that limit the throughput of a data channel:

- Modem speed. Within a single voice channel, modems ranging from 50 to 9,600 b/s can be accommodated.
- Half- or full-duplex mode of operation.
- Circuit error rate.
- Circuit reliability.
- Modem reversal time, which is the time required for a half-duplex modem to change from the sending to the receiving state.
- Protocol, including quantity of overhead bits and method of error handling such as selective retransmission or "go back N."
- Overhead bits such as start and stop bits, error-checking or forward error correction bits.

- Size of data block. The shorter the data block, the more significant the protocol overhead as a percentage of information bits. When the data block is too long, each error requires retransmitting considerable data. Optimum block length is a balance between time consumed in overhead and in error retransmission.
- Propagation speed of the transmission medium.

The throughput of a data channel is optimized by reaching a balance between the above variables. Because of the volume of calculations required to optimize the network, a computer program is generally required to generate throughput curves similar to those in Figure 3.15.

Line Conditioning

As discussed in Chapter 2, ordinary voice channels have limitations that prevent the transmission of data above 4,800 b/s. With *line conditioning,* voice channels can transport data up to 9,600 b/s. Two types of line conditioning, designated as types C and D, are available from telephone companies.

Type C conditioning is designed to minimize the effects of amplitude distortion and envelope delay distortion. See Chapter 2 for a description of these impairments. The amount of distortion is conditioned to various levels designated as C1, C2, C3, C4, or C5, in the increasing order of control over distortion. Type D conditioning controls the amount of harmonic distortion on a data channel and also controls noise to tighter limits than C conditioning.

STANDARDS

Data communications standards in the United States are principally established by ANSI, EIA, and CCITT. Standards activity in the United States is not highly developed. Many data communications services operate to proprietary protocols such as IBM's Binary Synchronous Communication (BSC) and Synchronous Data Link Control (SDLC). Inter-

FIGURE 3.15
Effect of Line Error Rate on Throughput

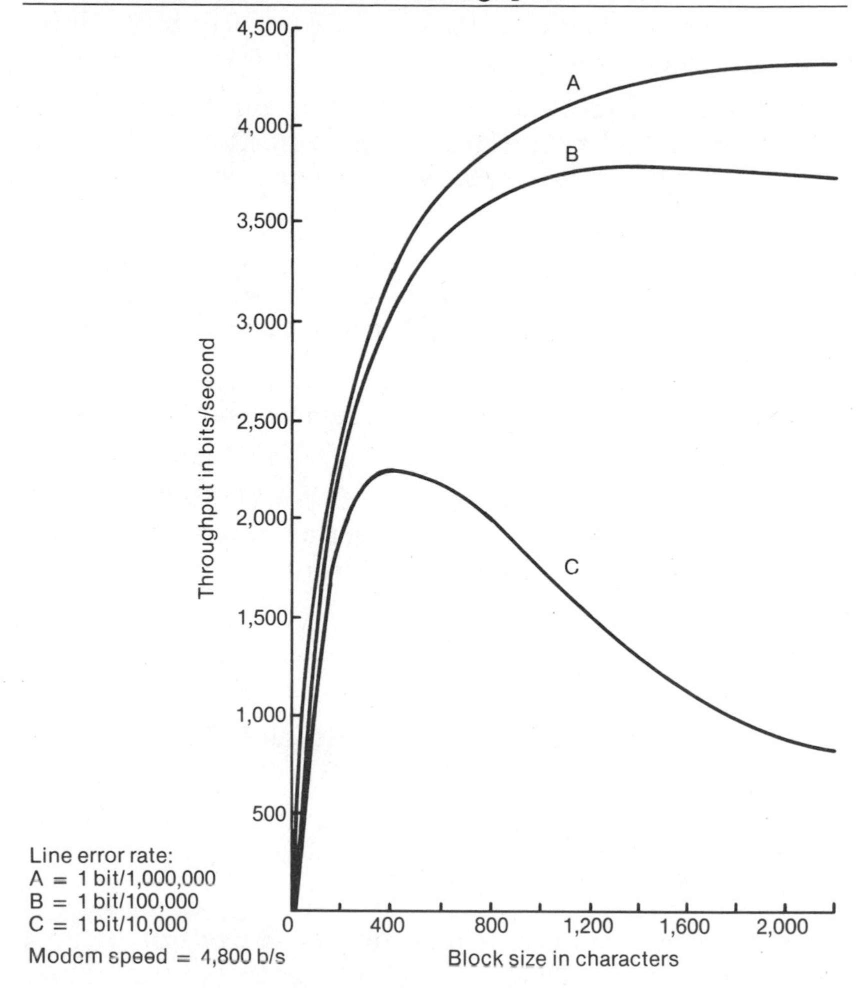

national standards are slowly being adopted, with many computers supporting CCITT X.25.

ANSI

X3.1 Synchronous Signaling Rates for Data Transmission.

X3.15 Sequencing of the American National Standard Code for Information Interchange in Serial-by-Bit Data Transmission.

X3.16 Character Structure and Character Parity Sense for Serial-by-Bit Data Communication in the American National Standard Code for Information Interchange.

X3.25 Character Structure and Character Parity Sense for Parallel-by-Bit Data Communication in the American National Standard Code for Information Interchange.

X3.28 Procedures for the Use of the American National Standard Code for Information Interchange in Specified Data Communications Links.

X3.36 Synchronous High-Speed Data Signaling Rates between Data Terminal Equipment and Data Communication Equipment.

X3.41 Code Extension Techniques for Use with the Seven-Bit Coded Character set of American National Standard Code for Information Interchange (ASCII).

X3.44 Determination of the Performance of Data Communications Systems.

X3.57 Structure for Formatting Message Headings for Information Exchange Using ASCII.

X3.64 Additional Controls for Use with the American National Standard Code for Information Interchange.

X3.66 Advanced Data Communication Control Procedure (ADCCP).

X3.79 Determination of Performance of Data Communications Systems that Use Bit Oriented Communications Control Procedures.

X3.100 Interface between Data Terminal Equipment and Data Circuit-Terminating Equipment for Packet Mode Operation with Packet Switched Data Communications Networks.

X3.102 Data Communications User Oriented Performance Parameters.

CCITT

V.2 Power levels for data transmission over telephone lines.

V.5 Standardization of data signaling rates for synchronous data transmission in the general switched telephone network.

V.6 Standardization of data signaling rates for synchronous data transmission on leased telephone-type circuits.

V.7 Definitions of terms concerning data communication over the telephone network.

V.15 Use of acoustic coupling for data transmission.

V.20 Parallel data transmission modems standardized for universal use in the general switched telephone network.

V.22 1,200 b/s duplex modems standardized for use on the general switched telephone network.

V.22 *bis* 2,400 b/s duplex modems standardized for use on the general switched telephone network.

V.24 List of definitions for interchange circuits between data terminal equipment and data circuit-terminating equipment.

V.29 9,600 b/s modem standardized for use on point-to-point leased telephone-type circuits.

V.32 9,600 b/s modem standardized for duplex use in the general switched telephone network.

V.50 Standard limits for transmission quality of data transmissions.

V.53 Limits for the maintenance of telephone-type circuits used for data transmission.

V.54 Loop test devices for modems.

X.2 International user services and facilities in public data networks.

X.3 Packet assembly/disassembly facility (PAD) in a public data network.

X.15 Definitions of terms concerning public data networks.

X.20 Interface between data terminal equipment (DTE) and data circuit-terminating equipment (DCE) for start-stop transmission services on public data networks.

X.20 *bis* Use on public data networks of data terminal equipment (DTE) that is designed for interfacing to asynchronous duplex V-Series modems.

X.21 Interface between data terminal equipment (DTE) and data circuit-terminating equipment (DCE) for synchronous operation on public data networks.

X.21 *bis* Use on public data networks of data terminal equipment (DTE) that is designed for interfacing to synchronous V-Series modems.

X.24 List of definitions for interchange circuits between data terminal equipment (DTE) and data circuit-terminating equipment (DCE) on public data networks.

X.25 Interface between data terminal equipment (DTE) and data circuit-terminating equipment (DCE) for terminals operating in the packet mode on public data networks.

X.28 DTE/DCE interface for start-stop mode data terminal equipment accessing the packet assembly/disassembly facility (PAD) in a public data network situated in the same country.

X.29 Procedures for the exchange of control information and user data between a packet assembly/disassembly facility (PAD) and a packet mode DTE or another PAD.

X.60 Common channel signaling for circuit switched data applications.

X.75 Terminal and transit call control procedures and data transfer system on international circuits between packet switched data networks.

X.121 International numbering plan for public data networks.

X.150 DTE and DCE test loops for public data networks.

EIA

RS-232-C Interface between data terminal equipment and data communication equipment employing serial binary data interchange.

RS-269-B Synchronous signaling rates for data transmission.

RS-334-A Signal quality at interface between data terminal equipment and synchronous data circuit-terminating equipment for serial data transmission.

RS-363 Standard for specifying signal quality for transmitting and receiving data processing terminal equipments using serial data transmission at the interface with nonsynchronous data communication equipment.

RS-366-A Interface between data terminal equipment and automatic calling equipment for data communication.

RS-384 Time division multiplex equipment for nominal 4 kHz channel bandwidths.

RS-404 Standard for start-stop signal quality between data terminal equipment and nonsynchronous data communication equipment.

RS-422-A Electrical characteristics of balanced voltage digital interface circuits.

RS-423-A Electrical characteristics of unbalanced voltage digital interface circuits.

RS-449 General-purpose 37-position and 9-position interface for data terminal equipment and data circuit-terminating equipment employing serial binary data interchange.

RS-484 Electrical and mechanical interface characteristics and line control protocol using communication control characters for serial data link between a direct numerical control system and numerical control equipment employing asynchronous full-duplex transmission.

RS-491 Interface between a numerical control unit and peripheral equipment employing asynchronous binary data interchange over circuits having RS-423-A electrical characteristics.

RS-496 Interface between Data Circuit-Terminating Equipment (DCE) and the Public Switched Telephone Network.

APPLICATIONS

Just as AT&T's telephone network design is used throughout the world, IBM's Systems Network Architecture (SNA) is the most successful computer network architecture in the world. In this section, SNA is described as an example of how a practical data communications network is implemented.

IBM Systems Network Architecture

SNA is a tree-structured architecture, with a mainframe host computer acting as the network control center. The boundaries described by the host computer, front-end processors, and cluster controllers and terminals are referred to as the network's *domain.* Figure 3.16 is a diagram of two such domains linked by a communications path between two front-end processors. Unlike the switched telephone network that establishes physical paths between terminals for the duration of a session, SNA establishes a logical path between network nodes, and it routes each message with addressing information contained in the protocol. The network is therefore incompatible with any but approved protocols. SNA uses the SDLC data link protocol exclusively. Devices using asynchronous or binary synchronous can access SNA only through protocol converters.

Network Addressable Units

The major components in the network are known as *Network Addressable Units* (NAUs). Four types are defined by SNA:

FIGURE 3.16
IBM Systems Network Architecture

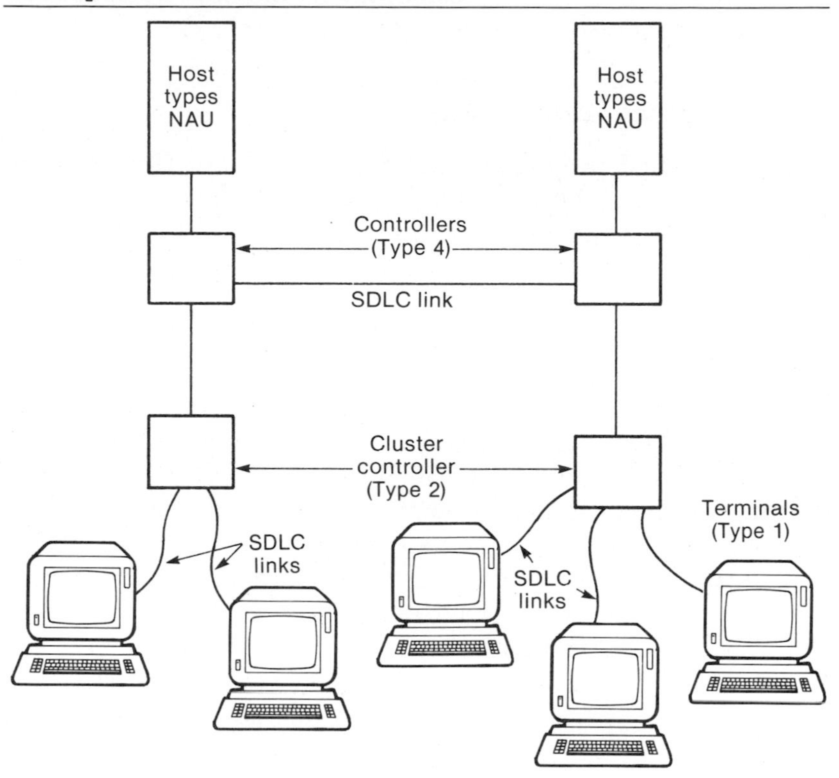

- Type 1—terminals.
- Type 2—controllers.
- Type 3—(not used).
- Type 4—front-end processors.
- Type 5—hosts.

The host NAU is categorized as a *System Service Control Point* (SSCP). Each network contains at least one SSCP in an IBM mainframe computer. The SSCP exercises overall network control, establishing routes and interconnections between logical units. The end user interfaces the network through a *Logical Unit* (LU). The LU, generally a terminal or a PC, is the user's link to the network. The next higher unit in the hierarchy is a *Physical Unit* (PU). Although its name implies that it is a piece of hardware, a PU is a control program executed in software or firmware.

SNA's Layered Architecture

SNA is defined in layers roughly analogous to the layers in ISO's OSI model. Unlike OSI, however, SNA is fully defined at each level. SNA was first announced in 1974 and is the basis for much of the OSI model, but it differs from OSI in several significant respects.

Level 1, Physical, is not included as part of the SNA architecture. The physical interface for analog telephone circuits is CCITT V.24 and V.25. The digital interface is X.21.

Level 2, Data Link Control, uses the Synchronous Data Link Control (SDLC) protocol. Figure 3.7 shows the SDLC frame, which consists of six octets or bytes of overhead. The first octet is a flag to establish the start of the frame. This is followed by a one-octet address and a one-octet control field. Next is a variable length data field, followed by a two-octet cyclical redundancy check field and an ending flag. The control field contains the number of the packets received to allow SDLC to acknowledge several packets simultaneously. SDLC permits up to 128 unacknowledged packets, which makes it suitable for satellite transmission. This layer corresponds closely to ISO's data link layer and the LAPB protocol used in X.25 networks.

Level 3, Path Control, is responsible for establishing data paths through the network. It carries addressing, mapping, and message sequencing information. At the start of a session the path control layer establishes a virtual route, which is the sequence of nodes forming a path between the terminating points. The circuits between the nodes are formed into transmission groups, which are groups of circuits having identical characteristics (i.e., speed, delay, error rate, etc.). The path control layer is also responsible for address translation. Through this layer, LUs can address other LUs by terminating address without being concerned with the entire detailed address of the other terminal. This layer is also responsible for flow control, protecting the network's resources by delaying traffic that would cause congestion. The path control layer also segments and blocks messages. Segmenting is the process of breaking long messages into manageable size so errors do not cause excessive retransmission. Blocking is the reverse process of combining short messages so the network's resources are not consumed with small messages of uneconomical size.

Level 4, Transmission Control, is responsible for pacing. At the beginning of a session the LUs exchange information about factors such as transmission speed and buffer size that affect their ability to receive information. The pacing function prevents an LU from sending more data than the receiving LU can accommodate. Through this layer, other such functions as encryption, message sequencing, and flow control are also provided.

Level 5, Data Flow Control, conditions messages for transmission by chaining and bracketing. Chaining is the process of grouping messages with one-way transmission requirements, and bracketing is grouping messages for two-way transmission.

Layer 6, Function Management Data Services, has three primary purposes. Configuration Service is concerned with activating and deactivating internodal links. Network Operator Services is the interface through which the network operator sends commands and receives responses. The Management Services function is used in testing and troubleshooting the network.

Layer 7, NAU Services, is responsible for formatting data between display devices such as printers and CRTs. It performs some of the functions of the ISO presentation layer such as data compression and compaction. It also synchronizes transmissions.

SNA has no applications layer as such, but IBM is defining three standards that allow for document interchange and display between SNA devices. *Document Interchange Architecture* (DIA) can be thought of as the envelope in which documents travel. DIA standards cover editing, printing, and displaying documents. The document itself is defined by *Document Content Architecture* (DCA), which is analogous to the letter within the envelope. The purpose of the DIA/DCA combination is to make it possible for business machines to transmit documents with formatting commands such as tabs, indents, margins, and other such format information intact. Documents containing graphic information are defined by *Graphic Codepoint Definition* (GCD), which defines the placement of graphic symbols on printers and screens.

SNA Usage Considerations

SNA is not the only alternative that IBM users have for data communications between devices. Other alternatives are the use of public data networks or private networks using packet switching, circuit switching, or other architectures. SNA, however, offers several advantages:

- It is mature. Developed in 1974, the architecture has undergone several evolutions and is reliable.
- The architecture is flexible. It can accommodate simple or complex networks and can communicate between networks.
- Because SNA is the most widely used data network architecture in the world, it enjoys a wide base of support.
- The network has IBM's power behind it, therefore assuring users of continuing support and compatibility with IBM's future enhancements.

Along with these advantages, SNA also has several limitations:

- It is not supported by all other vendors' products. Although IBM has taken steps to open the network to other vendors, many manufacturers have developed their own architectures instead.
- Installation and change procedures and protocols are cumbersome to administer.
- The cost per device is high compared to other alternatives.
- SNA is suited only for data and not for voice communications. This limits the use of SNA in integrated voice/data communications.

GLOSSARY

Adaptive equalizer: Circuitry in a modem that allows the modem to compensate automatically for circuit conditions that impair high speed data transmission.

American Standard Code for Information Interexchange (ASCII): A seven-bit (plus one parity bit) coding system used for encoding characters for transmission over a data network.

Analog: A transmission mode in which information is transmitted by converting it to a continuously variable electrical signal.

Asynchronous: A means of transmitting data over a network wherein each character contains a start and stop bit to keep the transmitting and receiving terminals in synchronization with each other.

Bandwidth: The range of frequencies a communications channel is capable of carrying without excessive attenuation.

Baud: The number of data signal elements per second a data channel is capable of carrying.

Binary: A numbering system consisting of two digits: zero and one.

Bit: The smallest unit of information that can be processed or transported over a circuit. Contraction of the words *BInary digiT.*

Bit error rate (BER): The ratio of bits transmitted in error to the total bits on the line.

Bridge: Circuitry used to interconnect networks with a common set of higher level protocols.

Byte: A set of eight bits of information equivalent to a character. Also sometimes called *octet.*

Circuit switching: A method of network access in which terminals are connected by switching together the circuits to which they are attached. In a circuit switched network, the terminals have full real-time access to each other up to the bandwidth of the circuit.

Central processing unit (CPU): The control logic element used to execute instructions in a computer.

Cyclical redundancy checking (CRC): A data error-detecting system wherein an information block is subjected to a mathematical process designed to ensure that undetected errors cannot occur.

Data circuit-terminating equipment (DCE): Equipment designed to establish a connection to a network, condition the input and output of DTE for transmission over the network, and terminate the connection when completed.

Data compression: A data transmission system that replaces a bit stream with another bit stream having fewer bits.

Data over voice (DOV): A device that multiplexes a full-duplex data channel over a voice channel using analog modulation.

Data terminal equipment (DTE): Any form of computer, peripheral, or terminal that can be used for originating or receiving data over a communication channel.

Datagram: A single unacknowledged packet of information that is sent over a network as an individual unit without regard to previous or subsequent packets.

Digital: A mode of transmission in which information is coded in binary form for transmission on the network.

Echo checking: A method of error checking in which the receiving end echoes received characters to the transmitting end.

Error: Any discrepancy between a received data signal from the signal as it was transmitted.

Expanded Binary Coded Decimal Interexchange Code (EBCDIC): An eight-bit coding scheme used for encoding characters for transmission over a data network.

Forward error correction (FEC): A method of correcting errors in a data channel by transmitting overhead bits that enable the receiving end to correct error bits.

Front-end processor: An auxiliary computer attached to a network to perform control operations and relieve the host computer for data processing.

Full-duplex circuit: A data communications circuit over which data can be sent in both directions simultaneously.

Gateway: Circuitry used to interconnect networks by converting the protocols of each network to that used by the other.

Half-duplex circuit: A data communications circuit over which data can be sent in only one direction at a time.

Line conditioning: A service offered by telephone companies to reduce envelope delay, noise, and amplitude distortion to enable transmission of higher speed data.

Mean time between failures (MTBF): The average time a device or system operates without failing.

Mean time to repair (MTTR): The average time required for a qualified technician to repair a failed device or system.

Message switching: A form of network access in which a message is forwarded from a terminal to a central switch where it is stored and forwarded to the addressee after some delay.

Modem: A contraction of the terms *MOdulator/DEModulator.* A modem is used to convert analog signals to digital form and vice versa.

Multidrop: A circuit dedicated to communication between multiple terminals that are connected to the same circuit.

Multiplexer: A device used for combining several lower speed channels into a higher speed channel.

Node: A major point at which terminals are given access to a network.

Octet: A group of eight bits. Also known as a *byte.*

Overhead: Any noninformation bits such as headers, error-checking bits, start and stop bits, etc. used for controlling a network.

Packet: A unit of data information consisting of header, information, error detection and trailer records.

Packet assembler/disassembler (PAD): A device used on a packet switched network to assemble information into packets and to convert received packets into a continuous data stream.

Packet switching: A method of allocating network time by forming data into packets and relaying it to the destination under control of processors at each major node. The network determines packet routing during transport of the packet.

Parity: A bit or series of bits appended to a character or block of characters to ensure that either an odd or even number of bits are transmitted. Parity is used for error detection.

Polling: A method of extracting data from remote terminals to a host. The host accesses the terminal, determines if it has traffic to send, and causes traffic to be uploaded to the host.

Propagation delay: The absolute time delay of a signal from the sending to the receiving terminal.

Protocol: The conventions used in a network for establishing communications compatibility between terminals and for maintaining the line discipline while they are connected to the network.

Serial interface: Circuitry used in DTE to convert parallel data to serial data for transmission on a network.

Split-channel modem: A modem that divides a communication channel into separate send and receive directions.

Statistical multiplexing: A form of data multiplexing in which the time on a communications channel is assigned to terminals only when they have data to transport.

Store and forward: A method of switching messages in which a message or packet is sent from the originating terminal to a central unit where it is held for retransmission to the receiving terminal.

Synchronous: A method of transmitting data over a network wherein the sending and receiving terminals are kept in synchronism with each other by a clock signal embedded in the data.

Throughput: The effective rate of transmission of information between two points excluding noninformation (overhead) bits.

Time division multiplexing: A method of combining several communication channels by dividing a channel into time increments and assigning each channel to a time slot. Multiple channels are interleaved when each channel is assigned the entire bandwidth of the backbone channel for a short period of time.

Topology: The architecture of a network, or the way circuits are connected to link the network nodes.

Value added network: A network in which some form of processing of a data signal takes place, or information is added to the signal by the network.

Virtual circuit: A circuit that is established between two terminals by assigning a logical path over which data can flow. A virtual circuit can either be permanent, in which terminals are assigned a permanent path, or switched, in which the circuit is reestablished each time a terminal has data to send.

BIBLIOGRAPHY

Adams, Marshall, and Ira W. Cotton, eds. *Computer Networks: A Tutorial.* Silver Spring, Md.: IEEE Computer Society Press, 1984.

Chorafas, Dimitris N. *The Handbook of Data Communications and Computer Networks.* Princeton, N.J.: Petrocelli Books, 1985.

Chow, Washow, ed. *Computer Communications.* Englewood Cliffs, N.J.: Prentice-Hall, 1983.

Doll, Dixon R. *Data Communications Facilities, Networks, and Systems Design.* New York: John Wiley & Sons, 1978.

Folts, Harold C. *McGraw-Hill's Compilation of Data Communications Standards,* 2d ed. New York: McGraw-Hill, 1982.

Freeman, Roger L. *Telecommunications System Engineering.* New York: John Wiley & Sons, 1980.

Held, Gilbert, and Ray Sarch. *Data Communications: A Comprehensive Approach.* New York: McGraw-Hill, 1983.

Lenk, John D. *Handbook of Data Communications.* Englewood Cliffs, N.J.: Prentice-Hall, 1984.

Martin, James. *Computer Networks and Distributed Processing.* Englewood Cliffs, N.J.: Prentice-Hall, 1981.

———. *Telecommunications and the Computer,* 2nd ed. Englewood Cliffs, N.J.: Prentice-Hall, 1976.

Reynolds, George W. *Introduction to Business Telecommunications.* Columbus, Ohio: Charles E. Merrill Publishing, 1984.

Sarch, Ray, ed. *Data Network Design Strategies.* New York: McGraw-Hill, 1983.

Sherman, Kenneth. *Data Communications: A Users Guide.* Reston, Va.: Reston Publishing, 1981.

Stuck, B. W. and E. Arthurs. *A Computer and Communications Network Performance Analysis Primer.* Englewood Cliffs, N.J.: Prentice-Hall, 1985.

Tannenbaum, Andrew S. *Computer Networks.* Englewood Cliffs, N.J.: Prentice-Hall, 1981.

MANUFACTURERS OF DATA COMMUNICATIONS EQUIPMENT

Modems

AT&T Information Systems
Case Rixon Communications, Inc.
Cermetek Microelectronics, Inc.
Codex Corp.
Databit Inc.
Datec Inc.
Digital Equipment Corporation
Gandalf Technologies, Inc.
General Datacomm, Inc.
Hayes Microcomputer Products, Inc.
IBM Corp.
Micom Systems, Inc.
NEC America Inc.
Paradyne Corporation
Penril DataComm
Prentice Corp.
Racal-Milgo, Inc.
Timeplex, Inc.
Universal Data Systems Inc.

Packet Switching Equipment

BBN Communications Corporation
Digital Communications Assoc.
Gandalf Data, Inc.
GTE Telenet Communications Corporation
Hewlett-Packard Co.
M/A COMM DCC Packet Switching
Micom Systems, Inc.
Timeplex, Inc.

Protocol Converters

Codex Corp.
Data General Corporation
Datapoint Corporation
Digital Communications Assoc.
Gandalf Data Inc.
IBM Corp.
Micom Systems, Inc.
Timeplex, Inc.

Time Division and Statistical Multiplexers

Case Rixon Communications, Inc.

Complexx Systems, Inc.

Digital Communications Assoc.

Digital Equipment Corporation

Gandalf Technologies, Inc.

General Datacomm, Inc.

Hewlett-Packard Co.

Micom Systems, Inc.

Penril DataComm

Prentice Corp.

Racal-Milgo, Inc.

Tellabs, Inc.

Timeplex, Inc.

Universal Data Systems Inc.

CHAPTER FOUR

Pulse Code Modulation and Digital Carrier

The pulse code modulation (PCM) method of multiplexing voice circuits on digital facilities was first patented in England in 1938 by Alec Reeves, an ITT scientist. Although the system was technically feasible at that time, pulse generating and amplifying circuits required vacuum tubes, and their size and power consumption consigned PCM to the shelf for another 20 years. Following the development of the transistor, PCM became commercially feasible in the 1960s, and with the development of large scale integration the following decade, costs, space, and power consumption continued to drop even in the face of high inflation.

Today, digital technology is rapidly replacing analog. Besides the lower costs of integrated circuitry, digital systems have the added advantage of directly interfacing digital switching machines and data circuits without an intervening voice frequency conversion. Digital carriers are more compact, with a per circuit density more than twice as great as equivalent analog systems. Digital circuits are also less susceptible to noise. In analog circuits, noise is additive, increasing with system length, but in digital carrier the signal is regenerated at each repeater. Over a properly engineered system, the signal arrives at the receiving terminal with virtually unimpaired quality.

Digital transmission has one primary disadvantage: a digital carrier requires several times more bandwidth than analog. Where an analog carrier transports a voice signal in 4 kHz of bandwidth, a digital voice channel requires 64 kb/s. Even in digital microwave radios with sophisticated modulation techniques, the density of digital signals is about half that of analog. Despite this disadvantage, digital transmission will gradually replace analog over the next two or three decades because of its advantages of higher signal quality and lower cost.

DIGITAL CARRIER TECHNOLOGY

In a T-1 digital carrier system, a voice signal is sampled, converted to an eight-bit coded digital signal, interleaved with 23 other voice channels, and transmitted over a line that regenerates the signal approximately once per mile. The signal is processed in a device called a *channel bank,* which is a 24-channel combination of voice circuits with an output bit stream operating at 1.544 mb/s. A photograph of a 48-channel bank, which is two banks packaged together, is shown in Figure 4.1.

Analog voice signals are sampled in a channel bank 8,000 times per second, slightly more than twice the highest frequency in the voice frequency range. The output of the sample is a *pulse amplitude modulated* (PAM) signal shown in Figure 4.2. If samples are taken frequently enough, the envelope of the PAM signal contains enough detail to reconstruct the original waveform.

The amplitude of the pulses from the sampling circuit is encoded into an eight-bit word by a process called *quantizing.* The eight-bit word provides 256 discrete steps, each step corresponding to the instantaneous amplitude of the speech sample. The output of the encoder is a stream of octets, each representing the magnitude of a sample.

The quantizing process does not exactly represent the amplitude of the PAM signal. Instead, the output is a series of steps as shown in Figure 4.3a. The error is audible in the

FIGURE 4.1
Wescom 360 D4 Channel Bank

Printed and used by permission of Rockwell Telecommunications, Inc., Wescom Telephone Products Division.

voice channel as *quantizing noise,* which is present only when a signal is being transmitted. The effects of quantizing noise are greater with low amplitude signals than with high. To overcome this effect, the encoded signal is compressed to divide low-level signals into more steps and high-level signals into fewer steps as shown in Figure 4.3b. When the

FIGURE 4.2

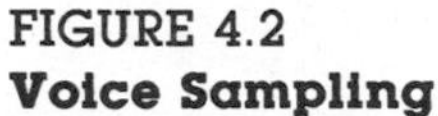

Voice Sampling

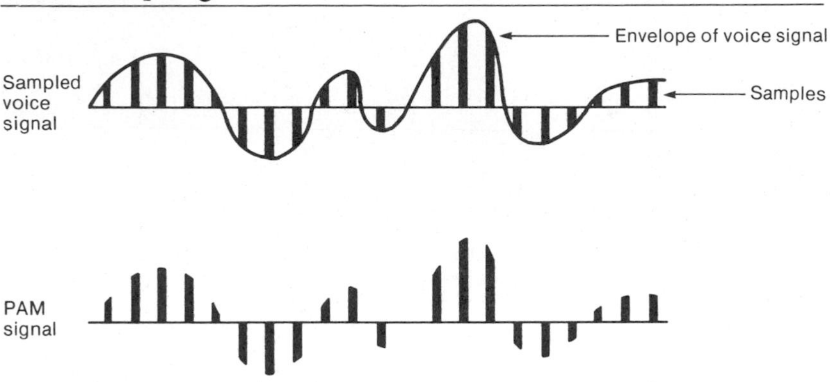

FIGURE 4.3
Companding in a PCM Channel Bank

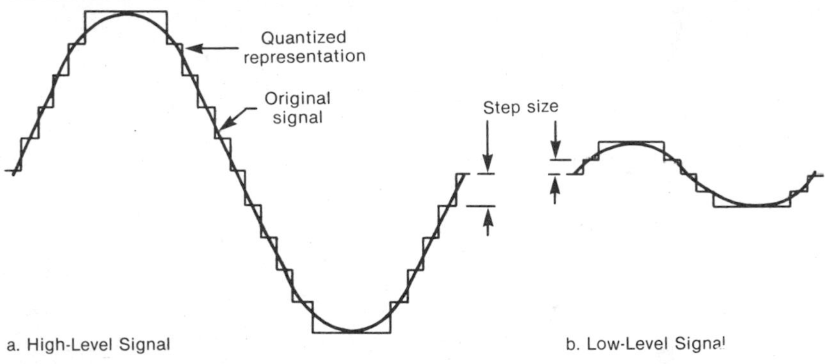

High-level signals are divided into steps of greater magnitude than low-level signal steps.

signal is decoded at the receiving terminal, it is expanded by reversing the compression steps. The combination of expansion and compression is called *companding.* In the United States, companding follows a formula known as "mu law" coding. In Europe, the companding formula is a slightly different form known as "A law" coding. Although the two laws are incompatible, they differ to only a slight degree.

The PAM voice signal is encoded in the channel bank and merged with 23 other voice channels. Each channel generates a bit rate of 64 kb/s (8,000 samples per sec × 8 bits per

FIGURE 4.4
PCM Frame

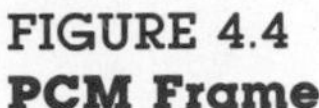

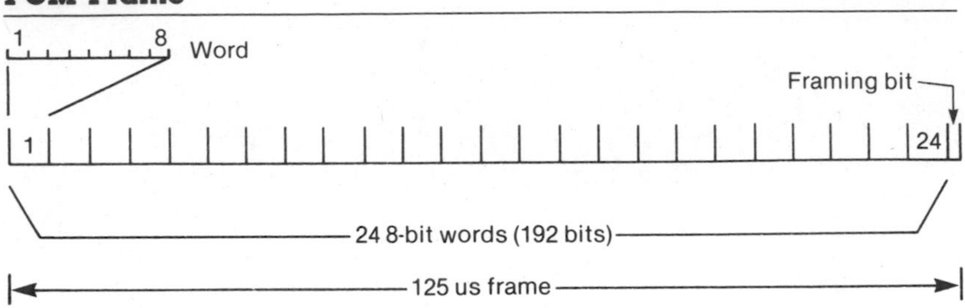

This frame is repeated 8,000 times per second to generate a line frequency of 1.544 mb/s.

sample). The 24 channels produce the frame format shown in Figure 4.4. A single framing bit is added to the 192 bits that result from the 24 eight-bit words, comprising a frame of 193 bits. The frame, 125 microseconds in duration, is repeated 8,000 times per second for a total line rate of 1.544 mb/s. European digital carrier systems use the same 64 kb/s channel bit rate, but multiplex to 32 rather than 24 channels for a 2.048 mb/s bit rate. Because of companding and bit rate differences, North American and European digital carrier systems are incompatible.

Standard digital channel banks use a technique called *bit robbing* to use the least significant bit out of every sixth frame for signaling. The distortion resulting from bit robbing is insignificant to voice signals and to digital signals using a modem. However, if a 64 kb/s data signal was directly coupled into a channel, the robbed bits would generate data errors. Therefore, a conventional channel bank is capable of handling direct digital input at a maximum of 56 kb/s per channel. Many manufacturers provide channel units called *dataport* units, which allow data to be applied directly to the digital bit stream without a modem.

Changes to the standard T-1 framing format have been proposed to provide 64 kb/s *clear channel* capability. Clear channel, also known as *extended super frame* (ESF), capability requires a change in the line coding, using a coding known as bipolar with 8-zero substitution (B8ZS).

FIGURE 4.5
Wescom 3440-01 T-1 Line Repeater

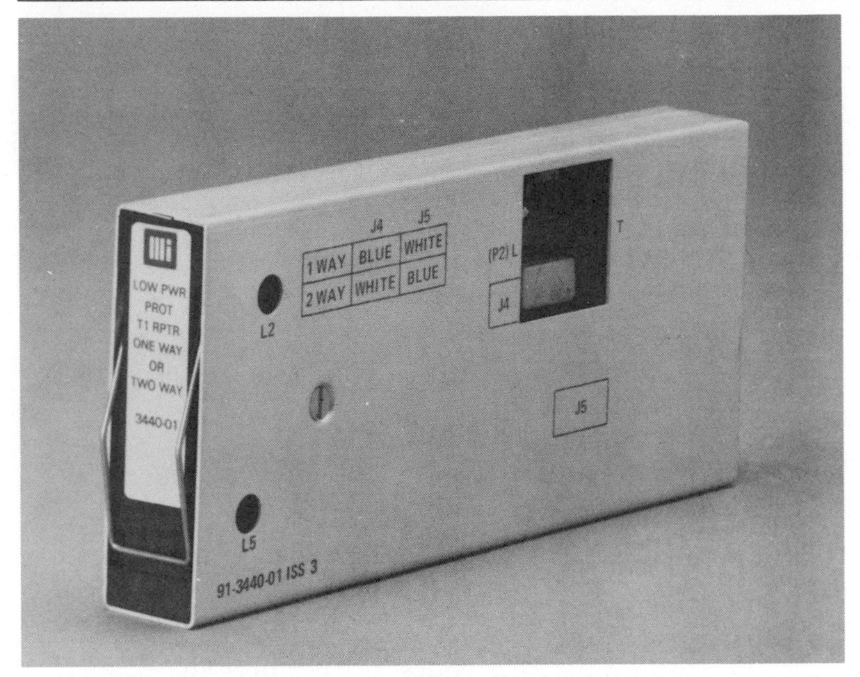

Printed and used by permission of Rockwell Telecommunications, Inc., Wescom Telephone Products Division.

Digital Transmission Facilities

The basic digital transmission facility is a T-1 line, which consists of an office repeater at each end feeding twisted pair wire, with digital regenerators spaced every 6,000 feet. The function of the office repeater is to match the output of the channel bank to the impedance of the line and to feed power over the line to the repeaters. The line repeaters regenerate the incoming pulses to eliminate distortion caused by the cable. A photograph of a line repeater is shown in Figure 4.5.

Digital signals are applied to twisted pair wire in groups of 24, 48, or 96 channels called T-1, T-1C, and T-2. The signals for all three originate in channel banks. The higher bit rates of T-1C and T-2 are developed by multiplexing as described later. Digital signals can also be applied to micro-

FIGURE 4.6
Block Diagram of T-1 Carrier System

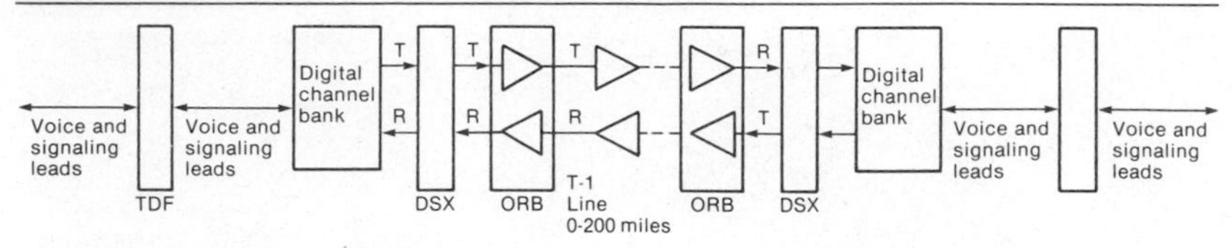

TDF - Trunk Distributing Frame
DSX - Digital Signal Cross-connect Frame
ORB - Office Repeater Bay
T - Transmit
R - Receive

wave radio, fiber optics, or coaxial cable for transmission over longer distances. Digital transmission over these facilities is described in Chapters 15, 16, and 20.

Digital Signal Timing

T-1 signals are kept in synchronization by *loop timing,* in which synchronizing pulses are extracted from the incoming bit stream. The PCM output of a channel bank is encoded in the *bipolar* format that is described later. The transition of each "one" bit is detected by the repeaters and the receiving terminals and used to keep the system in synchronization. If a signal consisting of more than 15 zeros is transmitted on a digital facility, the receiving end loses synchronization. To prevent this, the channel bank inserts a unique bit pattern that is detected by the receiving end and restored to the original pattern. This technique, called *bit stuffing,* is used by all digital carrier systems to prevent loss of synchronization.

THE T-1 CARRIER SYSTEM

A block diagram of a T-1 carrier system is shown in Figure 4.6. The primary elements of the system are the channel banks and repeaters. The other elements, distributing frame and digital cross-connect frame, are provided for ease of assignment and maintenance.

FIGURE 4.7
Block Diagram of a PCM Channel Bank

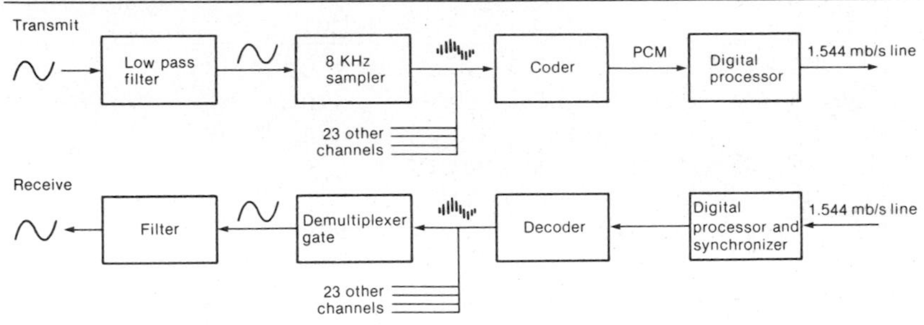

Channel Banks

A basic digital channel bank consists of 24 channels called a *digroup.* Most manufacturers package two digroups in a 48-channel framework. Five modes of operation are common in the industry:

- **Mode 1**—a 48-channel mode operating over a T-1C line.
- **Mode 2**—a 48-channel mode operating over a T-1C line but with the digroups separately timed for operation with an external multiplexer.
- **Mode 3**—independent 24-channel digroups operating over two T-1 lines.
- **Mode 4**—dual 48-channel banks combined to operate over a T-2 line.
- **Mode 5**—dual 48-channel banks combined to operate over a fiber optic pair.

The 48 channels share a common power supply and other common equipment, failure of which can interrupt all 48 channels simultaneously. A block diagram of a digital channel bank is shown in Figure 4.7.

The channel bank consists of a metal framework designed to accept plug-in common equipment and channel units. In the United States, there are no independent standards for digital channel banks. The electrical characteristics of a T-1 signal have been released in the public domain by AT&T through the U. S. Telephone Association. All T-1 channel

TABLE 4.1
Digital Carrier Typical Channel Unit Types

Code	*Function*
2W E&M	Two-wire E&M signaling trunk
4W E&M	Four-wire E&M signaling trunk
SDPO	Sleeve control dial pulse originating
DPO	Dial pulse originating
DPT	Dial pulse terminating
2W FXO	Two-wire foreign exchange office
4W FXO	Four-wire foreign exchange office
2W FXS	Two-wire foreign exchange subscriber
4W FXS	Four-wire foreign exchange subscriber
2W DX	Two-wire duplex signaling
4W DX	Four-wire duplex signaling
2W ETO	Two-wire equalized transmission only
4W ETO	Two-wire equalized transmission only
2W FXO/GT	Two-wire foreign exchange office with gain transfer
2W FXS/GT	Two-wire foreign exchange subscriber with gain transfer
4W SF	Four-wire single-frequency signaling
PLR	Pulse link repeater
PG	Program
RD	Ringdown
PLAR	Private line automatic ringdown
OCU DP	OCU dataport

banks sold for domestic use conform to the electrical characteristics of the AT&T plan and are therefore end-to-end compatible, but each manufacturer develops its own mountings. Channel and common equipment plug-ins are therefore not physically compatible unless the manufacturer has specifically designed them to plug into the mounting of another manufacturer.

A major advantage of digital channel banks over their analog counterparts is the availability of a wide variety of channel unit plug-ins. Integrated in these plugs are electrical functions that require external equipment in analog carrier systems. These functions include integrated signaling, voice frequency gain, wide band program transmission, and data transmission capability. Table 4.1 lists the channel units available from most major manufacturers. Most of these special service units have unique signaling options, concepts of which are explained in Chapter 8; the transmission functions are covered in this chapter.

Dataport Channel Units

As mentioned earlier, dataport channel units enable direct access to a T carrier bit stream without using a modem or digital-to-analog conversion. Dataport units compatible with the bit-robbed signaling of T carrier are available in speeds of 2.4, 4.8, 9.6, 19.2, and 56 kb/s. Each of the channel units occupies a mounting slot in the channel bank and uses the entire capacity of the bit stream. To use the entire 56 kb/s bandwidth of a voice channel for slower-speed data, external multiplexers can be used. Some manufacturers also produce channel units with built-in data multiplexers.

Dataport channel units are used in both common carrier and private networks. In private networks channel banks may be mounted at the user's premises. Over public networks digital lines are run to the DTE through a data service unit (DSU) or channel service unit (CSU), which provide a standard interface to the DTE. The CSU is a line driver and receiver that terminates a four-wire data loop at the user's premises. It equalizes thc cable, provides loop-around testing access, and includes electrical protection. A CSU is used when the DTE or a separate DSU provides zero suppression capability, clock recovery, and regeneration. The DSU performs all the functions of the CSU and is functionally equivalent to a modem in that it performs all the line functions needed to connect DTE over a data network. A block diagram of a T carrier dataport system is shown in Figure 4.8.

Special Transmission Functions

The plug-in units listed in Table 4.1 provide several special transmission functions besides the signaling functions that will be described in Chapter 8. *Foreign exchange* (FX) service is a combination of special signaling and transmission service. It is used by telephone companies to connect a telephone line in one exchange to a station located in another. FX channel units are also used in PBXs to connect the PBX to off-premise station lines or to foreign exchange trunks extended between PBXs. FX channel units are equipped for direct connection to metallic loops and have provisions for

FIGURE 4.8
Block Diagram of Dataport Application

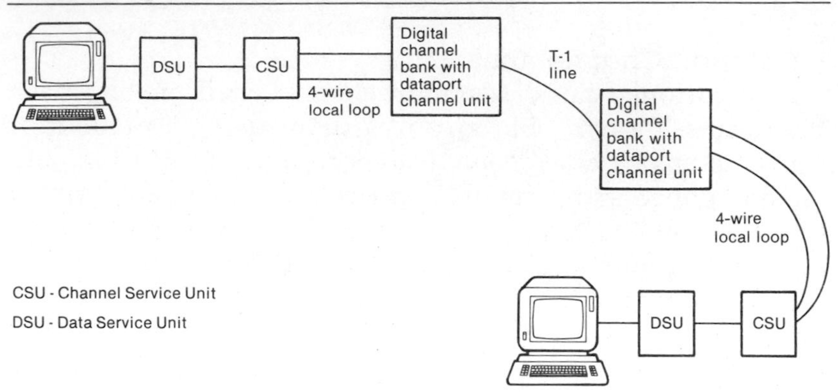

ringing telephones and for adding gain to long loops (gain transfer option).

Program (PG) channel units replace two or more voice channel units and use the added bit stream to accommodate a wider channel for use by radio and television stations or wired music companies, or other applications where a wide audio band is required. Program channels with 5 kHz bandwidth replace two voice channels, and 15 kHz units replace six voice channels.

Transmission only (TO) channel units are used for circuits that do not require signaling, or in which signaling is part of the application such as data services. TO channel units are also used to connect channels in channel banks wired back to back. Back-to-back wiring is used at intermediate points on a T carrier line when it is necessary to drop off individual voice channels.

Intelligent Channel Banks

Although the variety of channel units is a major advantage of digital channel banks, the use of this flexibility requires a large inventory of channel units to ensure that the required units are available for spares and growth. The latest generation of channel banks, typified by AT&T Network Systems' D5 Digital Termination System (DTS), includes intelligence in the channel bank and its channel units. The DTS consists of a System Controller and up to 20 D5 channel banks of 96 voice channels each.

The D5 channel bank can be accessed remotely over a data facility, and through its controller, the channel units can be remotely adjusted. Channel unit control settings and option selections are typed into a Craft Interface Unit that is either colocated with the DTS or located at a network control center. With this system, channel units can be remotely configured as one of several types by combining more than 25 channel unit types into 4, greatly simplifying the inventory management task. Furthermore, through this remote device, transmission levels are set and maintenance tests are made without requiring a technician at the terminal, which reduces the cost of maintaining the system.

Digital Cross-Connect Panel

In most multiple channel bank installations, a *digital signal cross-connect* (DSX) panel is provided. The transmit and receive circuits of the channel bank are wired through jacks in this panel to provide test access and to enable connecting the channel bank to another T-1 carrier line. The transfer of the T-1 bit stream to another line is called *patching* and is used for rearranging circuits and for manually transferring to a spare line in case of repeater failure.

The ability to patch to spare lines is a vital part of service restoration in common carrier applications and is provided in almost all instances. In private networks, the DSX panel is useful for temporary rearrangement of facilities. For example, a T-1 facility might be used for multiple low speed point-to-point terminal communications during working hours and patched to a high speed multiplexer for transfer of data after hours.

Automatic protection switching is often used in place of manual patching where rapid restoral is essential, and where one or both terminals are unattended. The automatic protection system monitors the bit stream of the protected channel banks and switches automatically to a spare line when the working line fails.

Distributing Frame

The voice frequency side of a channel is wired to *cross-connect* blocks on a distributing frame, which is discussed in

FIGURE 4.9
T Carrier Line Signals and Faults

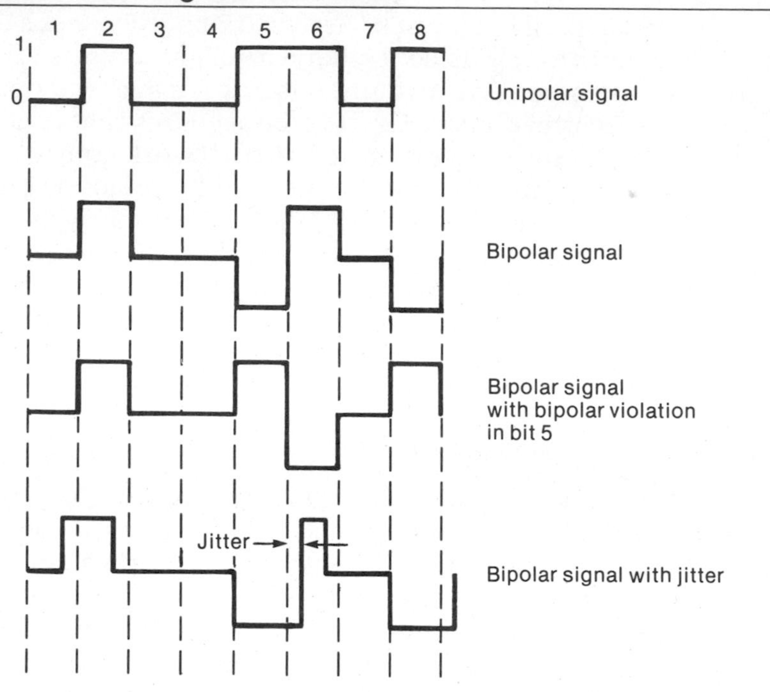

Chapter 14. A cross-connect block provides a convenient location for concentrating lines between the channel bank and its interfacing equipment. Where temporary rearrangements are needed, the voice frequency and signaling leads of a channel are wired through jacks. These jack panels are a convenient place for rapid restoral of circuits or for temporarily rerouting equipment to another channel.

T Carrier Lines

T-1 carrier lines can be extended for about 200 miles, although most private and common carrier applications are considerably shorter because longer circuits are usually deployed over radio or fiber optic facilities. A T carrier line accepts a bipolar signal from the channel bank as shown in Figure 4.9. An office repeater terminates the line at each end. In the receiving direction, the office repeater performs normal repeater functions, but it is passive in the transmit di-

FIGURE 4.10
Block Diagram of T Carrier Regenerator

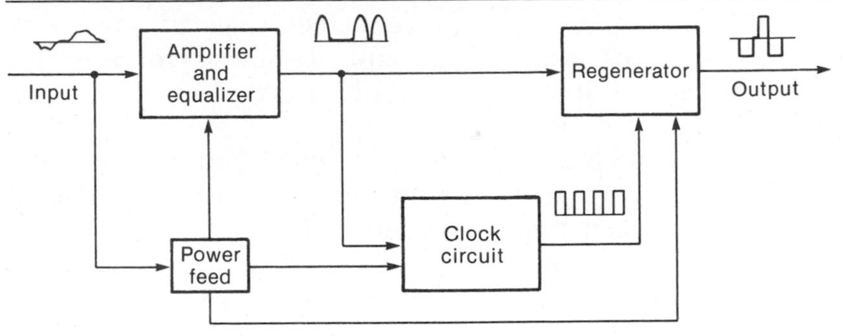

rection. Its transmit function is to couple the bipolar signal to the line and to feed power to the line repeaters. If interference or a failing repeater adds or subtracts one bits, a *bipolar violation* results, which indicates a fault in a T carrier system.

Line repeaters are mounted in an apparatus case holding 25 units. Apparatus cases are watertight for mounting on poles and in manholes. Repeaters, a block diagram of which is shown in Figure 4.10, perform these functions:

- Amplify and equalize the received signal.
- Generate an internal timing signal.
- Decide whether incoming pulses are zeros or ones.
- Regenerate pulses and insert in correct output time slot.

Incoming pulses are received in one of three states: plus, minus, or zero. If the incoming pulse, which has been distorted by the electrical characteristics of the line, exceeds the plus or minus threshold, the repeater generates a "one" output pulse. Otherwise, it registers a zero.

Phase deviations in the pulse, which are additive along a T carrier line, are known as *jitter*. Excessive jitter, as illustrated in Figure 4.9, can cause errors in data signals.

The transmit and receive paths of T-1 signals must be isolated to prevent crosstalk coupling. If excessive crosstalk occurs between the high-level pulses of a repeater's output and the low-level received pulses of adjacent repeaters, errors will result. These are prevented by assigning the transmit and receive directions to separate cables, to partitions within a specially screened cable, or to separate binder groups within a single cable. Cable binder groups are explained in Chapter 7.

T-1C and T-2 lines operate on the same general principles as T-1 lines except that their bit rates are higher and greater isolation between the pairs is needed to prevent crosstalk. Design rules for these systems require screened cable or separate transmit and receive cables.

DIGITAL CROSS-CONNECT SYSTEMS

Short-haul carrier systems converge in central offices in both private and common carrier networks for connection to long-haul facilities. If 24-channel digroups are connected through the office to a single terminating point, no channel bank is required. Instead, the incoming T-1 line is connected to the outgoing line with an *express office repeater* with channel banks needed only at the terminating ends. However, if fewer than 24 channels are needed, or if channels from a single originating point must be split to separate terminating points, *back-to-back* channel banks must be used to gain access to the bit stream for channel cross-connection.

Back-to-back channel banks are undesirable for several reasons:

- Cost of the channel banks.
- Channel banks are an added source of potential circuit failure.
- Labor cost of making channel cross-connections.
- Extra analog-to-digital conversions which are a source of distortion.

The *digital cross-connect system* (DCS) is a specialized electronic switch that terminates T carrier lines without channel banks and routes the individual channel bit streams to the desired output line. Unlike the electronic digital switches covered in Chapter 10, 11, and 12, DCS establishes a permanent path for the bit stream through the switch. This path remains connected until it is disconnected or changed by administrative action.

The DCS system is at least as costly as individual channel banks, but it eliminates most of the labor associated with rearrangement, eliminates extra analog-to-digital conversions, and offers a high degree of flexibility in rerouting circuits. Furthermore, routing changes can be controlled from a central location over a data link. If the organization uses a mechanized data base to maintain records of facility assignments, the same source that updates the data base can drive the DCS assignments also. DCS is a space-saving system because it eliminates the cross-connect blocks and wire required to interconnect the large numbers of voice frequency and signaling leads it replaces when back-to-back channel bank connections are used.

DCS is the key to implementing the intelligent digital network described in Chapter 25. Linked to the user's network control system, DCS offers the user the capability of reconfiguring the network to meet changes in demand or to accommodate changes that occur with time such as assigning network capacity to a voice switch during normal working hours and to a computer center for high-speed data transfer during off hours. However, users should be aware that DCS systems and IBM's Systems Network Architecture may be incompatible because SNA does not accommodate the rapidly shifting routes that DCS provides.

THE DIGITAL SIGNAL HIERARCHY

Digital signals from a T-1 source can be multiplexed to higher rates. Figure 4.11 shows the bit rates of the U. S. digital hierarchy. T-1 signals are applied to a standard repeatered line as described in the next section. Higher rate signals are applied to wire, coaxial cable, fiber optics, or digital microwave radio.

The basic digital group is raised to higher bit rates by a family of multiplexers. Multiplexers are designated according to the digital signal levels they interface. For example, a

FIGURE 4.11
U.S. Digital Signal Hierarchy

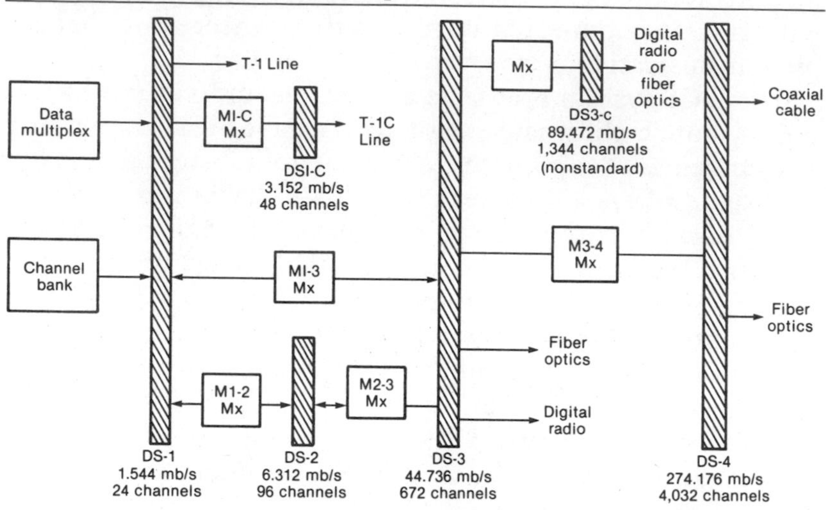

DS-1 to DS-3 multiplexer is designated as M1-3. A M1-3 multiplexer, which is shown in Figure 4.12, accepts 28 DS-1 inputs and combines them into a single 45 mb/s bit stream. (The bit stream is actually 44.736 mb/s, but it is commonly called 45 mb/s in the industry.) Multiplexer output can be directly connected to a digital radio or to a fiber optic system.

The primary use of the DS-4 signal level is to feed T-4 coaxial cable. The DS-4 signal speed is too high to apply to the limited bandwidth of a digital microwave, which is currently limited to carrying three DS-3 signals. Fiber optic cable, described in Chapter 16, can easily accommodate 274 mb/s, but its bandwidth can transport higher bit rates, so most manufacturers are producing equipment that operates at bit rates of 500 mb/s or more and are undefined in the digital hierarchy.

T-1 DATA MULTIPLEXERS

Digital transmission facilities are ideal for data transmission because of their inherent reliability and error immunity. A digital line can be interfaced directly by dataport channel units in the digital channel bank or by a T-1 digital multi-

FIGURE 4.12
M1-3 Multiplexer

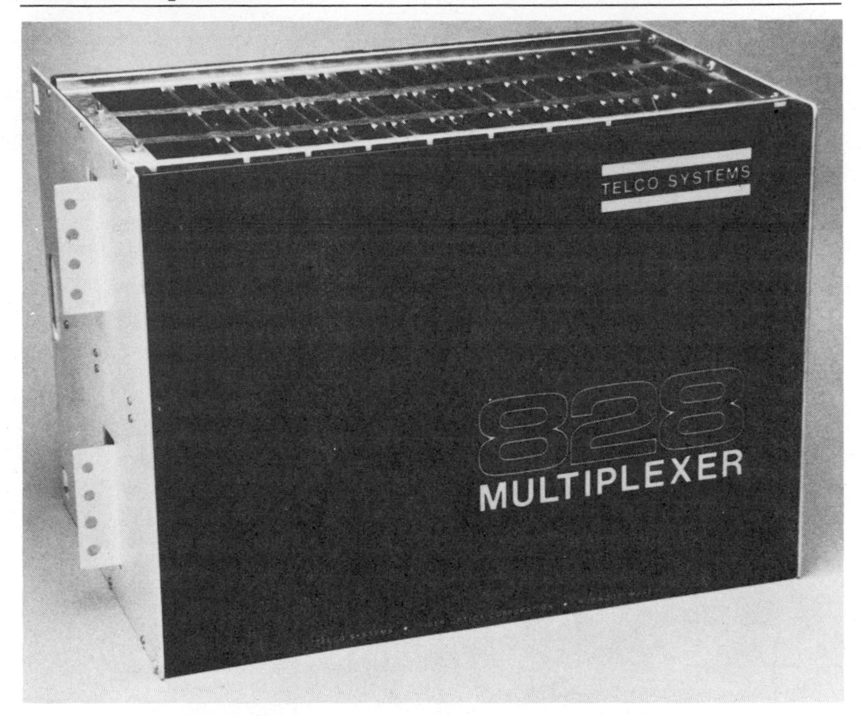

Courtesy Telco Systems Fiber Optics Corp.

plexer. T-1 multiplexers use TDM techniques to combine multiple low-speed bit streams into a 1.544 mb/s signal. A typical multiplexer is shown in Figure 4.13.

Both long-haul common carriers and telephone companies are offering point-to-point 1.544 mb/s service. This high speed bit stream can be used for any combination of voice, data, and video service providing a multiplexer is used to subdivide the bit stream. T-1 multiplexers connect to a T-1 line through a channel service unit (CSU) as described previously. Multiplexers contain control logic to provide clocking, to generate frames, and to enable testing. The control logic and the power supply are common to the entire multiplexer and may be redundant to improve reliability. The diagnostic capabilities include alarm and loop-back facilities, and may include a built-in test pattern generator to assist in

FIGURE 4.13
General Datacom Megamux Plus T-1 Multiplexer

Courtesy General DataComm Industries, Inc.

FIGURE 4.14
T-1 Multiplexer Layout Showing Drop-and-Insert Capability and Submultiplexing Low-Speed Channels

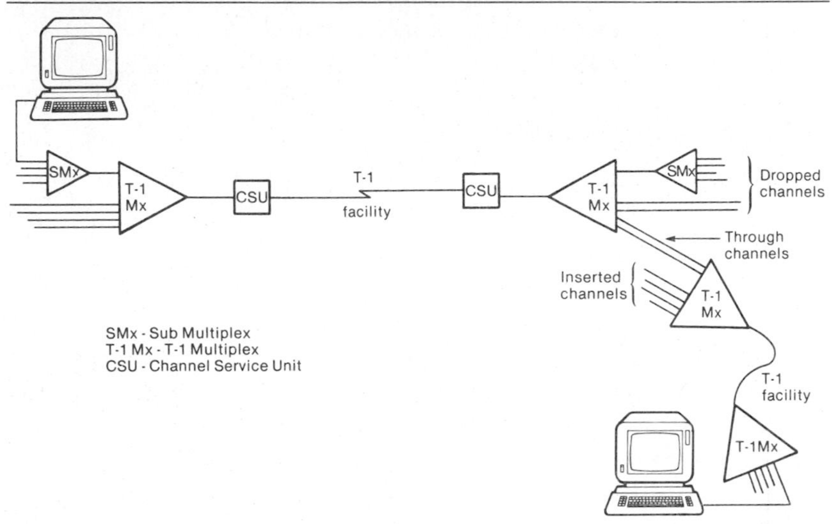

trouble diagnosis. Multiplexers can generally handle synchronous data from 50 b/s to 1.544 mb/s and asynchronous data from 50 b/s to 19.2 kb/s. Voice input is accommodated by using PCM or one of the modulation systems described in the next section; adaptive differential pulse code modulation (ADPCM), or delta modulation. ADPCM compresses the voice into 32 kb/s, and delta modulation compresses voice into either 32 kb/s or 16 kb/s.

T-1 multiplexers divide the bit stream into a series of subslots. Some provide slow speed subslots; others provide larger subslots, which require external multiplexers to use the capacity of the subslot for low-speed data. Figure 4.14 shows the methods of multiplexing data over T-1 lines and also shows another feature of multiplexers, *drop-and-insert* capability. This allows the multiplexer to extract selected channels while extending others on to another destination.

T CARRIER DATA COMPRESSION SYSTEMS

Numerous systems are proposed or already in use to increase the capacity of a carrier line beyond the basic 24 channels of T-1. These can be roughly grouped into two categories: those using different coding systems on the carrier line to increase its carrying capacity and those that use data compression techniques to encode voice in fewer than the 64 kb/s used by PCM.

T Carrier Line Encoding

The carrying capacity of a T-1 line can be doubled or quadrupled by using different coding schemes to replace the bipolar T-1 line signal. We have discussed T-1C, which employs bipolar line signaling at double the T-1 rate. In addition, other systems are employed that are proprietary to their manufacturers. A product called T-1G, recently announced by AT&T Network Systems, encodes the carrier line at 3.22175 mb/s to compress 96 channels into a cable designed for T-1C service.

Unlike T-1, which is standardized among all manufacturers, these systems are proprietary, and therefore are compatible only with like equipment. The increase in capacity is achieved at some disadvantage; to expand existing T carrier lines requires replacing repeaters, and in some cases, the apparatus cases that house them. Also, the design rules are more stringent because of crosstalk considerations. Furthermore, this line equipment is usable only on metallic cable facilities. It cannot be used to expand the carrying capacity of radio and fiber optic facilities.

Data Compression Equipment

Apparatus has recently been developed to compress two PCM bit streams into a single bit stream that can be transmitted over T-1 carrier lines or can be applied to M1-3 multiplexers and transmitted over fiber optic or radio facilities. These systems, known as low bit rate voice (LBRV) or *transcoders,* encode a digital signal at 32 rather than 64 kb/s, enabling transmission of 48 channels over a 1.544 mb/s line.

Although several methods for low bit rate voice such as Nearly Instantaneous Companding and Continuously Variable Slope Delta Modulation have been proposed, only Adaptive Differential Pulse Code modulation (ADPCM) is accepted as an international standard by CCITT. ADPCM

FIGURE 4.15
Comparison of Pulse Code Modulation and Adaptive Differential Pulse Code Modulation

PCM

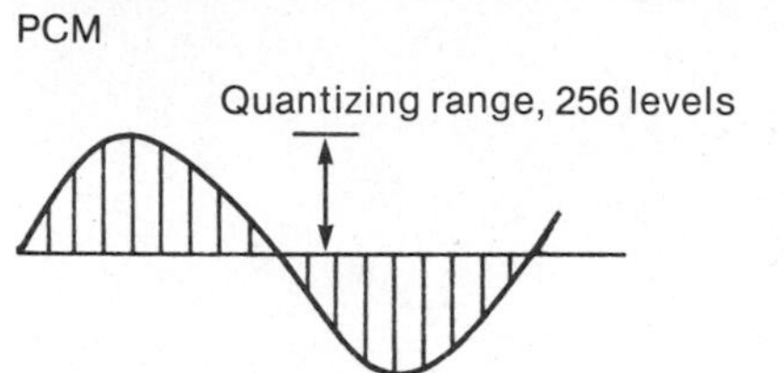

Sampling: 8,000 times per second
Quantize: 256 levels/sample
Code: 8 bits/sample

8,000 samples/sec
x 8 bits/sample
x 24 circuits
1,536 mb/s
\+ 8 mb/s framing
1,544 mb/s

ADPCM

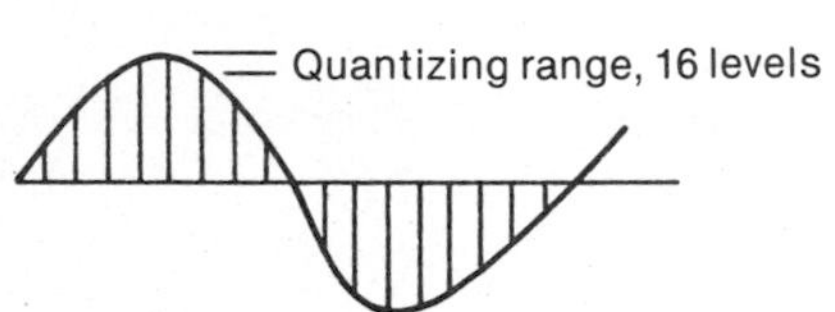

Sampling: 8,000 times per second
Quantize: 16 levels/change
Code: 4 bits/change

8,000 samples/sec
x 4 bits/sample
x 24 circuits
1,536 mb/s
\+ 8 mb/s framing
1,544 mb/s

samples the voice at 8,000 times per second as with PCM, but instead of quantizing the entire voice signal, only the changes between samples are quantized. The changes are predicted by a circuit known as an *adaptive predictor* that examines the incoming bit stream and predicts the value of the next sample. The difference between the actual sample and the predicted sample is quantized into 16 levels, which can be coded with four bits. The encoder adapts to the speed of change in the difference signal, fast for speech-like signals and slow for data signals. Figure 4.15 shows the difference between PCM and ADPCM, both of which result in a 1.544 mb/s line signal.

Signaling in ADPCM can use the bit robbing techniques of PCM, but to do so presents two problems. First, the voice compression techniques of ADPCM make it necessary to rob a bit from every fourth rather than every sixth frame. This makes ADPCM incompatible with the DS-1 signal format. Second, the robbed bit can degrade 4.8 kb/s data on a voice channel. To address this problem, a 44 channel format has been proposed to CCITT, in which 1 channel of every 12 is devoted to signaling. Some manufacturers provide either 44- or 48-channel operation as an option. An example is the

FIGURE 4.16
Adaptive Differential Pulse Code Modulated Transcoder

Courtesy Granger Associates.

Granger Associates TC7900-M1 Transcoder shown in Figure 4.16.

The primary disadvantages of ADPCM are the extra expense of the encoder, which is required in addition to the PCM channel banks, the inability of ADPCM to handle data above 4.8 kb/s, and the loss of two voice channels per digroup. Several products on the market are designed to provide 64 kb/s clear channel capability by using two voice channels for high-speed data transmission.

Delta Modulation

A less sophisticated method of signal compression is *delta modulation,* which registers changes in a voice signal. Delta modulation uses a one-bit code to represent the voice frequency waveshape. If a sample is greater in amplitude than the previous sample, a one is transmitted. If it is less, a zero is transmitted. These signals result in a code that represents the instantaneous slope of the voice frequency waveshape.

The primary advantage of delta modulation is its ability to compress more voice channels into a bit stream at a lower cost than PCM, ADPCM, or other adaptive low bit rate systems. A reasonably good quality signal can be obtained at 16 kb/s per channel. Subscriber carrier systems in use today apply 40 channels to a T-1 line for nearly twice the capacity of PCM.

The main weakness of delta modulation is its inability to follow rapid changes in the voice signal. Fortunately, however, voice signals are predictable in their behavior, so the effect is not noticeable in most conversations. A second weakness of delta modulation is the inability of the system to handle direct data transmission through data port channel units or to transport high-speed data using modems.

STANDARDS

T carrier standards are primarily set by manufacturers in North America with voluntary adherence to the standards. Many of the North American standards have been adopted by CCITT and are included in this section. The following are the primary standards and commercial publications affecting digital carrier:

Bell Communications Research

CB113 The Low Power T-1 Line Repeater Compatibility Specification.

CB118 The D3 Channel Bank Compatibility Specification.

CB119 Interconnection Specification for Digital Cross-Connects.

CB123 Digroup Terminal and Digital Interface Frame Technical Reference and Compatibility Specification.

CB126 D3 and D4 Subrate Dataport Channel Unit Technical Reference and Compatibility Specification.

CB127 M1C Multiplex Compatibility Specification.

CB128 M12 Multiplex Compatibility Specification.

CB129 M13 Multiplex Compatibility Specification.

CB130 M23 Multiplex Compatibility Specification.

CB131 MC2 Multiplex Compatibility Specification.

CB132 T1C Digital Line Compatibility Specification.

CB133 MC3 Multiplex Compatibility Specification.

CB134 T1D Digital Line Compatibility Specification.

CB141 D3 and D4 56 KB Dataport Channel Unit Technical Reference and Compatibility Specification.

CB142 The Extended Superframe Format Interface Specification.

CB143 Digital Access and Cross-Connect System Technical Reference and Compatibility Specification.

CB144 Clear Channel Capability.

TR TSY-000021 Item 363 Synchronous DS3 Format Interface Specification.

CCITT

G.702 Vocabulary of pulse code modulation (PCM) and digital transmission terms.

G.703 General aspects of interfaces.

G.704 Maintenance of digital networks.

G.705 Integrated services digital network (ISDN).

G.711 Pulse code modulation (PCM) of voice frequencies.

G.712 Performance characteristics of PCM channels at audio frequencies.

G.731 Primary PCM multiplex equipment for voice frequencies.

G.733 Characteristics of primary PCM multiplex equipment operating at 1,544 kb/s.

G.735 Characteristics required to terminate 1,544-kb/s digital paths on a digital exchange.

G.736 Characteristics of synchronous digital multiplex equipment operating at 1,544 kb/s.

G.7ZZ Proposed 32 kb/s low bit rate voice algorithm.

G.791 General considerations on transmultiplexing equipments.

G.792 Characteristics common to all transmultiplexing equipments.

G.793 Characteristics of 60-channel transmultiplexing equipments.

G.901 General considerations on digital line sections and digital line systems.

G.911 Digital line sections and digital line systems on cable at 1,544 kb/s.

G.917 Digital line sections and digital line systems on cable at 44,736 kb/s.

Q.46 and Q.47 Characteristics of primary PCM multiplex equipment operating at 2,048 kb/s and 1,544 kb/s.

Q.314 PCM line signaling.

APPLICATIONS

Digital carrier systems have wide application in both common carrier and private networks. Virtually all metropolitan common carrier networks use digital carrier today, with analog carrier confined to growth of existing systems. Digital carrier is not widely used in long-haul systems, but its use will increase as both satellite and terrestrial fiber optic facilities are deployed with digital technology.

Satellite common carriers offer both analog and digital circuits. As transcontinental fiber optics becomes available, long-haul digital circuits from terrestrial carriers will become more common. The application of T-1 facilities to the Rockwell International Corporation's private network is described in Chapter 12.

Evaluating Digital Transmission Systems

Originally, digital transmission equipment was used almost exclusively by common carriers. With deregulation of the telecommunications industry, private networks are increasingly composed of digital facilities. Terminal equipment is owned by the network user, and digital lines are implemented over common carrier or privately owned facilities. The criteria for evaluating digital transmission equipment are the same for both private and public ownership.

Reliability

The failure rate of terminal and line equipment is the primary concern in any network. Because the common equipment in the channel bank, T-1 multiplexer, and line repeaters is common to many channels, a single failure disrupts a considerable amount of service. Therefore, strict attention should be paid to the failure rate of these components.

Manufacturers quote failure rates as mean time between failure (MTBF) as discussed in Chapter 2. MTBF is the number of hours between service-affecting interruptions. Although the MTBF figure is an average, it offers a convenient way of calculating the expected frequency of failure by weighting the failure rates of the individual components. For example, if a manufacturer quotes 10,000 hours MTBF on the common equipment for a channel bank, and since a failure of a channel bank at either end of the circuit disrupts service, the MTBF for both channel banks together will be half the quoted rate, or 5,000 hours. If repeater failures are estimated at 20,000 hours per repeater and the line consists of four repeaters, the MTBF of the line will be 20,000/4 = 5,000 hours. If the channel banks and the line each have an MTBF of 5,000 hours, the failure rate of the entire system is 2,500 hours. This does not, of course, include failures of the transmission medium or failures from causes external to the equipment.

Availability of Special Service Features

T carrier channel banks should be equipped with a range of special services channel units to match the user's requirements. Most private networks will require dataport and foreign exchange channel units plus a variety of voice channel units to match the PBX or station equipment they interface. Compatibility with the extended superframe should be considered if the channel bank will be used extensively for point-to-point high-speed data transmission.

Maintenance Features

The design and layout of a T carrier system has a substantial effect on the cost of administering and maintaining it. Channel banks and line equipment should be equipped with an alarm system that registers local alarms and can be interfaced to telemetry for unattended operation. Test equipment needed to keep the system operative should also be considered. Typically this consists of a bit error rate monitor, repeater test sets, a T-1 signal source, and extender boards to obtain access to test and level monitor points on the plug-in units.

The physical layout of the system should be designed to facilitate maintenance. Voice frequency and signaling leads should be terminated on a distributing frame. If more than one T carrier line is terminated, DSX jacks should be provided. If temporary rearrangements will be made, voice frequency and signaling leads should be equipped with jacks.

Spare channel and common equipment plug-in units should be provided for rapid restoral, with quantities related to the number of units in service and the failure rate of the system. Carrier group alarm (CGA) may be required to lock out switching systems from access in case of channel bank failure. CGA should restore automatically when the failure is corrected.

Power Consumption

Differences in power consumption will be found between products of the various manufacturers. Digital carrier is normally left operating continually, so any saving in power will involve a substantial cost saving over the life of the system. Power consumption varies with the mix of channel units installed in the channel bank, so if alternative products are being evaluated, economic comparisons should be made assuming a similar mix of channel units.

Backup Power

Digital channel banks require −48 volts DC to match the storage batteries in telephone central offices. Converters are available to power the equipment from commercial AC sources. If continuous service during power outages is required, a battery plant should be provided. T-1 multiplexers that use random access memory to store the multiplexer configuration should be equipped with battery power to prevent loss of the configuration when the power fails. See Chapter 14 for power plant and common equipment considerations.

Operating Temperature Range

The operating temperature range of a digital channel bank is rarely a consideration in either private or common carrier networks. Most digital channel banks are capable of operation in the range of about 0 to 50 degrees C., which is well within the limits of most operating environments. T-1 multiplexers, which are not usually designed for a telephone central office environment, may have narrower limits that should be evaluated against the operating environment they will be installed in. Forced air circulation may be required for reliable operation at higher temperatures.

Density

Digital channel banks vary in the number of channels that can be installed in a given mounting space. If floor space is plentiful, this is not a matter of great concern, but when numerous channel banks or multiplexers are being installed, the density and mounting method are factors to consider.

Compatibility

T-1 channel banks built for use in the United States are generally compatible with each other and with the standard T-1 line format. T repeaters are not only compatible with the signal but are also plug compatible so that any manufacturer's repeater fits in the same slot in an apparatus case. Other than these two elements, compatibility is a matter of concern. Any T-1 multiplexer will be compatible only with itself unless it was specifically designed to be compatible with that of another manufacturer. Voice compression system that follow CCITT standards are compatible; other systems are proprietary and will be found incompatible with one another.

Evaluating T-1 Multiplex

Several features are unique to T-1 multiplexers and should be considered in addition to the above:

- Minimum bandwidth of the multiplexer slot. Wide bandwidths may require submultiplexers to use lower speed DTE.
- The ability to reroute traffic automatically when the primary link fails.
- Channel bypass and drop-and-insert capability to enable flexible routing of slots.
- Amount of redundancy in the power supply and central logic to improve reliability.
- The type of memory in which multiplexer configuration is contained. PROM memory tends to be inflexible; RAM memory can be lost during power failures. If RAM memory is used, a backup method of restoring the configuration such as booting from floppy disk or tape should be provided.
- Type of modulation used for voice channels. The use of ADPCM limits analog data to 4.8 kb/s; the use of delta modulation limits analog data to 2.4 kb/s.
- Amount of self-diagnostic capability in the unit. The more self-diagnostic capability, the more rapid trouble restoral will be.

GLOSSARY

Adaptive differential pulse code modulation (ADPCM): A method endorsed by CCITT for coding voice channels at 32 kb/s to increase the capacity of T-1 to either 44 or 48 channels.

Back-to-back channel bank: The interconnection of voice frequency and signaling leads between channel banks to allow dropping and inserting channels.

Bipolar coding: The T carrier line coding system that inverts the polarity of alternate "one" bits.

Bipolar violation: The presence of two consecutive "one" bits of the same polarity on a T carrier line.

Bit robbing: The use of the least significant bit per channel in every sixth frame for signaling.

Bit stuffing: Adding bits to a digital frame for synchronizing and control. Used in T carrier to prevent loss of synchronization from 15 or more consecutive "zero" bits.

Channel bank: Apparatus that converts multiple voice frequency signals to frequency or time division multiplexed signals for transmitting over a transmission medium.

Channel service unit (CSU): Apparatus that interfaces DTE to a line connecting to a dataport channel unit to enable digital communications without a modem. Used with DSU when DTE lacks complete digital line interface capability.

Clear channel: The elimination of bit-robbed signaling in a digital channel to enable use of all 64 kb/s for digital transmission.

Companding: The process of compressing high-level voice signals in the transmitting direction and expanding them in the receiving direction with respect to lower level signals to improve noise performance in a circuit.

Cross-connect: The interconnection of voice or signal paths between separate equipment units.

Data service unit (DSU): Apparatus that interfaces DTE to a line connecting to a dataport channel unit to enable digital communications without a modem. Used with CSU when DTE lacks complete digital line interface capability or alone when DTE includes digital line interface capability.

Dataport: A PCM channel unit that provides direct access to a digital bit stream for data transmission.

Delta modulation: A system of converting analog to digital signals by transmitting a single bit indicating the direction of change in amplitude from the previous sample.

Digital cross-connect system (DCS): A specialized digital switch that enables crossconnection of channels at the digital line rate.

Digital Signal cross-connect (DSX): A physically wired cross-connect frame to enable connecting digital transmission equipment at a standard bit rate.

Digroup: Two groups of 12 digital channels integrated to form a single 24-channel system.

Extended superframe (ESF): T-1 carrier framing format that provides 64 kb/s clear channel capability, error checking, 16 state signaling, and other data transmission features.

Foreign exchange (FX): A special service that connects station equipment located in one telephone exchange with switching equipment located in another.

Jitter: The phase shift of digital pulses over a transmission medium.

Loop timing: A digital synchronizing method that operates by extracting a synchronizing clock signal from incoming pulses.

Patch: The temporary interconnection of transmission and signaling paths. Used for temporary rerouting and restoral of failed facilities or equipment.

Pulse amplitude modulation (PAM): A digital modulation method that operates by varying the amplitude of a stream of pulses in accordance with the instantaneous amplitude of the modulating signal.

Pulse code modulation (PCM): A digital modulation method that encodes a PAM signal into an cight-bit digital word representing the amplitude of each pulse.

Quantizing: The process of encoding a PAM signal into a PCM signal.

Quantizing noise: Noise that results from the inability of a PAM signal to represent each gradation of amplitude change.

Transcoder: A device that combines two 1.544 mb/s bit streams into a single 1.544 mb/s bit stream to enable transmission of 44 or 48 channels over a DS-1 medium.

BIBLIOGRAPHY

American Telephone & Telegraph Co. *Telecommunications Transmission Engineering.* Bell System Center for Technical Education, vol. 2, 1977.

American Telephone & Telegraph Co. Bell Laboratories. *Engineering and Operations in the Bell System.* Murray Hill, N.J.: AT&T Bell Laboratories, 1983.

Bellamy, John C. *Digital Telephony.* New York: John Wiley & Sons, 1982.

Freeman, Roger L. *Telecommunication System Engineering.* New York: John Wiley & Sons, 1980.

———. *Reference Manual for Telecommunications Engineering.* New York: John Wiley & Sons, 1985.

MANUFACTURERS OF DIGITAL CARRIER EQUIPMENT

Channel Banks

AT&T Network Systems

Aydin Monitor System Digital Communications Group

GTE Communication Systems

ITT Telecom Network Systems Division

Lynch Communications Systems Inc.

Northern Telecom Inc., Network Systems

Wescom, Telephone Products Div. Rockwell Telecommunications, Inc.

Low Bit Rate Voice Equipment

AT&T Network Systems

Aydin Monitor System Digital Communications Group

Granger Associates

Tellabs Inc.

Wescom, Telephone Products Div. Rockwell Telecommunications, Inc.

T-1 Multiplexers

Case-Rixon Communications, Inc.

Coastcom

Datatel Inc.

Digital Communications Associates, Inc.

Ericsson Inc., Communications Division

General Datacomm, Inc.

Paradyne Corporation

Tellabs Inc.

Timeplex, Inc.

CHAPTER **FIVE**

Frequency Division Multiplex

The first transcontinental telephone circuits were made possible by Lee DeForest's invention of the vacuum tube in 1906. These voice frequency circuits were carried on open wire, using repeaters every 300 to 350 miles to overcome loss. Vacuum tubes also accelerated the development of radio, and soon thereafter led to carrier telephony using frequency division multiplex (FDM). FDM is a method of deriving multiple voice frequency channels over a single analog transmission medium such as wire, coaxial cable, or microwave radio.

The first carrier systems were designed for open wire because its heavy-gauge conductors introduced only about one-fortieth the loss of cable. Western Electric (now AT&T Network Systems) C carrier, some of which remained in service past the middle of the century, added three carrier channels to the voice channel over a single open wire pair. The Western Electric J carrier system, introduced in the late 1930s, was a 12-channel system that could be applied to frequencies above C carrier, thus deriving 15 carrier channels and 1 voice channel over a single pair of wires.

Open wire, which has all but disappeared from the nation's telecommunications networks, suffers from the severe disadvantage of susceptibility to weather. In warm weather, wires sag and loss increases. In cold weather, loss decreases but icing and winds cause circuit outages, resulting in unacceptable reliability.

With improvements in vacuum tube technology, the closer repeater spacing required by cable carrier became feasible and cable carrier came into general use to replace open wire carrier. One of the first, Western Electric K carrier, was a 12-channel system that operated over separate transmit and receive cables to prevent crosstalk. The more controlled environment of K carrier cable resulted in reliability superior to J carrier even though they shared the same basic channel bank. The primary limitation of K carrier was the small number of circuits it could accommodate. To prevent crosstalk between systems operating in the same direction, an elaborate network was required to balance every cable pair against every other pair. The physical size of the balancing network placed an upper limit on the number of systems that could be carried in one pair of cables.

The L carrier system, which was introduced just before World War II, is the immediate ancestor of today's long haul carrier systems. The basic frequency plan is followed by all manufacturers in the United States, and with minor variations, has been adopted as an international standard by CCITT. L carrier was originally designed for use on coaxial cable. As microwave radio became technically feasible in the late 1940s, L carrier was used to channelize the radio using the same modulation plan used for cable. The first L carrier system, L1, developed 480 channels over a pair of coaxial tubes. This was later expanded to 600 channels. Continual improvement has resulted in the current L5E system that supports as many as 13,200 channels over a pair of coaxial tubes.

These carrier systems, designed for transcontinental service, were either too large, as with L carrier, or too costly over "short-haul" distances of about 100 miles or less. However, the increasing demand for long-distance telephone service required quantities of short-haul circuits that could no longer be met economically with voice frequency cables and open wire. With continuing advances in carrier telephony and new, reliable miniaturized tubes, short-haul carrier was introduced in the 1950s. Western Electric N1 and Lenkurt (now GTE) 45BN carrier are early examples of these systems. The primary differences between short-haul and long-haul carrier are:

- Short-haul systems use less expensive modulation techniques.
- Voice channels over short-haul carriers generally have higher noise levels.
- Higher-order modulation plans are available only with long-haul carrier, limiting short-haul systems to a capacity of approximately 24 channels.
- Long-haul systems are designed to control level variations in transcontinental circuits to within 0.25 dB. Short-haul regulation systems also control level variations, but not over long distances.

Both long-haul and short-haul carrier have undergone numerous advances since their introduction; however, the frequency and modulation plans remain largely unchanged. As transistors and integrated circuits became feasible they were adopted to reduce cost and power consumption, and to increase the circuit density of carrier systems. Short-haul analog carrier is being replaced by digital carrier, and except for growth of existing routes, its use is rapidly declining in favor of digital carrier. L carrier, on the other hand, continues as an important source of long-haul circuits because terrestrial transcontinental digital circuits are not yet available except for a limited number of data under voice (DUV) circuits applied to microwave radio as described in Chapter 15.

ANALOG CARRIER TECHNOLOGY

Long-haul and short-haul analog carriers use the same basic techniques of generating a carrier frequency for each channel and modulating the amplitude of the carrier with a voice signal. Amplitude-modulation techniques are described in Appendix A for those readers who are unfamiliar with the methods. Single sideband suppressed carrier (SSBSC) modulation is used for all long-haul and most short-haul carrier. To reduce costs, some versions of short-haul carrier use double sideband modulation because it is less expensive and easier to synchronize.

A block diagram of an SSBSC channel is shown in Figure 5.1. The voice frequency signal is limited by filters to a range of about 300 to 3,300 Hz. The carrier frequency supply and the voice frequency signal are inserted into a balanced modulator that modulates the amplitude of the carrier and suppresses the carrier in the output signal. The output of a balanced modulator consists of only the upper and lower sidebands. Since the two sidebands are redundant, one is filtered out, leaving a single carrier frequency modulated by the voice signal and occupying 4 kHz of bandwidth. The voice channel bank combines 12 channels to produce an overall band of frequencies 48 kHz wide. In long-haul carriers the output of the channel bank is 60 to 108 kHz; in short-haul carriers the channel output is modulated directly to the carrier line frequency.

In the receiving direction, a carrier frequency identical to the transmit frequency is injected in one input to the demodulator. The received SSB signal is injected into the other input, and the voice frequency envelope of the carrier signal is extracted at the output.

With few exceptions analog carrier systems require external signaling and are designed with fixed transmit and receive levels. The standard levels are −16 dBm input into the channel modulator and +7 dBm output from the demodulator; a carrier system inserts 23 dB of gain into a circuit. Unlike digital carrier systems, signaling and gain transfer functions are provided in FDM carrier with external signaling and terminating units.

The L Multiplex System

The frequency and modulation plan of L multiplex (LMX) is shown in Figure 5.2. The 12-channel output of a channel bank is called a *group*. Five groups are combined to form a *supergroup* of 60 channels in a frequency range of 312 to 552 kHz. Ten supergroups are combined to form a 600-channel *master group*. Six master groups are in turn com-

FIGURE 5.1
Single Sideband Suppressed Carrier (SSBSC) Channel Block Diagram

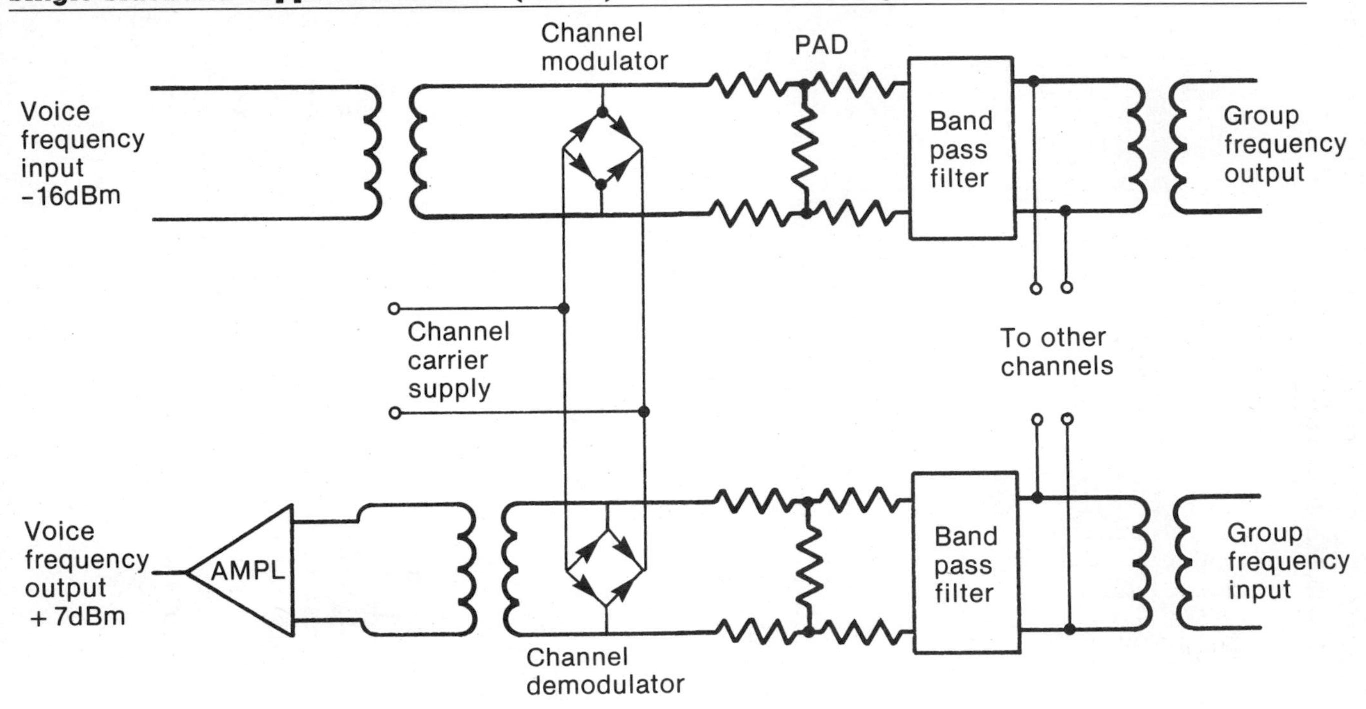

FIGURE 5.2
Basic L-600 Mastergroup

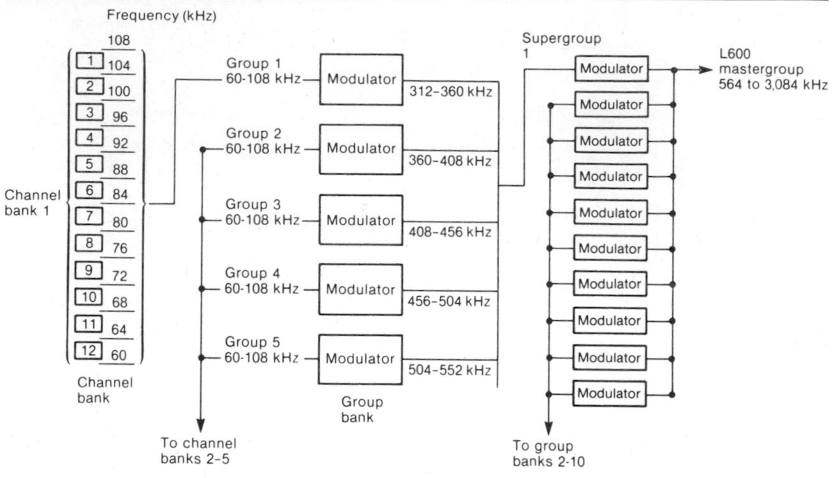

Six mastergroups are combined to form a 3,600-channel jumbo group.

bined into *jumbo groups* of 3,600 channels. A total of 13,200 channels can be supported on L-5E coaxial cable.

Some manufacturers produce a direct-to-line multiplex that modulates the voice channels directly to their final frequency. For example, the Granger DTL 7300 system shown in Figure 5.3 translates a band of 12 voice channels to any group frequency in the 12 to 2,540 kHz range.

The carrier supply is a critical element in an L carrier system. In the receiving terminal, a demodulating carrier identical in frequency to the modulating frequency is injected into the demodulator. If the receive carrier frequency is slightly different from the transmit frequency, the voice signal is shifted by the difference between the frequencies, resulting in an unnatural sound. Frequency stability is essential for data transmission because any offset may make it impossible for the modem to detect the incoming data signal. In the United States, analog carrier systems connecting to the AT&T Communications network are synchronized over a network that consists of 2.048 mHz signals transmitted over microwave radio and coaxial cable and 20.48 mHz signals transmitted over L5E coaxial cable. All carrier

FIGURE 5.3
Direct-to-Line Multiplex Unit

Courtesy Granger Associates.

systems connecting to this network are phase locked to these frequencies to ensure stability.

L Coaxial Transmission Lines

The signal from an LMX system can be applied to a coaxial cable. The coaxial cable consists of from 4 to 22 solid copper tubes 0.375 inches in diameter. The inner conductors are solid copper with insulating disks spaced at intervals of about 1 inch along the inner conductor to maintain separation from the outer tube. The spaces between the tubes are filled with twisted pair wire that can be used for short haul carrier or voice frequency circuits. Separate tubes are used in the transmit and receive directions. Equalizers are used to compensate for the high-frequency loss of the cable. In the transmitting direction low frequencies are attenuated and high frequencies boosted. In the receiving direction the equalization is reversed so the response curve of the signal applied to the amplifiers is flat. Repeaters are spaced at intervals ranging from 1 mile for L5E to 8 miles for L1. They are powered by voltages applied across the inner conductors of the transmit and receive coaxial tubes.

The gain of amplifiers is regulated by *pilots,* which are frequencies inserted on the line and separated from the voice channels by filters. The purpose of pilots is to adjust the amplifier gain automatically to compensate for variations caused by temperature fluctuations and aging components. As the pilot level varies, regulating circuits detect the variation and feed back a correcting signal to the amplifiers. Because the magnitude of the variations is unequal across the frequency range of the carrier line, pilots are spaced throughout the frequency range to equalize the amplifiers as well as to adjust their overall gain. Pilots are also used to detect faults in the system. When an amplifier begins to fail, the drop in pilot level triggers an alarm and may initiate a switch to spare facilities.

Spare coaxial lines and repeaters are provided to protect working lines. When the protection equipment in the receiving terminal detects a drop in pilot level, it signals the transmitting terminal to feed its signal on the spare line. Pilots inserted on the spare line keep the gain of the amplifiers regulated so channel levels do not vary when the switch occurs.

L Carrier on Microwave Radio

Channels derived with LMX can be applied to the input of microwave radio as described in Chapter 15. The number of channels the radio supports is a function of the bandwidth and modulation method used. Microwave radios with capacities as high as 5,400 voice channels are available.

Short-Haul Carrier Systems

The first short-haul carrier used in the United States by the Bell System was Western Electric's N1, which transmits 12 voice frequency channels over separate transmit and receive pairs in a single cable. Early versions of N carrier use a double sideband transmitted carrier modulation plan. This plan has the advantages of low cost, of synchronizing the demodulator in the receiving channel from the transmitted carrier, and of regulating the gain of the line amplifiers from the carrier power without the use of pilots. The disadvantages of this technique are the additional bandwidth required and the additional load the transmitted carriers place on the amplifiers. As discussed in Chapter 2, intermodulation noise results from the additional load, which makes this type of carrier unsuitable for radio transmission. The frequency plan of N1 carrier is shown in Figure 5.4.N carrier has undergone several improvements, resulting in the current N4 carrier system. N4 carrier, and its predecessor N3, use SSBSC modulation and transmit 24 channels in the same frequency range as N1 and N2.

N Carrier Lines

N carrier lines use separate transmit and receive frequencies as shown in Figure 5.4 to prevent crosstalk. To improve separation, separate transmit and receive cable pairs were also required until recent technology made it possible to use filters to combine the transmit and receive directions on the same cable pair. Frequency *frogging* is used to invert the transmit and receive frequencies to equalize loss at the different line frequencies. Because cable loss increases with frequency, by inverting the channels at each repeater, each channel is transmitted first at a high and then at a low frequency. This technique makes it possible to avoid the expensive equalizers used in other carrier systems.

Repeaters are designed for ground or pole mounting in weatherproof cabinets or in buildings. Repeaters, powered from DC voltage applied to the cable pairs at the terminals, are spaced at intervals of approximately eight miles.

ANALOG TO DIGITAL CONNECTORS

Because most of the long-haul facilities in the United States are analog, and because most metropolitan transmission systems are evolving toward all-digital, a requirement exists for analog-to-digital connectors. These generally fall into two categories: *L-to-T connectors* and *transmultiplexers.*

An L-to-T connector is a unit that contains back-to-back digital and analog channel units packaged in a single assem-

FIGURE 5.4
N-1 Carrier Modulation Plan

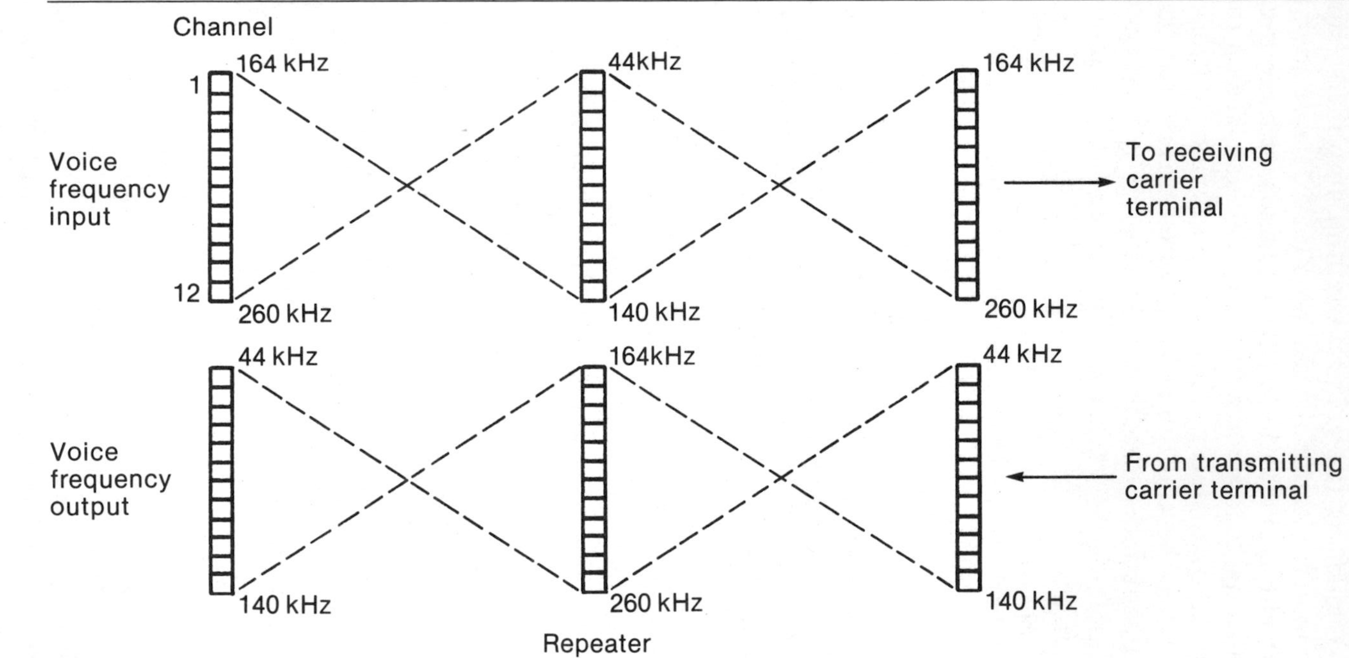

bly. The T carrier bit stream is demultiplexed to voice frequency and is coupled to the channel input of an analog channel bank. The built-in signaling bits from the digital channel bank are converted to single-frequency tones for transmission over the analog side of the connector. The costly external wiring interfaces associated with back-to-back channel banks are eliminated with this equipment, in which voice frequency and signaling interfaces are internally connected.

A second approach, is the transmultiplexer, which digitally processes the 60 to 108 kHz signals of two analog groups into a 1.544 mb/s bit stream without reducing either the analog or the digital signals to voice frequency. All pilots, signaling, and alarm functions of both analog and digital multiplex are accomplished in the transmultiplexer. Also, individual channel level controls are required to set analog channel levels. The unit must include pads to set the inserted connection loss of the circuit as described in Chapter 2. Transmultiplexers are packaged in groups of 24 channels that convert two analog groups to one DS-1 signal, or, as shown in Figure 5.5, in groups of 120 channels, which converts two supergroups to five DS-1 signals.

TIME ASSIGNMENT SPEECH INTERPOLATION

Voice communication circuits typically have large amounts of idle time because of the nature of human conversation. Long haul circuits are inherently capable of carrying information in both directions simultaneously. However, people cannot talk and listen simultaneously, which means the typical telephone circuit is idle at least 50 percent of the time. Furthermore, when one party to a conversation is speaking, there are often pauses in which no information passes in either direction.

Time Assignment Speech Interpolation (TASI) is equipment designed to concentrate conversations from multiple input circuits over a group of output circuits to take advantage of the pauses. Users speaking over a group of TASI-equipped circuits do not have a permanently assigned circuit

FIGURE 5.5
120-Channel Transmultiplexer

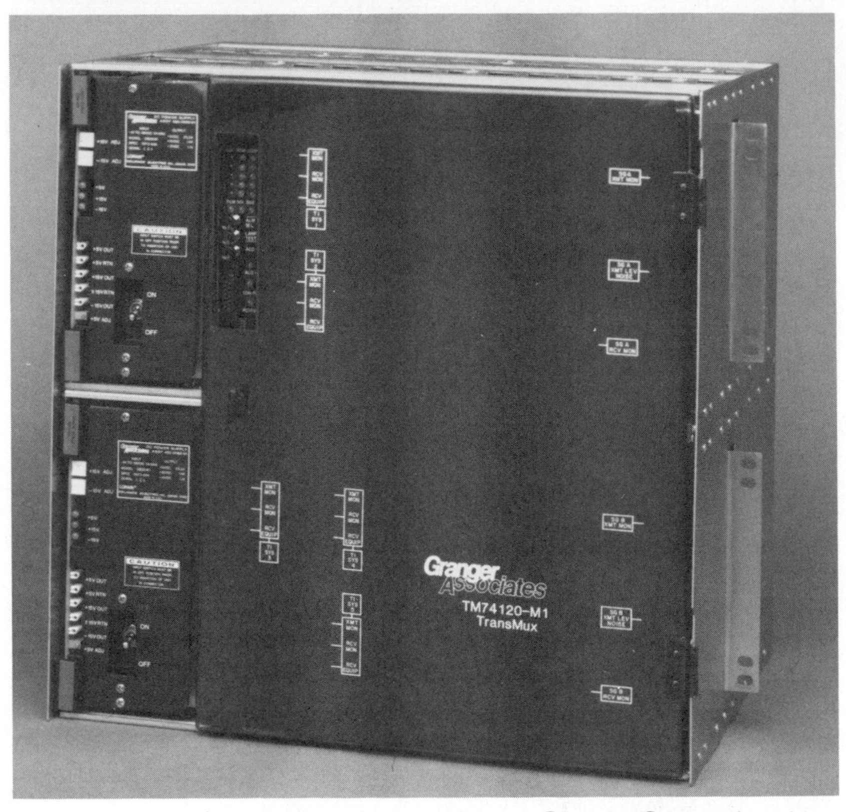

Courtesy Granger Associates.

but instead have a virtual circuit that is established by the system in response to commands from the control circuits.

TASI was originally developed by Bell Laboratories for use in transatlantic cables where it is important to make the maximum use of high-cost cable facilities. Although TASI was originally designed for overseas use, it is also finding domestic use in both common carrier and private networks.

The digital equivalent of TASI is Digital Speech Interpolation (DSI) equipment that operates on digital circuits. The predominant use of speech interpolation is on analog circuits at present because the majority of circuits long enough to justify TASI equipment are analog. As cross-country digital facilities become common, DSI equipment will be used to concentrate voice circuits on these.

The amount of circuit gain that can be derived with speech interpolation is a function of the amount and characteristics of the messages riding on the circuit group. Unlike switching equipment, TASI is unable to tolerate delays in circuit availability. If a path is not available at the time a talker begins, syllables are lost, which generates a repeat of the conversation and negates the benefits of the equipment. Therefore, it is essential that the system be properly engineered to prevent overloads.

Speech interpolation has two primary benefits in private networks. First, it can increase the carrying capacity of voice circuits without increasing leased circuit costs. If the costs of the equipment are less than the lease costs of the circuits, TASI results in a cost saving. Second, TASI equipment can be installed and rearranged by the user without involving the common carrier that provides the backbone circuits. This gives the network user independence from common carrier costs and service delays.

STANDARDS

Most standards in FDM carrier have been developed by AT&T and Bell Laboratories and have become *de facto* standards. The European plan for analog transmission varies from the North American plan only in minor detail. The following standards and publications cover analog transmission.

Bell Communications Research

CB122 Specification for the LT-1 connector which forms an interface between L-type carrier and the No. 4 ESS digital switching system.

CB139 LT-1B facility connector requirements and specification.

CCITT

G.231 Arrangement of carrier equipment.

G.232 12-channel terminal equipments.

G.241 Pilots on groups, supergroups, and so forth.

G.242 Through-connection of groups, supergroups, and so forth.

G.243 Protection of pilots and additional measuring frequencies at points where there is a through-connection.

G.332 12 mHz systems on standardized 2.6/9.5 mm coaxial cable pairs.

G.333 60 mHz systems on standardized 2.6/9.5 mm coaxial cable pairs.

G.334 18 mHz systems on standardized 2.6/9.5 mm coaxial cable pairs.

G.791 General considerations on transmultiplexing equipments.

G.792 Characteristics common to all transmultiplexing equipments.

G.793 Characteristics of 60-channel transmultiplexing equipments.

EIA

EIA RS-199-A Requirements for Solid and Semisolid Dielectric Transmission Lines.

RS-225 Requirements for Rigid Coaxial Transmission Lines—50 Ohms.

APPLICATIONS

Analog transmission is currently the dominant long-haul transmission method and will continue to predominate for several more years. However, the primary owners of analog facilities will be common carriers. Private networks will acquire large quantities of analog facilities from common carriers, but where the private network owns the facilities, digital equipment will predominate. Privately owned analog microwave radio will continue to be developed where its greater channel capacity is important.

The low level of usage of analog facilities by private networks stems from two factors. First are the technical advantages of digital facilities as described in Chapter 4. Second is the nature of services available over common carrier tariffs. Although common carriers offer high speed digital facilities directly to the end user, they do not provide broadband analog facilities on the same basis except for television transmission. This is the case because it is impossible to connect broadband analog facilities to the end user over twisted pair wire. Therefore, there will be relatively little use of analog channel banks and line repeaters by private network users.

Evaluating Analog Carrier Systems

The factors discussed in Chapter 4 for evaluating digital carrier systems are generally applicable to analog systems. Because special service channel units are not available with FDM carrier, these are not a consideration in evaluating systems. All the following factors should be considered in evaluating analog channel banks:

- Reliability including redundancy of carrier and pilot supplies, power supplies, automatic protection line switching, and an adequate alarm system.
- Floor space or channel density.
- Power consumption.
- Compatibility with the system at the other end including frequency plan, pilot frequencies and levels, and channel levels.
- Modulation plan and frequency plans must be compatible, and the same sideband, upper or lower, must be used.

GLOSSARY

Balanced modulator: An amplitude modulating circuit that suppresses the carrier signal, resulting in an output consisting only of upper and lower sidebands.

Direct-to-line multiplex: A system that modulates individual channel frequencies directly to their final carrier line frequencies.

Frogging: The process of inverting the line frequencies of a carrier system so that incoming low-frequency channels leave at high frequencies and vice versa. Frogging equalizes transmission loss between high- and low-frequency channels.

Group: A 12-channel band of frequencies occupying the frequency range of 60 to 108 kHz.

Jumbo group: A 3,600-channel band of frequencies formed from the inputs of six master groups.

L multiplex (LMX): A system of analog multiplex consisting of combinations of groups, supergroups, master groups, and jumbo groups to form a hierarchy of channels that can be transmitted over radio or coaxial cable.

L-to-T connector: A unit that interfaces two FDM groups with one TDM digroup to enable a 24-channel band of analog frequencies to interface a DS-1 line.

Master group: A 600-channel band of frequencies formed from the inputs of 10 supergroups.

Pilot: A single frequency that is transmitted on an L carrier line or microwave radio system to regulate amplifier stability and to actuate alarms.

Supergroup: A 60-channel band of frequencies formed from the inputs of five channel banks or groups.

Transmultiplexer: A device that connects digital and analog transmission lines by signal processing without the need for intermediate voice frequency conversions.

BIBLIOGRAPHY

American Telephone & Telegraph Co. *Telecommunications Transmission Engineering.* Bell System Center for Technical Education, vol. 2, 1977.

American Telephone & Telegraph Co. Bell Laboratories. *Engineering and Operations in the Bell System.* Murray Hill, N.J.: AT&T Bell Laboratories, 1983.

Freeman, Roger L. *Telecommunication System Engineering.* New York: John Wiley & Sons, 1980.

MANUFACTURERS OF ANALOG MULTIPLEX EQUIPMENT

Analog Carrier Equipment

AT&T Network Systems

GTE Communications Systems

ITT Telecommunications

NEC America, Inc.

Rockwell International Collins Transmission Systems Division

Analog to Digital Connectors

AT&T Network Systems

Granger Associates

Rockwell International Collins Transmission Systems Division

Direct-to-Line Multiplex

Granger Associates

Rockwell International Collins Transmission Systems Division

CHAPTER SIX

Station Equipment

The least exotic and technically sophisticated element of the telecommunications system is the ordinary telephone. Yet, its importance in the design of the network should not be underestimated. Because of the enormous numbers of telephone instruments in service, much of the rest of the network is designed to keep the telephone simple, rugged, and economical. Over the 100 plus years of telephony, the telephone set has undergone improvement, but the fundamental principles have changed little since Alexander Graham Bell's original invention. The primary changes have been improvements in three characteristics: in packaging, to make telephones esthetically appealing and easy to use; in signaling, which improves the methods used to place and receive telephone calls; and in transmission performance, which improves the quality of the talking path between users.

Station apparatus or customer premises equipment (CPE) includes not only the telephone set but also the bells and lights that alert the user to an incoming call, the key equipment that selects multiple telephone lines, the wiring that connects instruments on the user's premises, and auxiliary devices such as recorders and dialers. CPE also includes terminating and signaling equipment that is used for special telephone services.

TELEPHONE SET TECHNOLOGY

The telephone is inherently a four-wire device. Transmit and receive paths must be separated to accommodate the user's anatomy, but they must be electrically combined to interface the two-wire loops that serve all but a small fraction of the telephone services in the country. A functional diagram of the telephone set is shown in Figure 6.1.

Elements of a Telephone Set

The user's voice is converted into a pulsating direct current by a transmitter in the telephone handset. The transmitter in most telephones consists of a housing containing tightly packed granules of carbon that are energized by a DC voltage. The voice waves impinging on the transmitter compact the granules, changing the amount of current that flows in proportion to the strength of the voice signal. This fluctuating current travels over telephone circuits to drive the distant telephone receiver, which consists of coils of fine wire wound around a magnetic core. This current causes a diaphragm to move in step with changes in the line current. The diaphragm in the receiver and the carbon microphone in the transmitter are *transducers* that change fluctuations in sound pressure to fluctuations in electrical current, and vice versa.

The telephone set contains a hybrid coil that performs the four-wire to two-wire conversion as described in Chapter 2; it couples the two-wire line to the four-wire telephone handset. By design, the isolation between the transmitter and receiver is less than perfect. It is desirable for a certain amount of the user's voice to be coupled into the receiver in the form of sidetone, the feedback effect that regulates the volume of the user's voice. With too little sidetone, users tend to speak too loudly; with too much sidetone, they do not speak loudly enough.

The telephone set includes a switch hook that isolates all the elements except the ringer from the network when the

FIGURE 6.1
Functional Diagram of a Telephone Set

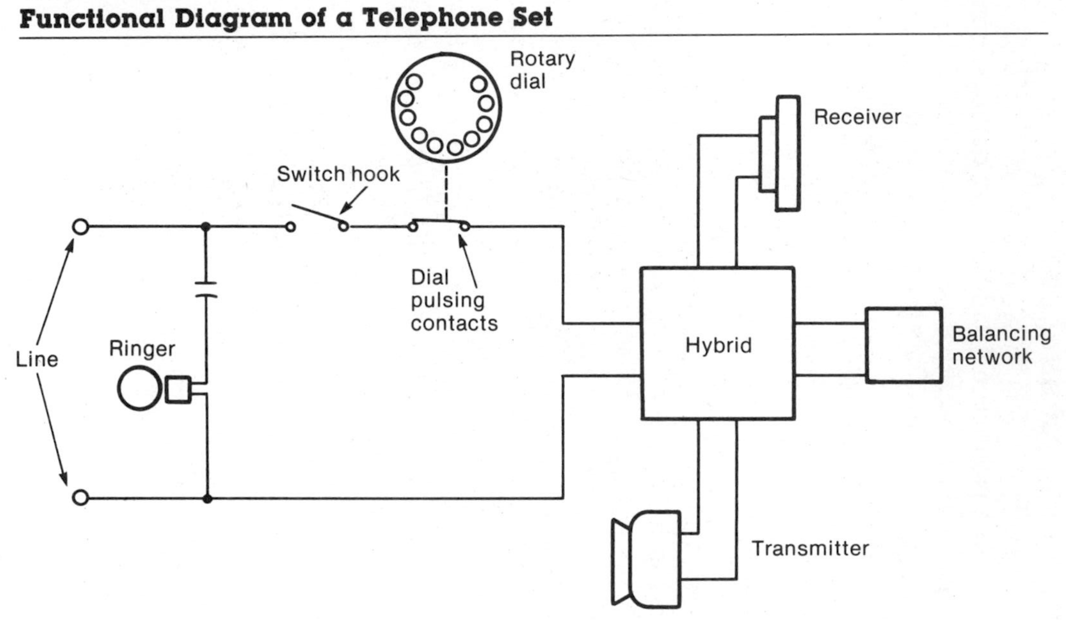

FIGURE 6.2
DTMF Dialing Frequency Combinations

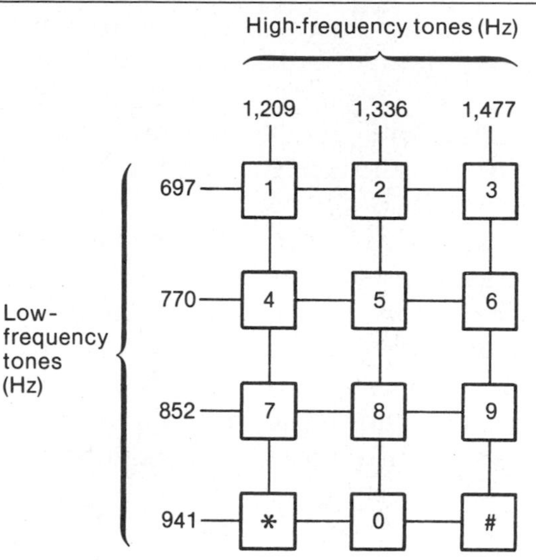

telephone is idle or *on hook*. When the receiver is lifted *off hook,* the switch hook connects the line to the telephone set and furnishes the battery needed to energize the transmitter. When the telephone is on hook, the ringer is coupled to the line through a capacitor that prevents the DC talking battery from flowing but permits the AC ringing voltage to actuate the bell. The telephone bell is an electromagnet that moves a clapper against a gong to alert the user to an incoming call. In many modern telephone sets the bell is replaced by an electronic tone ringer.

The dial circuit is connected to the telephone line when the receiver is lifted off hook. Dial circuits are of two types: rotary dials that operate by interrupting the flow of line current and tone dials that operate by sending a combination of frequencies over the line. Tone dialing, known as *dual tone multifrequency* (DTMF), uses a 3 × 4 matrix of tones, as shown in Figure 6.2, to transmit pairs of frequencies to a tone receiver in the central office. Each of the 12 buttons in a telephone set generates a unique pair of frequencies that are detected by a DTMF receiver.

The two wires of a telephone circuit are designated as the *tip* and *ring,* corresponding to the tip and ring of the cord plugs used by switchboard operators. Negative polarity talking battery is fed from the central office over the ring side of the line. When the receiver is off hook, current flows in the line. The amount of current flow is limited by the resistance of the local loop, which is a function of the wire gauge and length. For adequate transmission, at least 23 milliamps (ma.) of current is needed. If too little current flows, the transmitter is insufficiently energized and the telephone set produces too little output for good transmission. If more than approximately 60 ma. of current flows, telephone output will be uncomfortably loud for many listeners. Furthermore, DTMF dials need at least 20 ma. of current for reliable operation. The telephone set, the wiring on the user's premises, the local loop, and the central office equipment interact to regulate current flow in the line and the quality of local transmission service.

The contacts of a rotary dial are wired in series with the loop. When the dial is rotated to the finger stop, the loop is opened by a set of off-normal contacts. When the dial is released, the contacts alternately close and open to produce a string of square wave pulses. These pulses are detected by equipment in the central office to actuate switches that route the call.

Before the FCC opened the telephone network to connection of customer apparatus, all telephone sets were owned by the telephone companies, and because of the large quantities of sets in service, the network was designed for reliable operation with a minimum investment in station apparatus. The basic telephone set is a rugged and inexpensive device that is designed to provide satisfactory transmission over properly designed loops. If the current flowing through the loop is enough to provide between 23 and 60 ma. of current to the telephone set, satisfactory service is assured if the telephone is in good working order.

Telephone Sets for Party Lines

The previous discussion assumes the use of single-party telephones. In a single-party line, the ringer is bridged across the line and detects an AC ringing signal between the tip and ring of the local loop. If more than one user is connected to such a line, coded ringing is required. Early party lines

FIGURE 6.3
Multiparty Signaling in a Telephone System

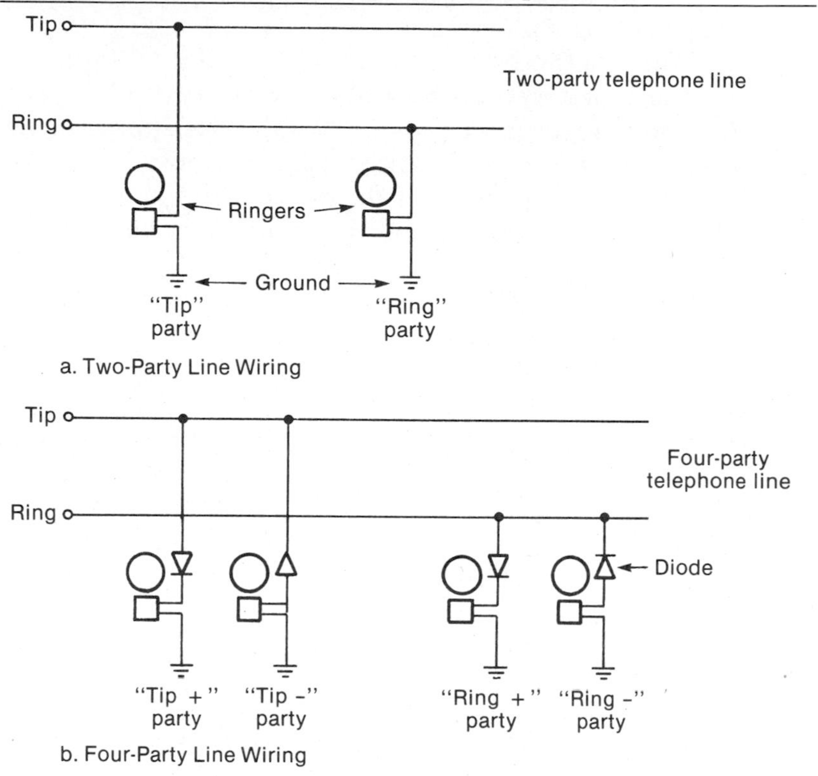

used combinations of long and short rings to signal the other party. The disadvantage, of course, is the annoyance of hearing unwanted rings, but to save costs, party lines still exist in many parts of the United States. Three methods are used to prevent rings from reaching any but the intended recipient.

The first method, used on two-party lines, requires wiring the ringer from one side of the line to ground as shown in Figure 6.3a. The party is wired in the central office as a "tip party" or a "ring party," and the ringer is wired from either the tip or the ring of the line to ground. The ringer wiring also actuates the automatic number identification equipment in the central office when either of the two parties places a long-distance call. This method of connecting ringers may unbalance the line and, as described in Chapter 2, increase the noise level. To alleviate the noise, *ringer isolators* are often used to disconnect the ringer except in the presence of a ringing signal.

The second method, used on four-party lines, employs a special tube that responds to only one polarity of ringing voltage inside the telephone. The central office applies either a positive or negative DC voltage superimposed over the ringing voltage to cause the tube to conduct as shown in Figure 6.3b. Bells are wired from tip or ring to ground through a tube that is polled positive or negative, yielding four ringing combinations. Because the tube conducts only in the presence of DC voltage superimposed over the ringing signal, the tube isolates the line from ground to improve balance.

The third system is *harmonic ringing.* In this system the ringers are tuned to respond to only one ringing frequency. This system is rarely used in the United States because the ringers can easily be detuned by dropping the telephone. Number identification for toll calls is manual on four-party lines.

Protection

Telephone circuits are occasionally subjected to high voltages that can be injurious or fatal to the user in the absence of electrical protection. Lightning strikes and crosses with high-voltage power lines are mitigated with a station *protector,* which is diagramed in Figure 6.4. Protectors use either an air gap or a gas tube to conduct high voltage from either side of the line to ground in case of hazardous voltages. The telephone is insulated so that any voltage getting past the protector will not injure the user. Protectors are placed by the telephone company and may also form a demarcation point with customer-owned wiring as shown in Figure 6.4. Protectors are connected to a ground rod, metallic water pipe, or other low-resistance ground.

The protector is connected to the telephone set by jacketed wiring, called *inside wiring,* placed by the user or the telephone company. Inside wiring terminates on the protector on the end nearest the central office and on a connecting jack designated by the FCC as RJ11 at the telephone end. FCC

FIGURE 6.4
Diagram of a Station Protector

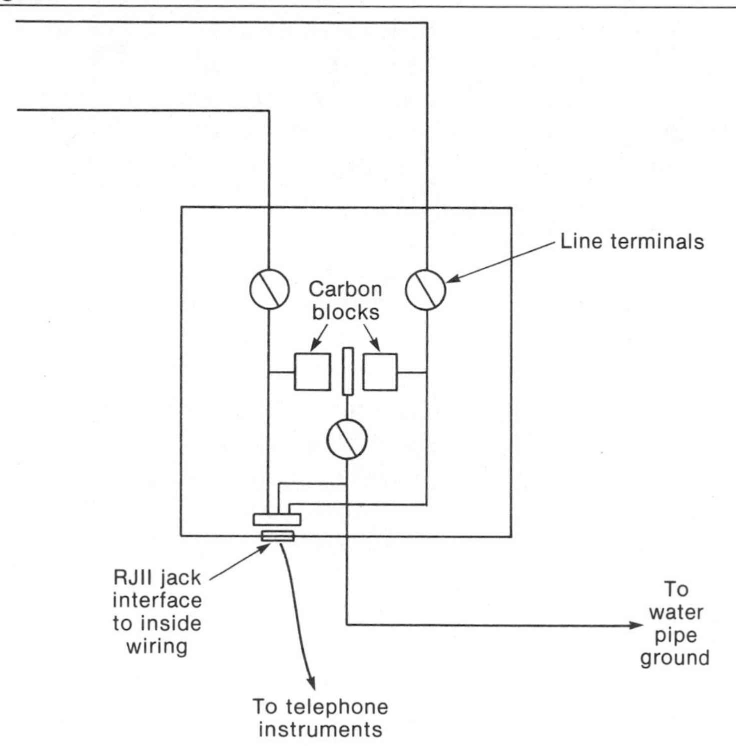

regulations require registration of the telephone set or other apparatus such as modems, PBXs, and key equipment. Registration indicates that apparatus has been approved for connection to the telephone network.

Coin Telephones

The coin-operated telephone business is worth about $4 billion in gross revenues per year. In 1984, the FCC authorized private ownership of coin telephones, paving the way for an entirely new breed of coin instrument. Coin telephones used by the telephone companies require a combination of circuitry in the telephone set and in the central office to operate coin collect and return apparatus. The reason for this is twofold: (1) the supervisory signals that indicate when the called party answers are not extended to the calling telephone.

Without supervisory equipment in the central office, the telephone would be unable to tell when to collect and refund coins. And (2), the system is designed to collect and return coins deposited for long-distance calls under operator control. This function is described in more detail in Chapter 10.

Privately owned coin telephones must be designed to operate independently of the central office. Some telephones are designed for coinless operation and are used only for collect, credit card, or third number calls. The majority require the calling party to signal the called party's answer by pushing a talk button or by depositing the coin after the called party answers. A microprocessor in the telephone handles the control functions.

Privately owned coin telephones are subject to state regulation and are permitted only where the state regulatory agencies have authorized their use. At the time of this writing, 46 states have either authorized or are considering private ownership of coin telephones.

Cordless Telephones

In recent years, cordless telephones have appeared on the market. These instruments use a low-powered radio link between a base unit and the portable telephone. The units are designed with enough range to use them on an average residential lot. As with all radio systems, privacy cannot be assured with cordless telephones. Early units were subject to interference and could be rung by any base unit operating on the same frequency. A more serious problem is the fact that anyone with a telephone on the same frequency can place unauthorized long-distance calls or eavesdrop on private calls. As only five frequencies in the 49 to 50 MHz range were initially allocated by the FCC, false rings and unauthorized long-distance calls are a problem on these early units.

Some systems operate on 49 MHz frequencies in both directions, while in others the base station transmits on a frequency between 1.7 and 1.8 MHz. These base stations use the AC power wiring in the house for an antenna. The frequencies are fixed, and interference may be encountered with another nearby unit. When interference occurs, the unit must usually be exchanged for a unit on another frequency. Be-

FIGURE 6.5
Code-A-Phone Model 2570 Answering Set

Courtesy of Code-A-Phone Corp., a subsidiary of Conrac Corp.

cause sets are designed with limited range, some units include a warning sound when the portable unit is carried outside the range of the base station.

The FCC has authorized 10 new cordless telephone channels in the 46 to 50 MHz range, and the new generation of telephones contains safeguards against false rings and unauthorized calls. The base-to-portable link is authenticated with a code from the portable unit so the base responds only to a unit with the correct code. This prevents unauthorized calls. The ringer in the portable unit is coded to prevent false rings. Encoders affect only the signaling and do not improve privacy. Anyone with a telephone tuned to the same frequency can listen to the call. For this reason, extended range is not necessarily an advantage.

Voice Recording Equipment

Voice recording equipment varies from simple telephone answering sets such as the one shown in Figure 6.5 to elaborate voice mail equipment that provides service similar to electronic mail. Answering sets, once provided exclusively by the telephone company, are widely available, and no more difficult to install than an ordinary telephone. Units such as the one shown in Figure 6.5 use separate cassette tapes for recorded announcements and messages from users. A separate unit is often provided so the recorder can be remotely accessed to retrieve recorded messages.

Voice mail is a service that is just entering the market. It is normally attached to a PBX or key telephone system. A caller dials the called party's mailbox number and leaves a message. The called party can retrieve messages by dialing the mailbox from any DTMF telephone and entering an authentication code. By entering instructions from the dial, the message can be repeated, erased, or retained in file. Compared to an ordinary answering device, voice mail gives the caller a choice whether to dial the called party directly or to dial his mailbox to leave a recorded message. A large number of calls, some estimate it to be as high as 50 percent, are placed for the sole purpose of transmitting information. The interruptions of many telephone calls are unneeded when voice mail is available.

Voice mail is similar in concept to electronic mail, but the two domains overlap very little. Electronic mail requires an expensive terminal, where voice mail can be sent and retrieved from an ordinary DTMF telephone. Voice mail is more personal. The caller's voice serves to authenticate the message, whereas electronic mail messages give no positive identification of the sender. On the other side of the coin, long messages are unsuited for voice mail. A complete message can be sent by electronic mail and reproduced or edited, but with voice mail, a verbatim transcript of long messages is impractical. Both services have their applications without a great deal of overlap.

Multiple Line Equipment

When more than one central office line is terminated in an office, equipment called a *key telephone system* (KTS) is used to pick up and hold multiple lines from a telephone. All KTS systems have several features in common:

- **Call pickup**—the ability to access one of several lines from a telephone.

- **Call hold**—the ability to place an incoming line in a holding circuit while the telephone is being used for another call.
- **Supervisory signals**—the lamps that indicate when a line is ringing, off hook, or on hold.
- **Common bell**—the ringing of a single alerting device to indicate an incoming call.

Key systems are rated according to the number of stations and lines they can accommodate. For example, a 6 × 24 system could terminate six telephone lines and 24 stations. Most key systems also include one or more intercom lines for conversations within the key telephone system. Intercom lines use either the telephone set or a speaker/microphone for the intercom talking path.

Electromechanical equipment known as 1A2 KTS, which is still in use in about 2.5 million locations, brings central office lines to the stations over multipair cable. The system uses a common bell for all lines and indicates which line is ringing by lighting a lamp on the telephone button. Separate leads are used for talking and illumination. A hold button transfers the line from the telephone set to a holding relay that terminates the line in a resistor and applies a flashing signal to the lamp on the telephone set. Figure 6.6 is a simplified diagram of how 1A2 KTS operates. Although this equipment is obsolescent, it is instructive to show the functions of KTS.

The control equipment is mounted in a cabinet and powered from commercial AC. The circuitry to supply illumination and to respond to pickup and hold signals is contained in plug-in units called *400 type* line cards. The common equipment and line cards light a steady lamp on busy lines and apply 60 flashes per minute to ringing lines and 120 flashes per minute to lines on hold. The common equipment also isolates lines in the holding mode so they cannot talk to each other when on hold.

The 1A2 system has a significant drawback—it requires large multipair cables that are expensive and costly to install. The number of conductors varies with the number of lines, but most stations are wired with a minimum of 25-pair cable.

FIGURE 6.6
Diagram of 1A Key Telephone System

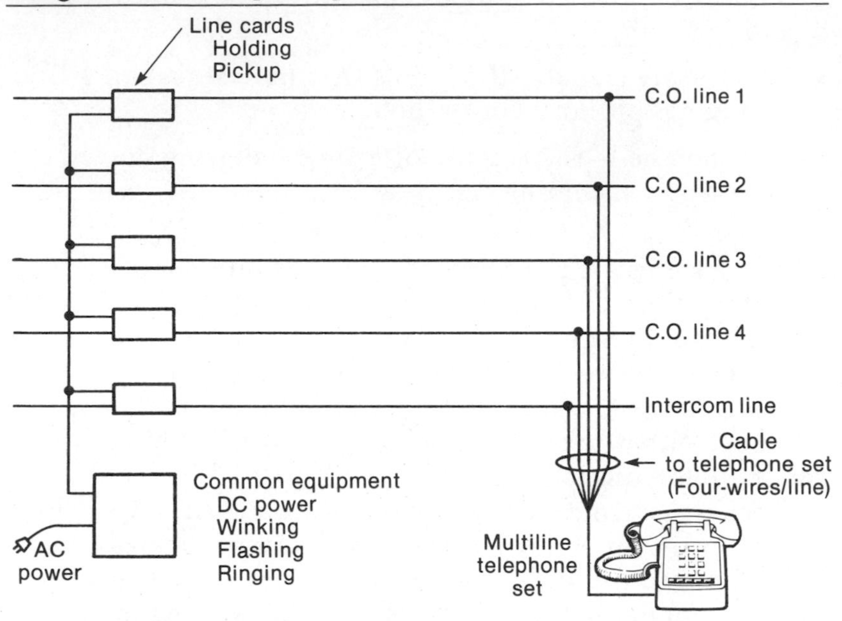

Line cards and common equipment are mounted in a central location.

It also lacks many of the enhancements of more modern equipment. For example, special attention must be paid to 1A2 KTS to ensure compatibility with DTMF dialing on the intercom line. Many systems use an electromechanical intercom line selector that is not compatible with DTMF dials. Auxiliary intercom line converters are required to select stations from DTMF pulses. Other features such as station message detail recording (equipment that records which station placed long-distance calls) and call management systems (which provides such functions as toll restriction and least cost routing) are not available with 1A2 KTS. Many special features may, however, be obtained with external add-on equipment.

The most recent generation of electronic KTS replaces electromechanical control with processor control and greatly reduces the number of cable pairs required for operation. A typical system uses two- or three-pair cable per station with separate paths for talking and control. Electronic KTS uses

FIGURE 6.7
Electronic Key Telephone Instrument

Courtesy of TIE/Communications, Inc.

a microprocessor to scan incoming lines. When an incoming ring is detected, the processor signals the attendant. The incoming call is picked up by depressing a button, the same procedure used in electromechanical KTS. However, instead of directly accessing the incoming line, the telephone set sends a data message to the controller, which connects the incoming line to the station. Calls can be held by depressing the hold button, which applies a flashing lamp signal to the line button.

Telephone sets in electronic KTS have numerous features that are unavailable on electromechanical systems. For example, the instrument shown in Figure 6.7 includes push button access to special features such as parking, call forwarding, and paging that are accessed in other systems by dialing access codes. A digital readout displays date and time, message waiting, called number, and other such messages. Effectively, electronic KTS is equivalent to electromechanical KTS except that lines and trunks are switched in a central switching matrix rather than in the telephone set. KTS systems that use electronic telephones are compatible only with telephones designed to interface the control equipment. In such systems, telephone sets are generally available only from the manufacturer.

The type and size of switching matrix vary with the manufacturer. Systems accommodating as many as 50 trunks and 100 lines are classed by the manufacturers as KTS, but the difference between such a system, called a *hybrid,* and a small PBX is not distinct. The provision of processor control allows KTS to provide many features that are similar to PBX features. Table 6.1 lists features provided by many popular KTS systems. These features, which are identical in most respects to similar PBX features, are discussed in Chapter 11. The switching matrixes used by KTS are either reed relays or time division networks. These technologies are described in Chapter 9.

KTS systems usually include one or more private communication or intercom lines used for station-to-station communication, primarily for conversations between the attendant and the called party. In large systems, however, the intercom line takes on the characteristics of the intrasystem talking paths of a PBX. Some systems provide more than one intercom line so several intrasystem conversations can be held simultaneously. Most systems provide a built-in speaker so the intercom line can be answered without using the telephone handset. Optionally, the handset can be lifted for privacy. The quantity of intercom lines provided is a distinguishing feature between a PBX and a hybrid KTS. Most hybrid KTS systems provide 10 or fewer intercom paths, which limits their usefulness for intrasystem calling.

While calls can be answered from any station in many systems, an attendant's console is usually provided. The attendant has all the features of regular stations, and may also be provided a busy lamp field to show which stations are occupied. Another common feature is direct station selection, which allows the attendant to transfer calls to stations

TABLE 6.1
Key Telephone Systems Typical Features

Attendant—provides telephone instrument and features for use by centralized attendant.

Automatic recall—calls that have been transferred to a station number return to the attendant after a given number of rings.

Automatic route selection—selects most economical route for long-distance calls.

Busy lamp—provides an array of lamps to indicate the busy/idle status of lines in a key system.

Camp-on—allows attendant to queue an incoming call on a busy number so it rings when the number is idle.

Call transfer—allows a user to transfer calls to another station without assistance from the attendant.

Call waiting—indicates by sending short tone over line that another call is attempting to access a busy line.

Call forward—allows user to transfer incoming calls to a second number if busy or after a given number of rings.

Conferencing—provides circuit for tying multiple lines together directly or through an amplifier.

Direct station selection—allows attendant push-button access to any station.

Distinctive ringing—incoming calls are identified as intercom or outside line based on the type of ring received.

Hands-free answer on intercom—allows user to talk to attendant on intercom trunk over speaker/microphone.

Last number redial—allows user to redial last number by pressing a button on the telephone set.

Message waiting—allows attendant to indicate to a busy or absent user with a lamp or other signal that a message is waiting.

Music on hold—inserts user-provided music on line while on hold.

Power fail transfer—transfers central office trunks to a standard telephone during power failure conditions.

Protected line—provides for locking out line that is being used for data transmission or other service so it cannot be picked up by another user.

Speakerphone—allows user to talk over trunk lines over speaker/microphone.

Speed calling—allows user to assign frequently called numbers to one- or two-digit access from telephone.

Station message detail recorder—registers long-distance calls against calling station for distributing system costs.

Toll restriction—prevents long-distance calling from other than authorized telephones.

by pushing a button instead of dialing the station number. To support the attendant, many systems include paging. The paging system is accessed by pushing a button or dialing a code and can be divided into zones if the building is large enough to warrant it. Many systems provide for "parking" a call so that a paged user can go to any telephone, dial his own extension, and pick up the incoming call.

Network Channel Terminating Equipment

Network channel terminating equipment (NCTE) is any apparatus mounted on a telephone user's premises that is designed to amplify, match impedance, or match network signaling to the signaling of the interconnected equipment.

Modems technically meet this definition, but for the purpose of this discussion, NCTE is confined to devices that process signals outside the range of ordinary telephone sets. NCTE includes the CSU and DSU equipment that terminate digital lines as discussed in Chapter 4. It also includes the customer premise end of the signaling equipment discussed in Chapter 8 and the terminating equipment for local area data transport as discussed in Chapter 18.

At the time of this writing, the FCC and CCITT disagree on the demarcation between the network and equipment on the customer premises. At issue in the United States is whether NCTE is part of the network and therefore provided by the telephone company or part of customer premise equipment provided by the user. Under CCITT definition, NCTE is considered part of the network and is provided by the telephone company. The FCC has ruled otherwise. This ruling presents design and compatibility hazards that users should understand before obtaining private line services from telephone companies.

The characteristics of private line signaling are discussed in Chapter 8 and are not repeated here. In a nutshell, private line signals travel over the voice telecommunications network and are actuated by combinations of tones inside the voice passband. At the local loop, these signals can be carried to the user's premises as tones, or converted to DC signals in the telephone central office. In the majority of cases, DC signaling is used. Often, the cable between the central office

FIGURE 6.8
NCTE on User's Premises

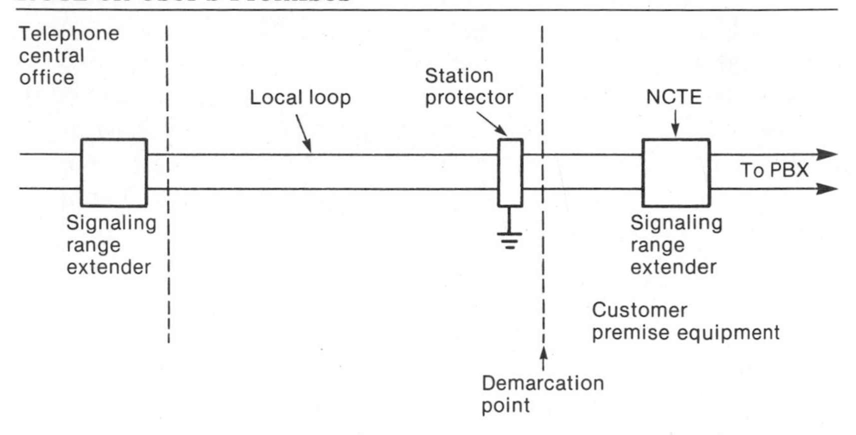

and the user's premises has too much resistance and loss to support reliable signaling. NCTE at the user's premises amplifies the voice, boosts the DC voltage on the line, or both. Figure 6.8 shows the layout of a typical circuit. In some cases, the form of the signal is altered to interface the terminal equipment. For example, a single-frequency signaling tone may be converted to 20 Hz ringing to ring a telephone on the end of a circuit. Like any other equipment connected to the telephone network, NCTE must be registered with the FCC.

NCTE is usually constructed on plug-in circuit packs that mount in shelves. The connector on the NCTE shelf is wired to a demarcation point to meet the telephone company cable pair on the input side and to the terminal equipment on the output side. Individual shelves can be wall mounted. Large concentrations of NCTE are usually mounted in racks or cabinets as discussed in Chapter 14.

STATION WIRING PLANS

A critical issue for most organizations is the form of station wiring to use. Older key telephone systems such as the 1A use 25-pair station wire, which is expensive, costly to install, and congests ducts and raceways. Nevertheless, where this wiring is already installed, it provides a convenient multi-conductor path for telephone equipment. The latest generation of key telephone equipment and PBXs use "skinny" wire that contains from one to four conductors.

A related issue is data terminal wiring. Terminals using RS-232-C or RS-449 interfaces are wired using seven or more conductors. Terminals such as the IBM 3270 are connected to the terminal controller with coaxial cable. The problem with these wiring plans is their inflexibility. Normally a building is wired with telephone outlets at most locations so station users can move their own telephones by making a simple cross-connection change in a wiring closet and plugging the telephone into a new jack. The cost of coaxial and RS-232-C cables usually precludes this kind of flexibility for data terminals.

When a building contains existing cabling, it is usually preferable to retain the existing wiring plan if it is adequate to support telephone and terminal systems. When it is necessary to rewire an existing building or to initially wire a new building, a carefully designed wiring plan should be installed. AT&T and IBM have wiring plans that support both voice and data terminals. The primary characteristics of the plans are:

- A backbone route links outside telecommunications circuits, PBX, and computer to "wiring closets" that are placed strategically throughout the building as shown in Figure 6.9.
- Wire is distributed from the wiring closet to station locations and cross-connected to the backbone circuit on a panel in the wiring closet.
- Data terminals are connected (within range limits) to twisted pair wire. In the IBM system the wire is shielded; in the AT&T system it is unshielded.
- Wall connectors are provided for both voice and data terminals. Terminals are cross-connected in the wiring closet to provide a path to the PBX, computer room, or other terminating point.

The AT&T and IBM wiring plans are similar in concept except for the shielded wire used by IBM. The shielding is

FIGURE 6.9
Station Wiring Plan

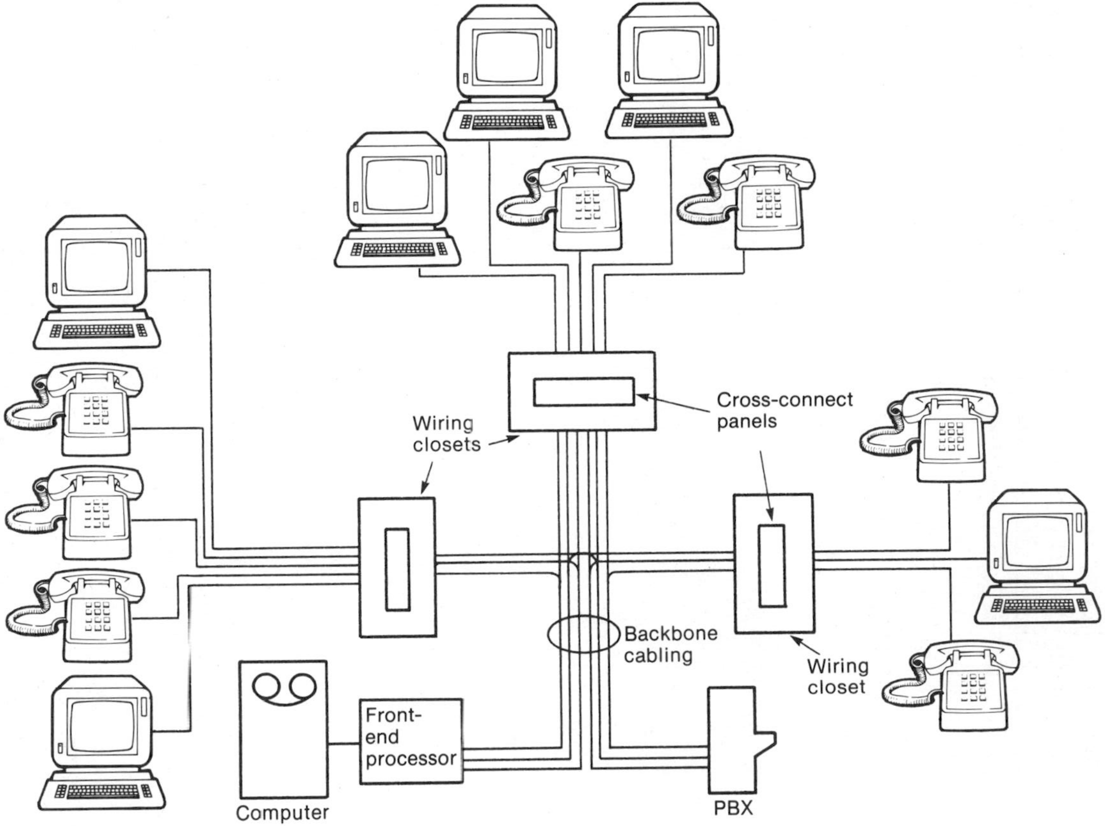

reported to be provided for the future use of a token ring local area network as described in Chapter 19. AT&T's system uses a wiring closet connector that is equipped with patch cords that are used to cross-connect station wiring to the backbone circuits. The chief advantages of these systems are the elimination of coaxial cable for runs of less than about 300 feet, the provision of a common wall outlet for both voice and data terminals, and the flexibility of a cross-connect panel for assigning circuits. Similar wiring plans can be assembled from components made by a variety of manufacturers.

STANDARDS

Telephone instruments, KTS, and auxiliary equipment such as recorders and dialers are not constructed to standards set by a standards agency. The FCC sets registration criteria, but these criteria relate to potential harms to the network or personnel from hazardous voltages, or to interference with other services from excessively high signal levels. The FCC also sets frequency requirements for cordless telephones and regulates the amount of electromagnetic radiation that processor-equipped devices can emit. FCC rules specify the amount of internal resistance of a telephone set, but otherwise do not regulate technical performance.

EIA has set certain criteria for telephone equipment, and CCITT has likewise standardized certain aspects of the telephone set and its operation. Primarily, telephone technical standards have evolved from practices of AT&T that were established when it had complete network design control. Other manufacturers have generally adopted those criteria.

ANSI

ANSI/IEEE 269 Method of Measuring Transmission Performance of Telephone Sets.

ANSI/IEEE 661 Method for Determining Objective Loudness Ratings of Telephone Connections.

ANSI/IPS FC-213 Performance Specifications for Flat Undercarpet Telephone Cable.

Bell Communications Research

Technical Reference 43001 Functional Criteria—Voice Frequency Terminating Equipment—Metallic Facilities Central Offices.

Technical Reference 43002 Functional Criteria—Voice Frequency Network Channel Terminating Equipment—Metallic Facilities Offices.

Technical Reference 43003 Functional Criteria—Voice Frequency Transmission Equipment—Maintenance Terminating Unit—Two-Wire Special Services.

Technical Reference 43004 Functional Criteria—Voice Frequency Transmission Equipment—Maintenance Terminating Unit—Four-Wire Special Services.

Technical Reference 43005 Functional Criteria—Voice Frequency Transmission Equipment—Impedance Compensator with Gain.

Technical Reference 43101 Voice Grade Entrance Facilities for Extending Customer-Provided Communications Channels.

Technical Reference 43201 and 43201A Private Line Interconnection Voice Applications.

TR-EOP-000001 Lightning and 60 Hz Disturbances at the Bell Operating Company Network Interface.

Technical Reference 43701 Private Line Interconnection—Connection to a Channel of a Communications System.

Technical Reference 47101 Standard Plugs and Jacks.

Technical Reference 47101A Description of Standard Registration Program Connection—Configurations—Supplementary Subpart F of Part 68 of FCC Rules and Regulations.

Technical Reference 47102 Bell System Miniature Plugs and Jacks.

Technical Reference 62113 Network Channel Interface Specifications for Off-Premises Station Lines (PBX End).

Technical Reference 62114 Network Channel Interface Specifications for Tie Trunks that Accommodate Registered Terminal Equipment Having Facility Interface Codes TL31M and TL32M.

Technical Reference 62115 Network Channel Interface Specifications for Tie Trunks that Accommodate Registered Terminal Equipment Having Facility Interface Codes TL31E and TL32E.

Technical Reference 62116 Network Channel Interface Specifications for Tie Trunk-Like Channels Accommodating Two-Wire Lossless Registered Terminal Equipment that Originates on M Lead.

Technical Reference 62117 Network Channel Interface Specifications for Tie Trunk-Like Channels Accommodating Two-Wire Lossless Registered Terminal Equipment that Originates on E Lead.

Technical Reference 62118 Network Channel Interface Specifications for Tie Trunk-Like Channels Accommodating Conventional Term Set Registered Terminal Equipment that Originates on M Lead.

Technical Reference 62118 Network Channel Interface Specifications for Tie Trunk-Like Channels Accommodating Conventional Term Set Registered Terminal Equipment that Originates on E Lead.

CITT

E.161 Arrangement of figures, letters, and symbols on rotary dials and pushbutton telephone sets.

E.180 Characteristics of the dial tone, ringing tone, busy tone, congestion tone, special information, and warning tone.

K.7 Devices for protection against acoustic shock.

P.10 Vocabulary of terms on telephone transmission quality and telephone sets.

P.33 Subscriber telephone sets containing either loudspeaking receivers or microphones associated with amplifiers.

P.34 Sensitivities of loudspeaker telephones.

P.62 Measurements on subscribers' telephone equipment.

P.75 Standard conditioning methods for handsets with carbon microphones.

Q.258 and Q.258 Telephone signals.

EIA

RS-453 Dimensional, Mechanical, and Electrical Characteristics Defining Phone Plugs and Jacks.

RS-470 Telephone Instruments with Loop Signaling for Voiceband Application.

RS-478 Multiline Key Telephone Systems (KTS) for Voiceband Application.

RS-487 Line Circuit (Card) for 1A2 Generic Multiline Key Telephone Systems.

RS-504 Magnetic Field Intensity Criteria for Telephone Compatibility with Hearing Aids.

FCC Rules and Regulations

Part 15 Radiation Limits for Class A Computing Devices.

Part 68 Telephone Set Registration.

APPLICATIONS

Telephone and key equipment are so common as to require little applications information aside from a discussion of the primary factors in their evaluation. Systems are simple enough that they can be installed in a short time in most offices.

Evaluation Considerations

Telephone sets, key telephone equipment, answering sets, NCTE, and all other equipment connected to the network must be registered with the FCC to guard against harms to the network. Although it is unlikely that equipment offered for customer premises will be unregistered, it should be noted that it is illegal to connect such apparatus without registration. All telephone apparatus must be protected from hazardous voltages as discussed in Chapters 7 and 14. The telephone company equips its lines with carbon block protectors. These are adequate for ordinary telephone sets, but some KTS may not be adequately protected. The manufacturer's recommendations should be consulted, and if necessary, gas tube protection should be provided as discussed in Chapter 7.

Telephones

The primary consideration in obtaining a telephone should be in the intended use of the instrument itself. The following criteria are important in evaluating a telephone instrument:

- **Durability and reliability.** Telephone sets are often dropped. It is difficult, in many cases, to obtain repair services, so the ability of the telephone to withstand wear and tear is of prime importance.
- **Type of dial.** Specialized common carriers and other telephone-related services require a DTMF dial to enter personal identification number and call details. Some telephones with pushbutton dials have a rotary dial output and are incompatible with these services.
- **Number of telephone lines served.** The number of telephone lines has a significant effect on whether single-line, multiline, or key telephone systems are acquired.
- **Transmission performance.** Some inexpensive telephones have inferior transmission performance and may give unsatisfactory service. This should be evaluated before purchase.
- **Additional features.** Such features as last number redial and multiple number storage are often desirable features in selecting a telephone.

Special Feature Telephones

The garden variety telephone set has gone the way of Henry Ford's Model T. Now, telephone sets can be obtained with dozens of optional features and with auxiliary equipment such as clock radios that have nothing to do with the telephone itself. Special features are available as either part of the telephone or as an add-on adapter.

Dialers. These units store a list of telephone numbers that are outpulsed by selecting a button. Dialers are particularly advantageous for accessing special common carriers who require dialing 23 digits to access long-distance numbers. Evaluation considerations include capacity, ability to handle 23-digit numbers, and the ability to pause for second dial tone.

Speaker Telephones. These units provide speaker and microphone for hands-free operation. The primary considerations are range of coverage (i.e., an office or a conference room), satisfactory voice quality, and the ability to cut off the transmitter during conversations.

Cordless Telephones. These units should include circuitry to prevent false rings and to restrict call origination to authorized telephones. The range of the base station and remote should be considered. Extended range may be an advantage in some applications, but may result in loss of privacy.

Memory. Telephones that store up to 10 digits and contain last number redial capability are often advantageous. Some telephones visually display the number dialed and allow correction of dialed digits before outpulsing begins.

Telephones for the Handicapped. Telephones are available with a variety of aids for handicapped users. These include special dials, amplified handsets, visual ringing equipment, and other such features. Of special concern are "hearing-aid compatible" telephones. Some hearing aids rely on magnetic pickup from the handset and are incompatible with some types of electronic handsets. Special telephone sets are equipped with keyboards and single-line readout for communication by the deaf. Compatibility between devices is important.

Transmission Quality

Telephones can be selected that match the transmission characteristics of a "500 type" telephone set, around which the network is designed. Telephones should operate satisfactorily at loop currents as low as 23 ma. Most key telephone systems have limitations on their signaling range, and some with time division networks introduce loss into the connection. In evaluating KTS, the following considerations should be investigated with respect to transmission:

- What is the maximum system to station set range in feet?
- How are off-premise or extended-range telephones handled?
- Is there a limitation on the range between the KTS and the telephone company? If so, can this range be extended with outboard apparatus?
- Does the system introduce transmission loss between lines or between lines and trunks?
- Is multiport conferencing amplified or is conferencing accomplished by switching lines together without amplification?

Answering Sets

Answering sets are available with numerous special features that should be considered before purchasing a unit. Among the most important features are:

- **Battery backup** for continued operation during power failures.
- **Call counter** to display how many calls are recorded.
- **Call monitoring** capability so incoming calls can be screened over a speaker.
- **Dual tape** capability so it is unnecessary to listen to the recorded message when playing back recorded calls.
- **Remote control recording** so the announcement can be changed remotely.
- **Ring control** to allow the user to adjust the number of rings before the line is answered.
- **Selective call erase** to allow selectively erasing, saving, or repeating incoming calls.

KTS versus PBX

Although the distinction between KTS and PBX is not clear in the larger KTS line sizes, if the organization requires more than 50 trunks, a PBX will undoubtedly be required because of its greater line, trunk, and intracalling capacity. Because the cost of common equipment is distributed among all stations, and because the common equipment for a PBX is more expensive, in smaller sizes KTS is definitely more economical. Between a lower range of about 20 trunks and an upper range of about 50 trunks, the decision can be based on cost, features, and technical performance. PBXs are designed to handle large amounts of calling within the organization, but intracalling on KTS often requires dialing over the local telephone network. This means dialing seven digits rather than three or four, and in locations with measured local service will result in higher service charges from the telephone company.

Blocking versus Nonblocking Switching Network

A nonblocking switching network is one that provides as many links through the network as there are input and output ports. For example, one popular key system has capacity for 24 central office trunks, 61 stations, and 8 intercom lines. The system provides 32 transmission paths, which supports calls to and from all 24 central office trunks. The eight intercom paths limit intrasystem conversations to eight pairs of stations. A nonblocking network provides enough paths for all line and trunk ports to be simultaneously connected. In this system, if all central office trunks are connected, of the remaining 37 stations, only 8 can be in conversation over the intercom paths. Although this system is not nonblocking, it meets an important test of having sufficient paths to handle all central office trunks and intercom lines.

Cost

The price of a KTS system is only part of the total cost incurred. As with all types of telecommunications apparatus, failure rate and the cost of restoring failed equipment are critical and difficult to evaluate. The most effective way to evaluate them on a key telephone system is by investigating costs of other users.

Installation cost is another important factor. In general, the larger the station cables required, the more expensive the installation. Another cost factor is the method of programming the station options in the processor. Most systems provide such options as toll call restriction, speed calling of a selected list of numbers, transfer of rings to another station, and other such features that are listed in a data base. If these features require a technician to program them, costs will be higher than a system with features that can be user programmed. The least costly systems allow a user to move a telephone to a vacant wall jack, plug it in, and change the data base from a central location. Maintenance costs are apt to be a significant item over the life of the system. Cost savings are possible with systems that provide internal diagnostic capability. Some systems provide remote diagnostic capability so the vendor can diagnose the system over an ordinary telephone line. These features offer cost savings.

Power Failure Conditions

During power outages, KTS is inoperative unless battery backup is provided. Some systems include emergency battery supplies, while with others the system is inoperative until power is restored. The system should include a power-fail transfer system that connects incoming lines to ordinary telephone sets so calls can be originated and answered during power outages. The method of restarting the system after power failures is also important because the method affects cost and the amount of time it takes to get the system back in operation. Some systems use nonvolatile memory that does not lose data when power is removed. Other systems reload the data base from a backup tape or disk, which results in a delay before the system can be used following restoral of power.

Wired versus Stored Program Logic

Many key telephone systems built after the mid-1970s use stored program logic. Older systems including 1A2 and equivalent KTS use wired logic. The primary advantage of wired logic is simplicity. Most failures in wired logic systems are in the plug-in 400 type cards, which can be replaced by the user for minimal cost. The primary disadvantages of wired logic systems are their lack of flexibility and the amount of cabling required. In stored program systems, new features can be added by changing the generic program. Features can be added and removed to customize stations by changing the feature data base.

The most important advantage of stored program systems is the wide range of features they offer. Wired logic systems offer essentially the features of the telephone system itself plus basic key features. Other features are added with outboard equipment. Stored program systems offer features that duplicate those of more expensive and complex PBXs.

System Size

Key telephone systems should be purchased with a view toward long term growth in central office lines and stations. Systems that are designed with plug-in common equipment are less costly than systems that must be purchased to their ultimate size at the outset. The number of internal or intercom calls should also be considered. The number of internal paths and central office lines required are a function of the calling rate.

GLOSSARY

Customer premises equipment (CPE): Telephone apparatus mounted on the user's premises and connected to the telephone network.

Dual tone multifrequency (DTMF): A signaling method consisting of a push-button dial that emits dual tone encoded signals used between the station and the central office.

Harmonic ringing: A method of preventing users on a party line from hearing other than their own ring by tuning the ringer to a given ringing frequency.

Inside wiring: The wiring on the customer's premises between the telephone set and the station protector.

Key telephone system (KTS): A method of allowing several central office lines to be accessed from multiple telephone sets.

Network channel terminating equipment (NCTE): Apparatus mounted on the user's premises that is used to amplify, match impedance, or match network signaling to the interconnected equipment.

Off hook: A signaling state in a line or trunk when it is working or busy.

On hook: A signaling state in a line or trunk when it is nonworking or idle.

Power fail transfer: A unit in KTS that transfers one or more telephone instruments to central office lines during a power failure.

Protector: A device mounted on the user's premises that serves as a demarcation between customer-owned and telephone company-owned equipment and protects equipment from damage by external voltages.

Registration: The process the FCC follows in certifying that customer premise equipment will not cause harms to the network or personnel.

Ring: The designation of the side of a telephone line that carries talking battery to the user's premises.

Ringer isolator: A device placed on a telephone line to disconnect the ringer when it is in an idle state. It is used for noise prevention.

Superimposed ringing: A method of preventing users on a party line from hearing more than their own ring by superimposing a DC voltage over the ringing signal and using it to fire a tube or semiconductor device in only the selected instrument.

Tip: The designation of the side of a telephone line that serves as the return path to the central office.

Transducer: Any device that changes energy from one state to another. Examples are microphones, speakers, and telephone handsets.

Voice mail: A system of storing messages in a private recording medium where they can later be retrieved by the called party.

BIBLIOGRAPHY

Clifford, Martin. *Your Telephone: Operation, Selection, and Installation.* Indianapolis, Ind.: Howard Sams & Co., 1983.

Datapro. *Reports on Telecommunications.* Delran, N.J.: Datapro Research Corporation, 1985.

Mingail, Harry B. *The Business Guide to Telephone Systems.* Seattle, Wash.: Self Counsel Press, 1983.

MANUFACTURERS OF TELEPHONE STATION EQUIPMENT

Telephone Instruments

American Telecommunications Corporation

AT&T Information Systems

GTE Business Communications Systems, Inc.

ITT Telecommunications

Northern Telecom, Inc.

Siemens Corporation

Cordless Telephones

AT&T Technologies Consumer Products

Cobra/Dynascan Corp.

ITT Telecommunications Business and Consumer Division

Panasonic Co. Telephone Products Division

Uniden

Telephone Answering Machines

Code-A-Phone Corporation

Dictaphone Corporation

Panasonic, Co.

Phone-Mate, Inc.

Key Telephone Equipment

AT&T Information Systems

Candela Electronics

Crest Industries, Inc.

Eagle Telephonics, Inc.

Executone, Inc.

IPC Technologies, Inc.

ITT Telecommunications

Melco Labs.

Northern Telecom, Inc.

Tel-Tone Corporation

TIE/Communications, Inc.

Tone Commander Systems, Inc.

Toshiba Telecommunications

Coin Telephones

AT&T Technologies

Cointel Communications

Digitech Communications

Northern Telecom, Inc.

Reliance Comm/Tec R-Tec Systems

U S Telecommunications

Network Channel Terminating Equipment

AT&T Network Systems

Lear Siegler, Inc.

Northern Telecom Inc.

Proctor & Associates Co., Inc.

Pulsecom Division, Harvey Hubbell, Inc.

Telco Systems, Inc.

Tellabs Inc.

Wescom, Telephone Products Div., Rockwell Telecommunications, Inc.

CHAPTER SEVEN

Outside Plant

The subscriber loop is supported by outside plant, which includes the wire or multiplex facilities that connect the customer's premises to the telephone central office, and the structures such as poles and conduit that support the physical attachment and routing of cable and wire. Although this chapter primarily discusses the application of outside plant by telephone companies, the same fixtures are used for wire facilities in private networks also. Moreover, because nearly every private network requires a local loop obtained from the telephone company, it is important for private network users to understand the characteristics of the loop and how it affects the performance of the network.

OUTSIDE PLANT TECHNOLOGY

Outside plant (OSP), diagramed in Figure 7.1, consists of the following components:

- Pole lines.
- Conduit.

FIGURE 7.1
Diagram of Outside Plant

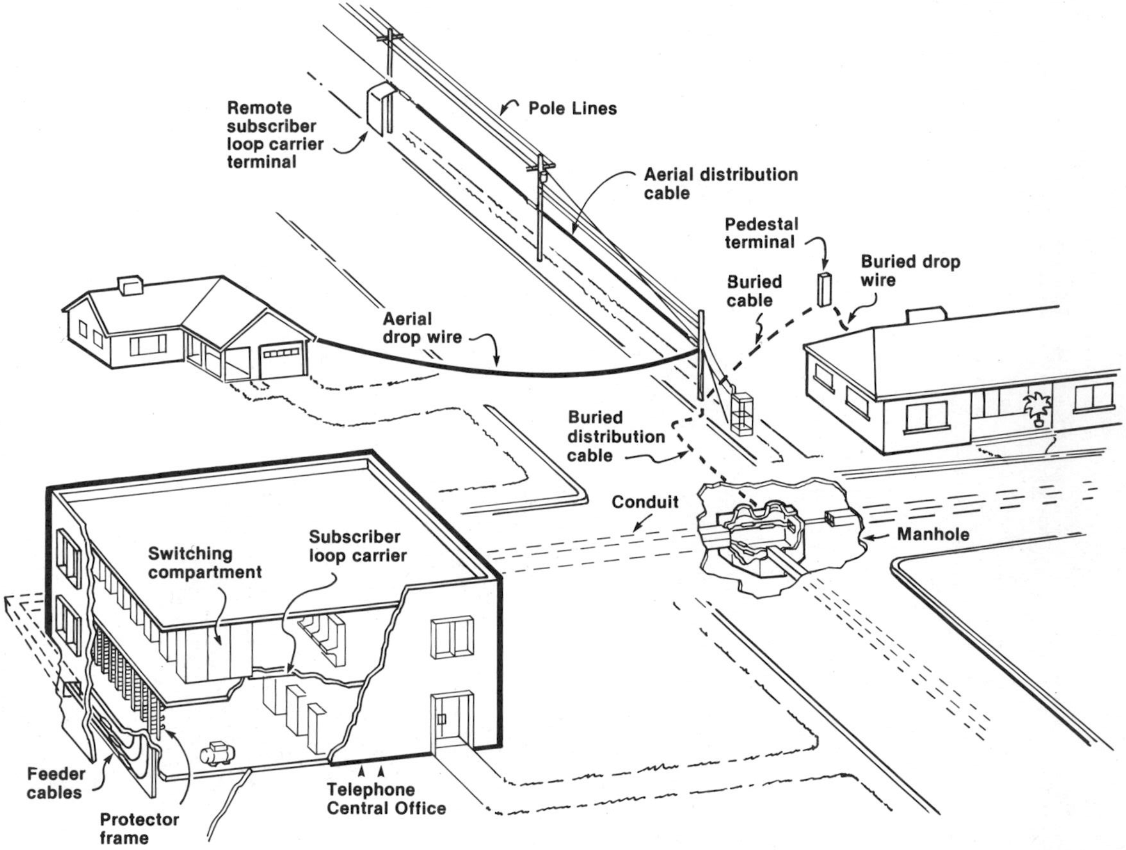

- Feeder cable.
- Distribution cable.
- Terminals.
- Subscriber loop multiplex equipment.
- Aerial and drop wire.

Protection equipment and range-extension devices located in the central office and on user's premises are also included in this discussion even though they are not normally considered part of outside plant by telephone companies.

Supporting Structures

The vast majority of subscriber loops are routed from the user's premises to the telephone central office over twisted pair cable, which is classified according to its supporting structure:

- Aerial cable, supported by pole lines.
- Underground cable, supported by conduit.
- Buried cable, placed directly in the ground without conduit.

Aerial cable is being discontinued as rapidly as economics permit because of environmental concerns and because of its vulnerability to damage. Aerial cable requires an external strength member to relieve tension on the conductors. Self-supporting aerial cable contains an internal strength member; all other cable requires an external *messenger* that is attached to poles. The messenger is a multistrand metallic supporting member to which cable is lashed with galvanized wire applied with a lashing machine. Down guys and anchors are placed at the ends and offsets in pole lines to relieve strain on the poles.

Direct burial is the preferred method for placing cable under ground because it is less expensive than conduit. Buried cable is either placed in an open trench or plowed with a special tractor-drawn plow that feeds the cable under ground through a guide in the plow blade. Where several cables are placed simultaneously, or where future additions will be required, conduit is placed to avoid the expense of opening streets more than once. Manholes are located in conduit runs at intervals corresponding to the maximum length of cable that can be handled physically. Technical considerations require manholes at 6,000-foot intervals to accommodate load coils and T carrier repeaters.

Cable Characteristics

Twisted pair cables are classified by their sheath material, their protective outer jacketing, the number of pairs contained within the sheath, and the wire gauge. Sizes available range from one- or two-pair drop wire to 3,600 pair paper-insulated cable used for central office building entrance. The upper limit of cable size, which is a function of wire gauge and the number of pairs, is dictated by the outside sheath diameter, which is limited by the size that can be pulled through conduit. Cables of larger sizes such as 1,200, 1,800, 2,400 and 3,600 pairs are primarily used for entrance into telephone central offices, which are fed by conduit in urban locations. Wire gauges of 26, 24, 22, and 19 AWG are used loop plant. Cost considerations dictate the use of the smallest wire gauge possible, consistent with technical considerations. Therefore, the finer gauges are used close to the central office to feed the largest concentrations of users. Coarser gauges are used at greater distances from the central office to reduce loop resistance.

Cable sheath materials are predominately high-durability plastics such as polyethylene and polyvinyl chloride. Cable sheaths are designed to guard against damage by lightning, moisture, induction, corrosion, rocks, and rodents. In addition to the sheath material, submarine cables are protected by coverings of jute and steel armor. Besides the outer sheath, cables are shielded with a serving of metallic tape to protect against induced voltages.

The twist of cable pairs is carefully controlled to preserve the electrical balance of the pair. As discussed in Chapter 2, unbalanced pairs are vulnerable to noise induced from external sources, so the twist is designed to ensure that the degree of coupling between cable pairs is minimized. This

FIGURE 7.2
Multipair Cable

Courtesy General Cable Company.

is accomplished by constructing cable in units of 12 to 100 pairs, depending on the size of the cable. Each unit is composed of several layers of pairs twisted around a common axis, with each pair in a unit given a different twist length.

Cable pairs are color coded within 50-pair *complements* as shown in Figure 7.2. Each complement is identified by a color-coded string binder that is wrapped around the pairs. At splicing points the corresponding pairs and binder groups are spliced together to ensure end-to-end pair identity and continuity. Cables can be manually spliced with compression sleeves or ordered from the factory cut to the required length and equipped with connectors.

Splicing quality is an important factor in preserving cable pair balance. Many older cables are insulated with paper and were spliced by manually twisting the wires together. These

FIGURE 7.3
Aerial Splice Cases

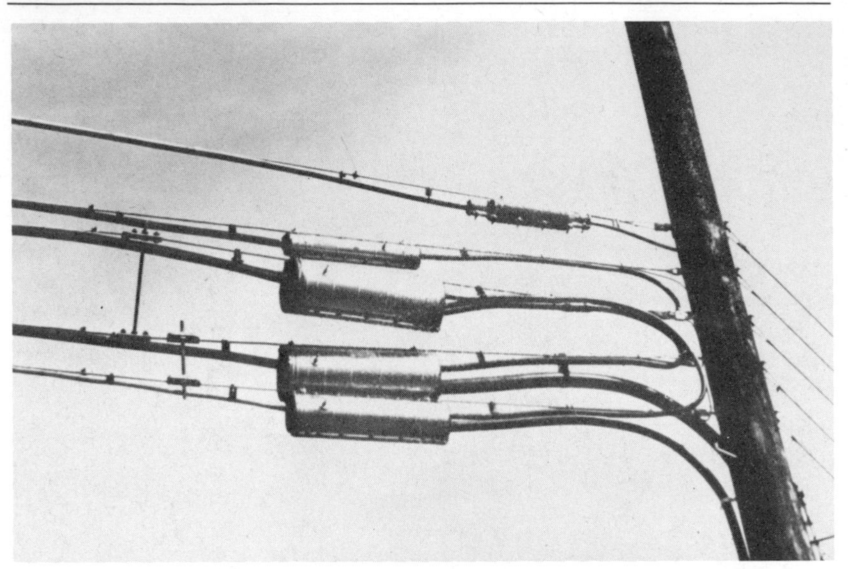

Courtesy Preformed Line Products Company.

older splices are often a source of imbalance and noise because of insulation breakdown or splice deterioration. To prevent crosstalk, it is also important to avoid *splitting* cable pairs. A split occurs when a wire from one pair is spliced to a corresponding wire in another pair. Although electrical continuity exists between the two cable ends, an imbalance between pairs exists, and crosstalk may result.

Cable splices are stored in above-ground closures in splice cases such as the ones shown in Figure 7.3. Cables must be manufactured and spliced to prevent water from entering the sheath because moisture inside the cable is the most frequent cause of noise and crosstalk. In a later section methods of keeping cables impervious to moisture are discussed.

Loop Resistance Design

The wire gauge is selected to achieve the desired loop resistance. All telephone switching machines, PBXs, and key telephone systems are limited in the loop resistance range they can tolerate. The loop resistance a machine can support is specified in ohms and consists of the following elements:

- Battery feed resistance of the switching machine (usually 400 ohms).
- Central office wiring (nominally 10 ohms).
- Cable pair resistance (variable to achieve the design objective of the central office or PBX).
- Drop wire resistance (nominally 25 ohms).
- Station set resistance (nominally 400 ohms).

The cable gauge is selected to provide the desired resistance at the maximum temperature under which the system will operate. This method of design is called *resistance design.* Telephone central offices have total loop resistance ranges from 1,300 to 1,500 ohms; most PBXs are less, with many supporting ranges of 400 to 800 ohms. The range can be extended with subscriber carrier or range extension devices, which are described later. The range limitation of subscriber loops is a function of the current required to operate DTMF dials and the telephone transmitter, the supervisory range of the central office, the range over which ringing can be supplied and tripped on answer (ring trip range), and the transmission loss of the talking path. Depending on the type of switching machine, one of these factors becomes limiting and determines the loop design range.

A further consideration in selecting cable is the capacitance of the pair, expressed in microfarads (mf) per mile. Ordinary subscriber loop cable has a high capacitance of 0.083 mf per mile. Low-capacitance cable, used for trunks because of its improved frequency response, has a capacitance of 0.062 mf per mile. Special 25-gauge cable used for T carrier has a capacitance of 0.039 mf per mile.

Special types of cable are used for cable television, closed circuit video, local area networks, and other such applications. Some types of cable are constructed with internal screens to isolate transmitting and receiving pairs. These types of cable are beyond the scope of this book; however, the reader should be aware that they exist and may be required for certain telecommunications applications such as high-capacity T carrier systems.

Feeder and Distribution Cable

Cable plant is divided into two categories: *feeder* and *distribution.* Feeder cable consists of cable pairs that are routed directly to a serving area without intervening branches to end users. Feeder cable is of two types, main and branch feeders. Main feeders are large backbone cables that exit the central office and are routed, usually through conduit, to intermediate branching points. Branch feeders are smaller cables that route pairs from the main feeders to a serving area. Distribution cable extends from a serving area interface to the user's premises. Figure 7.4 shows the plan of a typical serving area.

Where enough pairs are needed in a single building to justify wiring an entire complement into the building, the interface between feeder and distribution plant is a direct splice. Otherwise, the interface is a cross-connect cabinet where provisions are made to connect cable pairs flexibly between feeder and distribution plant.

Distribution cable is terminated in *terminals* similar to the one shown in Figure 7.5 to provide access to cable pairs. Terminals may be mounted on the ground in pedestals, in buildings, on aerial cable messenger, or underground. Aerial or buried *drop wire* is used to connect from the terminal to the protector at the user's premises.

To the greatest degree possible, feeder cables and distribution cables are designed to avoid *bridged tap,* an impairment shown in Figure 7.6. *Bridged tap* is defined as any portion of the cable pair that is not in the direct path between the user and the central office. It has the electrical effect of a capacitor across the pair and impairs the high-frequency response of the circuit. Bridged tap can render DTMF dials and data modems inoperative because of amplitude distortion, primarily at high frequencies. It can be detected by measuring the frequency response of a metallic circuit.

The frequency response of long subscriber loops is improved by loading as discussed in Chapter 2. Load coils are small inductors wound on a powdered iron core as shown

FIGURE 7.4
Feeder and Distribution Service Areas

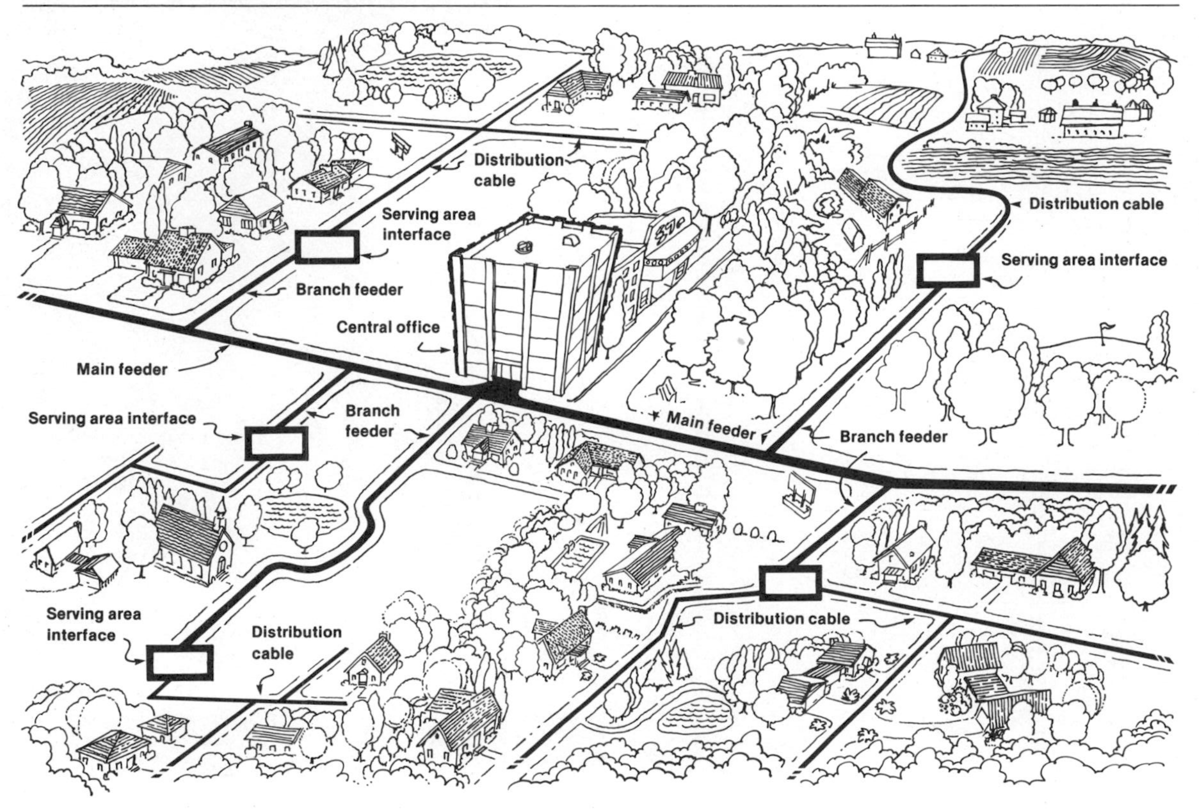

FIGURE 7.5
Terminals

a. Ground-Mounted Pedestal Terminal (distribution wire to residences is connected to cable with pressed-on sleeve).

b. Aerial Terminal (drop wires to residences emerge from bottom of terminal; photos by author).

FIGURE 7.6
Bridged Tap in Cable Pairs

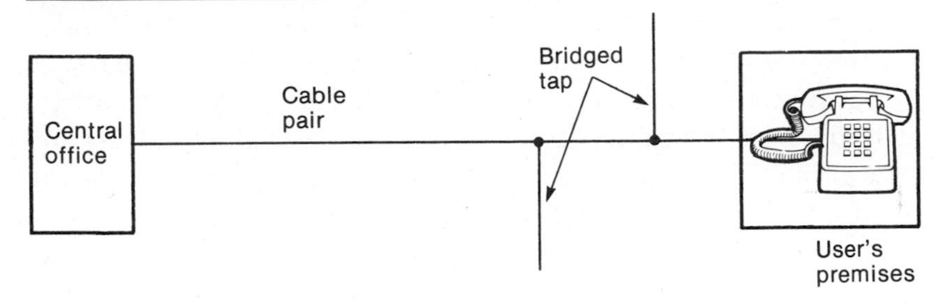

FIGURE 7.7
Toroidal Load Coil

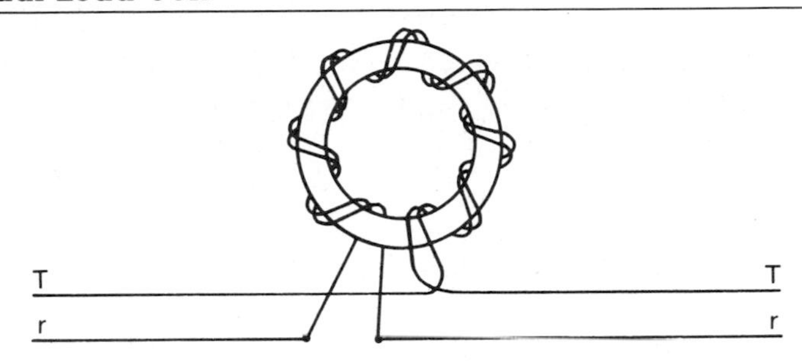

in Figure 7.7. They are normally placed at 6,000 foot intervals on loops longer than 18,000 feet. Load coils are contained in weatherproof cases that are mounted on poles or in manholes.

Subscriber Carrier

Loading is being displaced as a transmission-improving technique in many long subscriber loops by the use of subscriber loop carrier. Analog subscriber carriers provide from one to eight subscriber circuits on a single cable pair. Single-channel carrier is used to derive one carrier channel in the frequency range above the voice channel. Because of the high-frequency cutoff of load coils, single-channel subscriber carrier cannot be used on loaded cable. Therefore, single-channel carrier must be used within the 18,000-foot range of nonloaded cable to obtain satisfactory transmission from the voice channel.

Multiple-channel analog carriers sacrifice the voice channel to gain as many as eight carrier channels on a cable pair, depending on the manufacturer. Multiple-channel analog subscriber carriers are similar in concept to the trunk carriers described in Chapter 5, except they use double sideband modulation to reduce costs. Digital subscriber carriers operate over T-1 lines using either PCM or delta modulation techniques, as described in Chapter 4, to derive from 24 to 40 voice channels over two cable pairs. T-1 repeaters are placed at 6,000-foot intervals to provide a line equivalent to that used for trunk carrier.

Some subscriber carriers use *concentration* to increase the number of voice channels that can be transmitted over a T-1 line. Concentration operates on the probability that not all users will require service simultaneously. Users are not permanently assigned to digital time slots in concentrated systems. Instead, when the user requests service the system selects an idle time slot and identifies the line to the distant terminal with a data message. Concentrated carriers allow the termination of 48 subscribers on a 24-channel T-1 line.

A *line concentrator*, diagramed in Figure 7.8, is similar to subscriber carrier except that it is specifically designed for concentrated service and terminates multiple T carrier trunks to a larger number of subscriber lines to reduce the probability of a trunk being unavailable when a user requests service. Contrasted to concentrated subscriber carrier with a 2:1 concentration ratio, a remote line concentrator may assign, say, 250 users to five T-1 lines, a concentration ratio of 5:1. Concentrators are equipped with circuits to collect usage information that an administrator can use to avoid overloads. They must be engineered and monitored in the same manner as central office switching equipment, administration of which is described in Chapter 13.

Concentrators also may include circuitry known as *intracalling*, which offers the capability for users within the concentrator to connect to one another without using channels to the central office and back. Intracalling features require only one channel for the central office to supervise the connection. With intracalling a concentrator may be able to provide service between the users it serves even though the

FIGURE 7.8
Block Diagram of Remote Line Concentrator

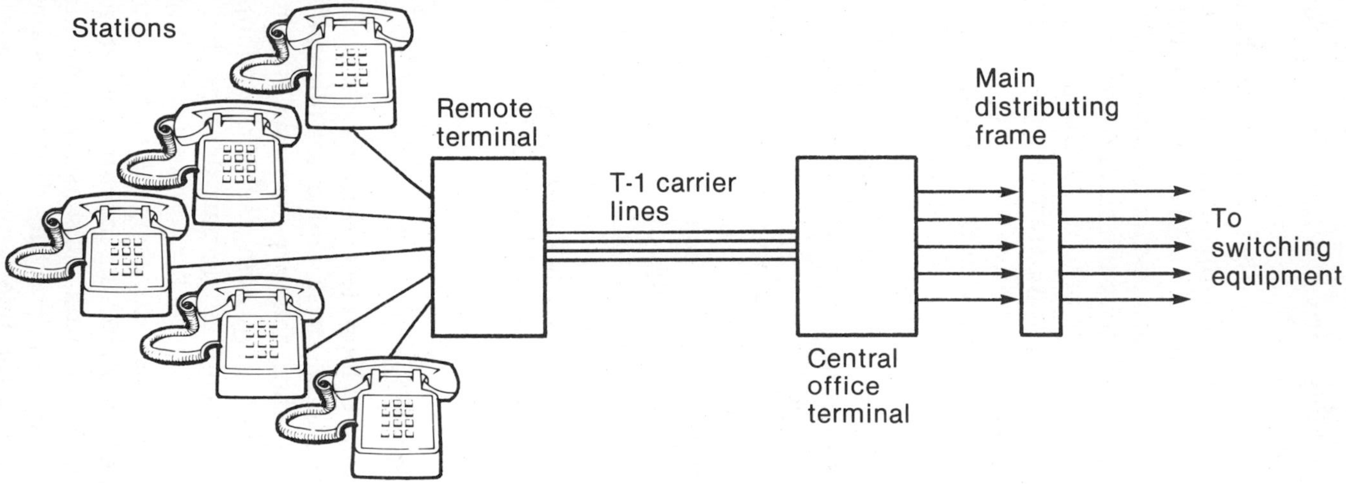

Stations

Remote terminal

T-1 carrier lines

Central office terminal

Main distributing frame

To switching equipment

carrier line is inoperative. Without this feature a carrier line failure disrupts all service to its users.

The term *pair gain* is used to describe the effectiveness of subscriber carriers and concentrators in increasing the channel-carrying capacity of a cable pair. The pair gain figure of a carrier or concentrator is the number of voice circuits that are added over and above the single-voice circuit that a cable pair supports. For example, a 24-channel digital subscriber carrier requiring separate transmit and receive pairs has a pair gain of 22. The families of single- and multiple-channel subscriber carriers and concentrators are called *pair gain devices.* A pair gain device consists of a central office terminal and a matching remote terminal with intermediate repeaters if required. The remote terminal is contained in a pole-mounted cabinet or a ground-mounted enclosure.

Pair gain devices provide better transmission quality than cable facilities. The transmission loss is fixed, normally at 5 dB, regardless of the length of the system. Pair gain devices also provide quieter channels than the cable pairs they replace. An advantage of digital subscriber carrier is its ability to use special service channel units to provide the range of special services listed in Table 4.1. Pair gain devices must be powered from an external source, so backup battery power is required to maintain telephone service during power outages.

Range Extenders

Another family of loop electronic devices is classified as *range extension.* Although these devices are normally mounted only in the central office, they are discussed here because of their exclusive use for improving subscriber loop performance. Range extenders are single-pair devices that boost the line voltage and may also include voice frequency gain. Battery-boost range extenders are designed to overcome the DC loop limitations of the switching machine and station equipment. Range extenders increase the sensitivity of the switching machine line circuits in detecting dial pulses and the on-hook/off-hook state of the line.

A second type of range extender increases the central office sensitivity and also feeds higher voltage to the station. This latter type boosts the normal −48-volt central office battery to −72 volts, which increases the line current when the station is off hook, supporting greater DTMF dial range and providing greater line current to the telephone transmitter. If the voice frequency transmission range of the cable pair is limiting, range extenders are available with built-in amplification to boost the voice level. These devices either contain a fixed amount of gain, or gain is automatically adjusted in proportion to the line current.

Cable Pressurization

Cable pressurization is used by telephone companies to keep moisture out of the cable and should also be considered in private networks when cable is exposed for long distances. In a cable pressurization system, a compressor pumps dehydrated air into the cable. At terminals where pairs are exposed, the cable sheath is plugged with a watertight dam and air bypasses the dam through plastic tubing. A flow meter at the source indicates the amount of leakage. When leakage exceeds a specified amount, it indicates damage to the sheath, which must be located and repaired to ensure watertight integrity. When a cable run is long with multiple branches, low air pressure alarms are installed to assist in locating trouble.

ELECTRICAL PROTECTION

Aerial cables frequently share supporting structures with power lines and can inadvertently become crossed with a high voltage that can injure users and damage equipment. Cables are also susceptible to lighting strikes that can be equally damaging. *Protectors* are devices designed to prevent both injury and damage from external electrical sources. Protectors must guard against unwanted voltage and current without impairing the talking and ringing voltages that are present on a normal circuit. Similar techniques are used to protect both the central office and the station ends of a circuit.

When lightning strikes, high voltages may exist for only a fraction of a second, but the magnitude may be harmful

FIGURE 7.9
Station and Central Office Protection Equipment

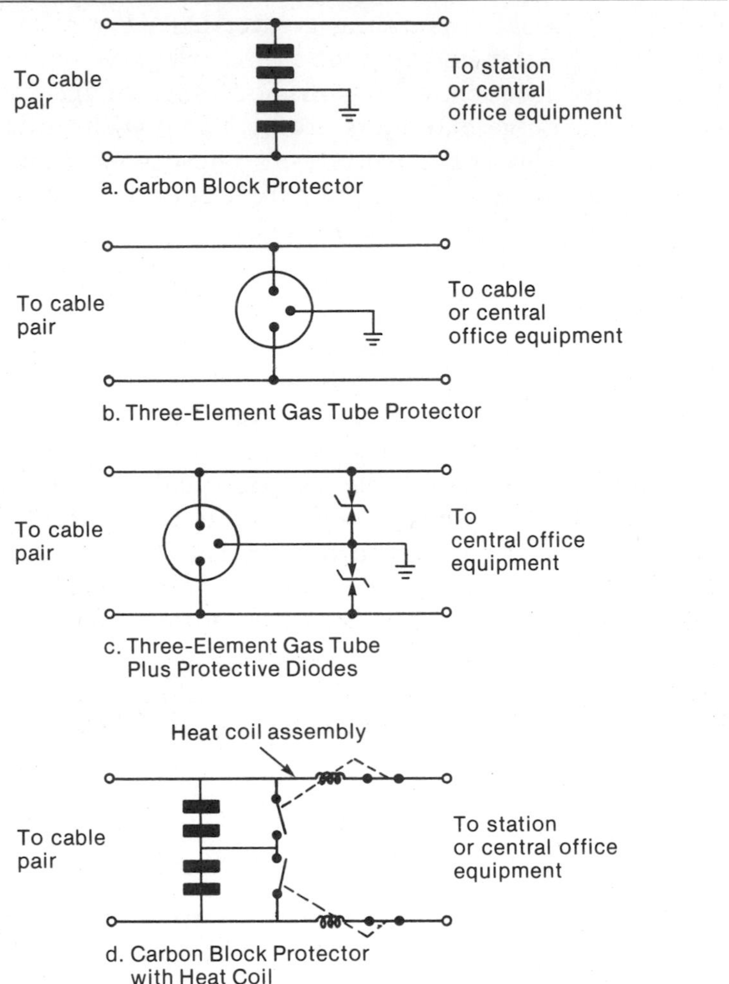

to both personnel and equipment. Lower voltages may present little hazard to people, but may cause equipment-damaging currents. Protectors are designed to limit both voltage and current. Not only is the magnitude of the current important, but the duration of the unwanted current must also be considered.

The most basic type of protector is a *carbon block* as illustrated in Figure 7.9a. The carbon block consists of two electrodes spaced so that any voltage above the design level arcs from the line to ground. Carbon block protectors are used in both the station and the central office. They are effective but have the drawback of being destroyed when high voltage is applied. Carbon blocks are replaced in many modern applications by *gas tubes,* shown in Figure 7.9b, which are glass capsules about the size of a pea that are connected between the circuit and ground. When a voltage higher than the design voltage strikes the line, the gas ionizes and conducts to ground. When the voltage is removed, the protector self-restores to normal. The integrated circuits used in many digital central offices and PBXs are highly sensitive to external currents and may not be protected adequately by gas tubes, which require a small but finite time to ionize. Some gas tube protectors are equipped with diodes, as shown in Figure 7.9c, to clamp the interfering voltage to a safe level until the gas tube ionizes.

Protectors are effective only if they are connected to a low-resistance ground. The most effective grounds are metallic water pipes connected to a public water supply, or direct connection to the ground used by the power company. Lacking these, an external grounding system must be used.

Sometimes voltages too low to operate the carbons or gas tubes flow through central office equipment to ground and eventually damage equipment. To guard against these unwanted currents, *heat coils* are used in central office protector frames as shown in Figure 7.9d. The heat coil is mounted in a spring-loaded mechanism that contains contacts in series with the telephone line. When excessive current flows, the heat melts solder surrounding a plunger, and the spring mechanism opens the circuit to the equipment and grounds the cable pair.

Each cable pair in a central office is protected in a frame. (This will be described in Chapter 14.) At the user's end of the circuit, protectors range from a simple one-pair device to multiple-pair protected terminals such as the one shown in Figure 7.10. Although station protectors are adequate to prevent injury to users, they are often inadequate to prevent damage to delicate electronic equipment. The owners of all devices connecting to the network, including modems, PBXs, key telephone systems, and answering recorders must be aware of the degree of protection offered by the telephone

FIGURE 7.10
Multipair Protected Building Entrance Terminal

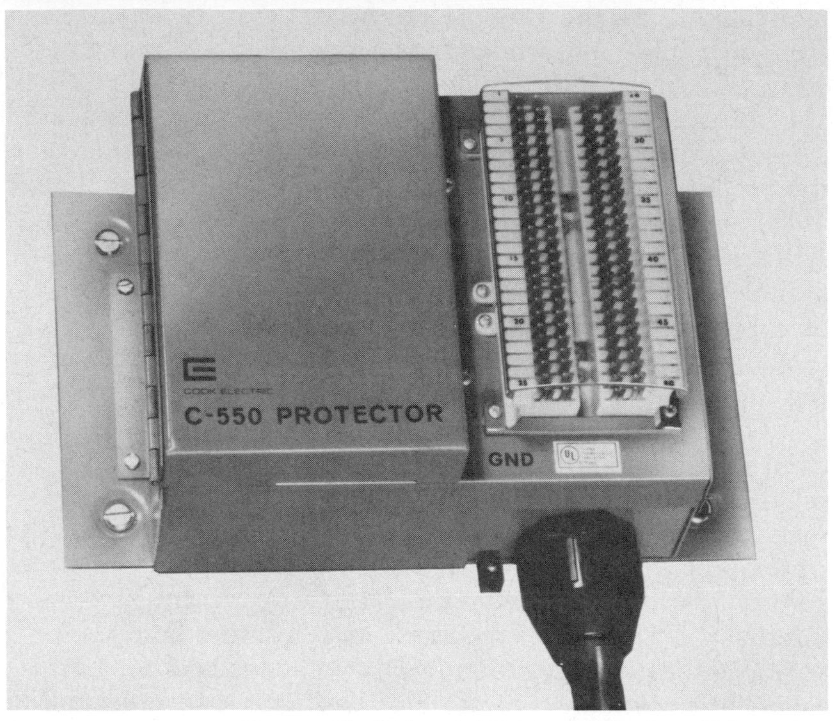

Courtesy Cook Electric Division of Northern Telcom, Inc.

line and the ability of their equipment to withstand external voltage and current.

STANDARDS

Outside plant is not manufactured to the standards of an independent agency except for wire gauge, which meets American Wire Gauge standards. Instead, manufacturer's specifications are used to determine cable size, construction, and sheath characteristics. Outside plant must be selected according to its specifications to meet the requirements of the application.

ANSI

ANSI/IEEE 487 Guide for the Protection of Wire-Line Communication Facilities Serving Electric Power Stations.

ANSI/IEEE 820 Telephone Loop Performance Characteristics.

Bell Communications Research

TR TSY-000172 Item 367 General Compatibility Information—Loop Fiber System.

CCITT

G.611 Characteristics of symmetric cable pairs for analog transmission.

G.612 Characteristics of symmetric cable pairs designed for the transmission of systems with bit rates of the order of 6 to 34 mb/s.

K.11 Protection against overvoltages.

K.12 Specification clauses for the requirements to be met by gas discharge protectors for the protection of telecommunication installations.

K.13 Induced voltages in cables with plastic-insulated conductors.

K.19 Joint use of trenches and tunnels for telecommunication and power cables.

APPLICATIONS

Loop plant is part of every network application. Even in private networks that bypass the local telephone company by routing circuits directly to an interexchange carrier, a connection is made from the network terminal to the station over metallic cable facilities that must be designed as part of the overall network. This section includes only electrical considerations. The evaluation of supporting structures involves mechanical considerations that are beyond the scope of this book.

Evaluating Subscriber Loop Equipment

This section includes the principal considerations for both telephone companies and private network users for selecting metallic outside plant facilities.

Cable Structural Quality

Cable is selected to match the pair size and gauge required by the network design. The sheath must be impervious to the elements if it is mounted outside. Crosstalk and balance characteristics are of paramount concern. When cables are used for special applications such as local area networks and high speed data transmission, the cable must meet the specifications of the equipment manufacturer.

Insulation resistance and DC continuity measurements should be made on all new cables. On loaded cables, structural return loss, gain frequency response, and noise measurements as described in Chapter 23 should be made to ensure the electrical integrity of the cable. When cable facilities are obtained from a common carrier, these measurements should also be made when trouble is experienced.

Air Pressurization

Air pressurization should be considered on long cables and any cable that carries essential services and is exposed to weather. The system should be equipped with a dehydrator, compressor, flow meter, and a monitoring and alarm system to detect leakage.

Protection

All metallic circuits, both common carrier and privately owned, are subject to lighting strikes and crosses with external voltage. It is unsafe to assume that the protection provided by the telephone company is enough to prevent damage to interconnected equipment. Private network users should determine the characteristics of the input circuits of their equipment and should obtain external protection if needed.

GLOSSARY

Bridged tap: Any section of a cable pair that is not on the direct electrical path between the central office and the user's premises.

Branch feeder: A cable that is used between distribution cable and the main feeder cable to connect users to the central office.

Carbon block: A form of cable protection that consists of a carbon conductor isolated from ground by an air gap.

Complement: A group of 50 cable pairs (25 pairs in small cable sizes) that are bound together and identified as a unit.

Concentration: The process of connecting a group of users to a smaller number of trunks between a remote terminal and the central office.

Concentration ratio: The ratio between lines and trunks in a concentrated carrier system or line concentrator.

Distribution cable: Cable that connects the user's serving terminal to an interface with a branch feeder cable.

Drop wire: Wire leading from the user's serving terminal to the station protector.

Gas tube protector: A protector containing an ionizing gas that conducts external voltages to ground when they exceed a designed threshold level.

Gauge: The physical size of an electrical conductor, specified by American Wire Gauge (AWG) standards.

Heat coil: A protection device that opens a circuit and grounds a cable pair when operated by stray currents.

Intracalling: The capability of a remote line concentrator to interconnect users served by the same concentrator without providing two trunks to the central office.

Main feeder: Feeder cable that transports cable pairs from the central office to branching or taper points.

Messenger: A metallic strand attached to a pole line to support aerial cable.

Outside plant: A collective term describing the cable and all supporting structures used to provide subscriber loops.

Pair gain: The increase in subscriber line capacity provided by a pair gain device over and above the capacity of the supporting cable pairs.

Pressurization: Apparatus that feeds dehydrated air under pressure to a cable to prevent entry of moisture.

Protector: A device that prevents hazardous voltages or currents from injuring a user or damaging equipment connected to a cable pair.

Range extender: A device that increases the loop resistance range of central office equipment by boosting battery voltage.

Remote line concentrator: A device that switches a number of users' lines to a smaller number of trunks to the central office.

Resistance design: An outside plant design concept that selects wire gauge based on the length of subscriber loops and the characteristics of the switching equipment.

Sheath: The outer jacket surrounding cable pairs to prevent water from entering and to protect against external damage.

Split pair: A situation that occurs in cable splicing when one wire of a cable pair is spliced to a wire of an adjacent pair.

Subscriber loop: The circuit that connects a user's premises to the telephone central office.

Terminal: A fixture attached to distribution cable to provide access for making connections to cable pairs.

BIBLIOGRAPHY

American Telephone & Telegraph Co. Bell Laboratories. *Engineering and Operations in the Bell System.* Murray Hill, N.J.: AT&T Bell Laboratories, 1983.

American Telephone & Telegraph Co. *Telecommunications Transmission Engineering.* Bell System Center for Technical Education, vol. 1, 1974; vol. 2, 1977; and vol. 3, 1975.

Freeman, Roger L. *Telecommunication System Engineering.* New York: John Wiley & Sons, 1980.

MANUFACTURERS OF OUTSIDE PLANT PRODUCTS

Cable and Wire Products

AT&T Network Systems
Alpha Wire Corp.
Anaconda Ericsson Inc. Wire and Cable Division
Belden Corp.
Brand-Rex Co.
General Cable Co.
Siecor Corporation
Standard Wire and Cable Co.

Cable Air Pressurization Equipment

Chatlos Systems, Inc.
General Cable Co.

Concentrators and Subscriber Carrier Equipment

AT&T Network Systems
Ericsson, Inc. Communications Div.
Northern Telecom, Inc.
Seiscor Inc.

Protectors (see Chapter 14)

Terminals and Cross-Connect Boxes

AT&T Network Systems
General Cable Co.
3M Co. Telcomm Products Div.
Northern Telecom, Inc.
Reliance Comm/Tec
Siecor Corporation

CHAPTER EIGHT

Signaling Systems

The objective of any network is to establish a communication path between end users, to supervise the path while it is in use, and to disconnect it when the users are finished. Users need to know nothing of how the path is established; they supply only the destination address and let the machine select the route. Routing is determined by controllers that are the brains of the network, over signal paths that are the nerve system. Signals travel between controllers either over the talking path or over separate data networks. In whatever manner the signals are exchanged, in the majority of cases they are ultimately converted to analog form at the user interface. Within the network, signals are converted to many different forms to accommodate differences in transmission facilities and equipment vintages.

These multiple-signaling state conversions are the principal complicating factors in network signaling. Signals can be roughly grouped into three functions:

- **Supervising** is monitoring the status of a line or circuit to determine if it is busy, idle, or requesting service. Supervision is a term derived from the function telephone operators perform in monitoring manual circuits on a switchboard. On switchboards, supervisory signals are displayed by an illuminated lamp indicating a request for service on an incoming line or an on-hook condition of a switchboard cord circuit. In the network supervisory signals are indicated by the voltage level on signaling leads, or the on-hook/off-hook status of signaling tones or bits.

- **Alerting** indicates to the addressee the arrival of an incoming call. Alerting signals are audible bells and tones or visual lights.
- **Addressing** is transmitting routing and destination signals over the network. Addressing signals are in the form of dial pulses, tone pulses, or data pulses over loops, trunks, and signaling networks.

SIGNALING TECHNOLOGY

Nearly every switched connection over the telecommunications network involves analog signaling; even if the bulk of the facility is digital, the subscriber loop usually remains analog and requires analog signals. Digital signaling equipment is simple and inexpensive. As indicated in Chapter 4, it involves little more than "robbing" the lowest order bit from every sixth frame and using the binary status of this bit to drive signaling leads in the channel units. Analog signaling is considerably more complex. Separate signaling units are required in virtually all analog carrier systems, which adds to the cost and complexity of the circuit.

Signaling System Overview

Figure 8.1 illustrates the signaling functions in the telecommunications network, showing the signals exchanged when one user calls another over a long-distance trunk. Switching has been omitted from this diagram, but a switching operation occurs at each circuit interface. In the idle state, subscriber loops have battery on the ring side of the line and an open circuit on the tip. No loop current flows in this state. Signaling equipment attached to the long-distance trunks between the local offices and the toll office furnishes a 2,600

FIGURE 8.1
Signaling on an Interoffice Connection

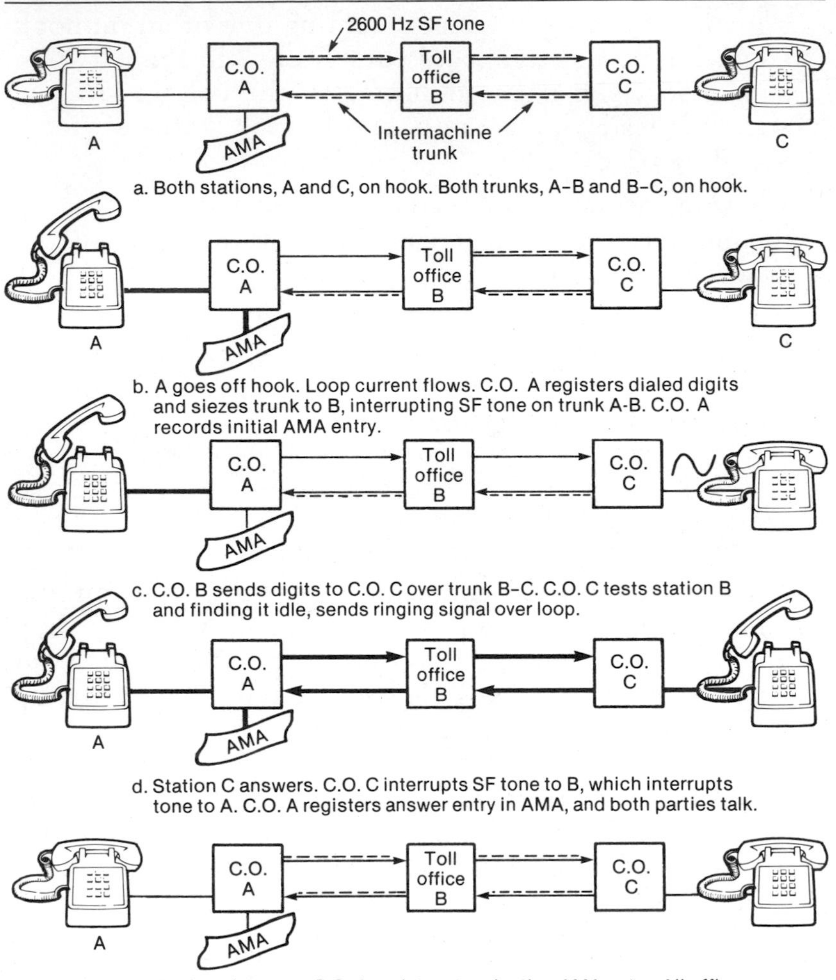

Hz signaling tone, indicating idle circuit status. The tone operates auxiliary single-frequency (SF) signaling sets that indicate the line status by the DC voltage of their signaling leads.

The switching machine continually scans the subscriber line and trunk circuits to detect any change in their busy/idle status. When station A lifts its receiver off hook, current flows in the subscriber loop, signaling the local central office of A's intention to place a call. The central office responds by marking the calling line busy (another status indication) and by returning dial tone to the calling party. Dial tone is one of several *call progress* signals that telephone equipment uses to communicate with the calling party. It indicates the readiness of the central office to receive *addressing* signals.

Station A transmits digits to the central office using either DTMF or dial pulses. The machine registers the digits and translates them into the address of the terminating station C. Included in central office A's address translation is a routing table that tells the switching machine the destination of the call. From the address, the machine determines that the call must be passed to the toll office B. Central office A checks the busy/idle status of trunks to office B and seizes an idle trunk. If no trunks are idle, central office A returns a *reorder* or fast busy call progress tone to station A. If an idle trunk is seized, A hears nothing except, perhaps, the clicking of operating relays. If central office A has automatic message accounting (AMA) equipment, it registers an initial entry to identify the calling and called number, and to prepare the AMA equipment to record the details of the call when C answers.

The trunk seizure removes the 2,600 Hz SF tone from the channel to indicate the change in status. Toll office B, detecting the change in status, returns a signal, usually a momentary interruption in the signaling tone toward central office A. This signal, called a *wink,* signifies that B is ready to receive digits. Detecting the wink, central office A sends its addressing pulses toward B. These pulses are either dial pulses, conveyed by interrupting the SF tone, or *multifrequency* pulses, conveyed by coding digits with combinations of two out of five frequencies. Office B continues to send an on-hook tone toward central office A, and will do so until station C answers. At this point, central office A has completed the originating functions and awaits the completion of the call.

Toll office B translates the incoming digits to determine the address of the destination, selects an idle trunk to central office C and seizes it. Central office C, detecting the seizure, sends a start signal to toll office B and prepares to receive digits. B sends the digits forward. Central office C tests the called station for its busy/idle status and, if busy, returns a busy tone over the voice channel. The calling party, recognizing the call progress tone, hangs up; the connection is taken down, and no completion entry is added to the AMA record. If station C is idle, central office C sends a 20 Hz alerting signal to ring the bell in C's telephone. It also returns an *audible ring* (another call progress tone) over the transmission path to the originating party. The line continues to ring until C answers, A hangs up, or the equipment times out.

When station C answers, central office C detects the change in status as line current begins to flow. This trips the ringing signal and discontinues audible ringing. Central office C changes the status of its signaling set toward B from on hook to off hook by interrupting the SF tone. B transmits the on-hook signal to A. Call completion is registered in the AMA equipment, indicating the time of day that charging begins.

When either party hangs up, the change in line current indicates a status change to its central office, which forwards the change to the other end by restoring the SF tone. Central office A registers a terminating entry in the AMA equipment to discontinue charging. The SF tones are restored to all circuits to indicate idle circuit status. All equipment is then prepared to accept another call.

This system, with minor variations that are discussed later, is used for signaling over dedicated as well as switched circuits. Not all circuits use SF signaling, however. For example, digital circuits usually employ internal bit-robbed signaling in place of the SF tones. SF signaling can be transmitted over a digital circuit, and often is when a digital link is connected to an analog link.

SF and MF signaling have two drawbacks that are leading to their replacement in toll circuits. First, circuit time that produces no revenue is used for signaling functions. Approximately 40 percent of all toll calls are uncompleted because the terminating station does not answer or is busy, or because of blockage or equipment irregularities, and the circuit time consumed in signaling is substantial. The second drawback is the vulnerability of this type of signaling to fraud. Devices have been constructed to defeat the operation of AMA equipment by inserting signaling tones on the circuit at the appropriate time.

To prevent these problems, the AT&T Communications network and many other carriers' networks use *common channel signaling,* which is described later. Common channel signaling uses a separate data communications network to exchange signals, establishing the connection after routing is complete. Common channel signaling has not been extended to local central offices, but developments now underway will include end offices in the common channel network.

E and M Signaling

By long-standing convention, signaling on interoffice circuits uses two leads designated as the E or recEive and the M or transMit leads for conveying signals. All external signaling sets and the built-in signaling of T carrier channels use E and M signaling to communicate status to attached central office equipment. Signaling equipment converts the binary state of line signals (tone on or off for analog and zero or one for digital equipment) to actuate the E and M leads.

There are five different types of E and M signaling interfaces, but in the most common type the M lead is grounded when on hook. An off-hook seizure is indicated by applying −48-volt battery to the M lead. The E lead is open when on hook; the signaling set applies ground to the E lead when it receives an off-hook signal from the distant end.

Signaling Irregularities

Signaling systems must be designed to transmit both status and addressing signals reliably enough to avoid errors. Signaling errors are rare with the internal signaling of digital carrier systems, but dial pulses transmitted over metallic facilities can be distorted by the line characteristics or by excessive noise.

Another irregularity occurs when both ends of a circuit are simultaneously seized. This condition, known as *glare,* is prevented by the use of one-way signaling on trunks. However, on small trunk groups the use of one-way trunks is uneconomical for reasons explained in Chapter 13. Therefore, to accommodate two-way trunks, signaling and switching systems must be designed to prevent glare. In the worst case, when glare occurs the equipment is unable to complete the connection, the circuit times out, and reorder is returned to the user.

DIRECT CURRENT SIGNALING SYSTEMS

DC signaling can be employed on metallic facilities, which include most subscriber loops and voice frequency interoffice trunks. The use of DC signaling facilities is not mandatory on metallic facilities. In many applications it is desirable to use SF signaling over metallic facilities.

The simplest status signal occurs when the telephone receiver is taken off hook, closing a DC path between tip and ring and allowing loop current to flow. This system is called *loop start.* Loop start is used on all subscriber loops that terminate in station sets.

Most PBXs connect to the central office over trunks equipped with two-way signaling and are thus subject to glare. Glare is particularly severe with loop start operation because the only indication the PBX has of a call incoming from the central office is the ringing signal, which occurs at six-second intervals. For up to six seconds the PBX is blinded to the possibility of an incoming trunk seizure and may seize a circuit for outgoing traffic when the trunk is carrying an incoming call the PBX has not yet detected. To provide an immediate trunk seizure signal toward the PBX, central office line circuits can optionally be wired for *ground start* operation. With this option the central office grounds the tip side of the line immediately upon seizure by an incoming call. By detecting the tip ground, the PBX is able to recognize the line seizure before ringing begins. Central office line circuits must be converted whenever this signaling method is used. PBX users must specify to the telephone company when ground start operation is required.

Metallic trunks and many special services use *duplex* (DX) signaling. DX signaling uses relays or electronic circuits that are sensitive to line status beyond the range of loop signaling. DX signaling equipment consists of either separate signaling

FIGURE 8.2
Single Frequency Signaling Simplified Block Diagram

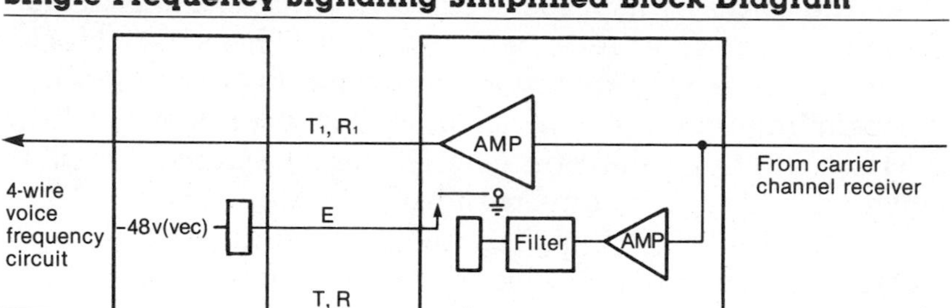

sets, circuitry built into carrier channel units as discussed in Chapter 4, or network channel terminating equipment (NCTE) as discussed in Chapter 6. An older system, composite (CX) signaling, is similar to DX signaling, except that it includes filters to separate the voice frequency path from the signaling path.

TRUNK SIGNALING SYSTEMS

Trunk signaling requires the built-in signaling of digital carrier, a separate SF set, or in some types of analog carrier, *out-of-band* signaling. The bit-robbed signaling of digital carrier is described in Chapter 4. Out-of-band signaling, rarely used in North America, passes a 3,700 Hz signaling tone over the channel. Narrow band filters separate the signaling tone from the voice frequency pass band.

Single-Frequency Signaling

The most common analog trunk signaling system is 2,600 Hz single frequency (SF) illustrated in Figure 8.2. The voice frequency leads from a carrier channel are wired directly into the SF set. The SF set contains circuitry to change the state of the E and M leads in response to the presence or absence of the SF tone, and to turn the signaling tone on and off when the status of its leads is changed by the switching machine or other central office equipment. The SF set blocks the voice frequency path toward the switching machine while the signaling tone is on the channel. The user does not hear the tone, although short tone bursts are occasionally audible when one party to a conversation hangs up.

One hazard of SF signaling is the possibility of *talk-off,* which can occur when the user's voice contains enough 2,600 Hz energy to actuate the tone-detecting circuits in the SF set. Voice filters are provided to minimize the potential of talk-off, but the problem may occur, particularly to people with high-pitched voices.

Addressing Signals

Addressing signals between station and central office equipment use either dial pulse or DTMF signals as described in Chapter 6. DTMF pulses require a DTMF receiver in the central office to convert the tones to the addressing signals. Because DTMF pulses travel over the voice path, they can be passed through the switching machine after the connection has been established. This capability is required to send addressing and identification information to interexchange carriers that use exchange access Feature Group A. (Feature groups are explained in Chapter 9.) DTMF is also useful for converting the telephone to a simple data entry device. All modern electronic central offices are capable of receiving DTMF dialing. Electromechanical central offices require externally mounted DTMF receivers to convert the tones to dial pulses.

Addressing signals are transmitted over trunks as dial pulse or multifrequency (MF) signals, or over a common channel as a data signal. MF signals are more reliable and considerably faster than dial pulse signals but require a *sender* to transmit the pulses. Most modern central offices are capable of MF signaling, but older electromechanical central offices can receive only dial pulses. Dial pulse or DTMF signals are also used between PBXs and their serving central office.

MF senders use a two-out-of-five tone method to encode digits as shown in Table 8.1. Digits are sent at the rate of

TABLE 8.1
Single-Frequency Signaling Tone Frequencies

Frequencies in Hz	*Digit*
700 + 900	1
700 + 1,100	2
900 + 1,100	3
700 + 1,300	4
900 + 1,300	5
1,100 + 1,300	6
700 + 1,500	7
900 + 1,500	8
1,100 + 1,500	9
1,300 + 1,500	0

about seven digits per second, compared to the 10 pulses per second of dial pulsing. Therefore, MF pulsing requires substantially less time to set up a call than dial pulsing.

Common Channel Signaling

Common channel signaling is an even more reliable and rapid method of establishing connections between processor-equipped switching machines. In North America, AT&T's common channel interoffice signaling (CCIS) system is used, although it will eventually be replaced with an international standard, CCITT No. 7. The CCIS network is a packet switched network operating at 4,800 b/s, using the architecture shown in Figure 8.3. CCIS replaces both SF and MF signaling equipment by converting dialed digits to data messages.

CCIS-equipped switching systems are interconnected with direct data links if the traffic volume warrants, but most signals are directed to a signal transfer point that acts as a concentrator for signaling traffic. CCIS signaling offers the following advantages over SF/MF signaling:

- Voice and signaling paths are separated, which reduces the potential for fraud and eliminates talk-off.
- Signaling speed is faster, reducing call setup time and allowing circuits to be disconnected faster than with SF signaling.

FIGURE 8.3
Common Channel Signaling System

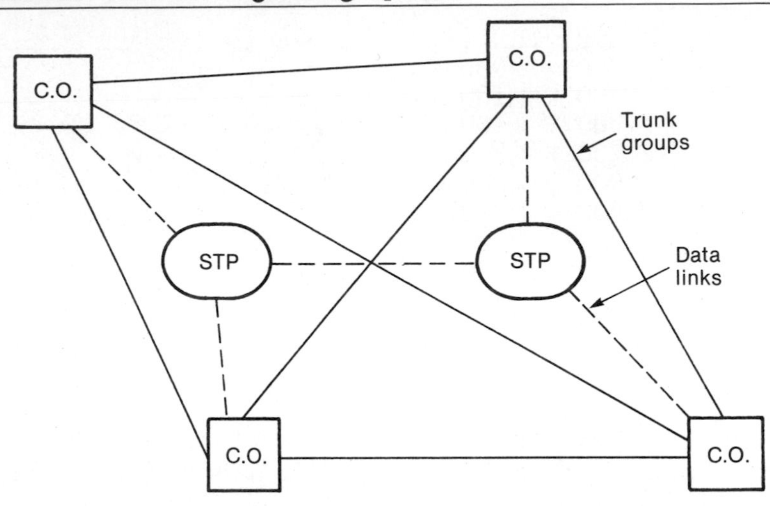

- Signals can be sent in both directions simultaneously and during conversation if necessary.
- Network management information can be routed over the CCIS network. For example, during trunk failure conditions, switching systems can be instructed with data messages to reroute traffic.

Private Line Signaling

Private or dedicated lines use all types of in-band signaling discussed so far plus *selective signaling,* an in-band system for operation of certain private line switching systems. Some dedicated circuits use signaling identical to that used by the telephone network. PBX tie trunks and large private switched networks require all the signaling capabilities of the telephone network. Special dedicated circuits require signaling arrangements that use the same techniques and equipment as the telephone network, but have no direct counterparts in switched systems. Examples are:

No Signaling. Some private lines require no signaling. Examples are data circuits that include signaling in the DTE and circuits that use microphones and speakers for alerting. Other examples of circuits requiring no signaling are program and wired music circuits.

Ringdown Circuits. In ringdown circuits a generator is used to ring the bell of a distant station. The 300 Hz cutoff frequency of carrier channels prevents 20 Hz ringing signals from passing over the channel. Ringdown circuits require equipment to convert the 20 Hz ringing supply to SF or E and M signals and vice versa. A similar circuit actuates ringing when the receiver is taken off hook. This type of circuit requires a loop-to-SF converter. The signal at the far end may actuate a light instead of a bell by using additional converter circuits.

Selective Signaling. Some private line networks use a four-wire selective signaling system to route calls without the use of switching machines. Dial pulses generate in-band tones to drive a simple switch to build up a connection to the desired terminating point.

Coin Telephone Signaling

Coin telephones use the dialing and ringing signals of ordinary telephones plus DC signals that operate apparatus within the telephone to collect and return coins. Coin tones are also generated in the telephone to enable the operator to distinguish between nickels, dimes, and quarters.

STANDARDS

Few published standards exist on the conventional single-frequency signaling system used in North America. Most standards have evolved by practice and are followed by signaling equipment manufacturers. Internationally, the CCITT No. 7 standard for common channel signaling and CCITT No. 4 and 5 line signaling systems are used. These standards are of little concern to users because they are administered by interexchange carriers. In North America, AT&T's CCIS system is used for common channel signaling. Similar systems for SF and MF signals have been adopted in North America.

ANSI

ANSI/IEEE 753 Standard Functional Methods and Equipment for Measuring Performance of Dial-Pulse (DP) Address Signaling Systems.

CCITT

Q.9 Vocabulary of switching and signaling terms.

Q.254 and Q.258 Telephone signals.

Q.311 2,600 Hz line signaling.

Q.314 PCM line signaling.

Q.320 Signal code for register signaling.

Q.330 Automatic transmission and signaling testing.

Q.331 Test equipment for checking equipment and signals.

Q.704 Signaling network functions and messages.

Q.705 Signaling network structure.

Q.723 Formats and codes.

EIA

RS-210 Terminating and signaling equipment for microwave communication systems.

APPLICATIONS

Signaling systems and equipment are used by all public and private networks except for some data communications networks that supply their own signals. Equipment used in these networks is identical and performs the same functions of alerting, addressing, supervising, and indicating status in both private and common carrier networks.

Evaluating Signaling Equipment

The same criteria discussed for evaluating digital carrier equipment also applies to signaling equipment. Reliability

TABLE 8.2
Voice Frequency Terminating and Signaling Units

Two-wire repeater
Four-wire repeater
Four-wire terminating set
Automatic ringdown unit
Conference bridge
DX signaling module
Data channel interface
Dial long line unit
Dial pulse correcting unit
Echo suppressor
Line transfer relay
Loop extender
Loop signaling repeater
Program amplifier
Pulse link repeater
Repeat coil
Signaling converter, 20 Hz to E and M
Signaling converter, loop start to ground start
Signaling range extender
Single-frequency signaling unit
Toll diversion unit
Voice frequency equalizer

is the primary concern, with power consumption, circuit density, and cost being other factors to consider.

A circuit design is essential before signaling equipment is selected. So many alternatives exist for interconnecting signaling equipment that it is necessary to determine the most economical design to minimize signaling costs. The most economical configuration in digital systems is the use of the built-in equipment in the T carrier channel. With the wide range of special channel units listed in Table 4.1, it is feasible to provide nearly any conceivable combination of signaling services to operate station signaling equipment.

Where built-in signaling is not included as part of the channel unit, external signaling converters are required. Table 8.2 lists the most common converters. These units are often available with amplifiers contained in the same package, and are built into plug-in units that mount in a special shelf. The voice frequency and signaling leads are cabled to

FIGURE 8.4
End-to-End versus Link-by-Link Signaling

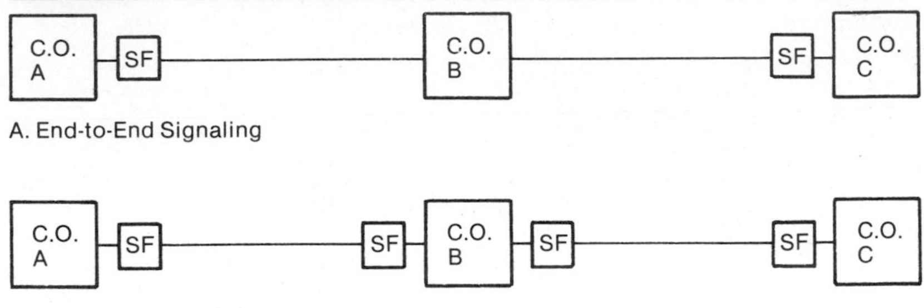

distributing frames as discussed in Chapter 14 for connection to carrier and subscriber loop equipment.

Signaling compatibility is also an important consideration in acquiring signaling equipment. Compatibility is rarely a problem with respect to signaling equipment of different manufacturers. Single frequencies, ringing frequencies, and E and M lead connections are universal with all manufacturers. However, the timing of signals, which is controlled by switching and transmission equipment connected to signaling sets, is a frequent cause of incompatibility, but this has little to do with the signaling equipment itself.

Plug compatibility is an issue that must be addressed by the manufacturer. Several types of plug-in shelves exist on the market; all are designed to manufacturer's specifications, and no standards exist. Some manufacturers make equipment that is designed to be plug-compatible with shelves of others.

Another important consideration in evaluating signaling equipment is testing capability. Circuits are designed with either *end-to-end signaling* or *link-by-link signaling,* as illustrated in Figure 8.4. End-to-end signaling is the easiest to test. If SF tones are used between both ends of a circuit, signaling status can be determined by listening to the tone over the voice channel or monitoring the signaling leads of a T carrier channel. The operation of the SF set is determined by measuring the electrical state of the E and M leads. With link-by-link signaling, the signals are extracted at intermediate points and connected by a *pulse link repeater,* which

interconnects the E and M leads. In most private networks with station, PBX, and NCTE equipment furnished by the user, the local loop by the telephone company, and the intercity circuits by an interexchange carrier, link-by-link signaling is the rule. For the user to diagnose signaling problems, testing capability is required to evaluate signaling in the station and NCTE.

GLOSSARY

Addressing: The process of sending digits over a telecommunications circuit to direct the switching equipment to the station address of the called number.

Alerting: The use of signals on a telecommunications circuit to alert the called party or equipment to an incoming call.

Audible ring: A tone returned from the called party's switching machine to inform the calling party that the called line is being rung.

Automatic message accounting (AMA): Equipment that automatically records long distance call billing details.

Call progress tones: Tones returned from switching machines to inform the calling party of the progress of the call. Examples are audible ring, reorder, and busy.

Common channel signaling: A separate data network used to route signals between switching systems.

Common channel interoffice signaling (CCIS): The AT&T common channel signaling system used in North America.

Composite (CX) signaling: A direct current signaling system that separates the signal from the voice band by filters.

Duplex (DX) signaling: A direct current signaling system that transmits signals directly over the cable pair.

E and M signaling: A common designation for the transmit and receive leads of signaling equipment at the point of interface with connected equipment.

End-to-end signaling: A method of connecting signaling equipment so it transmits signals between the two ends of a circuit with no intermediate appearances of the signaling leads.

Glare: A condition that exists when both ends of a circuit are simultaneously seized.

Ground start: A method of circuit seizure between a central office and a PBX that transmits an immediate signal by grounding the tip of the line.

Line status: An indication of whether a line or circuit is busy (off hook) or idle (on hook).

Link-by-link signaling: A method of connecting signaling equipment so signals are interfaced at intermediate locations between the two ends of the circuit.

Loop start: A method of circuit seizure between a central office and station equipment that operates by bridging the tip and ring of the line through a resistance.

Multifrequency signaling: A method of sending pulses over a circuit by using one pair of tones from a total set of five tones to encode each digit.

Out-of-band signaling: Signaling by means of a tone that is separated from the voice channel by filters.

Pulse link repeater: A signaling set that interconnects the E and M leads of two circuits.

Reorder: A fast busy tone used to indicate equipment or circuit blockage.

Selective signaling: An in-band signaling system used on private line networks to direct equipment to switch a connection.

Sender: A device used in central offices to store and forward addressing signals.

Single frequency (SF) signaling: The use of a single tone (usually 2,600 Hz) to indicate the busy/idle status of a circuit.

Supervision: The process of monitoring the busy/idle status of a circuit to detect changes of state.

Talk-off: A circuit irregularity that occurs when sufficient 2,600 Hz energy is present in the voice signal to cause the SF set to interpret it as a disconnect signal.

Wink: A momentary interruption in SF tone to indicate that the distant office is ready to receive digits.

BIBLIOGRAPHY

American Telephone & Telegraph Co. *Notes on the Network.* 1980.

———. *Telecommunications Transmission Engineering.* Bell System Center for Technical Education, vol. 1, 1974; vol. 2, 1977; and vol. 3, 1975.

Freeman, Roger L. *Telecommunication System Engineering.* New York: John Wiley & Sons, 1980.

Martin, James. *Telecommunications and the Computer.* Englewood Cliffs, N.J.: Prentice-Hall, 1976.

Sharma, Roshan La.; Paulo T. deSousa; and Ashok D. Inglè. *Network Systems.* New York: Van Nostrand Reinhold, 1982.

MANUFACTURERS OF SIGNALING EQUIPMENT

ADC Telecommunications

AT&T Network Systems

ITT Telecommunications

Lear Siegler, Inc.

Northern Telecom, Inc.

Pulsecom Division, Harvey Hubbel, Inc.

Telco Systems, Inc.

Tellabs, Inc.

Transcom Electronics, Inc.

Wescom, Inc.

CHAPTER NINE

Circuit Switched Network Systems

The word *network* has become an ambiguous term. It can describe the relationship of a group of broadcasting stations or a social fabric that binds people of similar interests. In telecommunications the usage is somewhat more specific, but the precise meaning of the term must be derived from context, and this can be confusing to those outside the industry. In its broadest sense a telecommunications network is the combination of all the circuits and equipment that enable users to communicate. All the switching apparatus, trunks, subscriber lines, and auxiliary equipment that support communication can be classified as elements of the network. In another sense, the word is narrowed to mean the switching circuits that interconnect the inputs and outputs of a switching machine. A third meaning is found at the circuit level where the term describes the interconnection of components to form a filter or level-reducing attenuator. The only way the ambiguity can be dealt with is to become familiar enough with the technology to take the meaning of the term from its context.

The direct interconnection of two stations meets the definition of a network in the strictest sense of the word, but this is a restrictive kind of private network because it lacks accessibility. In a broader sense, public networks provide the capability for a station to reach another station anywhere in the world without the need for complex addressing. Accessibility is achieved by introducing switching to control routing choices and to provide a point of entry to the network. An effective telecommunications network possesses these attributes:

- **Accessibility.** Any station can be connected to any other station if they are compatible with the network's protocols.
- **Ease of addressing.** Stations are accessed by sending a simple address code. From the address, the network performs the translation and code conversions to route the call to the destination.
- **Interconnectibility.** Network ownership rarely crosses national borders, and in the United States, multiple ownership is the rule. For the greatest utility, interconnection across sovereign or proprietary boundaries is required.
- **Robustness.** Networks must contain sufficient capacity and redundancy to be relatively invulnerable to overloads and failures, and to recover automatically from such failures that occur. They must offer some form of flow control to prevent users from accessing the network when overloads occur.
- **Capacity.** Networks must be capable of supporting enough users to meet service demands.

Some type of switching is required to meet all the above features. Networks employ three forms of switching: packet switching, in which traffic is divided into small segments, message switching, in which traffic is stored and forwarded when a path to the destination is available; and circuit switching, in which the users are directly interconnected by a path that is established for the duration of the session. The first two types of switching are practical only for data communications. Packet switched voice is technically feasible and packet switching may eventually be commonly used for voice traffic, but at present circuit switching is the only feasible form for voice. With some limitations in error-checking capability, circuit switching is also feasible for data traffic.

NETWORK TERMINOLOGY

Network terminology is often confusing because it is ambiguous. In this discussion of networks the following terms are used:

- A *node* is a network element that provides a point for stations to access the network, and it is the terminating point for internodal trunks. In circuit switched networks, nodes are always switching machines. In packet and message switched networks, they may be computers.
- *Trunks* are the circuits or links that interconnect nodes. In switching machines, the equipment that interfaces the internodal trunks to the switching machine is also called trunks, or sometimes *trunk relay equipment.*
- *Stations* are the terminal points in a network. Telephone instruments, key telephone equipment, data terminals, and computers all fall into the station definition for this discussion.
- *Lines* are the circuits or paths that connect stations to the nodes.

NETWORK ARCHITECTURE

Five basic network topologies—ring, bus, branching tree, mesh, and star—are discussed in Chapter 3. Circuit switched networks use the star and mesh topologies almost exclusively. Lines radiate from the central office to stations in a star topology, and the nodes are interconnected as a mesh.

The fundamental network design problem is determining how to assemble the most economical configuration of circuits and equipment based on peak and average traffic load; grade of service required; and switching, circuit, and administrative costs. With a small number of nodes it is prac-

FIGURE 9.1
Direct and Tandem Trunks in Single-Level and Hierarchical Networks

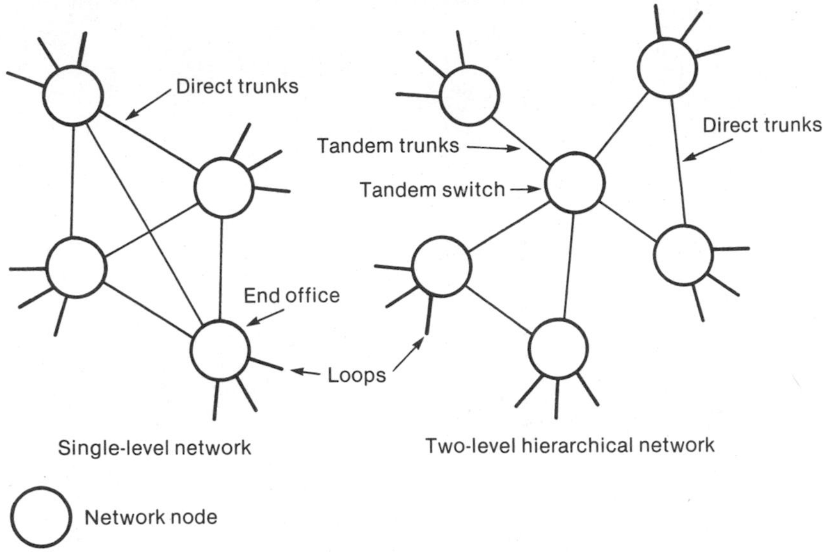

tical to connect them with *direct trunks*, which are circuits connected directly between nodes as shown in Figure 9.1. Direct connection is feasible up to a point, but as the number of nodes increases, the number of circuit groups increases as the square of the number of nodes, and the quantity of trunks soon becomes unwieldy. To conserve costs a hierarchical network can be formed, using *tandem switches* to interconnect the nodes. Figure 9.2 shows the North American telecommunications network structure, which is a hierarchy consisting of five levels or classes of switches.

The number of levels in a network hierarchy is determined by the network's owner and is based on a cost/service balance as explained in Chapter 13. The five-level structure of the telephone network is changing in response to new technology and changes in ownership mandated by regulatory agencies and the courts.

The Changing Network Environment

Telecommunications networks exist in an environment that is continually changing. Service demands are not constant—

FIGURE 9.2
Switching Hierarchy

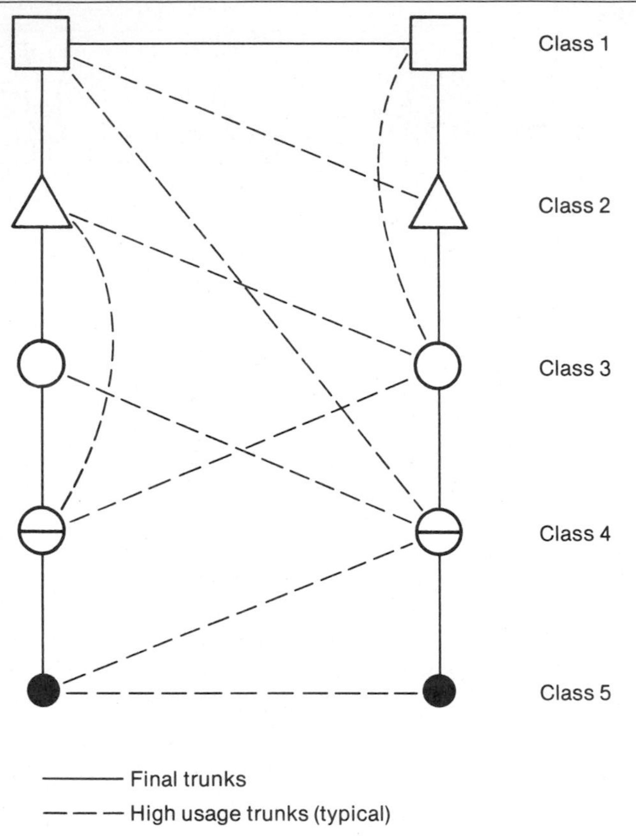

they vary by time of day, day of week, and season of the year. Demand is continually evolving in response to changing calling habits and business conditions. Competition and new technology have a substantial effect on cost and demand. Moreover, network design is always a compromise that seeks to use existing equipment as long as it is capable of providing satisfactory servicc. Because of these diverse forces, any network is a complex composite of modern and obsolescent equipment and is continually being shrunk or expanded to match demand. Even a new private network assembled with the latest technology is soon obsoleted in part by technical advances.

The remainder of this chapter is devoted to explaining the characteristics of switching equipment that serves North

FIGURE 9.3
Block Diagram of a Switching System

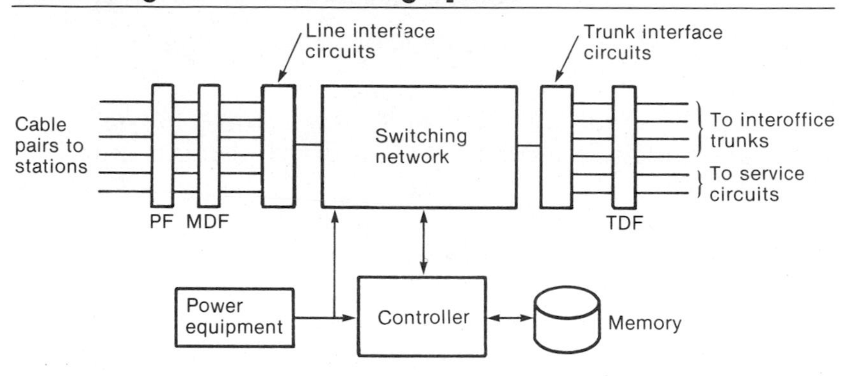

PF—Protector Frame
MDF—Main Distributing Frame
TDF—Trunk Distributing Frame

American telecommunications users. This technology is common to the three major classes of switching machines—local central offices, PBXs, and tandem switches—which are discussed in Chapters 10, 11, and 12. The distinction between these three types of switching machines is not absolute. A single machine can serve any or all functions.

Switching System Architecture

All switching systems contain the following elements as shown in Figure 9.3:

- A switching network or matrix that connects paths between input and output ports.
- A controller that directs the connection of paths through the switching network. Direct control switching machines that are discussed later do not employ a separate controller. The user controls the switch by dialing digits.
- Line ports that interface outside plant for connection to the end user. All local and PBX switching machines include line ports; tandem switches may have only a few specialized line ports.

- Trunk ports that interface interoffice trunks, service circuits, and testing equipment.
- Service circuits that provide call progress signals such as ringing and busy tones.
- Common equipment such as battery plants, power supplies, testing equipment, and distributing frames.

SWITCHING SYSTEM CONTROL

When a user signals a switching machine with a service request, the switch determines the terminating station's address from the telephone number dialed and *translates* the number to determine call routing. Translation tables specify the trunk group that serves the destination, an alternate route if the first choice route is blocked, the number of digits to dial, any digit conversions needed, and the type of signaling to use on the trunk. Some switches lack translation capability. These machines, called *direct control* machines, are capable of routing only in response to dialed digits. *Common control* machines include circuitry that enables them to make alternate routing choices; that is, when one group of trunks is blocked, another group can be selected. Electromechanical common controlled switching machines use wired relay logic. Modern electronic switching systems use *stored program control* (SPC) controllers to perform all call processing functions.

SWITCHING NETWORKS

Switching machines can be classified by their type of switching network. Direct control switching machines have inflexible networks that are directed by dial pulses to a single destination. Common control and stored program control use one of four types of switching network:

- Crossbar analog.
- Reed relay analog.
- Pulse amplitude modulated analog.

- Pulse code modulated digital.

The basic function of the switching network is to provide paths between the inputs and outputs. Like all other design tasks, the network design objective is to provide enough paths to avoid blocking users while keeping costs to a level users are willing to pay. The first two network types use electromechanical relays for the switching medium and are more expensive than the last two types which employ digital logic circuits to provide and control the network paths. Electromechanical networks are constructed with some restriction on the number of users that can be served at one time. A network that contains fewer paths than terminations is referred to as a *blocking network* because not all users can be served simultaneously.

With the lower costs of integrated circuits, nonblocking PCM networks are feasible and are provided in many digital central offices. A nonblocking network means that the number of paths through the network is equal to the number of possible users that can be connected at one time. It should be remembered, however, that a nonblocking network does not imply that a switching machine has unlimited call-carrying capacity. Some machine element other than the network will limit its capacity.

Modern switching networks are wired in grids as shown in Figure 9.4. Each stage of the grid is composed of a switching matrix that connects input links to output links. Links are wired between switching matrixes to provide a possible path from any input port to any output port. The network shown in Figure 9.4 is nonblocking because the number of input ports and output ports is equal and the number of possible paths equals the number of ports.

Many switching networks use concentration to reduce network cost. Assume, for example, that the primary switch on the input side of the network in Figure 9.4 has six input ports instead of the three ports shown. Such a network would have a two-to-one concentration ratio because only three of the six inputs could be serviced simultaneously. Local central offices and PBXs typically use a line switch concentration ratio of four to one or six to one. It should be understood that even though a switch may have a nonblocking network,

FIGURE 9.4
Switching Network Diagram

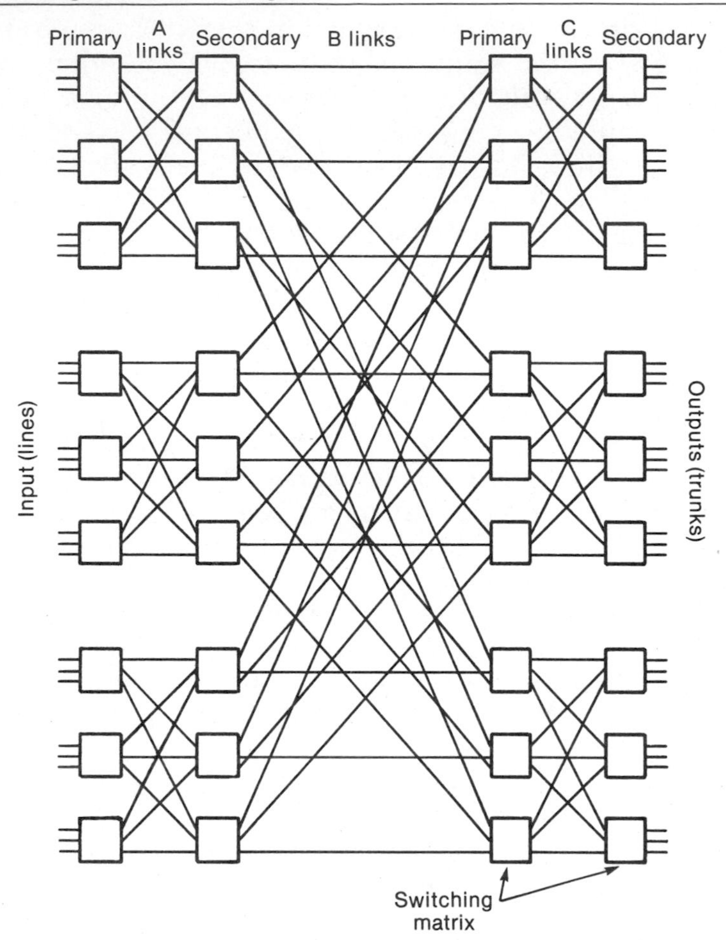

it can be blocked in the line switch networks. Trunk switches usually use no concentration.

The capacity of a switching network is directly related to its number of switching stages. The switching matrix is physically limited by the number of terminations it can support. To avoid blocking, the controller must have multiple choices of paths through the network. These paths are obtained by providing multiple switching stages so that each stage has enough choices that the probability of blocking is reduced to a level consistent with grade-of-service objectives.

FIGURE 9.5
Direct Control Switching System (Step-by-Step)

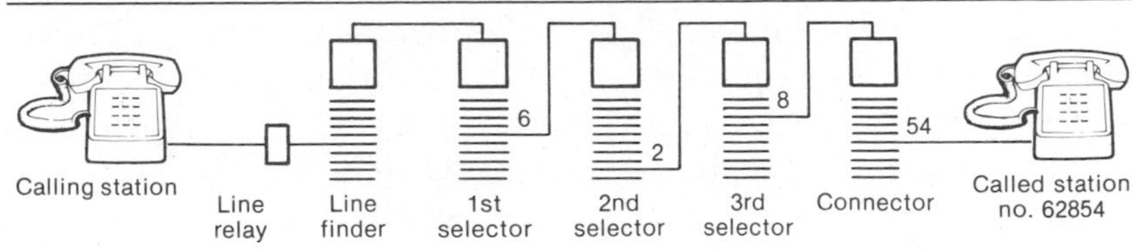

DIRECT CONTROL SWITCHING SYSTEMS

The earliest type of switching system employed operators to make manual connections. An incoming signal, actuated by taking the receiver off hook or turning a crank, operated a signal on the switchboard to notify the operator of an incoming call. The operator answered a call by inserting a cord in a jack, obtaining the terminating number, and inserting the matching cord in the jack of the called line or of a trunk to a distant office. Disregarding the cost of the manual switchboard, it was efficient because of the ability of the operator to make alternate routing decisions.

The first practical dial system was the Strowger method, which used an electromechanical switch to make connections. The Strowger, or step-by-step, switch operates through two axes; it is driven vertically to 1 of 10 levels and rotates horizontally to 1 of 10 terminals. It is actuated by pulls on the telephone dial or by an internal operation, depending on the function of the switch. A block diagram of a simple step-by-step central office is shown in Figure 9.5.

Subscriber lines are wired to line relays that operate when the user lifts the receiver. Line relays are wired to line finders that are connected in groups to serve as many as 200 lines with up to 20 switches, depending on the traffic volume. When all switches in a line group are busy, other users cannot get dial tone until someone hangs up. Line finders automatically step vertically and rotate horizontally to locate a calling telephone line. A first selector switch, permanently wired to the line finder, furnishes dial tone to the user. As digits are pulsed, selector switches are driven upward to a level corresponding to the digits dialed and automatically rotate horizontally to find the first path to the next selector.

The final two digits in the train actuate a connector switch, which is cross-connected to the user's line. The next-to-last digit drives the connector to the appropriate level; the last rotates the connector to the correct terminal. If the terminal is busy, the connector returns a busy signal. If it is idle, the connector attaches a ringing signal, which remains attached until the called party answers or the calling party hangs up.

A step-by-step central office can be visualized as a concentrator in which a large number of originating stations are routed over a smaller number of paths through the selector train. The selector switches find vacant paths to the expansion side of the office, which contains all the telephone numbers terminating in the central office. A step-by-step office has no translation capability. It routes to lines and trunks directly on the basis of the dialed digits. Different selector levels are wired to internal switches or to trunks to distant offices. The initial digit "1" is reserved for dialing to service circuits such as long-distance, directory assistance, and repair service.

Incoming calls from distant offices appear on incoming selectors, which have access to the selector switch train. The dial pulses drive the selector train exactly like an intraoffice call except for the fewer number of digits that are needed to drive the selector train and connector to the destination.

Most step-by-step central offices have no automatic message accounting (AMA) capability. Long-distance call details are recorded in a serving toll office that is equipped with centralized AMA (CAMA).

About half of the telephone switches in the United States use the step-by-step system today, but these are being replaced by stored program machines. Although the step-by-step office comprises a significant number of offices, the remaining machines are small and serve only a fraction of the total telephone users in the country. Several pertinent points should be understood about directly controlled central offices:

- Step-by-step switches are incapable of making routing decisions. When blockage is encountered, the switch can return only a reorder or busy signal to the user, who must hang up and try again.

- The switch can be driven only by dial pulses. To provide DTMF service, a separate converter is wired to the first selector to convert DTMF signals to dial pulses.
- The switching speed is regulated by the speed of the telephone dial and is a function of the quantity of digits in the dialed number and the dialing speed of the user.
- All equipment used for a call remains connected for the duration of the call. The system has no ability to release high-usage equipment to another user as long as it is occupied.
- Each call introduces wear to the mechanical equipment. Unless periodic adjustments and replacements are made, service deteriorates.
- The switches and relays in a step-by-step office introduce noise that is coupled into other circuits. Therefore, data transmission is likely to be impaired by errors in a step-by-step office, particularly when the equipment is improperly maintained.
- A step-by-step office is robust with no common circuits, aside from power and ringing supplies, that can interrupt all lines from a single failure. Redundancy is unneeded except in ringing machines.

COMMON CONTROL CENTRAL OFFICES

Common control switching systems employ electromechanical logic circuits to drive the switching network. These logic circuits are brought into a connection long enough to establish a path through the switching network, then are released to attend to other calls. A block diagram of a crossbar common control switching machine is shown in Figure 9.6.

Several types of switching networks are used with common control machines, but the most common is the crossbar switch that is shown in Figure 9.7. The crossbar switch is a matrix consisting of horizontal and vertical connections. Switches are manufactured in several different sizes, but for the purpose of illustration, a 10 × 20 switch will be considered. With this type of switch, 20 input paths assigned

FIGURE 9.6
Block Diagram of Crossbar Switching Machine

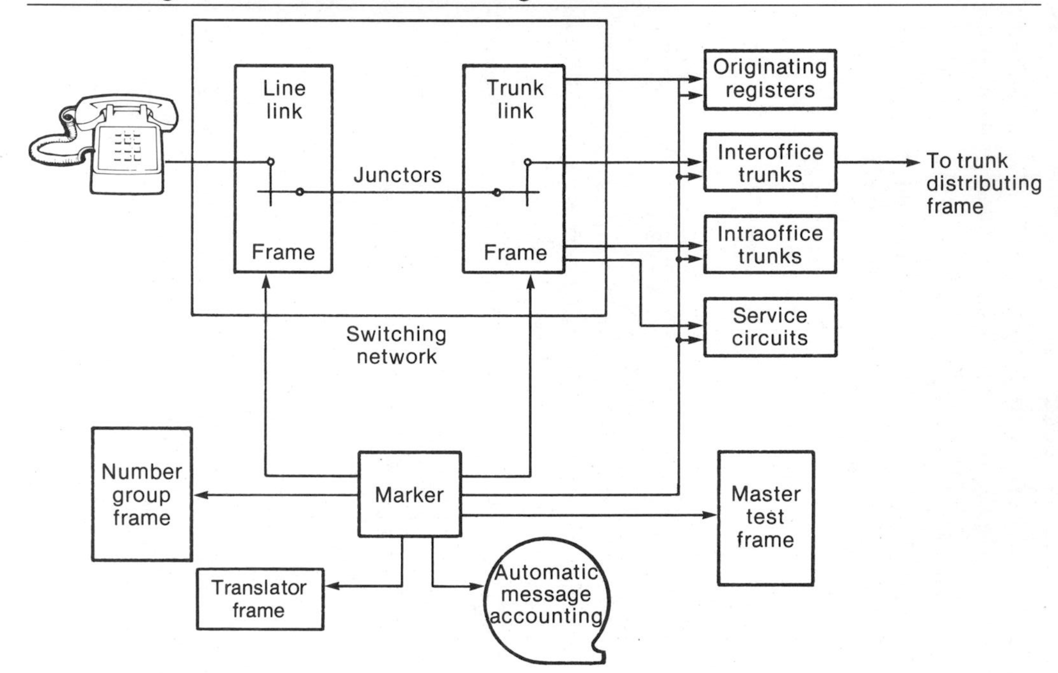

FIGURE 9.7
Crossbar Switch

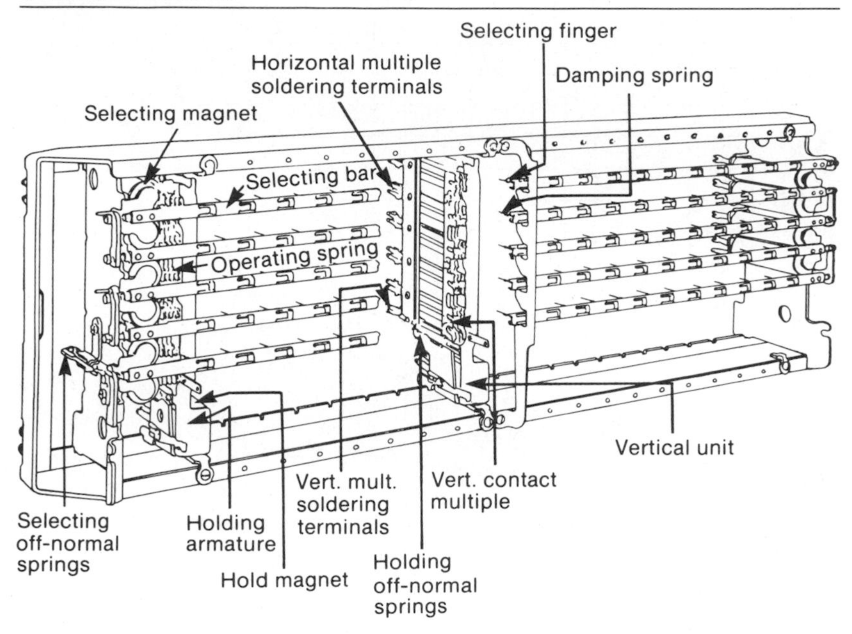

to the vertical portion of the matrix can be connected to 10 output paths assigned to the horizontal side of the matrix, providing a 2:1 concentration ratio.

A set of crosspoints is wired to the horizontal and vertical paths at each matrix intersection as shown in Figure 9.8. The common control equipment operates and releases the crosspoints. For example, to make the connection between vertical 3 and horizontal 5 as shown in the figure, an impulse from the common control operates a horizontal select magnet. This magnet moves a finger against an actuating card on the switch contacts, where it is clamped by operation of the vertical hold magnet. The select magnet can then be released. The hold magnet keeps the contacts operated until it is released by the common control when the call is terminated. The function of the common control equipment is to select an idle path through the switching network and to operate crosspoints in all switches simultaneously to establish path continuity.

Crossbar switches are constructed with two-wire contacts for local central office and PBX applications and with four-

FIGURE 9.8
Crossbar Switch Schematic Diagram

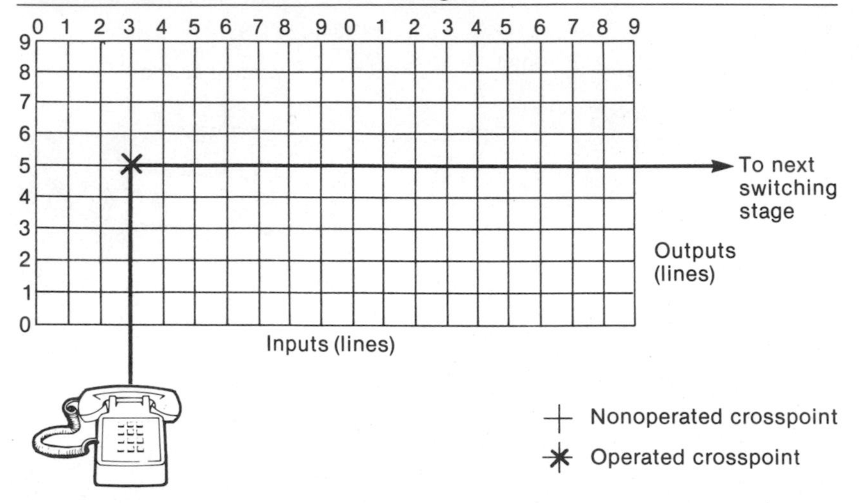

wire contacts for toll tandem operation. Some two-wire toll tandem machines are used for local tandem offices and lower classes of toll switching machines, but these are obsolescent and are being replaced because of the transmission deficiencies inherent in a two-wire switch as described in Chapter 2.

The brain of a crossbar switch is the *marker*, an electromechanical device that performs functions similar to a central processor in a computer. The switching network and auxiliary equipment in a crossbar office are driven by instructions from the marker. Markers are provided in a quantity sufficient to serve the traffic load in the office. A minimum of two is provided for redundancy.

When a line goes off hook, the marker detects the change of state and connects the line to an originating register trunk. The register provides dial tone and receives the dialed digits. Registers are of two classes, dial pulse and DTMF. Dial pulse registers can detect only dial pulses; DTMF registers can detect either type of signal. The register stores dialed digits in a relay circuit and signals the marker when dialing is complete. If dialing has not been completed in a specified interval, the register times out and the marker connects the line to a reorder trunk.

After dialing is complete, the marker calls a translator, a device that stores routing and signaling information. If the translator informs the marker that the call is to a chargeable destination, the marker calls in AMA equipment and records the initial entry. If the number is within the same office, the marker tests the status of the terminating number and, if busy, attaches the calling line to a busy signal trunk. If the called line is idle, the marker reserves a path through the network, connects the called number to a 20 Hz ringing trunk, and connects the calling number to an audible ringing trunk. When the called party answers, the audible and 20 Hz ringing are removed and a path is connected through the network. The first party to hang up signals the marker to take down all the connections.

When a call is incoming from a distant office, the marker recognizes the trunk seizure from the change in status of the signaling leads and attaches an incoming register to receive the digits. From the trunk classification the marker tells the register how many digits to expect. When dialing is complete, the register signals the marker, which connects a path from the trunk to the terminating number.

The most significant advantage of common control offices compared to direct control offices is the flexibility of common control. If, for example, an office has direct trunks to another office but they are all busy, it can attach an outgoing call to an alternate route to a tandem office. This capability is designed into most networks. Direct circuit groups called high-usage (HU) groups are established to terminating offices if the traffic volume is high enough to justify the cost. The capacity of HU groups is engineered to keep the circuits fully occupied during heavy calling periods. Overflow traffic is routed to tandem trunks that are more liberally engineered to support the overflow. Alternate routing concepts are discussed in more detail in Chapter 13.

The following characteristics are significant in common control switching machines:

- Alternate routing capability offers flexibility in handling overloads.
- Crossbar offices require less maintenance than step-by-step offices because their internal circuitry has no

wiping action to generate wear. Also, trouble indications are detected by the marker and recorded in a trouble recorder for analysis.

- Although crossbar offices are not as electrically noisy as step-by-step offices, relay and switch operations cause noise that can result in errors in switched data circuits.
- The structure of crossbar switches makes four-wire tandem operation feasible to improve transmission performance.
- Common control switches are vulnerable to total central office failure. When markers are all occupied or out of service no further users can be served even though idle paths are available.
- Common control offices are much faster than step-by-step offices. Dial pulses are registered in shared circuits to avoid tying up equipment with slow users or long dial pulse strings.

STORED PROGRAM CONTROL

Stored program control (SPC) central offices have been in operation in the United States since 1965 and are gradually replacing their electromechanical counterparts where it is economical to do so. The primary economies of SPC offices are their lower maintenance cost and their ability to provide enhanced features that are impractical with electromechanical central offices. SPC machines, diagrammed in Figure 9.9, use concepts similar to common control electromechanical offices except that electronic logic replaces wired relay logic.

Call processing is controlled by the central processor. When an off-hook signal is detected, the processor attaches a dial pulse or DTMF receiver to the originating line. The receiver supplies dial tone and registers the incoming digits. The processor stores the details of the call in a temporary call store memory. Translation tables are stored in semipermanent memory. The processor establishes a path through the switching network, attaches 20 Hz ringing to the called party

FIGURE 9.9
Block Diagram of Stored Program Switching Machine

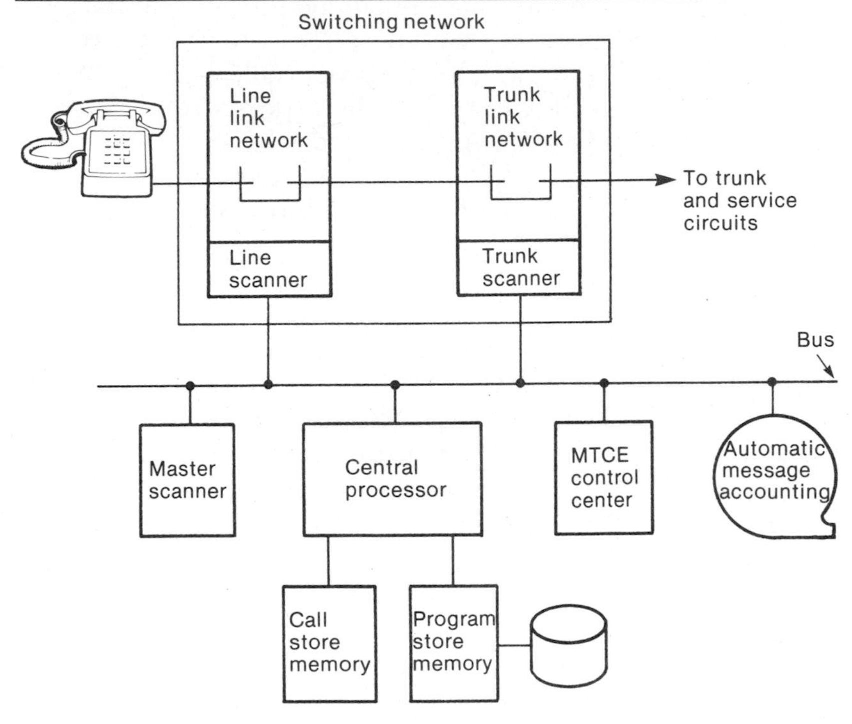

and an audible ringing trunk to the calling party. When the called party answers, the connection is completed.

Although the call processing is similar to that in a common control electromechanical office, the SPC processor offers much greater flexibility. The processor operates under the direction of a *generic program,* which contains the details of call processing. Features can be added by replacing the generic program with a new issue. Because of this factor, SPC machines are far more flexible than their electromechanical counterparts. Special features that are described in Chapters 10, 11, and 12 are contained in the generic program and can be activated, deactivated, or assigned to a limited group of users by making program changes. SPC machines are also capable of collecting statistical information and diagnosing circuit and machine irregularities to a much greater degree than electromechanical machines.

Electronic Switching Networks

Over the past two decades many changes have been made in SPC switching networks, which can be broadly classed as analog or digital. Some machines have used processor-driven crossbar switching networks, but the majority of analog machines use reed relay networks. A reed relay consists of contacts enclosed in a sealed glass tube and surrounded by a coil of wire. The contacts are closed by a short pulse of current and remain closed until opened by a second pulse. The switches are wired in a matrix of horizontal and vertical paths to establish a DC circuit through the network.

The majority of the electronic switching systems in use in the United States today use reed relay networks; however, this technology is obsolescent. Reed switches are more expensive to manufacture and require more maintenance than the digital networks that are replacing them. The latest generation of switching machines uses digital switching networks.

Pulse Amplitude Switching

The earliest all-electronic switching networks used pulse amplitude modulation (PAM), a concept that is described in Chapter 4. At the heart of the network is a high speed time-multiplexed bus that provides a talking path for all connected conversations. The bus is divided into time slots. Stations are interconnected by assigning them to the same time slot, during which they are allocated the full bandwidth of the bus long enough to send a single pulse. Line circuits sample the voice signal 8,000 times per second and generate a PAM signal as described in Chapter 4. Line and trunk ports are connected to the bus through gating circuits as illustrated in Figure 9.10. At the proper time slot instant, the gate opens to connect originating and terminating users to the bus for the duration of one pulse. By allowing a port to send during one time slot and receive during another, full-duplex operation is achieved.

PAM networks have the advantage of being electronic and thus less costly to manufacture than reed networks. Western Electric Dimension® PBXs manufactured during the 1970s and early 1980s use PAM switching networks. This type of

FIGURE 9.10
Time Multiplexed Bus System

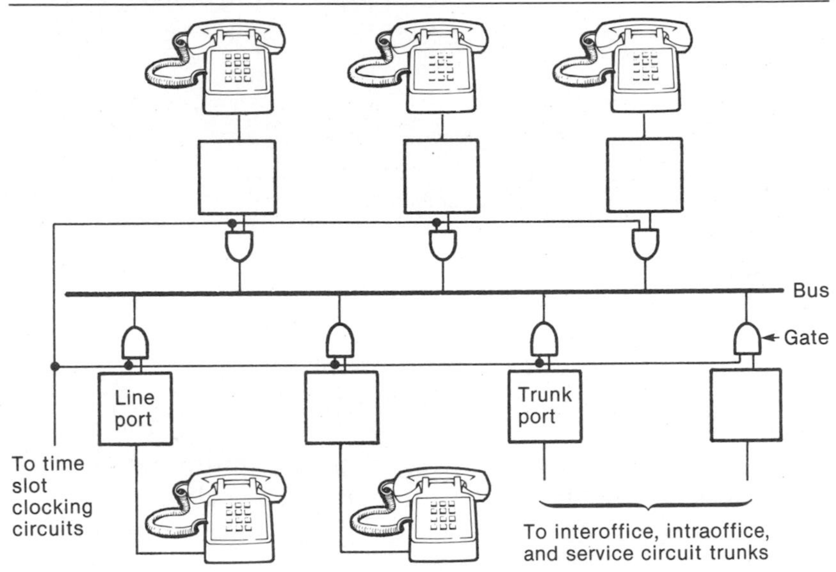

network has the disadvantage of passing only analog signals and is being replaced by switching machines that use PCM networks.

Pulse Code Modulated Networks

PCM switching networks are used in the latest generation of switching machines of all types. PCM networks are similar in concept to the analog matrix shown in Figure 9.4 with some important exceptions.

First, PCM networks connect the encoded signal over parallel paths. The incoming serial bit stream from the line or trunk circuits is converted to a parallel signal and assigned to a time slot. At the proper time slot instant, all eight bits of the input PCM signal are gated to the output port in parallel. In the output circuits they are converted to serial for application to trunk and line circuits.

The second exception is in the mode of switching employed in PCM networks. Analog networks use *space division* switching exclusively. That is, input paths are physically connected to output links. Digital networks use a combination of space division and time division switching.

FIGURE 9.11
Time Slot Interchange

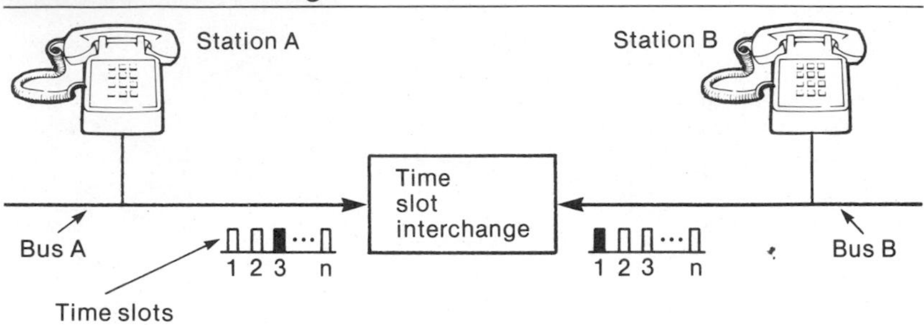

Although the term *space division* implies a relay operation, the switches used in digital networks contain no moving parts. Integrated logic gates form the switching element to direct the PCM pulses from one path to another.

To understand the digital switching concept, assume, as shown in Figure 9.11, that two separate time multiplexed buses are linked by a circuit called a *time slot interchange* (TSI). If station A wants to call station B, the TSI circuit receives a PCM word from A during time slot 3, stores it for a cycle, and puts it out on time slot 1 of bus B during the next cycle. Time slot assignments are made by the processor and released when the call terminates.

Networks of greater complexity are composed of combinations of space and time division switches. For example, the network shown in Figure 9.12 uses a "time space space time" (TSST) network. Although only one direction of transmission is shown, digital switching networks are bi-directional and provide a four-wire transmission path. Different manufacturers employ a variety of time and space switching configurations to achieve their network design objectives.

Other Switching Networks

Two other types of digital switching network are used in PBXs. Several systems use delta and pulse width modulated networks. Like the more common PAM network, these are technically digital switches but are incapable of directly interfacing digital lines.

FIGURE 9.12
Digital Switching Network

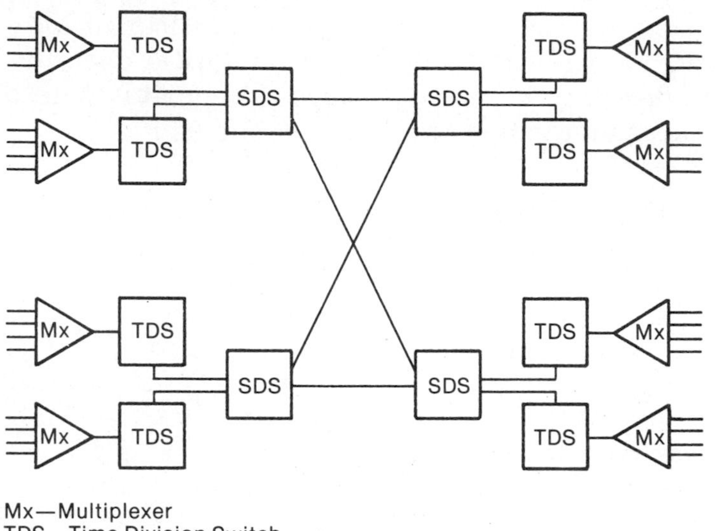

Comparison of Digital and Analog Switching Networks

Digital switches have several advantages over their analog counterparts:

- The switching networks are less expensive to manufacture because of the ability to use integrated components.
- T-1 circuits can interface the switching machine directly without using a channel bank to bring the circuits down to voice frequency.
- High-speed data can be switched without the use of modems if digital line interface circuits are provided.

Digital switches have offsetting disadvantages compared to analog networks that should be considered where the advantages are not compelling:

- Their line circuits are more complex and expensive than analog line circuits. In analog switches, the costs are concentrated in the switching network; in digital switches, costs are concentrated in the line circuits.

- Digital switches consume more power than analog switches because the switching network is operating and drawing current continually, although it is handling little or no traffic load.

LINE, TRUNK, AND SERVICE CIRCUITS

All switching systems are equipped with circuits to interface the switching network to stations, trunks, and service circuits such as tone and ringing supplies. In some machines these circuits are external devices. In other cases, they are integral to the switching equipment. For example, some digital central offices develop tones internally by generating the digital equivalent of the tone so when it is applied to the decoder in a line or trunk circuit, it is converted to an analog tone.

Line Circuit Functions

In a digital central office, line circuits perform seven basic functions that can be remembered by the acronym "BORSCHT." Analog central office line circuits require five of the seven functions; because they have two-wire switching networks, the hybrid and coding functions are omitted. The BORSCHT functions are:

Battery is fed from the office to the line to operate station transmitters and DTMF dials.

Overvoltage protection is provided to protect the line circuit from damaging external voltages.

Ringing is connected from a central ringing supply to operate the telephone bell.

Supervision refers to monitoring the on-hook/off-hook status of the line.

Coding is required in digital line circuits to convert the analog signal to a PCM bit stream.

Hybrids are required in digital line circuits to convert between the four-wire switching network and the two-wire cable pair.

Testing access is provided so an external test system can obtain access to the cable pair for trouble isolation.

In digital central offices, line circuits are contained on plug-in cards. Because much of the cost of the machine is embedded in the line circuits, shelves are installed, but to defer the investment, line cards are added only as needed. In analog central offices, line circuits are less expensive because they omit the analog to PCM conversion and the two- to four-wire conversion. Analog line circuits are permanently wired in frames that are connected to a distributing frame for cross-connection to the cable pairs.

Trunk Circuits

Trunk circuits interface the signaling protocols of interoffice trunks to the internal protocols of the switching machine. For example, in an SPC office, a trunk is seized by an order from the central control to a trunk distributing circuit in the trunk frame. This seizure causes the trunk circuit to connect battery to the M lead toward the carrier system. When a trunk is seized incoming, the ground on the E lead is passed from the trunk circuit to a scanner that informs the controller of the seizure.

Digital central offices interface analog trunks using external trunk circuits just as analog offices do. Digital trunks are interfaced without the use of trunk circuits, however. Digital offices have interface circuitry that provides the 1.544 mb/s bit stream direct access to the switching network.

Service Circuits

All types of switching systems require circuits that are used momentarily in routing and establishing connections. The hardware used to provide these circuits is discussed under common equipment in Chapter 14. In this section the applications are discussed so the reader will understand how the services are obtained and applied in all types of switching systems.

Ringing and Call Progress Tone Supplies

All switching machines that interface end users require 20 Hz ringing supplies generating approximately 90 volts to ring telephone bells. In addition, audible ringing supplies, busy tones operating at 60 interruptions per minute (IPM), and reorder operating at 120 IPM are required. In digital central offices, tones are generally digitally created in software.

Recorded Announcements

Recorded announcements are used to provide explicit information to the user when calls cannot be completed and tone signals are insufficient to explain the cause. For example, recorded announcements are used to intercept calls to disconnected telephone numbers. When a transfer of calls is required, the machine routes the incoming call to an intercept operator or automatic intercept announcement equipment, but otherwise calls to nonworking numbers are routed to a recorder. Announcements are also used on long-distance circuits to indicate temporary circuit or equipment overloads. In many cases, these are preceded by a two-tone code so automatic service observing equipment can collect statistics on ineffective dialing attempts.

Permanent Signal Tones

A *permanent signal* occurs when a line circuit is off hook because of trouble or because the user has left the receiver off hook. Permanent signals in trunk circuits occur because of equipment malfunctions or maintenance actions. Combinations of loud tones and recorded announcements are used on most line switching machines to alert the user to hang up the phone. Permanent trunk signals are indicated by interrupting the SF tone at 120 IPM, which flashes the supervision lights attached to E and M leads in some signaling apparatus.

Testing Circuits

Testing circuits are contained in all end offices and many PBXs to provide testing access to subscriber lines. These circuits connect the tip and ring of the line to a test trunk to allow a test position access to both the cable facility and, to a limited degree, the central office equipment.

All tandem switching machines and most end offices and PBXs also include trunk-testing circuitry to make transmission and supervision measurements on central office trunks. These circuits vary from 1,004 Hz tone supplies that can be dialed from telephones served by the switching machine, to trunk-testing circuits that enable two-way transmission and supervision measurements on trunks. Trunk-testing apparatus is covered in Chapter 24.

ACCESS TO THE LOCAL NETWORK

Until AT&T's divestiture of its Bell Operating Companies (BOCs) on January 1, 1984, the telephone network was designed for single ownership. Divestiture, intended to open long-distance telephone service to competition, has far-reaching effects that should be understood by telecommunications users. This section discusses the architectures of local and long-distance telephone networks and how the interexchange carriers (IECs) obtain local access.

Predivestiture Network Architecture

Until the mid-1970s, AT&T had a monopoly on switched long-distance telephone service. The local network up to that time was owned by the 22 BOCs and some 1,500 independent telephone companies (ICs). The local telephone companies furnished all local telephone service, most intrastate long-distance service, and a limited amount of interstate long-distance service. Most interstate service was furnished by AT&T's Long Lines division, which was renamed AT&T Communications after divestiture.

The network was segregated into five classes of switching machines and their interconnecting trunks as shown in Figure 9.2. Class 5 central offices were owned and operated by the BOCs and the ICs, a situation that remains unchanged. Ownership of Class 4 and higher offices was established by agreement between the parties; many of these machines were jointly owned.

The Class 5 offices are connected to higher class offices by toll connecting trunks. The higher class offices are interconnected by *intertoll trunks.* The toll connecting trunks are owned by the local telephone company; intertoll trunks are owned by AT&T Communications, the BOCs, and the ICs as negotiated by the parties and as regulated by the FCC and the courts.

Beginnings of Competition in the Toll Network

In 1978, the FCC decided to permit other common carriers to offer switched long-distance telephone service in competition with AT&T. A key issue in the decision was how the carriers would obtain access to the telephone user through the Class 5 office. Because the network had been designed for single ownership, it was impossible to give other carriers access similar to AT&T's without extensive redesign of the local central offices. AT&T's Long Lines circuits terminated on trunk ports in the central offices as did circuits to ICs. This trunk-side access was not suitable for multiple common carriers, however, because local central offices were not designed for customer-directed access to a carrier. Switching machines were designed to route only on the basis of destination, not on the basis of a customer-selected access code.

The BOCs filed tariffs to provide IECs other than AT&T access to the local network through line-side terminations in the central office. These tariffs were called the Exchange Network Facilities Interconnecting Arrangement (ENFIA) and have subsequently been replaced by Feature Group tariffs that are discussed in a later section.

The line-side access to the local switching machines offered under ENFIA A tariffs allowed customers of the IECs to access the network by the arrangement shown in Figure 9.13. Compared to the trunk-side access offered to AT&T, line-side access has the following disadvantages:

- Access to the IEC's network is achieved by dialing a seven-digit local telephone number. The IEC provides second dial tone to indicate when its switching and recording equipment is attached. Access to AT&T's network is by dialing "1."

FIGURE 9.13
Feature Group A and D Access to Local Telephone Network

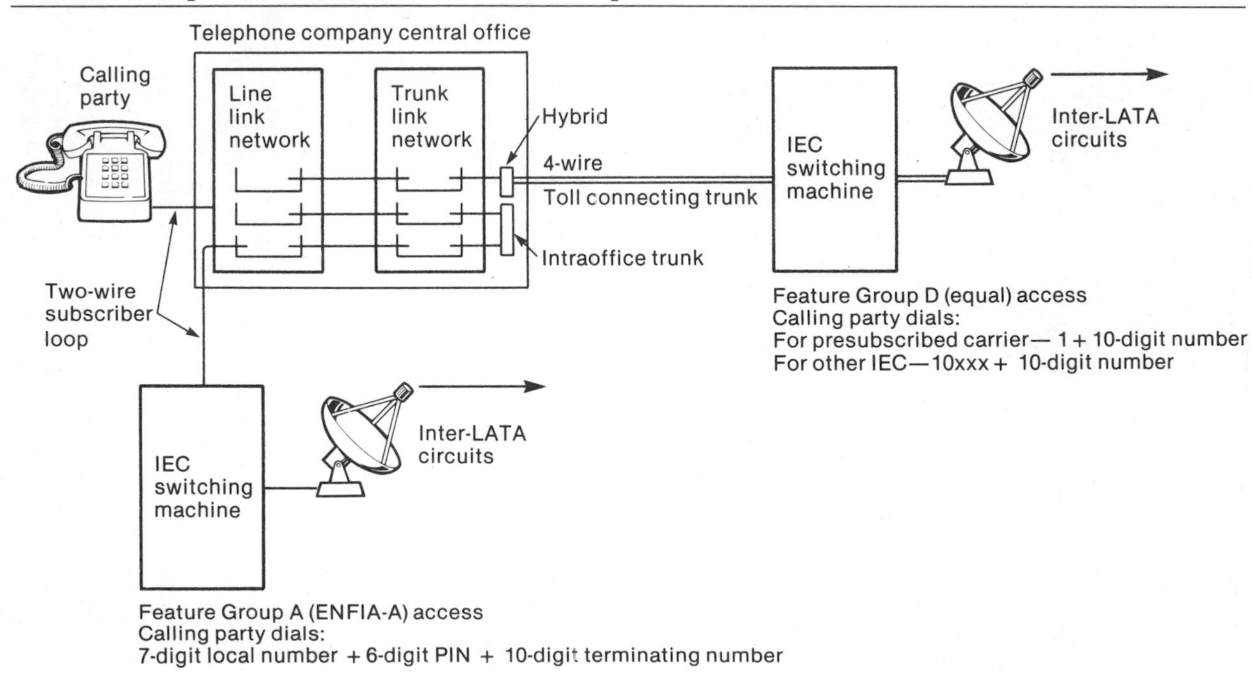

- Automatic number identification, a feature that is discussed in Chapter 10, is possible only over trunk-side connections. IEC users dial a personal identification number (PIN) for calling party billing identification. Rotary dials cannot be used for dialing PIN numbers; only DTMF dial pulses can pass through the class 5 office to the IEC's switching equipment.
- Line-side connections are inherently two-wire and provide poorer echo performance than the four-wire terminations of trunk side access.
- Answer and disconnect supervision are not provided over line side connections. Call timing can be determined only by monitoring the originating party's holding time.
- Multifrequency signaling, standard on trunks, is not available on line-side connections.

The BOCs filed two more ENFIA tariffs prior to divestiture to improve access. These tariffs, ENFIA B and C (now combined as Feature Group B), offer trunk-side access to the local switching machine and most of the features of AT&T's access. The most notable exception to equal access under Feature Group B is that single-digit access is provided only to AT&T's network. Feature Group B uses the code 950-10XX for access to the IEC's network, where XX is a two-digit code identifying the IEC.

Equal Access

In AT&T's agreement with the Department of Justice, the BOCs are required to provide access substantially equal to that given AT&T by September 1986, except where it is technically and economically infeasible to do so. Equal access, called "Feature Group D," gives all IECs access to the trunk-side of local switching machines. Users presubscribe to service from a preferred IEC and obtain access to that IEC by dialing "1." Other IECs are accessed by dialing 10XXX where XXX is a nationwide access code to the particular carrier. IECs choosing equal access receive automatic number identification, eliminating the need to dial a personal identification number.

Equal access requires intelligence in the Class 5 office to route the call to the required IEC trunk group. This requires changes to the generic programs in SPC offices. Electromechanical offices require extensive redesign to provide equal access. As many of these are being replaced, they will probably not be converted.

Many BOCs are providing equal access through a tandem switching machine as shown in Figure 9.14. An equal access tandem registers the dialed digits from the end office and, on the basis of originating telephone number, routes the call to the selected IEC. When equal access tandems are provided, IEC trunks interfacing electromechanical central offices terminate on the tandem rather than the end office. The equal access dialing plan is listed in Table 9.1.

Local Access Transport Areas

Under the terms of the agreement between AT&T and the Department of Justice, the BOCs are prohibited from transporting long-distance traffic outside geographical boundaries called *Local Access Transport Areas* (LATAs). LATA boundaries correspond roughly to Standard Metropolitan Statistical Areas defined by the Office of Management and Budget. Traffic crossing LATA boundaries cannot be transported by the BOCs. Traffic within LATA boundaries is regulated by state utilities commissions and varies with jurisdictions. Intra-LATA equal access may be authorized in some states, and in others may be assigned exclusively to the local telephone company. Where the local telephone company is assigned dial one access within the LATA, other carriers can be accessed only by dialing 10XXX.

STANDARDS

Switching machine standards are covered separately by type of machine in Chapters 10, 11, and 12. Also, refer to the standards section of Chapter 2 for standards on the transmission performance of networks.

FIGURE 9.14
Equal Access through Telephone Company Provided Tandem

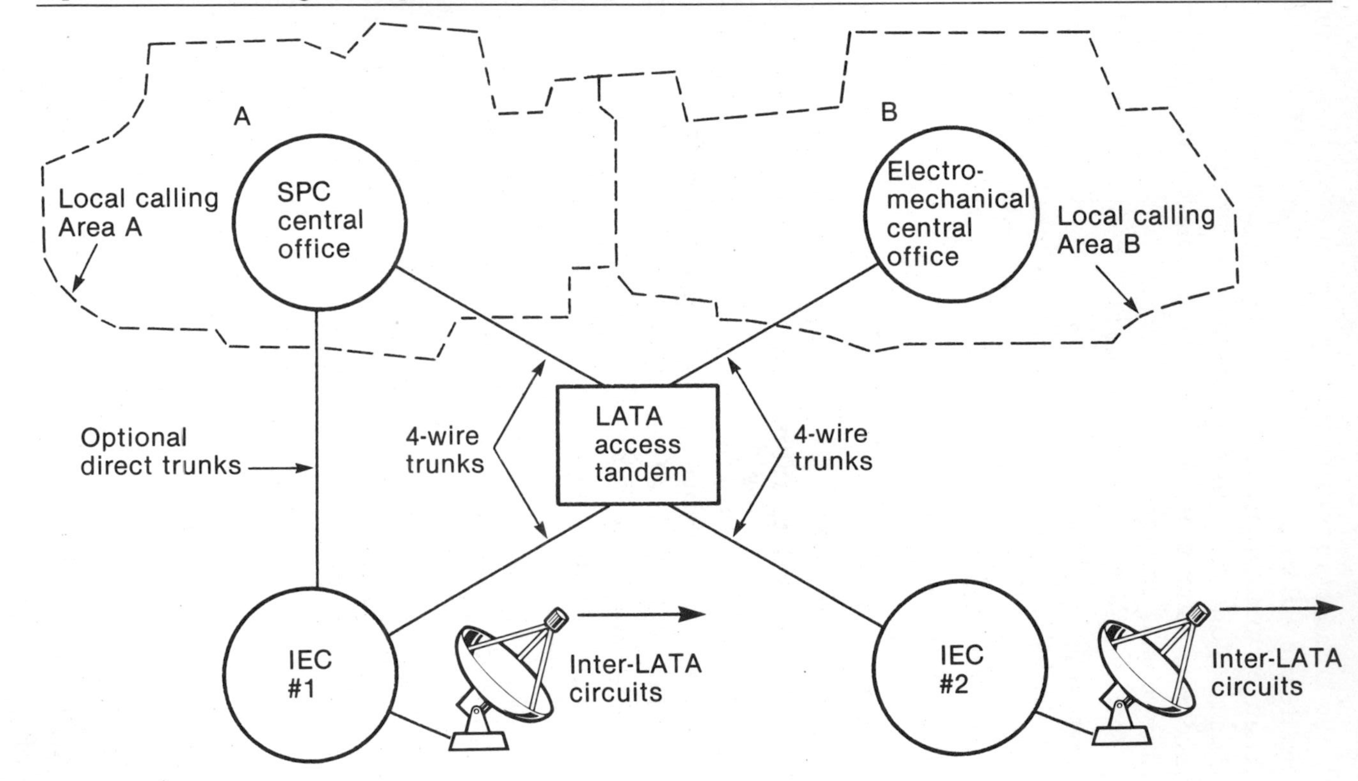

TABLE 9.1
Dialing Plan under Equal Access

Access	*Dialing Plan*	*Total Digits*
Local	NNX-XXXX	7
FGD Toll—Intra NPA (presubscribed)	1-NNX-XXXX	8
FGD Toll—Intra NPA (other carrier)	10XXX NNX-XXXX	12
FGD Toll—Inter NPA (presubscribed)	1-N0/1X NNX-XXXX	11
FGD Toll—Inter NPA (other carrier)	10XXX N0/1X NNX-XXXX	15
FGA Toll—Intra NPA	NNX-XXXX (DT) XXXXXX NNX-XXXX	20
FGA Toll—Inter NPA	NNX-XXXX (DT) XXXXXX N0/1X NNX-XXXX	23

Legend:

FGA = Feature group A

FGD = Feature group D

0/1 = Digits "0" or "1"

N = Any digit from 2 to 9 or 0

X = Any digit from 0 to 9

10XXX = Carrier access number

XXXXXX = Personal identification number (PIN)

(DT) = Dial tone

ANSI

ANSI/IEEE 312 Definitions of Terms for Communication Switching.

APPLICATIONS

Applications of switching apparatus are covered separately by type of machine in Chapters 10, 11, and 12.

GLOSSARY

Alternate routing: The ability of a switching machine to establish a path to another machine over more than one circuit group.

Automatic message accounting (AMA): Equipment that registers the details of chargeable calls and enters them on a storage medium for processing by an off-line center.

Call progress tones: Tones such as busy, reorder and audible ring that indicate the progress of a call to the user.

Common control switching: A switching system that uses shared equipment to establish, monitor, and disconnect paths through the network. The equipment is called into the connection to perform a function and then released to serve other users.

Crossbar: A type of switching system that uses a centrally controlled matrix switching network consisting of electromechanical switches connecting horizontal and vertical paths to establish a path through the network.

Direct control switching: A system in which the switching path is established directly through the network by dial pulses without use of central control.

Direct trunks: Trunks dedicated exclusively to traffic between the terminating offices.

Equal access: A central office feature that allows all interexchange carriers to have access to the trunk side of the switching network in an end office.

Generic program: The operating system in a SPC central office that contains logic for call processing functions.

High-usage groups: Trunk groups established between two switching machines to serve as the first choice path between the machines and thus handle the bulk of the traffic.

*Independent company (IC):*A non-Bell telephone company.

Interexchange carrier (IEC): A common carrier that provides long-distance service between LATAs.

Intertoll trunks: Trunks interconnecting Class 4 and higher switching machines in the AT&T network.

Local access transport area (LATA): The geographical boundaries within which Bell Operating Companies are permitted to offer long-distance traffic.

Marker: The logic circuitry in a crossbar central office that controls call processing functions.

Node: The switching machine or computer that provides access to the network and serves as the concentration point for trunks.

Permanent signal: A permanent seizure of a line or trunk in a switching machine.

Personal identification number (PIN): A billing identification number dialed by the user to enable the switching machine to identify the calling party.

Space division switching: Interconnecting links in a network by the physical connection of the links.

Step by step: A direct control central office that uses Strowger switches to establish a talking path through the network.

Stored program control (SPC): A common control central office that uses a central processor under direction of a generic program for call processing.

Tandem trunks: Trunks between an end office and a tandem switching machine or between tandem switching machines to provide alternate routing capability when direct trunks are occupied.

Tandem switch: A switching machine that interconnects local or toll trunks to other switching machines.

Time division switching: The connection of two circuits in a network by assigning them to the same time slot on a common bus.

Translation: The process of analyzing dialed digits and using them to select a route or equipment features based on information contained in the machine's data base.

BIBLIOGRAPHY

American Telephone & Telegraph Co. *Notes on the Network.* 1980.

Bellamy, John C. *Digital Telephony.* New York: John Wiley & Sons, 1982.

Briley, Bruce E. *Introduction to Telephone Switching.* Reading, Mass.: Addison-Wesley, 1983.

Freeman, Roger L. *Telecommunication System Engineering.* New York: John Wiley & Sons, 1980.

Hobbs, Marvin. *Modern Communications Switching Systems,* 2d ed. Blue Ridge Summit, Penn.: Tab Books, 1981.

Sharma, Roshan La.; Paulo T. deSousa; and Ashok D. Inglè. *Network Systems.* New York: Van Nostrand Reinhold, 1982.

CHAPTER TEN

Local Switching Systems

The switching machines discussed in this chapter are predominately used by telephone companies, although some large organizations use modified local central office machines for PBXs and tandem switching applications. The key to the application of a local switch lies in the features provided in its generic program. Hardware differences between local switches, PBXs, and tandem switches are not significant. Some manufacturers produce machines that are used in all three applications with software changes and minor hardware variations.

The field of local switching systems is broad; so much so that some simplification is needed to condense it to a single chapter. This chapter will focus primarily on digital central offices (DCOs) because they are the technology that will last into the foreseeable future. Where substantial differences exist between DCOs and their analog electronic counterparts, the differences will be mentioned but will not be covered in depth because the future of analog switching is predominately one of growth additions to existing machines. The features covered in this chapter are available in most DCOs. Some features may be available in electromechanical offices also, but because of the obsolescence of this technology, the differences are not discussed in this chapter.

Central office switching machines are occasionally used as PBXs or tandem switches. With the right software, a local CO can function in any of the three applications. The primary differences between these applications are:

- **Line circuits.** Tandem switches have few or no line circuits. PBX line circuits often omit some of the BORSCHT functions, have less loop range, and use less expensive technology than CO line circuits.
- **Trunk circuits.** CO and tandem switch trunk interfaces must meet identical requirements. PBX trunk interfaces are built for private network applications and have a narrower range of features than CO trunks.
- **Maintenance features.** Local COs include several features for subscriber line testing and maintenance. These features are usually omitted from PBXs and tandem switching machines. Trunk maintenance features are usually more sophisticated in tandem switches than either local COs or PBXs. Both tandem and local switching machines have administrative, self-diagnostic, and internal maintenance features that exceed the capability of all but the most sophisticated PBXs.
- **Capacity.** With some exceptions, local COs have greater capacity than all but the largest PBXs. The switching network capacity of local and tandem switches is generally equivalent except for the No. 4 ESS tandem switches used in the AT&T Communications network.

DIGITAL CENTRAL OFFICE TECHNOLOGY

DCOs are divided into two categories; community dial offices (CDOs) that are used for unattended operation serving up to about 10,000 lines and COs designed for urban applications up to about 60,000 lines. The distinction between these categories is not absolute. Both use similar technology, but CDOs use an architecture that limits their line size. When this capacity is exceeded, the machine must be replaced. Urban COs, on the other hand, have a size that is limited primarily by the calling rate of the users. If the calling rate is low, some DCOs can terminate as many as 100,000 lines. With a high calling rate, the central processor or switching network limit the capacity to fewer lines. Capacity considerations are discussed in Chapter 13.

Switching machines are enclosed in cabinets or are mounted in relay racks. CDOs tend to be cabinetized, and larger COs are usually rack mounted. A small machine can be installed with little labor, although skilled personnel and special equipment are required to install and test the machine. The interconnecting circuits in cabinetized machines are prewired, and installation involves cabling the machine to protector and distributing frames, and to power and alarm equipment. Large COs are installed in floor-mounted relay racks that are interconnected with plug-ended or wire-wrapped cables. The major components of a DCO are shown in Figure 10.1.

Distributed Processing

The earliest SPC switches used a central processor for all call-processing functions. With the advent of microprocessors, many SPC systems now use distributed processing, in which a central processor is linked to distributed microprocessors over a data bus. The major call-processing functions, such as marking a path through the switching network, are still controlled by the central processor. Functions that require no access to the machine's data base, such as line scanning and supervision, digit reception, and ringing, are controlled by processors located in the line switch units.

SPC Central Office Memory Units

DCOs require three types of memory. The *generic program* provided by the manufacturer is common to all switching machines of the same type. It resides in program store memory and directs call processing. *Parameters* also reside in a program store data base and are unique to the particular CO. They are used by the generic program to determine the quantities and addresses of peripheral equipment. The data from

FIGURE 10.1
Major Components of a Digital Central Office

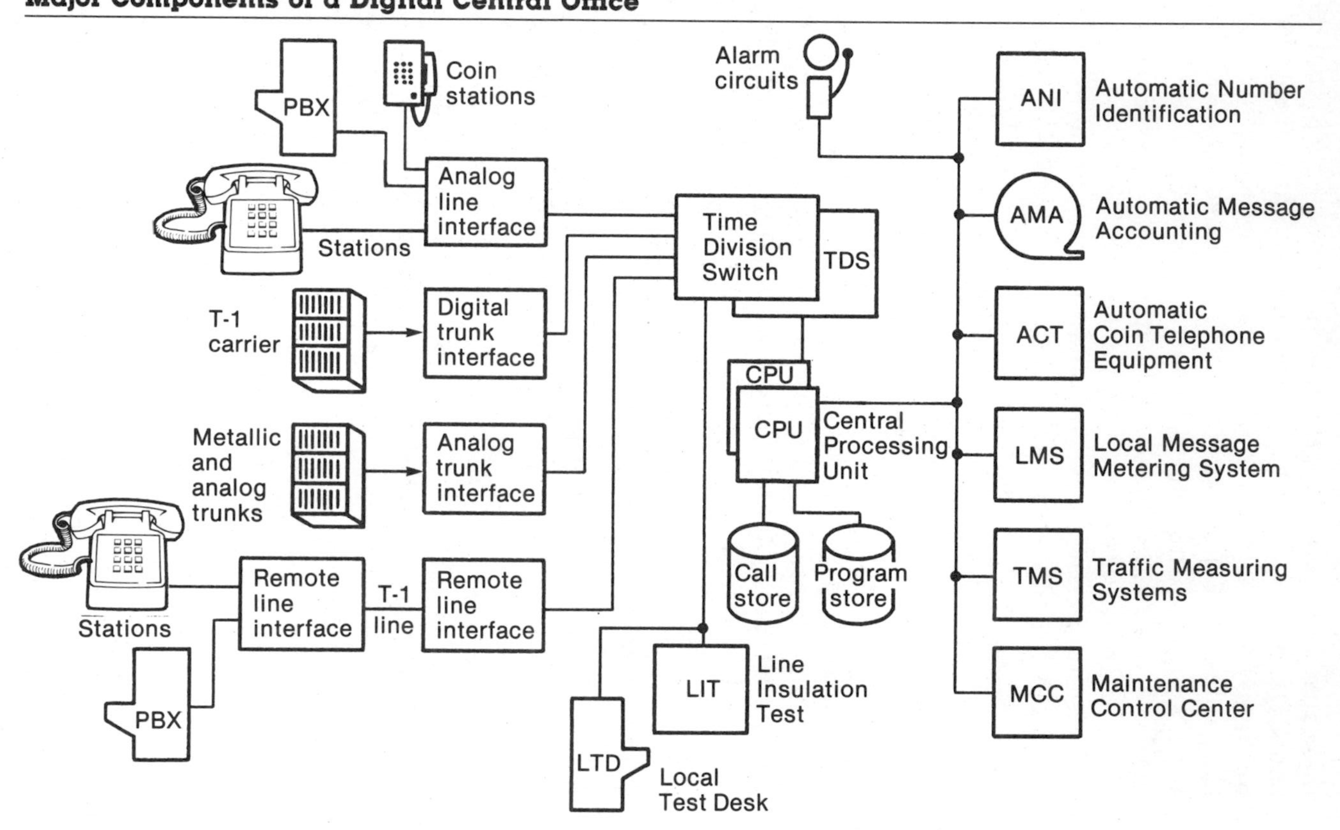

which parameters are developed is engineered at the time the application is designed and remains constant until the office characteristics are changed or major equipment items are added.

Translations are unique to the office and are created by the user or the manufacturer to enable the generic program to identify working lines and trunks, to determine the features associated with lines and trunks, and to provide trunk routing information for interoffice calls. Translations are also stored in the program's semipermanent data base.

Each line is assigned a record in the line translation memory. The line translation includes information about each user such as the following:

- Class of service (i.e., one-party, two-party, or four-party service).
- Telephone number associated with the line.
- Optional features such as call waiting, call forwarding, three-way calling, DTMF, etc.
- Status of the line (i.e., working, temporarily disconnected, out of service, etc.).

Trunk translations identify signaling and terminating characteristics of the trunk such as:

- Method of pulsing (i.e., dial pulse or multifrequency).
- Terminating office identity.
- Type of signaling on trunk (i.e., loop, E and M).
- Class of trunk (i.e., local, toll, service circuit, etc.).

The third type of memory is the *call store,* which is a temporary memory that the program uses to store details of calls in progress. Temporary memory is also used to store *recent change* information, which is line and trunk translations that have been added to the system but that have not yet been merged with the semipermanent data base.

Redundancy

Local DCOs are designed to provide reliability in the order of one hour's outage in 40 years of operation. This is not to imply that individual users experience that degree of reliability because individual component failures can disrupt individual lines without failure of the entire switching machine. Redundancy of critical circuit elements is required to provide this degree of reliability.

All local DCOs have duplicate central processors. In addition, other circuit elements that can cause significant outage, such as scanners and signal distributors, are duplicated. Redundant switching networks are economically feasible and often provided in digital machines. In analog machines the high cost and low probability of total network failure make redundant switching networks unnecessary. Other than network redundancy, the degree of duplication is similar between digital and analog machines.

Redundancy is provided on one of three bases:

- **Shared load** redundancy provides identical elements that divide the total load. When one element fails, the others can provide service to all users, but during heavy loads a poorer grade of service will be provided.
- With **synchronous** redundancy, both regular and duplicate elements perform the same functions in synchronism with each other, but only one element is on-line. If the on-line element fails, the standby unit accepts the load with no loss of service. Either unit is capable of carrying the entire office load alone.
- With **hot standby** redundancy, one unit is on-line with the other waiting with power applied, but in an idle condition. When the regular unit fails, the standby unit is switched on-line with some momentary loss of service.

The critical service questions posed by the type of redundancy are the degree to which the machine diagnoses its own problems and initiates a transfer to standby, and the degree to which a transfer results in loss of service. With the first two forms of redundancy, little or no loss of in-progress calls should be experienced. With the third form of redundancy, calls that have been connected are usually unaffected by a failure, but calls being established are usually lost and must be redialed. When a large number of calls in progress are lost, a number of immediate reattempts can be expected with the possibility of temporary processor overload.

Maintenance and Administrative Features

DCOs contain numerous features designed to monitor the machine's health from a colocated maintenance control center (MCC) and from a remote location. These features enable the machine to respond automatically to abnormal conditions. These features can be classified as fault detection and correction, essential service and overload control, trunk and line maintenance features, and data base integrity checks.

Fault Detection and Correction

The central processor continually monitors all peripheral equipment to detect irregularities. When a peripheral fails to respond correctly, the processor signals an alarm condition to the MCC and switches to a duplicated element if one is provided. The MCC interfaces the fault-detecting routines of the generic program to maintenance personnel. At the MCC, the central processor communicates its actions with messages on a CRT, printer, or both. Depending on the degree of sophistication in the program, the machine may register the fault indication or may diagnose the fault down to a list of suspected circuit cards. For unattended operation, the fault information is transmitted to a central location over a data link.

The processor also monitors its own operation through a series of built-in diagnostic routines. If irregularities are detected, the on-line processor calls in the standby and goes off-line. All such actions to obtain a working configuration of equipment can be initiated manually from the MCC or from a remote center. The ultimate maintenance action, which can be caused by an inadvertently damaged data base or a program loop, is a restart or initialization. Initialization of the machine is usually initiated only manually because it involves total loss of calls in progress and loss of recent change information.

Essential Service and Overload Control

Switching machines are designed for traffic loads that occur on the highest normal business days of the year. Occasionally, peaks higher than normal yearly peaks occur. Heavy calling loads can occur during unusual storms, political disorders, and other such catastrophes. During these peaks the switching machine may be overloaded to the point that service is delayed or denied to quantities of users. COs are designed with "line load control" circuitry that makes it possible to deny service to nonessential users so that essential users such as public safety and government can continue to place calls.

COs may also be equipped with features that control overloads in the trunk network. These features are discussed in Chapter 12.

Trunk Maintenance Features

Local COs contain varying degrees of trunk maintenance capability. The system monitors trunk connections in progress to detect momentary interruptions or failures to connect. For example, if a trunk fails during outpulsing, the unexpected off hook from the far end causes dialing to abort. The failure is registered by the system and entered on a trunk irregularity report, which is used to locate trunk trouble. The system marks defective trunks out of service and lists them on a trunk out-of-service list. When all trunks in a carrier system fail, the system detects a carrier group alarm, marks the trunks out of service in memory, and through its alarm system reports the failure to the MCC.

DCOs are also equipped with apparatus for off-line trunk diagnosis. Trunk test systems, which are discussed in Chapter 12, interface with distant COs to measure transmission performance and to check trunk supervision.

Line Maintenance Features

Switching machines contain circuits to detect irregularities in station equipment and outside plant. Like trunk tests, these are made on a routine or per call basis. On each call many machines monitor the line for excessive external voltage (foreign EMF), which indicates cable trouble.

Line insulation tests (LIT) are made routinely during low-usage periods and are intended to detect incipient trouble. The LIT progresses through lines in the office on a pre-programmed basis and measures them for foreign voltage or low insulation resistance, which is low resistance between tip and ring or from each side of the pair to ground. These tests detect outside plant troubles such as wet cable, and terminal, drop wire, and protector troubles.

Data Base Integrity

Changes to a switching machine's data base of line, trunk, and parameter translations are made only after checks to ensure the accuracy of the input record and to assure that existing records will not be damaged. These checks are particularly important with remote data base input from asynchronous lines that lack error-checking capability. Update of the data base may be allowed only from authorized input devices, and then with password control to ensure that only qualified personnel are given access to the files. A copy of the data base may also be kept off-line in disk storage so it can be reinserted if the primary file is damaged or destroyed. The manner of assuring data base integrity varies with the manufacturer and is an important consideration in evaluating local switching systems.

Line Equipment Features

Line switch frames in DCOs are constructed modularly with a number of line cards concentrated into a smaller number of links to the switching network. The ratio of lines to links is the *concentration ratio* of the office. Two different architectures are employed in the line interface as shown in Figure 10.2. In the coder/decoder (*codec*) per line architecture, a separate analog-to-digital converter is contained in each line card. The 64 kb/s bit streams from multiple cards are combined in a multiplexer into a high-speed bit stream and routed to the switching matrix. In the shared codec architecture, the output of the analog line circuits is switched to a group of shared codecs. The former method, while more expensive, reduces the service impact of a single codec failure. The line card also provides testing access to the local test desk through a relay that transfers the cable pair to a testing circuit.

The integrated circuits used in DCO line cards are highly sensitive to foreign voltage and may be damaged if not pro-

FIGURE 10.2
Digital Central Office Line Circuit Architecture

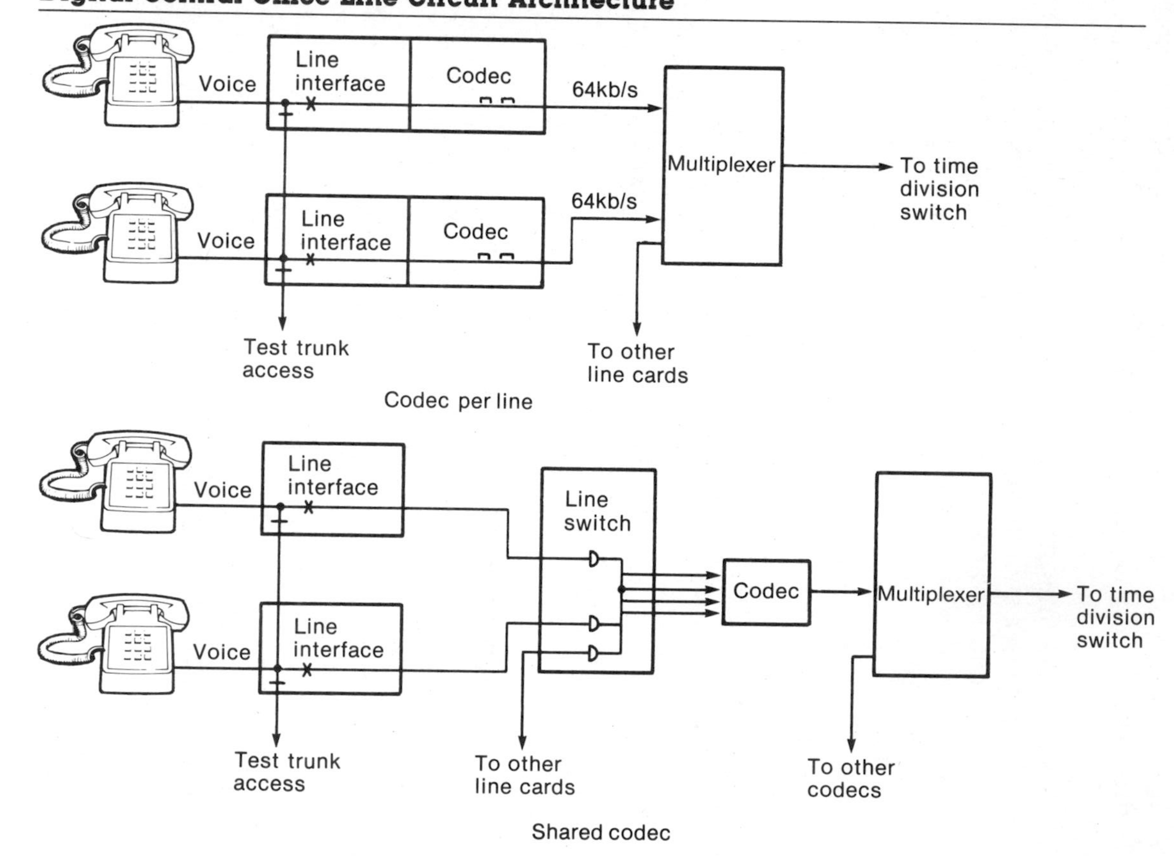

tected with gas tubes and clamping diodes as discussed in Chapter 7. The manufacturer's protection recommendations should be followed.

Transmission Performance

Because every line circuit in a digital switching machine contains a hybrid circuit, line circuit balance is of particular importance in a DCO. The variability of outside cable plant makes difficult the design of an economical line circuit to balance a wide range of cable pairs. Some manufacturers compensate by designing loss into the line circuit hybrid. The addition of loss to the line circuit is undesirable because it degrades transmission performance.

Distributed Switching

Most DCOs have the capacity to distribute switching to a remote location close to a cluster of users. As most line switches have a concentration ratio of four or more to one, distributed switching reduces the quantity of circuits needed between the CO and the users. Use of a remote line switch reduces the need for range extension and gain devices because the line circuit is moved close to the user and linked to the CO with low- or zero-loss trunks.

Remote line equipment comes in two general forms as illustrated in Figure 10.3. The first, a remote line switch, contains a switching matrix. Calls within the module are switched through an intracalling link. A single circuit to the CO supervises the connection. The second form, a remote line module, contains no intracalling features. A call between two stations within the module requires two CO links. If the "umbilical" to the CO is disrupted, calling within a line switch is still possible. A remote line module can neither place nor receive calls when the umbilical is inoperative.

Subscriber carrier, as described in Chapter 7, can also be interfaced to digital COs. Subscriber carrier differs from a remote line module in that carrier does not include concentration unless it is connected to a central office carrier terminal. The quantity of lines served at the remote location is equal to the number of trunks from the remote to the CO. Digital COs are capable of direct interface to digital subscriber carrier without use of a CO terminal.

FIGURE 10.3

Digital Remote Line Equipment

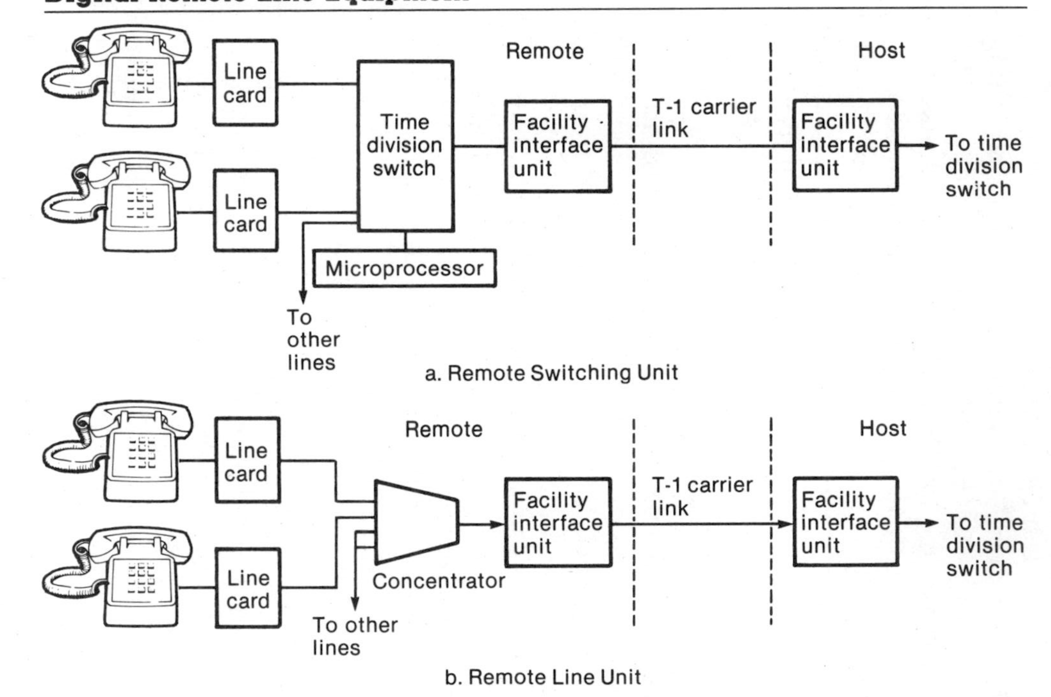

a. Remote Switching Unit

b. Remote Line Unit

Trunk Equipment Features

DCOs are designed for trunk compatibility with all kinds of pulsing and signaling employed in the trunk side of the network. Both analog and digital trunk interfaces are available. Analog trunk interfaces, used with trunks derived over FDM carrier and metallic facilities, are mounted in shelves similar to line interface mountings. Digital trunks interface the switching network in groups of 24 trunks through a digital trunk interface circuit that requires no channel banks. Special trunks are provided to interface service circuits such as directory assistance, repair service, local test desk, and disconnected number intercept.

Compatibility with CCIS or CCITT Signaling System No. 7 (SS7) is optional. Bell Communications Research is currently developing a specification for applying SS7 to Bell operating telephone company networks including local COs. Deployment is expected to begin in the second half of 1986.

LOCAL CENTRAL OFFICE EQUIPMENT FEATURES

Local COs contain a number of peripheral equipment units that facilitate maintenance and special software features. The most significant of these features are listed below.

Alarm and Trouble Indicating Systems

As COs have progressed, alarm systems have evolved from simple visual and audible alarms to present-day systems that include internal diagnostics. In early switching machines, trouble was located by alarm lights located on the ends of equipment frames and on the individual equipment shelf. Crossbar offices also punch alarm indications into trouble cards that provide information on status of the circuits in use when an alarm occurs.

SPC offices have internal diagnostic capability; the degree of sophistication varies with the manufacturer. Trouble indications are registered on a console and printed on a teletypewriter. The console is used to send orders to the machine to switch equipment, make circuits busy, and other such actions designed to diagnose trouble or obtain a working configuration of equipment. Some machines carry the diagnostic capability down to directing which circuit card should be replaced to clear the trouble.

Automatic Number Identification (ANI)

ANI automatically identifies the calling party for billing purposes. Single-party lines are identified from their line circuit or telephone number equipment. In electronic offices, this is a table lookup function. In electromechanical offices, separate equipment is required to translate the billed telephone number from the line equipment. On two-party lines the ANI equipment determines which party to bill by determining whether the station ringer is wired from the tip or the ring of the line to ground.

ANI equipment is interrogated by automatic message accounting equipment to determine the identity of the calling party. Where automatic identification is not provided, as with most four-party lines, the switching machine bridges an operator on the line to receive the calling party's number and key it into the AMA equipment.

Automatic Message Accounting (AMA)

AMA equipment records call details at each stage of a connection. The calling and called party numbers are registered initially. An answer entry registers the time of connection, and the terminating entry registers the time of disconnect. These entries are linked by a common identifying number to distinguish them on the storage medium from other calls.

The storage medium is paper tape (now obsolete), magnetic tape, magnetic disc, or solid state memory. Tapes are sent to distant data processing centers for assembling the entries into completed messages. AMA systems using disk or solid state storage transmit the call details to the processing center over a data link. Fragments of incomplete calls are analyzed for administrative purposes and fraud detection. AMA equipment is also used to register local measured service billing details.

Coin Telephone Interface

As described in Chapter 6, the ownership of coin telephones is being opened to private access, in which case all coin functions are contained within the telephone itself. In the majority of coin telephones, however, CO equipment is required to control the flow of coins. Three classes of coin operation are in use in the United States today:

- **Dial tone first.** This class offers dial tone to the user without a coin so calls can be placed to operators and emergency numbers. When a call is placed to a chargeable number without coin deposit, the CO signals the user that coins are required to complete the call.
- **Prepay or coin first.** This class requires a coin deposit before dial tone is supplied. Calls to the operator and emergency numbers require a coin, which is returned at the end of the call.
- **Postpay.** The CO supplies dial tone to the phone, and calls can be placed to any number without a coin. However, the coin must be deposited before the parties can talk. This system is widely used in Europe. Its use was disappearing in the United States, but it is being reintroduced with customer-owned coin telephones that are not connected to central office coin control equipment.

Coin telephones are assigned to a separate class of service that provides access to special coin trunks. The coin trunk is designed for one of the three classes of coin operation. Its function is to send coin collect and return signals to operate relays in the telephone to route the coin to the collection box or to the return chute. In an SPC office the processor controls the coin trunk; in an electromechanical office the coin trunk contains the logic to supply dial tone and to collect and return coins.

For long-distance calls the coin interacts with the operator or automatic coin equipment to forward tones corresponding to the value of the coins deposited. The operator controls the collecting and refunding of coins with switchboard keys that control the coin trunk. This gives the operator the ability to collect coins before the call is placed and to refund them if the call is not completed. This is done by holding the coins in the coin phone and collecting them only when the called party answers.

Common Equipment

Chapter 14 is devoted to a discussion of equipment common to all types of CO equipment. This section describes how common equipment items are used to provide external interfaces for the CO.

Power equipment consists of a commercial AC entrance facility, a backup emergency generator, a −48 volt DC storage battery string, and battery charging equipment. Emergency generators are provided in all metropolitan COs. CDOs may be wired with an external plug to couple to a portable generator. In either case, the battery string is designed with sufficient capacity to operate the office until emergency power generators can assume the load during a power failure.

The battery plant is connected to the switching equipment with heavy copper or aluminum bus bars. Where higher voltages are required, separate battery strings are provided in some offices, but in most applications, DC converters are used to raise the battery voltage to the higher voltage required by the equipment.

Trunks, subscriber lines, loop electronic equipment, and other miscellaneous CO equipment are wired to terminal strips mounted on protection and distributing frames. Distributing frames provide access points to the transmission and signaling leads and are used for flexible assignment of equipment to user services and to trunks. Equipment is interconnected by running cross-connect wire or "jumpers" between terminals.

Local Measured Service (LMS)

Many telephone companies base their local service rates on usage. Flat rate calling is usually available, with LMS as an optional service class. With the LMS class of service, calls inside the local calling area are billed by number of calls, time of day, duration of the call, and distance between the parties. The method is similar to long-distance billing except that calls are usually bulk billed with no individual call detail.

In electronic and crossbar offices equipped with AMA equipment, LMS calls are usually registered on the AMA record, separated from long-distance calls by an identifying mark. In offices lacking AMA equipment, auxiliary registers are wired to subscriber lines.

Traffic Measuring Equipment

All COs are sized on the basis of usage. For example, in step-by-step offices the number of switches in a line group and the number of users that can be assigned to a group are determined by collecting statistical information. In common control and SPC offices the concentration ratio and quantities of trunks and service circuits are also based on usage.

Usage information is based on the number of times a trunk or circuit is seized (attempts) and the average holding time of each attempt. This information is collected by attaching two registers to the equipment: one that measures the number of times the circuit is seized and one that measures elapsed time the circuit is busy. The method of evaluating this information is discussed in Chapter 13. Electromechanical offices require external traffic usage registers; SPC offices use software registers. The registers are periodically unloaded to a processing center for summary and analysis.

Most COs also provide dial tone speed registers, which are devices that are attached to line circuits, periodically go off hook, and measure the number of times dial tone is delayed more than three seconds. Dial tone speed is an important measure of the quality of local switching service.

Traffic measuring equipment is capable of providing a variety of other data for administering the CO. For example, data base statistical information is provided to update the availability of vacant lines and trunks. Point-to-point data collection enables the machine administrator to detect the calling pattern of various subscriber lines to improve the utilization of the network by reassigning lines to different network terminations. Subscriber line usage (SLU) measurements enable the administrator to distribute heavy usage lines among different terminations on the line switch frame to avoid overloading the line switch with several heavy users. Service measurements are made by the machine to determine key indicators such as *incoming matching loss*, which is the failure of an incoming call to obtain an idle path through the network to the terminating location.

Network Management

Many local offices contain network management provisions to prevent overloads. For example, *dynamic overload control* automatically changes routing tables to reroute traffic when the primary route is overloaded. Trunks can automatically be made busy to a congested central office. This feature enables the blocked system to take recovery action without being overwhelmed by ineffective attempts from a distant CO.

Local Central Office Service Features

Local COs provide a variety of service features that enhance call processing. In SPC offices these are provided primarily by software. In electromechanical COs they are sometimes provided by hardware, in some cases by auxiliary equipment, and in other cases they are not provided at all. The following is a brief description of the principal features provided by most electronic end offices:

Call-Processing Features

Call-processing features allow the customer to recall the CO equipment by a momentary on-hook flash. The machine responds with "stutter" dial tone to indicate that it is ready to receive the information. Each of these features is optional to the user, generally at extra cost.

Three-way calling is a feature that allows the user to add a third party to the conversation by momentarily holding the other party while a third number is dialed. **Call transfer** is a similar feature that is available only on Centrex-equipped lines. The operation is exactly like three-way calling except the originating party can hang up, leaving the other two parties in conversation.

If a line equipped with **call waiting** is busy, the switching machine sends a tone signal to indicate that another call is waiting. The user can place the original call on hold and talk to the waiting call by flashing the switch hook. **Speed calling** is different from the other call-processing features in that a switch hook flash is not used. This feature gives the user the ability to dial other numbers with a one- or two-digit number.

Call forwarding enables the user to forward incoming calls to another telephone number. While this feature is activated, calls to the user's number are automatically routed by the CO to the alternate telephone number.

Centrex Features

Centrex is a PBX-like service furnished by telephone companies through equipment located in the CO. In most cases, the switching equipment is a partition in the end office. Centrex features allow **direct inward dialing** (DID) to a telephone number and **direct outward dialing** (DOD) from a number without operator intervention. For calls into the Centrex, the service is equivalent to individual line service. Outgoing calls differ from individual line service only by the requirement to dial an outgoing access code, usually "9." Calls between stations in the Centrex group require four or five digits instead of the seven digits for ordinary calls.

An attendant position located on the customer's premises is linked to the CO over a separate circuit. Centrex service provides PBX features without locating a switching machine on the user's premises. Centrex features are similar to the PBX features that are discussed in more detail in Chapter 11.

Emergency Reporting

In the United States, the code 911 is dedicated to reporting to fire, police, ambulance, and other emergency numbers. The local CO switches a 911 call to a trunk to a 911 center. Several desirable 911 features can be provided by SPC offices, but they are impractical or impossible with electromechanical offices. For example, often the jurisdictional boundaries of the 911 bureau do not coincide with the CO exchange boundary. If the SPC data base has been so constructed, the CO can route calls to the proper 911 bureau. Another valuable feature is the ability of the CO to identify and report the calling telephone number to the 911 bureau in case the caller becomes confused during an emergency and is unable to give his or her address.

Multiline Hunt

This feature, often called *rotary line group,* connects incoming calls to an idle line from a group of lines allocated to a user. In older COs, the numbers had to be in sequence; in electronic offices, any group of numbers can be linked by software into a multiline hunt group.

STANDARDS

Few local switching machine standards have been set by standards agencies. Subscriber line and trunk interface standards are published by EIA; the bulk of DCO performance criteria is a matter of matching the manufacturer's specifications to the user's requirements. A comprehensive list of requirements is published by Bell Communications Research on behalf of the seven Bell regional operating companies. This publication, *Local Switching System General Requirements* (LSSGR), defines the features and technical specifications required of local COs used by the Bell regions. LSSGR is not a standard; compliance by manufacturers is voluntary. Nevertheless, it is the most comprehensive publication available for local CO operation.

Bell Communications Research

CB 119 Interconnection Specification for Digital Cross-Connects.

CB 123 Digroup Terminal and Digital Interface Frame Technical Reference and Compatibility Specifications.

CB 137 Local Digital Switching System Transmission Performance Standards.

CB 151 General-Purpose Central Office Concentrator (GPCOC) Service Network Interface Specification.

CB 154 Specifications for Special Information Tones (SIT) for Encoding Recorded Announcements.

PUB 48501 Local Switching System General Requirements (LSSGR).

PUB 4850A LSSGR Update.

PUB 53300 Stored Program Control (SPC) Software Documentation Standards.

PUB 55025 Central Office and No. 5 ESS Gas Tube Protective Devices.

PUB 61201 Compatibility Information for Feature Group D Switched Access Service.

CCITT

E.180 Characteristics of the dial tone, ringing tone, busy tone, congestion tone, special information tone, and warning tone.

G.705 Integrated services digital network (ISDN).

Q.9 Vocabulary of switching and signaling terms.

Q.35 Characteristics of the dial tone, ringing tone, busy tone, congestion tone, special information tone, and warning tone.

EIA

RS-464 Private Branch Exchange (PBX) Switching Equipment for Voiceband Applications.

RS-470 Telephone Instruments with Loop Signaling for Voiceband Applications.

APPLICATIONS

DCOs are applied almost exclusively by telephone companies. The selection of a digital switch is a complex process that is a function of the companies' maintenance strategy and service offerings as well as cost objectives. This section briefly describes the primary criteria used in selecting a DCO.

Evaluating Digital Switching Systems

The primary criteria in evaluating a DCO are cost, features, compatibility, maintenance features, and the ability of the office to provide the desired grade of service. The complexity of a DCO makes the provision of maintenance features of particular significance. A list of the primary DCO evaluation criteria follows.

Maintenance Features and Reliability

DCOs should be designed to provide a high degree of reliability through the use of duplicated critical circuit elements and high-grade components. All processors should be duplicated, preferably on a synchronous basis. The processor should be capable of full self-diagnosis of trouble. In case of overload, automatic recovery is essential, including discontinuing nonessential call-processing operations if necessary.

The machine should be capable of complete operation from a remote location with the exception of changing defective circuit cards and running cross connects. Manufacturers should provide an on-line technical assistance center to aid in solving unusual maintenance conditions.

The system should be fully documented with maintenance practices, trouble locating manuals, software operation manuals, office wiring diagrams, and a complete listing of data base information.

Transmission Performance

The CO should insert no more than 1 dB of loss into a line-to-line connection and 0.5 dB into a line-to-trunk or trunk-to-trunk connection.

Line Concentration

The machine should be capable of providing line concentration ratios to match the expected usage. Line concentration ratios as low as two to one may be required in heavy-usage machines, with ratios as high as eight to one in low-usage machines.

Environmental

The system should be capable of operation under the temperature extremes that could occur with power, heating, or air conditioning failure. The normal operating temperature range of most CO equipment is 0 to 50 degrees C. (32 to 120 degrees F.).

Multiclass Operation

A DCO should contain the software needed for all Class 5 office functions. In addition, some applications require Class 4 software. If the DCO is used as a PBX, it must include the PBX features described in Chapter 11. Of major significance in PBX use is the ability to interface a Class 5 CO through the line side rather than the trunk side of the machine.

Remote Line Capability

A DCO should include remote line capability to minimize range extension and subscriber loop costs. The remote should contain intracalling capability if this feature is required by the owner. The user will experience no difference in operation if a machine lacks intracalling capability unless the link to the CO is severed. This capacity is important when the remote switch is placed in a locale with a strong community of interest. The DCO should also be capable of interfacing subscriber carrier at the T1 level.

Capacity

The system should have the capacity to handle the traffic load for the expected life of the machine. The machine should also be capable of terminating the required number of lines and trunks. All central offices must have the capability of measuring and recording usage, whether with built-in or external equipment.

GLOSSARY

Automatic number identification (ANI): Circuitry in a switching machine that automatically identifies the calling party's telephone number on a billable call.

Call store: The temporary memory used in an SPC switching system to hold records of calls in progress and pending changes to permanent memory.

Capacity: The number of call attempts and busy hour load that a switching machine is capable of supporting.

Centrex: A class of central office service that provides the equivalent of PBX service from a telephone company switching machine.

Coder/decoder (codec): The analog-to-digital conversion circuitry in the line equipment of a digital CO.

Code conversion: The process of registering incoming digits from a line or trunk and converting them to a different code required for call routing.

Community dial office (CDO): A small CO designed for unattended operation in a community, usually limited to about 10,000 lines.

Concentration ratio: As applied to CO line equipment, it is the ratio between the number of lines in an equipment group and the number of links or trunks that can be accessed from the lines.

Digital central office (DCO): a local central office using PCM switching.

Distributed processing: The distribution of call-processing functions among numerous small processors rather than concentrating all functions in a single central processor.

Distributed switching: The capability to install CO line circuits close to the served subscribers and connect them over a smaller group of links or trunks to a CO that directly controls the operation of the remote unit.

Dynamic overload control (DOC): The ability of the translation and routing elements in a CO to adapt to changes in traffic load by rerouting traffic and blocking call attempts.

Line insulation test (LIT): CO equipment that is capable of automatically measuring resistance and voltage on subscriber lines for the purpose of detecting faults.

Maintenance Control Center (MCC): A center in an SPC CO where system configuration and trouble testing are controlled.

Multiline hunt: The ability of a switching machine to connect calls to another number in a group when other numbers in the group are busy.

Parameters: The record in an SPC office's data base that specifies equipment and software quantities, options, and addresses of peripheral equipment for use in call processing operations.

Program store: Permanent memory in an SPC CO that contains the machine's generic program, parameters, and translations.

Recent change: Changes to line and trunk translations in an SPC switching machine that have not been merged with the permanent data base.

Redundancy: The provision of more than one circuit element to assume call processing when the primary element fails.

Remote line switch: A line unit mounted near a cluster of users and equipped with intracalling capability.

Remote line unit: A line unit without intracalling capability mounted near a cluster of users.

BIBLIOGRAPHY

Briley, Bruce E. *Introduction to Telephone Switching.* Reading, Mass.: Addison-Wesley Publishing, 1983.

Freeman, Roger L. *Telecommunication System Engineering.* New York: John Wiley & Sons, 1980.

Hobbs, Marvin. *Modern Communications Switching Systems,* 2d ed. Blue Ridge Summit, Penn.: Tab Books, 1981.

Sharma, Roshan La.; Paulo T. deSousa; and Ashok D. Inglè. *Network Systems.* New York: Van Nostrand Reinhold, 1982.

MANUFACTURERS OF LOCAL SWITCHING SYSTEMS

AT&T Network Systems

CIT-Alcatel Inc.

Ericsson, Inc. Communications Division

GTE Communications Systems

ITT Telecommunications

NEC America, Inc.

Northern Telecom, Inc.

Stromberg Carlson Corporation

CHAPTER ELEVEN

Private Branch Exchanges

In the past, PBXs were designed and used almost exclusively for voice communication. Any data communications requirements within the organization were met by equipping a PBX line with a modem. Now the PBX is carving out a role for itself as the integrator of the automated office. Local area networks (see Chapter 19) also claim that distinction, but LANs are effective primarily in data handling; many modern PBXs are equally proficient at both data and voice. The driving force behind the integration of data and voice in the PBX is the proliferation of terminals needed in the automated office. In the automated office the telephone will be supplemented by a data terminal for sending and receiving messages, filing and retrieving data, creating and editing text, and supporting the office worker's productivity with computer aids such as schedulers, calendars, spreadsheets, and a multiplicity of other specialized tools.

The voice and data requirements of the automated office can be fulfilled in three ways: separate PBXs and LANs, a PBX that interfaces LANs, or a PBX that integrates the total voice and data requirements in a single machine. The telecommunications industry is unable to agree which approach is the most effective, and for good reason; the needs of organizations differ so widely that no single approach will meet the communication needs of all offices.

It is safe to assume that any organization large enough to justify a PBX needs voice communications capability; data communications requirements vary greatly with the organization. Where the need for voice communications predominates, the conventional solution of a PBX with modems for occasional data traffic is the least expensive alternative. Where the organization has data terminals that require frequent access to one another for heavy traffic loads, a separate LAN can probably be justified. If the organization has numerous lightly loaded terminals, the integrated voice/data PBX becomes attractive. One of the principal benefits of using the PBX to switch data is that management features such as class of service restriction, call detail recording, and least-cost routing can be extended to data terminals.

The PBX is rapidly becoming the precursor of future public network technology. The integrated services digital network (ISDN), which will be discussed in Chapter 25, is a concept of a future network that accommodates simultaneous voice and data transmission. The ISDN is evolving slowly in public networks because the costs of change are enormous and the needs are as yet undeveloped. However, ISDN concepts are emerging in the office. Digital PBXs are the present test bed of the future ISDN in an environment where experimentation is encouraged in the quest for improved productivity.

PBXs conform to the general purposes of the voice switching machines described in Chapters 9 and 10, but their architectures are disparate and their features vary widely. This chapter discusses features that are unique to PBXs, some ways that PBXs depart from conventional central office practice, and concludes with a discussion of factors for evaluating PBXs.

PBX TECHNOLOGY

PBX technology has progressed through three generations and is just entering the fourth. First-generation machines used wired solid state or relay logic and space division analog switching networks. The second generation introduced stored program control driving reed or PAM switching networks. The third generation, the first to facilitate data transmission, employs PCM switching technology. With stored program control, the PBX evolved from a mere call processor to a device that supports office management. For example, features to manage building energy are now included in some systems that were once limited to call processing.

The fourth generation of PBXs employs LAN-type architecture to provide a highway of paths for interconnecting the variety of terminals and mainframes found in the automated office as illustrated in Figure 11.1. Fourth-generation machines provide an interactive user interface that supports office automation functions as well as voice transmission. The user's terminal is a multipurpose work station that integrates voice, enhanced telephone features, and data applications into a single device. The fourth-generation machine is intended as a total office communication system. Unless it is applied as such, it will be found prohibitively expensive for voice alone.

Contrasts between PBXs and Local Central Offices

The first three generations of PBX used technology closely resembling that of central offices. However, local telephone service is a highly standardized product furnished by a switching machine that is designed for long life and low-cost operation within a carefully controlled environment of numbering plan, trunking plant, outside plant, and standard features. The modern PBX is intended for a specialized environment. The uniqueness of organizations suggests products that can be customized to the application. A standard user interface is of less concern in a PBX than in a public environment because within an organization, system operation can be taught and controlled. The distinguishing differences between a PBX and a local central office are:

- Because a PBX is intended for private use, it often omits service protection features that are mandatory in central offices. For example, battery power and processor redundancy may be omitted to save costs.

FIGURE 11.1
Conceptual Diagram of a Fourth Generation PBX

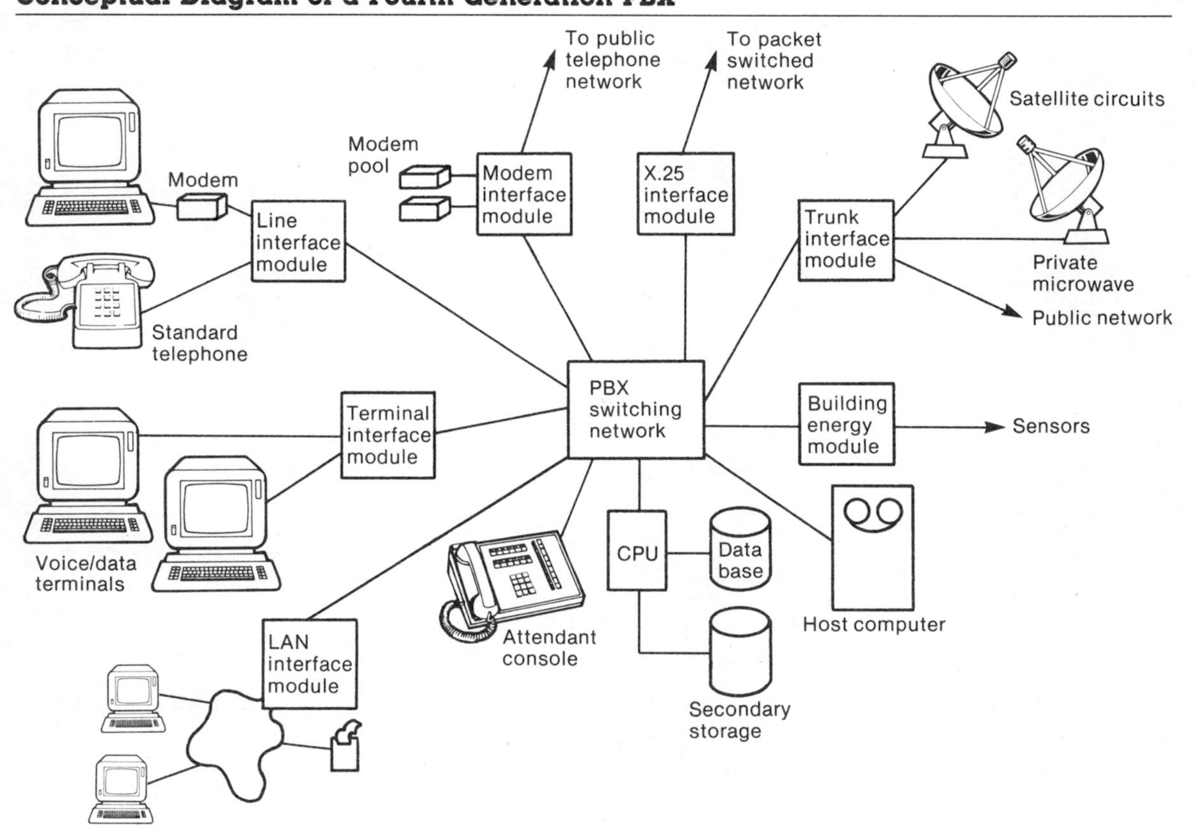

- PBXs are constructed to be economical in much smaller line sizes than central offices. Clusters of small users that would be handled by a remote line unit in central offices are served by small PBXs.
- PBXs are usually equipped with an attendant console for answering calls to the main telephone number. This feature is unnecessary in central offices except for those equipped for Centrex service.
- Central offices connect to the interexchange network with trunk-side connections to a higher class office. A PBX is designed to terminate on the line side of a Class 5 central office and therefore uses line signaling methods (loop or ground start and dial or DTMF signaling).
- PBXs are designed with more flexible numbering plans than central offices. Class 5 offices normally enforce seven-digit dialing. PBXs employ a flexible numbering system to suit the needs of the user.
- PBXs are increasingly equipped with voice/data integration features and with software to support office automation. They contain data communications features such as X.25 interface, digital addressing, and high-speed digital bus interface. These features are not provided with the current generation of local central offices, although extended data communications features are under development by most manufacturers.
- PBXs contain a provision for restricting stations from selected features such as off-net and tie trunk access. Central offices provide unrestricted access to other numbers.

Until recently, PBXs have used architectures similar to that used by central offices, that is, a switching network driven by SPC or electromechanical common control. Recently, however, PBXs have departed from traditional design and have begun to employ architectures more suited to integrating the multiple data terminals in the automated office in addition to voice. The design of these machines is modular. The switching elements are designed for remote operation, and the processing is distributed close to the end user.

Voice/Data Integration in the PBX

For PBXs to control the automated office, data transmission capability is required. Any machine can switch data from a modem-equipped line, but this does not answer the requirement of automated office control. Moreover, modem operation has several drawbacks such as:

- Many PBXs are incapable of passing data at speeds higher than 4,800 b/s.
- Call setup time in a conventional machine is excessive, sometimes exceeding the transmission time of the message.
- The costs of the modem and extra PBX line ports are high. Alternate voice/data capability can be used in some machines, but this takes the line out of service for voice while data is being transmitted and vice versa.
- The load imposed on the switching machine by data is much different from voice. Overload of the PBX is a hazard. See Figure 11.2 for differences in call attempts and holding time for voice and data.

The key to data transmission through a PBX lies in sizing the switching network and processor specifically for the organization's data traffic load and in providing line ports designed for data transmission. Three different techniques are common in the PBX market today for "simultaneous voice and data operation" and should be understood when a PBX is being evaluated for data transmission.

Data over voice (DOV) carrier applies a high-speed data signal, usually up to 19.2 kb/s, over existing voice wiring. Separate DOV modems are required at each end of the circuit. The data signal is applied to the analog data carrier and is assigned to a separate port on the PBX. While this method technically qualifies as simultaneous voice/data operation, it actually uses separate circuits that are the equivalent of the single channel subscriber carrier described in Chapter 7.

FIGURE 11.2
Comparison of Call Attempts and Holding Time for Voice and Data in a PBX

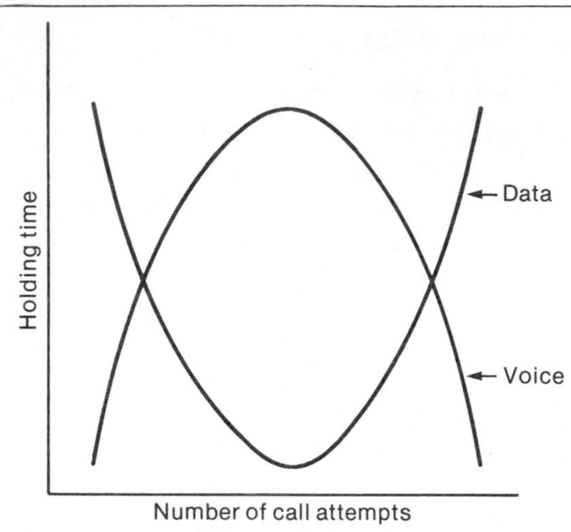

Separate voice and data pairs are used by some PBXs, requiring two to four pairs of wire to the station. The station set contains a line driver to couple the data signal to the data pairs. The voice pair carries an ordinary analog voice signal. Separate voice and data ports are required in the PBX. This system and DOV have the advantage of being compatible with ordinary inexpensive telephone sets.

Digitized voice and data PBXs employ a special station instrument to convert the voice to a 64 kb/s signal. The data signal is combined with the voice and linked to the PBX with one or two pairs. The signals are separated in the PBX line port. This system provides a good technical solution to the simultaneous voice/data problem but has the drawback of requiring an expensive telephone set. Where extensive use of data transmission is required, this factor is less important because the digital telephone set offsets the cost of an ordinary telephone plus a modem. Where data transmission is rarely required, this alternative is costly.

INTERFACES

PBXs are interfaced to local and interexchange telephone networks. In addition, many machines are equipped with special services such as WATS lines, tie lines, and foreign exchange trunks to local calling areas in distant cities. Machines serving automated office and data communications require interfaces to data terminals, gateway circuits, and other computer networks. The variety of interface circuits is one of the key distinguishing features between PBX generations.

Line Interfaces

PBXs often include key telephone-like features in their software and may employ line interface circuits that require more than one pair of wire for the station set. Also, integral data transmission features often require two- to four-pair station wiring. The station loop resistance range of a PBX is usually less than the 1,300- to 1,500-ohm range typical of a central office. Therefore, many PBXs are restricted to operation within a building or campus and must use special off-premise line circuits for longer loops, or must use remote line switch units to move the line circuits closer to the end user.

Voice-only PBXs have line interface circuits that resemble those used in digital central offices except that some CO features such as line testing access are usually omitted from PBX line circuits. Data PBX line circuits are of three types: separate voice and data ports, combined voice and data ports, or universal ports that can be used interchangeably for voice or data. The first two types of ports may result in unused port capacity if the PBX has the wrong ratio of data to voice stations.

Ping Pong

Full-duplex operation is achieved on some two-wire line circuits by using "ping pong" transmission. In this method the alternate ends of the circuit transmit bursts of data at twice the data transfer speed. For example, 64 kb/s transmission is obtained by the two ends alternately transmitting 128 kb/s bursts and rapidly switching between transmit and receive.

PBX Trunk Interfaces

Like central offices, PBXs interface the outside world through trunk circuits that exchange signals with other switching machines through a variety of signaling interfaces. Trunk and signaling compatibility are important considerations in selecting PBXs.

Many digital PBXs are capable of direct T-1 line interface with a local central office, but telephone company tariffs permitting T-1 interface to the central office are unavailable in many locations. It is important to remember that a PBX interfaces the line side of a local switching machine. Therefore, ground or loop start signaling is employed; the central office furnishes dial tone to the PBX when it is prepared to receive digits, and DTMF or dial pulse addressing is used. Most local central offices are not equipped for signaling and supervision over a T-1 line interface to a PBX.

Compatibility with central office line equipment is important to proper PBX operation. The interface standard is RS-464, published by EIA, which specifies technical and performance criteria for the interface between the two types of machines. The central office interfaces the PBX with the central office's own supervision; the supervision from a distant trunk is not transferred through the line circuit to the PBX.

PBXs require an access digit, usually "9," to connect station lines to central office trunks. When the user dials "9," the PBX seizes an idle central office trunk and connects the talking path through to the station if the station is permitted off-net dialing. The station hears central office dial tone as a signal to proceed with dialing.

Tie Trunks

Organizations operating multiple PBXs often link them through *tie trunks,* which are intermachine trunks terminating on the trunk side of the PBX. Trunk facilities may be privately owned or obtained from interexchange carriers. Signaling compatibility is a concern with tie trunks. The trunks normally use E and M signaling, but the trunk interface in the PBX may require the use of signaling converters.

If tie trunks terminate in a single location, they are accessed by dialing a digit (usually "8"), which connects them to the distant PBX. Dialing in the distant machine follows that machine's dialing plan. Many multimachine organizations have separate dialing plans for each machine and a single organizationwide dialing plan. The PBX is then programmed to provide the translations needed to reach the distant number over the tie trunk network.

Transmission is of concern in designing a tie trunk network. Transmission loss is designed according to the via net loss or fixed loss plans as discussed in Chapter 2.

Special Trunks

Many PBXs have a variety of special trunks to provide access to lower cost long-distance service such as WATS lines, foreign exchange, and 800 numbers. When several of these special trunks are connected to the PBX, it is impractical to expect the users to determine which class of trunk to use. Many PBXs offer *least-cost routing* (LCR) features to enable the machine to determine the most economical route based on the class of trunks terminated, their busy/idle status, and the station line class.

Gateways and External Interfaces

All PBXs provide interfaces to public voice networks. In addition, some PBXs offer gateways to public data networks such as Tymnet and Telenet. These gateways are usually implemented with an X.25 interface (see Chapter 3) or a proprietary equivalent. This interface allows a terminal to address an off-net station over a digital circuit. The protocol conversions are handled between the PBX and the public data network.

Interfaces to local area networks (LANs) are also provided by some PBXs as shown in Figure 11.3. Token rings and contention bus interfaces are available with some PBXs. See Chapter 19 for a description of LAN technologies. With a LAN interface, data terminals on the LAN can communicate with one another as usual. They also can gain access to terminals on the PBX, or can use the PBX as a gateway to public telephone or data networks. Through the LAN interface the PBX emulates a terminal on the LAN and communicates with its protocol so a terminal on the PBX can have access to terminals on the LAN.

FIGURE 11.3
PBX to LAN Interface

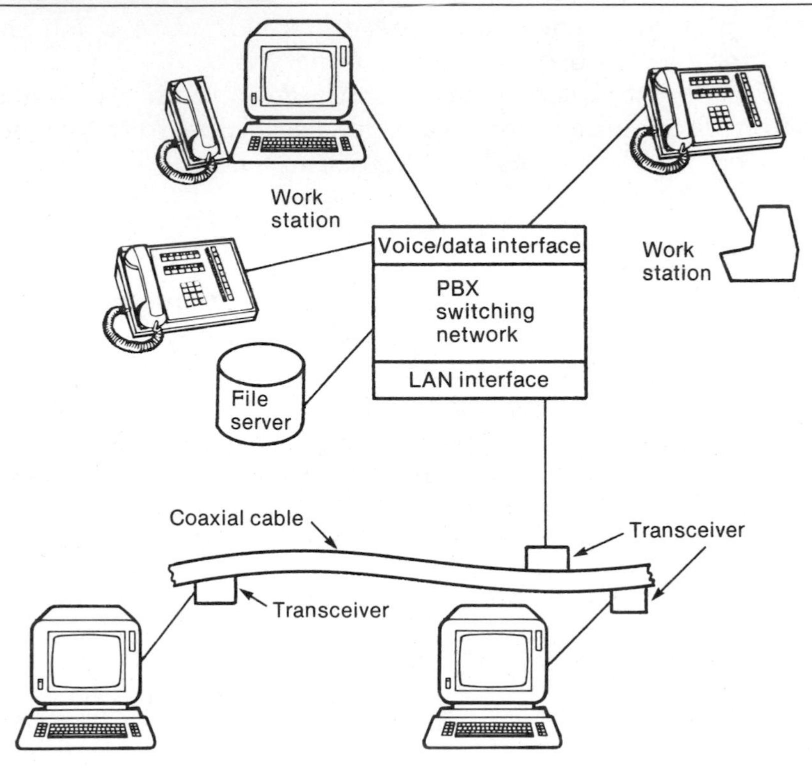

Interfaces to mainframe computers are also provided by some PBXs. For example, a PBX may be capable of emulating an IBM 3270 terminal and can support access by noncompatible terminals into a mainframe cluster controller.

Modem Pooling

Digital PBXs offer station-to-station data communication without a modem. However, modems are required for data communications outside the PBX. Many systems provide a pool of modems supporting a range of speeds and protocols. Modems are accessed by dialing a PBX station number. After the modem is attached, it is used like a dedicated modem. One important feature in a modem pool is the ability of the machine to hold users in queue until a modem is free.

Code, Speed, and Protocol Conversion

The greatest impediment to data communication is the lack of standard codes, speeds, and protocols. Some PBXs are capable of interfacing dissimilar terminals. The system automatically adjusts to the terminal, either after a short log-on sequence or from the machine's line translations. The machine likewise adjusts to the requirements of the called station, enabling dissimilar terminals to communicate through the system.

PRINCIPAL PBX FEATURES

Features common between PBXs and central offices are not repeated in this chapter. Refer to Chapter 10 for a discussion of voice switched network features not discussed here. Many features that are impractical or unnecessary in a local telephone network are both feasible and desirable in the private environment of a PBX. These are discussed in this section.

Least-Cost Routing (LCR)

Most PBXs terminate a combination of public switched and private trunks on the machine. For example, in addition to the long-distance service provided by the presubscribed interexchange carrier, the PBX may terminate WATS lines, foreign exchange lines, tie trunks to another PBX, and possibly access lines connected directly to an IEC. Educating users about which service to use is a difficult task, particularly as rates vary with time of day and terminating location and the dialing plan varies with the carrier called. However, it is a relatively simple matter to program route selection into the PBX's central processor. With LCR the user dials the number, and the machine determines the least expensive route and dials the digits to complete the call over the appropriate trunk group.

Many machines also provide a warning tone when calls are about to be routed over an expensive service so the user can hang up before the call is completed. Complete flexibility in route selection is a highly desirable PBX feature and is available on most stored program controlled machines.

Station Message Detail Recording (SMDR)

This feature provides the equivalent of a detailed toll statement for PBX users. If the users have access to a common group of trunks to the local CO, the telephone company provides a toll statement listed by local telephone number. The calling PBX station number is not identified to the local CO, so such a toll statement will not provide the identity of the calling station. Most businesses require station message detail to control long-distance usage and to spread costs among the user departments. The SMDR provides this capacity. An SMDR is limited by the fact that answer supervision is not returned over a local telephone loop. Therefore, the SMDR is unable to determine whether the called station answered or not. The determination of called station answer is based on the amount of time the calling station is off hook, or by monitoring for voice frequency energy in the talking path. Because of the lack of answer supervision the SMDR is unable to balance precisely with central office billing detail. However, it is accurate enough for most organizations.

Voice Store and Forward (Voice Mail)

Voice mail is available as an optional feature in some PBXs and can be added as an auxiliary service to most PBXs. When a station is busy or unattended, the caller can leave a message, which is stored in secondary storage, usually digitally. The station user can dial an access and identification code to retrieve the message. Voice mail is the analog equivalent of electronic mail, a service that allows a terminal-equipped station to leave a data message for another station.

Some voice mail systems include an option that is the functional equivalent of direct inward dialing. Incoming calls are greeted with an announcement that invites them to dial the extension number if it is known, to dial a number for extension information, or to stay on the line for an attendant. This feature saves money by deloading the attendant.

PBX Voice Features

As all PBXs are designed for voice switching service, they have a number of features intended for the convenience and productivity of the users. Not all the features listed below are universally available, and many machines provide features not listed. This list briefly describes the most popular voice features found in PBXs.

- The PBX can be equipped with **paging trunks** that are accessed by an attendant or dialed from a station. An option is zone paging, which allows the attendant to page in specific locations rather than the entire building.
- **Integrated key telephone** system features such as those described in Chapter 6 can be integrated into the PBX software and accessed from a nonkey telephone. Features such as call pickup, call hold, music on hold, and dial-up conferencing can be provided.
- **Camp-on** is a feature that allows an attendant to queue an incoming call to a busy station. When the station becomes idle, the call is automatically completed.
- The caller's name or **telephone number can be displayed** on special telephones equipped with an alphanumeric readout.
- **Special dialing** features are provided in many machines. In addition to ordinary digit dialing, some machines can interface a work station keyboard for addressing by a mnemonic or name.
- **Distinctive ringing** is provided by many machines so trunk calls can be distinguished from intramachine calls.
- The **executive override** feature allows a station to interrupt a busy line or preempt a long-distance trunk if the station class is higher than the class of the user.
- **Trunk answer any station** allows stations to answer incoming trunks when the attendant station is unoccupied or busy.

Attendant Features

Attendant consoles are provided in most PBXs for incoming call answer and supervision. Direct station selection (DSS) allows the attendant to call any station by pressing an illuminated button associated with the line. This feature is usually available only in small PBXs. Automatic timed reminders alert the attendant when a called line has not answered within a prescribed time. The attendant can also act as a central information source for directory and call assistance. Attendant controlled conferencing is available for multiport conference calls. Centralized attendant service (CAS) is a feature provided on many machines to enable a central group of attendants to perform attendant functions for remote PBXs.

Automatic Call Distributors

An automatic call distributor (ACD) is a device combining the features of PBXs and tandem switches for concentrating incoming lines or trunks to a group of service positions. An ACD is either a stand alone system or a partition in a tandem switch or PBX. Typical applications are the service positions of any large organization such as airlines, utilities, catalog houses, and department stores.

Incoming calls are routed to an attendant position on the basis of incoming line type. For example, airlines typically screen calls by providing one line to the public and a different number to travel agents. When attendant positions are idle, the call is routed immediately. If all positions are occupied, the calls are placed in queue and the caller is notified by recorded announcement or music that the call is being held.

ACDs are similar in architecture to PBXs except that they are designed only for incoming calls. ACDs are often designed for multiple-machine operation so calls can be transferred to a distant location during off-peak hours. This requires careful transmission design to prevent excessive loss or singing on trunks interconnecting the machines.

The attendant console for ACDs is usually linked to some form of data base in addition to the voice communications circuits. Thc most effective operation enables the operator to obtain access to a data base with the same console used for answering incoming calls.

Office Automation Features

The advent of the automated office has prompted an expansion of the PBX to provide a variety of features. Most organizations large enough to use a PBX today have some form of data communications between stations or over access lines outside the PBX. Such communications use the PBX to switch modem-equipped lines. In this application the PBX is transparent to data transmission. In the automated office, however, data communication becomes a substantial purpose of the PBX. Many station users are equipped with a work station that either supplements the ordinary telephone or integrates the telephone with a video display terminal.

The method of providing data communications over the PBX is a key factor in distinguishing among the alternatives. In some PBXs, data is an adjunct added to a voice PBX; in others, data is integrated into the design. Unlike a voice PBX, a PBX with integrated data features is interactive with the terminal. Standard physical interfaces such as RS-232-C and RS-449 are provided between the telephone and a desktop computer, or the telephone and terminal may be combined as a single instrument.

PBXs offer a wide range of data speeds through their networks. A PBX with an analog reed or electromechanical network places little limitation on the data speed because the transmission path is connected through the machine and is transparent to the underlying machine operations. PCM networks are capable of passing data up to 64 kb/s, provided the line port allows direct access to the switching network. Most digital PBXs provide high-speed data ports that allow transmission at common data speeds such as 19.2, 56, and 64 kb/s.

Some modern PBXs provide higher data transmission speeds by allowing direct access to a high-speed bus. In some machines, this is accomplished by replacing a group of line ports and giving the data terminal access to the combined bit stream. In other machines, line circuits are bypassed completely and access is provided directly to the bus.

Office automation features supported by many PBXs include:

- Electronic mail—a service that enables users to leave messages and send letters or documents from terminals to a central data file for later retrieval by the recipient.
- Electronic filing—a service that enables users to store

records in secondary storage in the PBX. This service usually includes file search and data retrieval capability.

- Interface between incompatible terminals may include the ability to communicate between incompatible word processors. For example, IBM and Wang word processors could exchange files through such an interface if supported by the PBX.

PBX STANDARDS

Few standards have been established for PBXs. The interface between a PBX and its serving local central office has been standardized by EIA, and trunk interfaces follow accepted industry practices for signaling and electrical interface. PBX features are generally accepted in the industry, but the method of actuating the feature is left to the manufacturer. Lines interfacing analog telephones follow accepted industry signaling practice, but the loop resistance range is left to the manufacturer. Proprietary station interfaces are determined entirely by the manufacturer with little uniformity among products.

All PBXs require FCC registration of line and trunk terminations to public telephone networks.

EIA

RS-232-C Interface between data terminal equipment and data communication equipment employing serial binary data interchange.

RS-449 General-purpose 37-position and 9-position interface for data terminal equipment and data circuit-terminating equipment employing serial binary data interchange.

RS-464 Private branch exchange (PBX) switching equipment for voiceband applications.

RS-464-1 Addendum to RS-464.

RS-470 Telephone instruments with loop signaling for voiceband application.

RS-478 Multiline key telephone systems (KTS) for voiceband application.

FCC Rules and Regulations

Part 68 Connection of Terminal Equipment to the Telephone Network
Subpart C Registration Procedures
Subpart F Connectors

APPLICATIONS

A "typical" PBX application is difficult to define because the very nature of the product is in its diversity. A large office equipment manufacturer recently installed a network of PBXs that are interconnected with tie lines to serve several plants. This network is presented as a case study, not because it is typical but because it represents an unusually complete job of identifying requirements and selecting a product that best fulfilled their needs from the many alternatives on the market.

The ABC Corporation PBX Network[1]

The ABC Corporation is a large manufacturer of office products. Prior to 1984, service was furnished by Centrex supplied by the local telephone company. After an exhaustive search of alternatives, service was changed to eight Rolm CBX II PBXs serving about 3,500 users. The headquarters location is a VL (very large) model with M (medium) PBXs at the seven satellite locations.

At the time of cutover, tie lines between the locations were provided over analog facilities. In 1985, the tie lines were converted to DS-1 facilities so that the network is entirely digital except for local central office trunks.

Although this network is far more elaborate than the typical PBX installation, the planning, product selection, and introduction activities that went into developing the net-

[1]Company identity disguised.

work are representative of the kind of activities required in even the smallest PBX.

Planning began with an identification of company objectives for a system to replace Centrex. The primary motivation for the change was to obtain integrated voice and data capability over a facility that would insulate the company from future cost increases. The company is a heavy user of internal data communications, so it was essential to obtain a vehicle that would accommodate both voice and data, and that could perform the functions of a local area network until LAN standards were developed.

The company developed a detailed list of requirements that it sent to eight major PBX vendors who had the potential capability of meeting them. The requirements included the following:

- The system had to meet future growth plans of the company without major additions or replacements.
- A full range of features were required, including direct inward dialing, least-cost routing, automatic call distribution, and the standard station features commonly provided with modern PBXs.
- A system of recording message detail was required for control of long-distance costs and to charge back costs to the organizational unit incurring them.
- A network management and control center was required to oversee fault isolation and restoral in both the switching and facility portions of the network as well as for centralized administration of moves, changes, and features.
- A uniform numbering plan was required among all sites.
- The system had to be capable of providing an automatic telephone directory to keep the company directory current with moves and changes.
- A wide range of data features were required including:

    Full integration of voice and data between the office work station and the PBX.

Modemless operation.

PCM-based architecture.

Modem pooling.

Speed and code conversion.

X.25 interface to public data networks.

The company also felt that the choice of vendor was equally as important as the choice of the PBX itself. A key factor used in evaluating the vendor was its reputation in the marketplace for producing a quality product with a continuing commitment to research and development and customer service. Because this system was more complex than the typical PBX, the vendors were required to demonstrate design capability and experience with large multisite networks.

Although the company intended to perform its own system maintenance and administration to the maximum degree feasible, the vendor was required to demonstrate the reliability of its products and make a commitment to continuing support. Also, the vendor's financial stability was evaluated.

The PBX equipment itself was evaluated on how well it met the requirements, its basic architecture, its warranty, and how well it fit the company's business plan for growth and expansion. The feature list was evaluated for completeness and ease of use. The architecture of the machine was evaluated for its restoral capability and for its protection against future obsolescence. As with any such system, cost was evaluated, although cost was a secondary consideration compared to the company's perception of value.

Responses from the initial eight vendors were narrowed to three products that appeared to meet the basic requirements. Company managers met with the vendors to discuss issues identified during the preliminary evaluation phase and to listen to more detailed vendor presentations on their products. After the Rolm system was tentatively selected, company financial and legal staffs were brought into the discussions to aid in final negotiations and in developing a purchase contract.

The selection from among the three finalists was based on these factors:

- The use of a distributed architecture.
- The system's capacity and probability of blocking.
- The maturity of the software.
- The effectiveness of SMDR and usage reports.
- Expandability.
- The vendor's installation plan.
- Qualifications of personnel who would be assigned to the project.
- The vendor's plan for continuing support.

The first part of the project, the installation of the switching systems, was completed in 1984. The second phase, the installation of T-1 internodal links, is being completed in 1985. Also, a Teleresearch SMDR system will be installed in 1985. In the future, the company plans a new headquarters building, which will contain a second Rolm VL CBX II.

In evaluating the experience of this installation, company personnel recommend that organizations addressing the question of a new PBX should ensure that several key issues are addressed during the project. First, the extent of voice and data integration should be determined. This factor is crucial in determining the architecture needed, which in turn plays a large part in determining the cost of the system and determines whether central office-based systems such as Centrex are suitable.

Closely related to this is the question of how many electronic telephones will be required and who they will be assigned to. In integrated voice/data PBXs, a significant portion of the total cost is concentrated in electronic telephones, but these are not usually cost justified for all users.

The amount of capacity required is a third critical factor because of the tendency of data features to overload PBXs. The system must be purchased with the correct type of line and trunk ports to correspond to the type of instruments and trunk terminations that will be used. If the PBX is being installed in an existing network, it is essential to consider the cutover method. Existing trunks can be transferred to the new system, but they will be difficult to test. A more costly, but more satisfactory alternative is to use a paralleling trunk route so the entire network can be tested prior to cutover.

A detailed training plan is required in a conversion of this sort to ensure that the users know how to use the features. Otherwise, there is a definite risk of features falling into disuse. This is particularly true of features that are implemented by switchhook flash and dialing a code as opposed to features that are activated on an electronic telephone by pressing a button.

After the cutover, the training should be validated with a follow up visit to be sure the users are satisfied with the service and that they know how to operate the features. Finally, it is essential that the work of the users and the vendors be closely coordinated. In any PBX conversion, the work of the telephone company must be coordinated with the work of one or more equipment vendors. Weekly meetings are necessary to ensure that this work is closely coordinated and that the cutover progresses smoothly.

Evaluation Considerations

The uniqueness of every organization makes a universal PBX specification impractical. Considerations that are important in some applications will have no importance in others. Therefore, the user should weigh them accordingly. It is, of course, essential that the machine contain the required features and that it meet the requirements for the number of line and trunk terminations. Reliability and cost are implicit requirements in any telecommunications system and are not separately discussed.

Office Automation

The need for office automation features is a primary consideration in evaluating a PBX. If a large percentage of the stations will be terminal equipped during the system's life, an integrated voice/data PBX should be considered. If few work stations are required, the cost of an integrated PBX may be difficult to justify.

Automated office features such as electronic mail, electronic filing, and voice mail require storage in excess of the amounts provided for call processing. Third- and fourth-generation PBXs may provide auxiliary disk storage to accommodate these features. Organizations implementing an automated office may rely on the PBX for some automated office features such as the following:

- Voice mail (voice store and forward).
- Electronic mail and messaging.
- Multiport voice conferencing.
- Work station interface.

External Interfaces

Closely related to office automation features is the need for external interfaces. Every PBX must conform to the standard RS-464 interface to a local telephone central office and must be registered with the FCC for network connection. In addition, interfaces such as these should be considered:

- X.25 interfaces to packet switching networks.
- RS-232-C or RS-449 data set or work station interface.
- IBM 3270 or equivalent mainframe terminal interface.
- Protocol and format conversion.
- Interface to a local area networks.
- 1.544 mb/s interface to external trunk groups.

Terminal Interfaces

A key consideration in evaluating a PBX is the type of terminals it supports. All PBXs have, at a minimum, a two-wire station interface to a standard telephone. Ordinary telephones are the least expensive terminal, and because of the quantities involved in a large PBX, inability to use standard telephones can add significantly to the cost. The telephone falls short, however, as a terminal for the automated office. Voice/data terminals are of two types, a stand-alone terminal with an EIA interface or an integrated terminal that interacts with the PBX's processor to actuate its features. Integrated terminals are usually available only from the PBX vendor unless specifications have been released to others.

If a proprietary integrated terminal is the only type supported by the PBX, station costs may be excessive. If the organization makes extensive use of data communications, the cost of these work stations is not significant compared to the cost of modems. However, if many stations use only voice traffic, the inability to interface a standard telephone will prove costly.

Integrated terminals may be the only way of accessing some PBX features such as integrated key telephone features. Some of the more complex key telephone-like features tend to fall into disuse because of the difficulty of using them from a standard telephone. A proprietary terminal may facilitate feature operations such as call pickup and hold by using buttons to replace switch hook flashes and special codes required with standard telephones.

Consideration should be given to the different types of ports the PBX employs. Where the port type is restricted to voice or data, it may be impossible to use the full capacity of the machine. The most flexible ports are capable of handling either voice or data with no restriction on the type of terminal connected to the port. This, of course, requires that voice be digitized in the terminal.

These features should be considered in evaluating a PBX terminal interface:

- Proprietary or nonproprietary telephone interface.
- Number of conductors to the station.
- Availability of RS-232-C or RS-449 interface to the terminal.
- Station conductor loop range.
- Number of line ports required for voice/data operation.
- Where data encoding is performed (i.e., in the line interface circuit or telephone set).
- Dialing interface (i.e., keyboard, telephone, mnemonic).
- Integrated key telephone system features.

Switch Network

First- and second-generation PBXs use analog switching networks. All of the types of analog networks described in Chapter 9 are employed by early vintage machines. Third-generation machines predominately use electronic switching networks with pulse code modulation. However, the distinction among the generations is not universally accepted. Some PBXs may have switching networks that use pulse amplitude, pulse width, and delta modulation.

PCM switching is the most desirable in full-featured PBXs because of the ability to handle high-speed bit streams and because of the compatibility with digital channel banks. Pulse amplitude modulation, pulse width modulation, and delta modulation networks are equally satisfactory for voice. Although a modemless interface to these non-PCM networks is possible, they are limited in data speed to 4,800 b/s or less in many PBXs, whereas PCM can handle 56 or 64 kb/s data. Where modems are employed, little difference will be perceived by the user among the four types of networks, and the PBX choice can be based on other factors.

A key evaluation consideration is whether the network is blocking or nonblocking. With nonblocking networks, a further consideration is whether line concentration limits the number of simultaneous connections to fewer than the number of line and data ports on the machine.

Program Storage

One distinguishing feature among PBXs is the method used for storing the generic program and updates. If the program is stored in a read-only memory (ROM), it is invulnerable to power outages but can be changed only by changing the card containing the program chips or by "burning" a new program into the ROM. If the program is stored in random access memory (RAM), the program is lost whenever power is removed. With RAM memory, a backup program must be provided on tape, floppy disk, or hard disk. With a volatile memory, even a momentary power interruption will require reinitializing or restarting the program, which may be time consuming. Many PBXs make some provision for battery backup of the memory to retain translations during power failure conditions. An uninterruptable power supply can also be used to provide service continuity during power outages.

Environmental Considerations

Except for the largest machines, PBXs are cabinet mounted and connected to external circuits with distributing frames or wall-mounted backboards. Power supplies are self-contained; backup battery power may be provided or not, depending on the application.

Most PBXs are capable of operating in an ordinary office environment without air conditioning. However, the operating temperature range should be evaluated because many systems require an air-conditioned environment. In any case, adequate air circulation will be required.

Data Base Updates

The ease of changing class of service and telephone numbers is an important evaluation consideration. If these can be controlled by the attendant, it is possible to place, remove, and move stations, and to change restrictions without using a trained technician. With a fully flexible system, station jacks are wired to line ports. A new station is added by plugging in the telephone and actuating the line from an attendant or maintenance console.

Diagnostic Capability

The degree to which a PBX is capable of diagnosing its own trouble and directing a technician to the source of trouble is an important factor in controlling maintenance expense. It is also important that a machine be equipped with remote diagnostic capability so the manufacturer's technical assistance center can access the machine over a dial-up port.

Transmission

Many digital PBXs insert loss into line connections to improve echo performance. A line-to-line loss of four or five dB is acceptable under many conditions. However, if the system makes extensive use of tie lines, WATS lines, and other external trunks, transmission performance may be unacceptable to some stations, particularly those with long station loops. The total network should be designed before a PBX is chosen to determine whether insertion loss of the switching system is acceptable.

Station Wiring Limits

The station loop range of both proprietary terminals and ordinary telephones must be considered in evaluating a PBX.

Most proprietary terminals and those terminals requiring an EIA interface can operate only over restricted range. Range limits can often be extended in some systems by using distributed switching, which moves the line circuits close to the stations.

Bandwidth Requirements

Bandwidth is of concern with machines interfacing high-speed digital lines. Digital voice PBXs limit the data speed to 4,800 b/s or less. With special data ports, higher speeds are supported, including, in some machines, enough bandwidth to switch a 1.544 mb/s bit stream.

GLOSSARY

Computer branch exchange (CBX): A computer-controlled PBX.

Data over voice (DOV) carrier: A system of multiplexing a data channel on the same circuit as a voice channel to enable simultaneous voice/data operation.

Electronic mail: A service that allows text-form messages to be stored in a central file, and retrieved over a data terminal by dialing access and identification codes.

Integrated Services Digital Network (ISDN): A local network for transporting digital voice, data, and video services. The ISDN is expected gradually to replace the present telephone network.

Integrated voice/data: The combination of voice and data signals from a work station over a communication path to the PBX.

Least-cost routing (LCR): A PBX service feature that chooses the most economical route to a destination based on cost of the terminated services and time of day.

Modem pool: A centralized pool of modems accessed through a PBX to provide off-net data transmission from modemless terminals.

Ping pong: A method of obtaining full-duplex data transmission over a two-wire circuit by rapidly alternating the direction of transmission.

Private automatic branch exchange (PABX): A term often used synonymously for PBX. A PABX is always automatic, whereas switching is manual in some PBXs.

Restriction: Limitations to a station on the use of PBX features or trunks on the basis of service classification.

Station message detail recording (SMDR): The use of equipment in a PBX to record called station, time of day, and duration on trunk calls.

Tie trunk: A privately owned or leased trunk used to interconnect PBXs in a private switching network.

Voice store and forward (voice mail): A PBX service that allows voice messages to be stored digitally in secondary storage and retrieved remotely by dialing access and identification codes.

BIBLIOGRAPHY

Datapro. *Reports on Telecommunications.* Delran, N.J.: Datapro Research Corp., 1985.

Green, James H. *Automating Your Office.* New York: McGraw-Hill, 1984.

———. *Local Area Networks.* Glenview, Ill.: Scott, Foresman, 1985.

Hobbs, Marvin. *Modern Communications Switching Systems,* 2d ed. Blue Ridge Summit, Penn.: TAB Books, 1981.

Martin, James. *Telecommunications and the Computer.* Englewood Cliffs, N.J.: Prentice-Hall, 1976.

Stallings, William. *Local Networks.* New York: Macmillan, 1984.

MANUFACTURERS OF DIGITAL PBXs AND RELATED PRODUCTS

PBX Manufacturers

AT&T Information Systems

Ericsson, Inc. Communications Division

GTE Communications Systems

Harris Digital Telephone Systems

Hitachi America, LTD

Intecom, Inc.

IPC Technologies, Inc.

Mitel, Inc.

NEC America, Inc.

Northern Telecom, Inc.

Oki Telecom

PBX Manufacturers (*concluded*)

Rolm Corporation

Siemens Communications Systems

Ztel, Inc.

Station Message Detail Recorders

Account-A-Call Corporation

Bitek International, Inc.

Contel Information Systems

Ericsson, Inc. Communications Division

Rolm Corporation

Sykes Datatronics, Inc.

Tekno Industries, Inc.

Voice Mail Equipment

AT&T Information Systems

BBL Industries, Inc.

Ericsson Information Systems

GTE Corporation Business Communications Systems, Inc.

IBM Corporation

Northern Telecom, Inc.

Rolm Corporation

Wang Laboratories

CHAPTER TWELVE

Tandem Switching Systems

Both local central offices and PBXs can be equipped to serve as tandem switches. If trunking requirements are light, the most economical way to switch trunks is through an existing Class 5 switch or PBX. However, the market offers switching machines designed specifically for trunk switching applications. These switching machines are the subject of this chapter.

The primary users of tandem trunk switching machines are:

- Local telephone companies that use tandem switches for switching local interoffice trunks and for interexchange carrier access.
- Interexchange carriers that use tandem switches for access to their networks.
- Large government and business organizations that use tandem switches for private message and data networks.

The technologies used in these machines are similar; however, the size varies considerably from a few dozen trunks to machines capable of terminating more than 100,000 trunks.

TANDEM TRUNKING FACILITIES

Long-haul trunk facilities are largely owned by private carriers and furnished to end users as individual circuits, bulk groups of circuits, or as bandwidth that can be channelized at the user's discretion. Some large organizations own private facilities, but the majority of long-haul facilities are obtained through common carrier tariffs.

Facilities are obtained over either terrestrial or satellite services. Both digital and analog circuits are available. Terrestrial circuits more than 500 miles long are largely provided over analog facilities, although transcontinental digital circuits are rapidly gaining acceptance in private networks. Most common carrier digital terrestrial facilities more than 500 miles long are derived through data under voice (DUV) facilities on analog microwave radio (see Chapter 15) or over analog facilities using a transmultiplexer or L-to-T connector to derive the digital channel (see Chapter 5 for a discussion of transmultiplexers).

The type of facility has an important effect on tandem switching economies. Where two-wire analog facilities are provided, two-wire switching is generally most economical. Where the facilities are four-wire, transmission considerations mandate the use of four-wire switching if at all possible. Many two-wire tandem switches are in use by telephone companies, but they are obsolete and are being replaced with digital switches where feasible. Digital switches can interface T-1 lines without using channel banks and can avoid balance problems generated by four- to two-wire conversions. In modern networks, two-wire switching may be feasible for long-distance service resellers that use WATS and Feature Group A circuits, but most other tandem switching machines are four-wire with direct digital trunk interfaces.

TANDEM SWITCH TECHNOLOGY

Tandem switch architecture is quite similar to Class 5 digital switch architecture except that a tandem switch has few, if any, line terminations. Machines are designed with a nonblocking network controlled by central or distributed processors. Digital trunks are terminated directly in a digital interface frame that couples incoming 1.544 mb/s bit streams directly to the switching network. Signaling is detected by the central processor. The processor sets up a path through the switching network from the incoming time slot to an outgoing time slot that is assigned to an outgoing digital channel. Calls that cannot be completed because of trunk congestion are routed to a tone or recorded announcement trunk. The digital switch acts as a large time slot interchange device that is transparent to the bit stream in the terminated circuits.

Most digital switches designed for the public telephone network are equipped for common channel signaling (CCS). The CCS interface is a circuit that interprets incoming data messages and communicates them to the call processor. When CCS signaling is employed the signaling bits in the terminated digital circuits are not used.

Wideband Switching Capability

Some tandem switches are capable of switching bit streams as high in speed as 1.544 mb/s, a capability that can prove useful in private networks for high-speed data transfer and video teleconferencing. With this capability, mainframe computers can be linked during off-peak hours to transfer data over backbone digital circuits that are normally used for voice during peak load periods. In this kind of application the digital switch takes the place of a manual patch that would otherwise be required to set up the broadband circuit.

With current technology, a video signal can be compressed into bandwidths of 515 kb/s to 1.544 mb/s as described in Chapter 20. A tandem switch can therefore be used to route video signals over a backbone network as shown in Figure 12.1. Because a video signal requires the use of contiguous channels in a T-1 bit stream, a preempting capability may be required to forcibly vacate channels that are occupied with voice transmissions. Some switches preempt by overriding busy channels, sending a warning tone, and cutting off the channel so it can be occupied by a video signal.

The ability to share a T-1 bit stream between voice and video offers an economical solution to setting up occasional

FIGURE 12.1
Use of a Digital Tandem Switch for Alternate Voice and Video Service

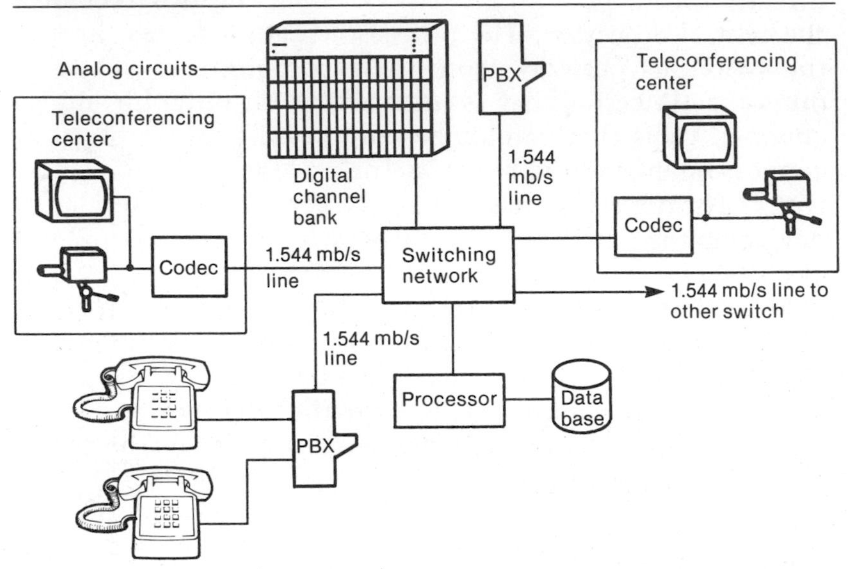

video conferences. For the duration of the conference, overflow voice calls can be denied or completed over a public network, depending on the restrictions in the originator's class of service. Overflowing calls on a public voice network is usually a less expensive strategy than establishing a separate video conferencing channel.

TANDEM SWITCH FEATURES

Most of the features in a tandem switching machine are implemented in software. Therefore, machines can be used in either public or private networks by changing the generic program. Although the architecture for private and public tandem switches is similar, feature differences between the two applications are significant.

Public Tandem Switch Features

Tandem switches for the public telephone networks are capable of terminating large numbers of both analog and digital trunks. The primary features employed are discussed in the following sections. Some of these same features are also used in private tandem switches as noted under the feature descriptions.

Testing Access

Trunk test positions are required in all tandem switches. Trunks are switched through the switching network to the test position for making continuity, transmission, and supervision tests. In a network of multiple tandem switches, communication between the test positions is provided by dialing a special access code, usually 101. This connects the two test positions so technicians can talk and test over the trunk. Several different codes are employed in the long-distance trunks of the IECs, local telephone companies, and some private networks for testing as shown in Table 12.1. These test lines can be used for both manual and automatic tests.

To test direct trunks between two offices automatically, computer-controlled test equipment dials a *remote office test line* (ROTL) over an ordinary telephone circuit as shown in Figure 12.2. The ROTL seizes an outgoing trunk and dials the test line number over the trunk. A *responder* at the distant office interacts with the ROTL and responder at the near-end office to make two-way transmission, noise, and supervision measurements. The results of the test are registered by the test system. Trunks exceeding maintenance limits are automatically taken out of service at the switching machines and marked for maintenance action.

Signaling

Common channel switching is the rule for public toll tandem switching machines and is increasingly being employed on private systems. Machines used in the AT&T Communications network employ AT&T's CCIS system; however, all networks are expected eventually to convert to CCITT Signaling System No. 7. A common channel interface circuit is connected between the processor and the data circuits that comprise the CCS network. The switching systems select an idle path by exchanging data messages, test the path for con-

TABLE 12.1
Trunk Maintenance Test Lines

Type	*Function*	*Purpose*
100	Balance	Provides off-hook supervision and terminates trunk in its characteristic impedance for balance and noise testing.
101	Communications	Provides talking path to a test position for communications and transmission tests.
102	Milliwatt	Provides a 1,004 Hz signal at 0 dBm for one-way loss measurements.
103	Supervision	Provides connection to signaling test circuit for testing trunk supervision.
104	Transmission	Provides termination and circuitry for two-way loss and one-way noise tests.
105	Automatic transmission	Provides access to a responder to allow two-way loss and noise tests from an office equipped with a ROTL and responder.
107	Data transmission	Provides access to a test circuit for one-way voice and data testing. Enables measurement of P/AR, gain-slope, C-notched noise, jitter, impulse noise, and various other circuit quality tests. See Chapter 23 for description of tests.

tinuity, and assign a path through all switches that are part of the connection.

Operator Service Positions

Several large switching systems such as AT&T Network Systems' No. 4 ESS and Northern Telecom's DMS 200 provide operator service positions for assistance in completing calls and for registering special billing arrangements such as collect and third party billing. Credit card billing is largely dialed by the user into a network that is described in a later section.

Two different architectures are employed for operator access. The earlier system uses a bridging circuit on every incoming operator trunk as shown in Figure 12.3. A separate

FIGURE 12.2
Automatic Remote Trunk Testing System

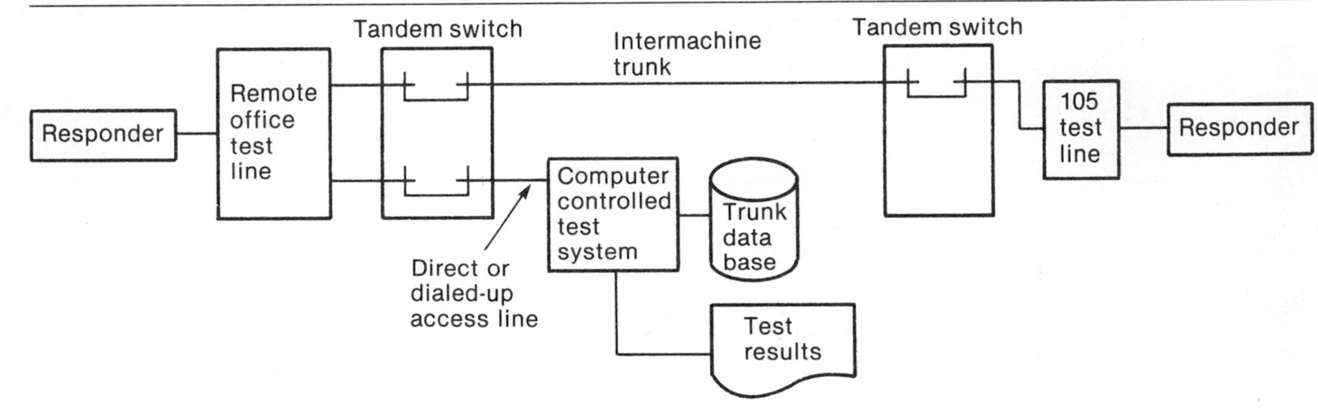

FIGURE 12.3
Traffic Operator Service Position Access

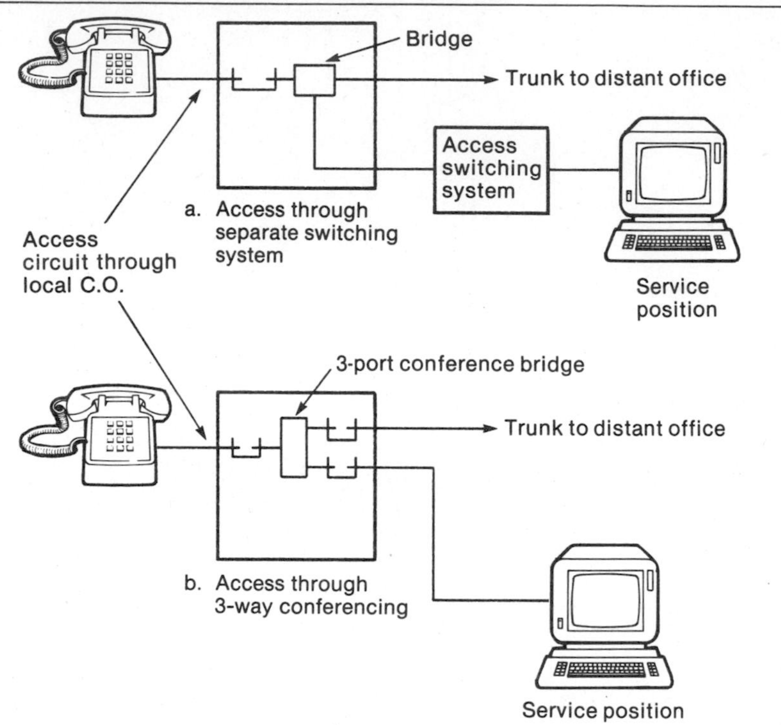

switching machine connects trunks to a group of operator service positions. The operator is bridged to the connection long enough to obtain the billing information, key it into the system, verify the billing, and drop off the call.

The operation of the other type of operator service unit is similar except the service positions are attached to ports in the tandem switch. The switch connects the operator, the incoming trunk, and the outgoing trunk to a three-way conference bridge while billing information is obtained. The call is switched directly to the outgoing trunk for completion.

User-Dialed Billing Arrangements

Most coin and credit card calls are accommodated in public tandem switches by special networks that register and verify charging information. Credit card calls are verified over a data link against a centralized data base. The customer dials a called number, and then when signaled by the machine, dials a credit card number. A message is sent to the data base to verify the validity of the card number, after which the call is switched through the machine.

Coin calls are connected to a circuit that computes the rate, informs the user of the charge with a voice announcement circuit, registers the values of the coins deposited, and connects the call. The equipment monitors the conversation time and reconnects the voice announcement circuit when additional coin deposit is required.

Private Tandem Switching Machine Features

Except for coin and operator services, private tandem switches require much the same features as public switches. Also, other features unique to a private network are provided. Many of these are the same as the PBX features described in Chapter 11. Circuit costs tend to control the total network costs in a private system; therefore, tandem switches are designed to make efficient use of circuit capacity. This section covers features that are not generally included on other types of switches except for those that use a significant part of their capacity for tandem switching.

Personal Identification Data Base

Tandem switches designed for common carrier use are usually equipped for personal identification number (PIN) access to the network. A file of authorized users and their restrictions is kept in the machine's data base. Users access the machine by dialing a seven-digit number used for Feature Group A access through the public telephone network, or by dialing an access code in a PBX that is directly trunked to the tandem switch. If the PBX and tandem switch are equipped for automatic number identification, PIN dialing is not required.

The user file lists the restrictions of each PIN. Most machines offer a full restriction range such as limiting calls to tie lines only or blocking calls to selected area codes, central office codes, or even station numbers. For example, a company could allow its accounting personnel to call the accounting department in another branch, but calls to all other numbers could be blocked. In addition to the PIN, most systems also accommodate accounting code information. This allows a firm to dial an accounting classification for the call and have charges automatically tabulated against that account.

Other methods of controlling costs are optionally provided in most machines. Some users may be permitted to access a tie line network, but when a call is about to advance to a high-cost facility because lower cost routes are occupied, a *call warning tone* sounds. The call will not route to the high-cost facility until the user dials a *positive action digit* to instruct the machine to proceed. When the call duration exceeds a threshold set in the user's file, a time warning signal can be set to sound. These and other similar features are normally not available in public networks because the network owner has no interest in limiting usage.

Other features that are available for certain line classes include *security blanking* which blanks the called telephone number on the billing record if the user wants to avoid leaving a record of calls to certain numbers. Some line classifications are given queuing priority to enable them to jump to the head of the queue when all circuits are busy. Most of these features are implemented in software and can therefore be provided for little or no cost, enabling an organization to customize a network to meet its individual needs.

Remote Access

The remote access feature allows a user to dial a local telephone call to the tandem switch in order to complete long-distance calls over the private network. PIN identification is required to control unauthorized usage. Even with PIN control, security is apt to be a significant problem with remote access. Many systems are designed to raise an alarm when repeated attempts are made to dial with invalid PINs.

Transmission is also a potential problem with remote access. As shown in Figure 12.4, the loss of the connection includes one or two additional subscriber loops, which increases loss by as much as 16 dB. If access to the tandem switch is through a PBX that inserts loss in the connection, poor transmission on many calls can be expected. If the call uses a Feature Group A trunk to an IEC, another subscriber

FIGURE 12.4
Sources of Transmission Problems with Remote Access to a Tandem Switch

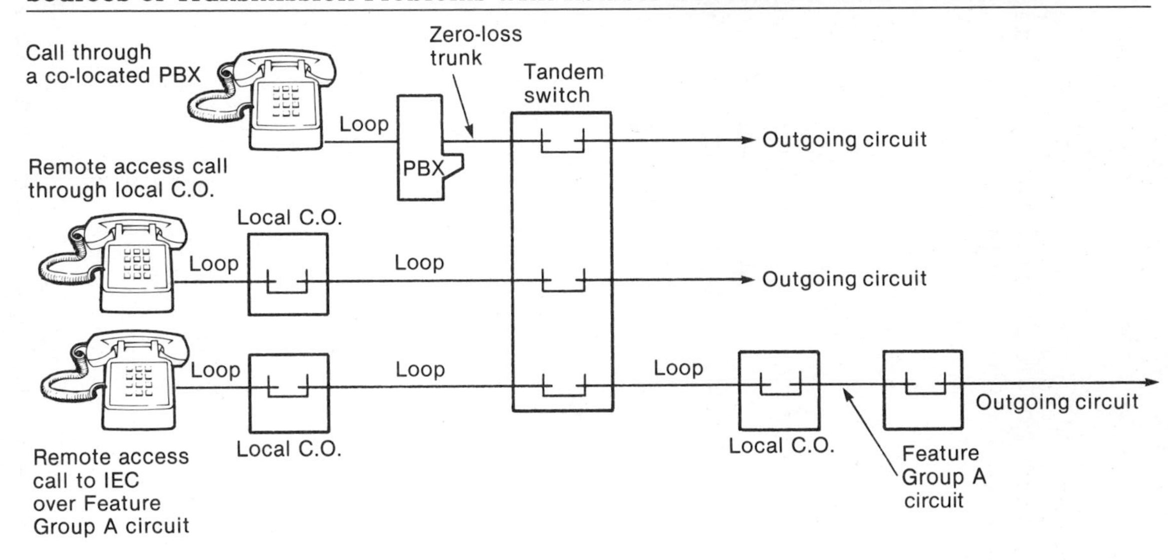

loop is inserted in the connection. A digital tandem switch in a remote access arrangement inserts little or no loss but requires two two-wire to four-wire conversions, which may result in poor echo performance or hollowness.

Queuing

Public networks are designed so that blockage occurs on 1 percent or fewer of the calls placed. When blockage occurs, the call is rerouted to a reorder trunk; and if the user wants to complete the call, redialing is required. In a private network environment, circuit usage can be increased and costs controlled by queuing users for access to outgoing trunks. If a few calls are always backed up in the queue, trunk utilization can approach 100 percent, with the only idle time that of disconnecting and reconnecting a call.

User dissatisfaction and lost productive time always result from queuing, so this technique is rarely used in public networks where the user can easily dial a competitor. In private networks, however, the organization can have control over the network and its users and may choose to pay the penalty of lost productive time by using queuing. An audible indication signals when a call is placed in queue. Some systems use a tone, others a recorded announcement, and others use music in queue. Any network that employs queuing should collect statistical information to indicate the average length of queue, number of calls abandoned from queue, and the average holding time in queue. This information indicates the number of productive hours lost because of queuing.

Routing

Tandem switches are capable of least-cost routing identical in concept to the LCR feature in PBXs as discussed in Chapter 11. Other routing features are also available with most tandem switches and some PBXs. *Code blocking* is used to prevent dialing of certain digit combinations. With this feature, an organization can block access to a given geographical area by blocking the area codes. This can be carried down to blocking individual terminating telephone numbers. For example, some organizations use this feature to block calls to "dial-a-joke" and "dial-a-porn" telephone numbers. The accounting information described in a later section can be sorted by the called telephone number to identify heavily used numbers that might indicate abuse of communication services. Where abuse is found, the unauthorized number is blocked.

Another feature, *code conversion*, is used to limit the number of digits a user must dial to make a connection. For example, a frequently used number could be dialed with seven digits. If the machine completes the call over a tie line, only the seven digits are required. If the number is completed over an IEC network, the switch appends the area code before forwarding the call. Off-net calls to an IEC over a Feature Group A connection may require dialing a PIN that differs from the user's on-net PIN. The user can dial his own PIN, but the switch dials the PIN required by the IEC.

Network Statistical Information

A considerable amount of information about calling patterns and habits is needed to administer a network. Information on the disposition of all originating calls is used to size circuit groups and switching equipment. Most systems provide the following types of information:

- Queuing statistics including average and peak number of calls in queue, duration of calls in queue, and abandoned calls.
- Trunk information for each trunk group on the network including peak and average number of calls, holding time of calls, and number of ineffective attempts because of blockage, cutoff, or equipment trouble.
- Service circuit statistics including number of attempts, holding time, and overflow to service circuits such as reorder, busy, and multifrequency receivers and senders.

The quantity of statistical information collected from a machine is apt to be overwhelming unless some form of mechanized analysis is provided. Some tandem switches are controlled by commercial processors that are also capable of analyzing usage information. Otherwise, external processors are usually needed to determine from the information collected what action should be taken to adjust load and capacity.

Accounting Information

Message accounting information is provided by tandem switches to enable allocating communications costs to the users. This information is provided either in machine readable format for separate processing or is processed by the switching system's processor to assemble completed message detail. Charges can be summarized by individual user, department, and accounting code. For most effective communications cost control, machine readable information should be provided so calling habits can be analyzed and so destinations can be evaluated to determine whether additional private trunks should be added to reduce overflow to more expensive common carrier circuits.

Network Management Control

Many switching systems provide a *network management control center* (NMCC) to administer service and performance on private networks. The center collects all network management information in a single location. Typical NMCC functions include:

- Manual and automatic testing of trunks using apparatus and techniques similar to those used in public networks.
- Compilation and analysis of statistical information to determine loads and service levels, and to determine when circuit types and quantities should be changed.
- Machine performance monitoring including diagnostics and maintenance control of all machine features and circuits.
- Alarm surveillance to detect and diagnose troubles and determine status of switching and trunking equipment.
- Performance and status logging to monitor and log machine history including records of trouble and out-of-service conditions.

STANDARDS

As with other types of switching systems, few standards are published with respect to tandem switching. Trunk interfaces in private network switches require FCC registration and must meet the standard technical interface information for compatibility. Compatibility information is largely adopted from manufacturer's specifications and has not been standardized.

APPLICATIONS

The Rockwell International Telecommunications Network[1]

Rockwell International serves as an excellent example of a highly diversified and complex communications environment. The company engages in a wide variety of commercial and government businesses and operates through an extensive geographic distribution of large and small operating facilities. In addition, the company, with a vast data communications network, is heavily dependent on remote computing services.

Business activities range from the manufacture of space shuttles and B-1B aircraft in the United States to the worldwide manufacture and distribution of automotive parts and electronic components and systems. With over 300 facilities in North America, dozens of overseas locations, 80,000 voice network users, and 35,000 computing service users, the voice, data, and message communications requirements are massive.

To support this operating environment, an extensive centralized computing services capability was established in the early 1970s. Today, four large regional computing centers currently support over 95 percent of the computing requirements of the company with the overall capability of 55 IBM 3033s or their equivalents. Over the years, this central ser-

[1]The information in this section was written by Dr. Joseph G. Robertson, Director of Telecommunications at Rockwell International, and is used with permission of Dr. Robertson and Rockwell International.

vice has been augmented by distributed data processing, dedicated Computer Aided Design and Manufacturing (CAD/CAM), and word processing and office automation capabilities at a majority of Rockwell facilities.

As might be expected, separate communications networks evolved for voice, data, dial-up data, facsimile, teletype, mail, and messaging. Some of these networks were part of the central service function while others were established and operated independently by the business units. As a result, by the end of the 1970s, aggregate telecommunications costs were large and growing, and efforts were duplicated. It was recognized that something must be done to control costs that were driven by inflation and subjected to the idiosyncrasies of state and local government regulations.

Advanced technology and the economy of scale resulting from the integration of the diverse networks represented the basis for remedying this growing problem. Ownership of more of the equipment in the telecommunications network offered an additional economic opportunity. Increasing the level of centralized management and control over these services appeared to be a necessary step in this process.

Voice Communications in the 1970s

Beginning in 1967, Rockwell's voice and dial-up data communications needs were supported by a Common Controlled Switching Arrangement (CCSA) network service provided by AT&T. Initially, five switching centers supported 16 company locations. By 1981, the network had been expanded to support 160 locations, and call volumes and costs had increased at an even faster rate. It became obvious that the economic performance and technical capabilities of a service designed in the 1960s would not sustain the company and its growth requirements for the 1980s.

Major limitations of the CCSA service included an outmoded electromechanical switch technology, degrading service quality, and absence of management information to evaluate and optimize the network configuration. There also were clear indications that costs of service would increase as AT&T migrated its network to modern electronic switching technology.

Data Communications in the 1970s

During the 1970s, dedicated data lines were leased, connecting each major business location to one or more of the four large regional computing centers. Since not all centers necessarily provided the same services, applications, or data bases, the nation was crisscrossed by hundreds of dedicated Rockwell data lines. Users frequently had multiple computer terminals on their desks, each connected to a different computing service or computing center. This pseudo network and operating environment was chaotic, and data communications costs were increasing at a rate disproportionate to service levels. The potential for bandwidths greater than 56 kb/s offered further incentive to redesign the data communications network. The IBM SNA architecture and communication approach gave us a migration strategy.

Time for a Change

It became clear in the late 1970s that it was time for a change in both voice and data communications. Such change would have to provide better visibility over network operations and costs, greater flexibility, management control, and wideband data services. Consideration would also have to be given to an eventual migration toward an integrated digital telecommunications environment, either public or private in nature. Any new network strategy would have to be compatible with the rapidly expanding telecommunications requirements for dial-up data and office automation as well as for such communications-based applications as electronic mail, voice mail, and video conferencing. Finally, the current data communications strategy of dedicated terminal access to single services would have to be drastically changed to allow true networking to take place, as well as to reduce the number of terminals and their costs. Rockwell had long committed to the IBM SNA strategy and was proceeding in that direction through planning and participation in IBM's early support programs for SNA architecture and evolving network support software. Improved network architecture and performance appeared to be achievable objectives in the early 1980s.

Alternatives

Recognizing the need for a change, a corporate study group was established to identify telecommunications requirements for the 1980s, assess new and emerging technologies, and recommend approaches to increase resource utilization, efficiency, and cost effectiveness. The results of this study led to an extensive analysis and evaluation of alternatives and the eventual recommendation of a specific approach.

Early in the analysis it was concluded that continuing to attempt to improve the operational and economic performance of the existing CCSA voice network was not a viable option, and this was excluded from further consideration. Continued migration to an SNA-based data communications environment was strongly endorsed, but changing the data network was not a part of this study. Since effort was already underway to rehome data lines to the nearest center and proceed to the destination through an SNA network, the two areas were treated as independent efforts with no intent at that time to integrate voice and data.

The alternatives for voice network services that survived an extensive technology and business case review included:

- Lease of an Enhanced Private Switched Communications System (EPSCS) from AT&T.
- Lease of a digital satellite service from SBS. Data transmission could be included in this alternative.
- Implementation of a private switched and transmission network involving microwave and satellite transmission in conjunction with local terrestrial service.
- Installation of private switching equipment in conjunction with lease of terrestrial transmission services from common carriers.

The latter two alternatives involved extensive use of products manufactured and distributed by Rockwell's Commercial Electronics Operations and its associated Collins divisions.

AT&T EPSCS

The AT&T EPSCS offering was highly attractive. Employing newer generation switching equipment, it had all the features of CCSA along with such additional capabilities as originating call screening, user-dialed authorization codes, call queuing, time variable routing, and a subscriber's network control center. Basically, it offered improved quality, management tools for network control, and low initial cash investment. It also offered significant savings over the CCSA service.

Basic limitations of the EPSCS alternative included:

- Susceptible to inflationary and tariff increases for both switching and transmission services.
- Provision of only limited access to network management information regarding reliability and optimization, and little user network control capability.
- Incapacity to incorporate wideband digital service.

SBS Satellite Service

The SBS offering involved four more dedicated leased earth stations and wideband digital transmission for voice, data, and video. While the service was priced attractively, there were limitations in the quality and means of handling voice communications, which represented over 85 percent of our transmission requirements. There also were problems in connecting the hundreds of local facilities to the backbone network and in providing cost effective off-net access. As a new service from a new company, it represented an undesirable degree of risk. Finally, it appeared unlikely that service could be provided in time to meet the desired CCSA replacement schedule.

Although the SBS offering eventually would be considered for wideband data telecommunications applications, it was rejected as an alternative for replacement of the CCSA voice service.

Private Switching and Transmission Network

The next alternative considered was a complete private network involving four Galaxy Digital Tandem Switches from Rockwell's Commercial Electronics Operations along with microwave equipment from the same supplier. The transmission strategy included the establishment of microwave networks in areas where a number of facilities were concentrated such as Los Angeles, California; Dallas, Texas;

and Pittsburgh, Pennsylvania. Long-distance transmission would be supported by a combination of owned or leased satellite earth stations and leased common carrier terrestrial transmission service.

This alternative offered significant economic advantage as it buffered switching costs and a large portion of transmission costs from anticipated inflationary and tariff rate increases. It also gave recognition to the large portion of telecommunications costs associated with off-net calls through centers of major voice traffic concentration. The switches offered least-cost routing of this traffic along with other means of network optimization. The total private network approach offered all the benefits of the EPSCS network along with:

- Real-time network availability and reliability statistics.
- Real-time diagnostic capability.
- Automatic network reconfiguration.
- Wideband transmission.
- User managed and controlled network management center.

This network architecture represented the ultimate system required to meet Rockwell's long-term telecommunications requirements in the most economic fashion. However, in the interest of reducing risk and minimizing initial capital investments, it was concluded that the private transmission portion of this proposal should be reevaluated for implementation at a later date on a phased basis.

Private Switching Network

The final alternative consideration was the "switches only" portion of the previous proposal. It involved the acquisition of four Galaxy Digital Tandem Switches from Rockwell's Commercial Electronics Operations and the lease of common carrier terrestrial transmission services. A network management center was included in the design along with tools for the collection and evaluation of performance statistics.

This alternative provided economic advantage by capping the cost of switching services through ownership of the switches. While it offered less direct control of the cost of leased transmission services, it still provided many economic and operational benefits. For example, performance statistics allowed frequent optimization of the network transmission architecture through a computer model developed by Rockwell. The network management center allowed real-time monitoring of circuit availability, precluding the need for extensive backup lines to cover unknown levels of circuit outages.

The intelligence of the DTS switches allowed least-cost routing and automated In-WATS calling. Both "hop-on" and "hop-off," the network capabilities could be provided. The extensive use of dedicated lines provided the potential for incorporation of voice multiplexing equipment for more efficient usage of transmission lines. Most important, this network architecture would position the company to migrate to digital wideband services provided over private microwave, satellite, and common carrier terrestrial digital services.

In summary, this private switched network offered substantial economic and operational benefits in the near term and a foundation for migrating to improved transmission services in the future. It minimized implementation risks and allowed the phasing of capital investment. For these reasons, it was selected for implementation.

Initial Network Configuration and Capabilities

The initial configuration of COMNET, the new Rockwell private switched voice network, involved the installation of four Digital Tandem Switches that were located at four major company facilities: Richardson, Texas; Pittsburgh, Pennsylvania; Cedar Rapids, Iowa; and Seal Beach, California. These switches were interconnected by 314 voice grade intermachine trunks (IMTs). The network management center was located at the Seal Beach facility in association with corporate management centers for computing and telecommunications operations.

An additional 1,800 voice circuits were associated with the four switches. The majority of these circuits were dedicated analog lines to the serving vehicles of approximately 150 company facilities. The remaining circuits were foreign exchange lines to high voice traffic off-net locations and WATS lines to support overflow from the backbone IMTs and to access off-net locations. Approximately 150 additional company facilities could be addressed through the network by four-digit abbreviated dialing features of the DTSs. In-WATS circuits provided access to the network from off-net locations and also allowed off-net to off-net access.

The DTSs, through software, supported least-cost routing, alternate routing, facility restrictions, personal authorization codes, traffic measurement, and 100 percent call detail reporting. They also supported a common numbering scheme where anyone on the network could be accessed by a seven-digit number.

In parallel with planning for the voice and dial-up data network, significant effort was applied to upgrading the data network. The object was to rehome the individual data circuits to the nearest regional computing center, and then, through network control software, to redirect the circuit over a backbone network to the target computing center and service. Upon the completion of this migration, the data communications network involved approximately 1,000 dedicated access lines to the regional computing centers, 37 backbone lines interconnecting these centers, and 500 dial-up access ports.

Implementation

A dedicated project team was established early in the process and stayed with the activity through the cutover period and until the network was turned over to the telecommunications organization for operation. This team developed a highly detailed installation and cutover plan that was subjected to frequent "red team" and management reviews to identify any problems or weaknesses in the plan.

As part of the implementation process, facilities had to be designed and constructed, power and backup power planned and installed, and normal and emergency operating procedures developed. Spare parts provisioning was a major part of the planning activities of the maintenance staff along with the acquisition of the necessary test equipment.

Software was developed in advance to allow for network performance monitoring and evaluation, transmission optimization modeling, cost collection, and user billing.

Training and documentation was another area requiring considerable effort. With a potential of 80,000 users of the voice service, a comprehensive user training program was critical. New phone books and operating instructions had to be in place well before the cutover. Video tapes and automated 35mm slide training programs were developed to simplify the process of training a large number of people to use the system in a very short period of time.

The switches were installed during the summer of 1981 and were thoroughly checked out. New transmission circuits were installed during the same period. The actual cutover of the new network took place during the Christmas holiday to avoid disrupting the corporation during full business operation. Although some problems were encountered, they were quickly resolved, and the network was declared operational the first week of January 1982 when business was resumed after the holidays.

The majority of the problems that occurred during the staging and cutover were associated with coordinating the installation of circuits by AT&T and the local telephone companies. To preclude the risks and potential disaster associated with any cutover, parallel circuits were maintained throughout this period and were phased out only when satisfactory service was confirmed.

Because of the comprehensive planning activity, the migration to a new corporate-wide network took place on schedule, within budget, and with virtually no serious incidents. The highly successful installation must be attributed to the quality of the project team and their dedication. However, the results would not have been achieved without the early establishment of the team, the detail of the planning effort, and the extensive project reviews.

Managing the Network

With a network of this complexity and geographic distribution, centralized management and control was critical. A Network Operations department was established within the

Corporate Telecommunications Services organization with total responsibility for:

- Network operation.
- Network management center staffing and operation.
- Central operator assistance staffing and management.
- Traffic analysis and network optimization through computer modeling.
- Circuit ordering and cancellation.
- Invoice review and approval.
- User billing administration.
- Authorization code administration.
- Maintenance management and vendor coordination.
- Performance reporting.

Much of this activity was conducted through the network management center where the network could be monitored in real time, standard and special performance reports could be generated, and coordination of all the involved staff and organizations could be accomplished.

Because of the availability of detailed calling information and circuit performance data, the network architecture could be reviewed and optimized every two weeks through computer modeling. With an established grade of service over each circuit link, the model assisted in identifying circuit requirements, least-cost routing, and least-cost vendor selections. This capability provided even greater economic benefits from the private switched network than was anticipated during the planning period.

Because of the detailed billing information, local management was better able to review and control their communications costs. In general, by establishing a private network, management gained a better understanding of the telecommunications requirements and gained better control over communications costs.

Maintaining the Network

On-site maintenance of the voice network was provided under contract to the Telecommunications Services organization by personnel from Rockwell's Commercial Electronics operation. This full-time staff was responsible for scheduled and unscheduled hardware and software maintenance. It also was responsible for quality assurance of all circuits in the voice network and for coordinating with hardware vendors and transmission service suppliers.

Comprehensive monitoring, trouble shooting, and problem escalation procedures were developed to ensure rapid problem identification, interim problem reduction actions, minimum downtime, and rapid full-service restoral. All of these activities involved working closely with the network management center staff, the hardware vendors, and transmission service suppliers. The success of this effort is demonstrated by the fact that the maintenance service organization has established an enviable record of system availability and rapid service restoral.

Network Evolution

Since its implementation in 1982, the network has undergone significant change and enhancements in the interest of supporting growth, providing a high level of service, and controlling costs. Since 1982–83 was a period of rapid tariff rate increases, the network management and optimization modeling tools became more valuable to counter the increasing economic burden.

As calling patterns on the new network became more stable, some of the economically more attractive transmission services of the other common carriers were incorporated, producing significant savings for both the backbone network and off-net calling services. With better understanding of calling patterns, voice multiplexer equipment utilizing time assignment speech interpolation (TASI) was installed on the backbone, further reducing the average cost of a call. This effort virtually halved the number of calls. Dial-up data calls were routed around the TASI equipment by separate dialing codes.

The most significant change to the network took place in January 1984 with the establishment of the Integrated Digital Backbone Network (IDBN). This involved the replacement of 175 analog IMT circuits that made up the voice backbone network with 13 of AT&T's Accunet® T1.5 digital wideband circuits. At that time, through the use of TASI equipment, 283 voice circuits were actually supported by the 175 IMTs. Implementation of these wideband links was almost as complex as the installation of the initial private switching network and required the same level of planning, scheduling, and project review.

Figure 12.5 shows the new voice/data/video network incorporating Accunet T1.5 circuits for the backbone. The four major computer centers and the four digital tandem switches are shown as rectangles connecting to the four nodal MUXs. Columbus and Tulsa along with seven other major facilities in the Los Angeles area are also connected by T-1 links to the nearest regional computing center.

Wideband circuits connect among these multiplexers, each supporting twenty-four 64 kb/s channels. Selected 64 kb/s channels carry two voice circuits digitized at 32 kb/s per circuit using the Variable Quantization Level (VQL) algorithm developed by Aydin Monitor Corp. Many of the voice circuits were also processed through TASI equipment with incorporated echo suppression capability, providing even greater transmission efficiency at an acceptable quality level. Dial-up data circuits and non-TASI voice circuits were routed around the TASI equipment and were processed through separate echo cancellation devices.

In addition to the voice service, this wideband backbone network supports all of the digital data circuits interconnecting the four major computing centers. The multiplexer ports can incorporate cards that support a wide range of data transmission services. Cards are available for (1) selectable 64, 56, or 40.8 kb/s circuits; (2) 56 kb/s circuits; (3) three 19.2 kb/s circuits; (4) five 9.6 kb/s or slower circuits; and (5) a high-speed card for up to 768 kb/s in 128 kb/s increments. The latter capability is intended to support bulk data transport and compressed video teleconferencing applications. Video rooms are located in Seal Beach and Newport Beach, California; Dallas, Texas; and Chicago, Illinois. The service can be provided to our Pittsburgh, Pennsylvania, and Cedar Rapids, Iowa, nodes when rooms are established.

Rockwell had gained considerable earlier experience with 1.544 mb/s services for data transmission and was well aware of their performance and economic advantage over multiple

FIGURE 12.5
Rockwell Integrated Digital Network

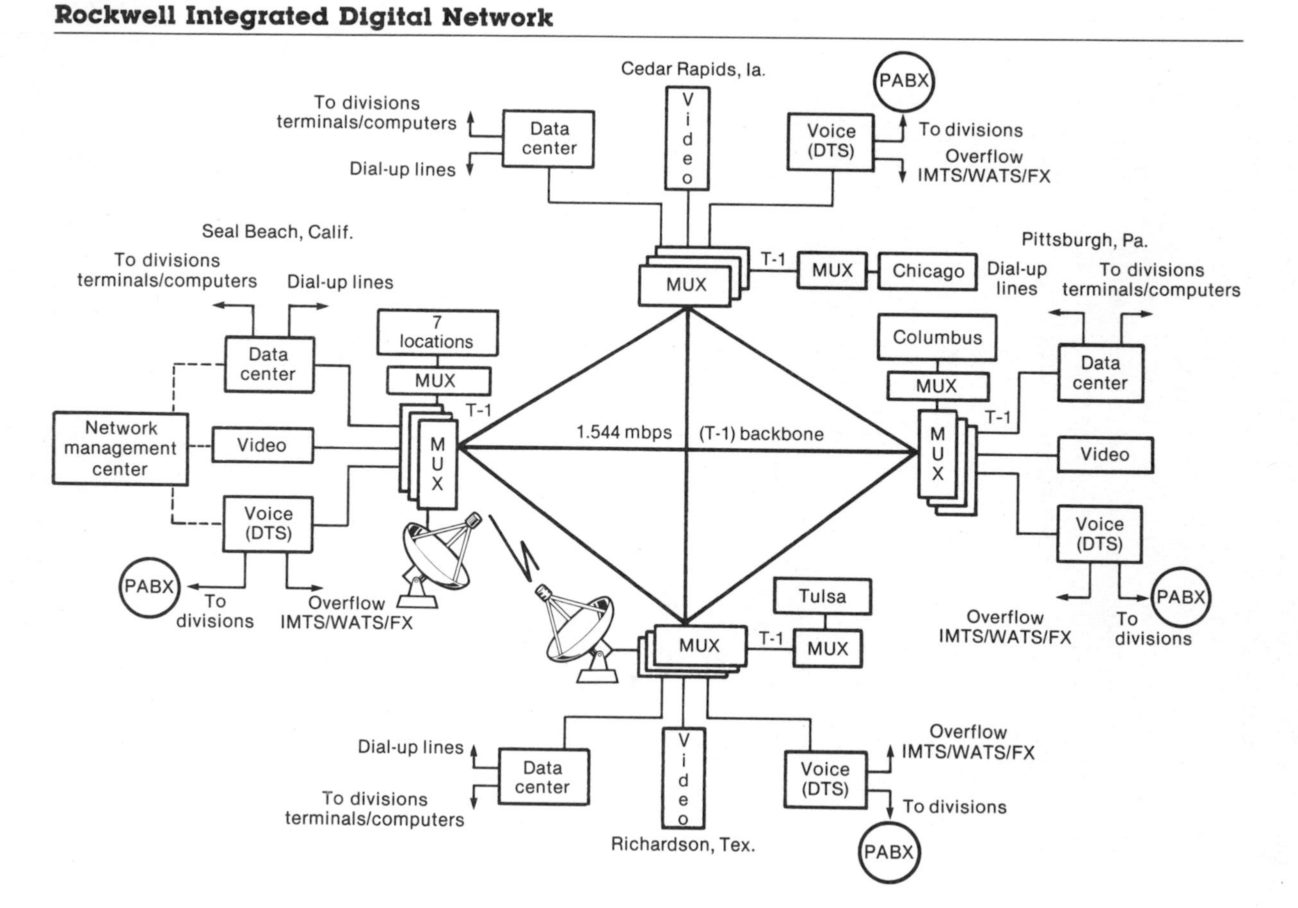

Printed and used by permission of Rockwell International Corporation.

circuits. With an initial installation of 1.544 mb/s lines in 1981 in the Los Angeles area, there now are 13 lines in operation, connecting our Western Region Computing Center to seven major Los Angeles facilities.

Since hundreds of company locations are connected to our regional computing centers and digital tandem switches, all data and voice calls go to the nearest node and then ride the wideband backbone network to a remote node. The call can then be switched to a nearby company facility or to the public network or foreign exchange lines for off-net calls.

Because of the attractive tariffs associated with the Accunet T1.5 service, considerable transmission savings were realized in the replacement of the multiple dedicated voice and data circuits that previously made up the backbone network. Additional benefits included:

- Improved bit error rate for data.
- Greater flexibility in dynamically reconfiguring the network.
- Inherent growth potential within most of the wideband links through the simple addition of multiplexer port cards.
- Avoidance of delays associated with acquiring individual service from the telephone companies.

We cannot fully describe the Rockwell telecommunications environment without giving recognition to the integrated SNA-based computer data network that involves selected circuits on the backbone network that connect the major computing centers. This capability allows any terminal in the country connected to a major computing center to access any authorized computer and service at any other computing center. It also allows easier problem recognition, supports alternate routing within the network in case of failure of any circuit, and supports the transfer from a regional computing data center to other locations internal and external to the company. The new wideband backbone network is fully compatible with the corporate data communications strategy based on SNA/SDLC architecture and will continue to provide both a performance and economic advantage over the previous multicircuit environment.

The Future

The new backbone network has been operational since January 1984, and the company has realized significant economic and operational benefits. Near-term plans involve adding 1.544 mb/s links and multiplexers to the backbone and to other large facilities having extensive data communications requirements. When economically advantageous, digital voice circuits also will be carried over these links. We found that in the interest of voice quality we had to reduce the number of circuits associated with each TASI device. We also replaced all existing echo suppressors with echo cancellers. We are investigating locations with sufficient voice communications requirements to warrant establishing a 1.544 mb/s link between their PABX and the digital tandem switches that provide access to the voice network. In some cases, smaller locations with PABXs will be connected to larger PABXs in their area to further reduce transmission line costs.

Rockwell has provided video conferencing service between Seal Beach, California, and Richardson, Texas, for the past 2½ years. Two new locations have been added and several others are being considered. Current plans involve the use of the backbone network to provide the bandwidth necessary to support compressed video conferencing at 768 kb/s. To provide the needed bandwidth, some voice calls will be alternate routed or transferred to Rockwell's WATS service. When fully implemented, virtually any location served by a 1.544 mb/s link could have video conferencing access to virtually any other company location having similar wideband transmission and conferencing service.

Rockwell has been operating a satellite communications network between Seal Beach, California, and Dallas, Texas, for several years. Initially the service was used for video conferencing. Now it is available for wideband data transfer and for emergency backup for the T-1 links.

Rockwell has made major changes in its overall telecommunications environment since 1982 and is continuing to plan and implement new capabilities and services. In part, these changes were made for economic factors and to control costs. Possibly even more important, these changes were made to increase service and operational flexibility, and to position the company to take advantage of the evolving all-digital telecommunications environment.

While this period of industry deregulation and AT&T divestiture is confusing and fraught with problems, we are beginning to see the opportunities that competition offers. The new products and services that are flooding the market are being evaluated and the impact on existing operations considered. We see great potential in some of the new intelligent wideband multiplexer products that offer remote diagnostics and software control. Eventual interface with some of the anticipated enhanced transmission services, including Software Defined Networks, may offer even greater opportunities for improved voice, data, and video telecommunications services in an economical fashion.

There probably never has been a similar period of opportunity to control cost and improve performance through technology. As a major supplier of telecommunications equipment and as a company heavily dependent on communications services, Rockwell intends to maintain a leading edge position in the development and use of new technology to support our competitive position.

Evaluation Considerations

Many of the same factors that are important for other switching systems are also important for tandem switches including reliability, capacity, compatibility of external interfaces, operational features, and internal diagnostic capability. This section covers the features that are included in tandem switches primarily to control private network costs and that are omitted from most other switching systems.

Queuing

Tandem switch manufacturers tend to emphasize queue efficiency as an important way of improving circuit occupancy and controlling circuit costs. The method of handling calls during blocked circuit conditions is an important factor to evaluate in comparing tandem switches. Queuing can definitely increase circuit occupancy during busy periods. However, the increase in efficiency must be evaluated against the decrease in productivity that results from users waiting in queue for calls to be completed and the time required to administer the system. The time spent in queue is generally nonproductive and can easily outweigh the cost of the alternatives. Even if users have speaker telephones so they can continue to work while waiting, which is not always practical, busy station lines tend to aggravate the "telephone tag" problem.

Three alternatives can be used to avoid queuing. The first is to return the call to reorder and force redialing. With a telephone that includes a last number redial feature, this approach is tolerable. The second alternative is to increase the quantity of circuits. With adequate data from the queuing and circuit usage statistics, the cost of this alternative can easily be calculated. The third alternative is to overflow to a higher cost facility. IEC networks are designed to a low blocking probability, and overflow to a common carrier is usually an effective way to limit delays.

Despite its hazards, queuing is an effective tool to use in managing a network. To be effective, a system should provide for variable queuing by class of service and time of day. The length of queue should be administratively variable. When a reasonable queue length is reached, additional callers should be turned back to avoid long queue holding times. The system should provide near-real time information about queue length so system administrators can take corrective action.

Network Management Control

An effective tandem switch should provide the tools needed to manage and control the network. The system should be capable of unattended operation from a remote location. It should have complete flexibility in changing line classifications, restriction levels, trunk classifications, queuing parameters, and other such features that affect line and trunk administration. The system should provide real-time information about trunk status. Defective trunks should be removed from service and referred to the control center for corrective action.

The system should be able to diagnose trunk performance automatically during light load periods. It should be equipped for transmission and supervision tests to distant tandem switches and, if permitted, to IEC and local telephone company responders. Direct access to individual circuits should be permitted for making transmission measurements and supervision.

A full range of statistical information should be available for evaluating service and for determining when trunk quantities should be adjusted. A method of automatically analyzing the raw statistical information should be included. The information should be formatted so the corrective action to be taken is apparent.

The system should diagnose its own troubles and direct technicians on what action to take, ideally to the point of specifying the circuit card to change. The manufacturer should provide remote diagnostic and technical assistance, and should provide a method of keeping both hardware and software current with changes.

Network Modeling

Network sizing, as discussed in Chapter 13, is based on the use of simulation or modeling. If the vendor provides a simulation or modeling service to determine equipment and trunk quantities and to predict the effects of changes in load, tandem switch administration will be facilitated. An effective tool should be able to accept usage information from machine-produced statistics and summaries of originating and terminating traffic to determine the most efficient use of equipment and facilities. The system should also be capable of accepting cost information to determine the best balance between alternatives such as WATS lines, tie lines, public switched networks, etc.

Routing, Blocking, and Translations

A tandem switch should be capable of fully flexible routing to trunk groups based on class of service. It should have the capability of blocking area codes, central office codes, and line numbers. It should also be capable of filtering so calls from some users can be terminated only to selected codes. The system should be capable of blocking access to circuit groups to prevent ineffective attempts to distant machines that are known to be temporarily experiencing service difficulties. This capability allows the distant machine to recover without being inundated with excessive call attempts. The system should also be capable of code translations so users dial the same code to reach a destination regardless of the routing.

Common Channel Signaling (CCS)

In a multitandem network, common channel signaling should be considered to achieve more effective circuit utilization. The costs of CCS equipment and the separate data network should be weighed against the improved circuit utilization from eliminating in-band signaling.

Remote Access

The value of remote access to enable use of the network from a separate location should be considered. Security and transmission performance must be weighed against the benefits of remote access. A switch should provide an effective screen against unauthorized access and should alert the network administrator to unauthorized attempts. The transmission characteristics of the access circuits should be evaluated to predict whether service will be satisfactory.

High-Speed Switching Capability

If video conferencing, closed circuit TV, or high-speed data transfer are anticipated, the system should be evaluated for its capability to allocate the bandwidth dynamically. The system should be capable of reserving any segment of capacity required and should be able to preempt occupied circuits when necessary to vacate the required bandwidth.

GLOSSARY

Call warning tone: A tone placed on a circuit to indicate that the call is about to route to a high-cost facility.

Code blocking: The capability of a switching machine to block calls to a specified area code, central office code, or telephone number.

Network management control center (NMCC): A center associated with a network to monitor network service and performance, analyze statistical information, and take maintenance action on failed equipment.

Positive action digit: A digit that a switching machine requires to be dialed before it will advance a call to a high-cost route.

Queuing: The holding of calls in queue when a trunk group is busy and completing them in turn when an idle circuit is available.

Remote access: The ability to dial into a switching machine over a local telephone number in order to complete calls over a private network from a distant location.

Remote office test line (ROTL): A testing device that acts in conjunction with a central controller and a responder to make two-way transmission and supervision measurements.

Responder: A test line that can make transmission and supervision measurements through its host switch under control of a remote computer.

Security blanking: The ability of a switching machine to blank out the called digits for certain lines so no called number detail is printed.

BIBLIOGRAPHY

American Telephone & Telegraph Co. *Notes on the Network.* 1980.

Bellamy, John C. *Digital Telephony.* New York: John Wiley & Sons, 1982.

Freeman, Roger L. *Telecommunication System Engineering.* New York: John Wiley & Sons, 1980.

Hobbs, Marvin. *Modern Communications Switching Systems.* 2d ed. Blue Ridge Summit, Penn.: Tab Books, 1981.

Sharma, Roshan La.; Paulo T. deSousa; and Ashok D. Inglé. *Network Systems.* New York: Van Nostrand Reinhold, 1982.

MANUFACTURERS OF TANDEM SWITCHING SYSTEMS

AT&T Network Systems

Action Communication, Division of Honeywell, Inc.

Datapoint Corporation

DSC Communications Corporation

Northern Telecom, Inc.

Rockwell International Switching Systems Division

CHAPTER THIRTEEN

Network Design Concepts

To integrate switching systems and trunks into a network requires determining the quantities of trunks and the amount of shared network equipment needed to reach a reasonable balance between service and costs. This function is called *network design* or *traffic engineering* in the telecommunications industry. Trunk groups and shared equipment quantities are based on the probability of the size of the offered traffic load. When the load is less than the network is designed for, money is wasted on unused capacity. When demand exceeds the designed load, service is affected by ineffective attempts and delays, and money is wasted on lost revenues or unproductive employee time spent on ineffective attempts.

Telecommunications traffic engineering is similar in many respects to highway traffic engineering. A circuit group provides a given amount of capacity. In highway engineering, the amount of capacity depends on the number of lanes. In a group of circuits, each circuit is analogous to a lane in a highway. The objective is to keep the lanes full without introducing so much load that traffic grinds to a halt. The amount of load offered to the path varies with time of day, day of the week, and season of the year. Within limits, these variations are predictable. The network is designed for anticipated peaks, called the *busy hour* in network design terms. When the offered traffic load exceeds the designed capacity, blockage results.

The following information about the character of the network is factored into an overall design:

- Network owner's grade of service objectives.
- Anticipated load measured in number of call attempts and call holding time distributed by hourly, daily, and seasonal variations.
- Behavior of users in placing and holding calls.
- Capacity of each of the elements of the network: trunks, switching network, common equipment, and processors.

THE NETWORK DESIGN PROBLEM

The essential problem in designing a telecommunications network is how much equipment and trunking is required to meet an objective balance between service and cost. The process used to reach this balance is rather complex and requires special formulas, software tools, tables, and training. This chapter explains only the concepts. Its purpose is to advise the reader what must be done to size the network, not how to do it.

Network design is the process of predicting future demand based on past results, evaluating the capacity of equipment and facilities, and providing the correct amount of capacity, in the proper configuration, in time to meet service objectives. The primary complication in network design is how to provide the right amount of equipment and facilities to meet a constantly fluctuating demand. In any part of the network, demand fluctuates from minute to minute as users originate and terminate calls. Hourly fluctuations also occur, as shown in Figure 13.1, because of changes in usage as the business day peaks and wanes during breaks and lunch hours. Furthermore, demand varies by season of year, by class of

FIGURE 13.1
Hourly Variation in Calls for a Typical Local Central Office

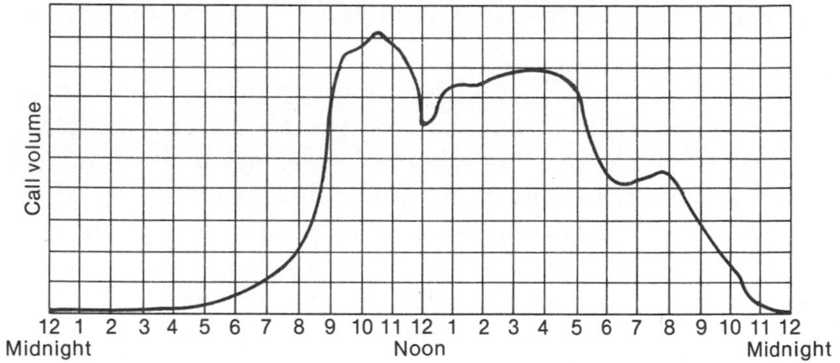

service, and by type of call, that is, whether the call is local or long haul.

Queuing Theory

The most common network design method involves modeling the network according to principles of queuing theory, which describes how customers or users behave in a queue. Three variables are considered in network design. The first variable is the arrival or input process that describes the way users array themselves as they arrive to request service. Examples are arrivals of users at a group of DTMF receivers to begin dialing or at a group of trunks to a distant office. The second variable is the service process, which describes the way users are handled when they are taken from queue and admitted into the service providing mechanism. The third variable is the queue discipline, which is the way users behave when they encounter blockage in the network. The network is designed by observing how users behave and selecting the appropriate design formula. Three reactions to blockage are possible:

- **Blocked calls held** (BCH).When users encounter blockage, they immediately redial and reenter the queue.
- **Blocked calls cleared** (BCC).When users encounter blockage, they wait for some time before redialing.

- **Blocked calls delayed** (BCD).When users encounter blockage, they are placed in a holding circuit until capacity to serve them is available.

Traffic engineers have different formulas or tables to apply, corresponding to the assumption about how users behave when blockage occurs. Service systems are grouped into two categories: loss systems and delay systems. In a loss system when blockage is encountered, the user is turned away. An example is the "fast busy" tone that signals that all trunks are busy. In a delay system the user is held in queue until a server is available. An example is the queue for an idle DTMF receiver. Until a receiver is available, the user does not receive dial tone, but if he waits long enough, a receiver will be attached.

Traffic Load

In a given hour, the load on any part of the network is expressed as the product of the number of call attempts and the average holding time of all attempts. For example, if a circuit experienced six call attempts that averaged 600 seconds (10 minutes) each, the group would have carried 3,600 call seconds of load. To express the load in more convenient terms, traffic engineers divide the number of call seconds by 100 and express the result as "*cent* call seconds" (CCS). A load of 36 CCS represents 100 percent occupancy of a circuit for one hour. Traffic loads are also expressed in *erlangs*, a unit named for A. K. Erlang, a Danish mathematician. One erlang is equal to 36 CCS; both represent full occupancy of a circuit for one hour.

It is important to distinguish between the carried and offered loads of a network. The difference between the two lies in the ineffective attempts that are offered but not carried because of blockage or failures. The offered load can be estimated, but it is impossible to determine accurately because the number of calls that are immediately redialed is unknown.

One hundred percent occupancy in a circuit is unachievable in the real world because of variations in timing and duration of access attempts. Calls do not align themselves

FIGURE 13.2
Waiting Time as a Function of Circuit Occupancy

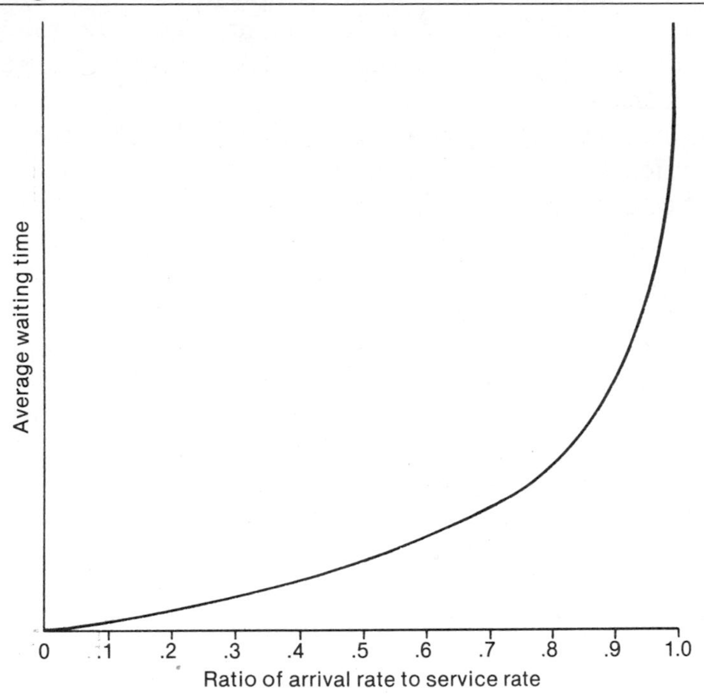

neatly so that when one terminates another is waiting to occupy the vacated capacity. As discussed in Chapter 12, a call queuing process can be used in a switch to ensure that a call is waiting when a circuit becomes idle, but even with this technique the full 36 CCS capacity of a circuit cannot be reached. The reason for this can be seen from the queuing formula:

$$\text{Average waiting time} = \frac{\text{Average arrival rate}}{\text{Service rate} \times (\text{Service rate} - \text{Arrival rate})}$$

When the service rate and arrival rate are equal, as they would be at 100 percent occupancy, the denominator of the equation is zero, which means the waiting time will be infinite. Figure 13.2 illustrates the relationship between percent occupancy and length of queue. Network design becomes a task of determining how long the queue can be allowed to extend before costs of delays outweigh the cost of adding more capacity. It is evident, therefore, that network design requires a service objective, of which more will be said later.

The network design problem is further complicated by the unpredictability of the users. Users vary widely in the number of calls they attempt per hour and in how long they hold a circuit after it is connected. In a network without blocking, call originations are random. That is, the attempts and holding time of any user are independent of other users. Users are far from uniform in their behavior. Some dial faster than others; some frequently dial a few digits and hang up (called *partial dial*); some redial immediately when a reorder or busy is encountered, while others wait for a time before redialing. If this randomness in user behavior is observed, it begins to fall into a pattern that can be used to predict how load in a network element will be distributed.

To illustrate how random behavior is used in network design, assume that numerous observations of the habits of customers attempting to access a group of DTMF receivers have been made. With enough observations it becomes possible to predict the probability that a given number of attempts will occur in the busy hour. Probabilities are stated as a decimal number between zero and one; the sum of the probabilities always equals 1.0. If probabilities are plotted in a bar chart such as Figure 13.3, a pattern begins to develop. Countless observations have shown that arrivals tend to array themselves according to the curve shown by the solid line in Figure 13.3. This curve can be described by a formula known as *Poisson* distribution, so named after the French mathematician S. D. Poisson. Although Poisson distribution is not a perfect match for the distribution of incoming service arrivals, its accuracy is sufficient for network designs.

With a good estimate of patterns of call attempts, the second question is how long the average holding time of the group of DTMF registers will be. If holding times are plotted, they tend to follow the distribution shown in Figure 13.4. This distribution is described by an exponential curve shown by the solid line in Figure 13.4.

FIGURE 13.3
Poisson Distribution of Call Arrivals

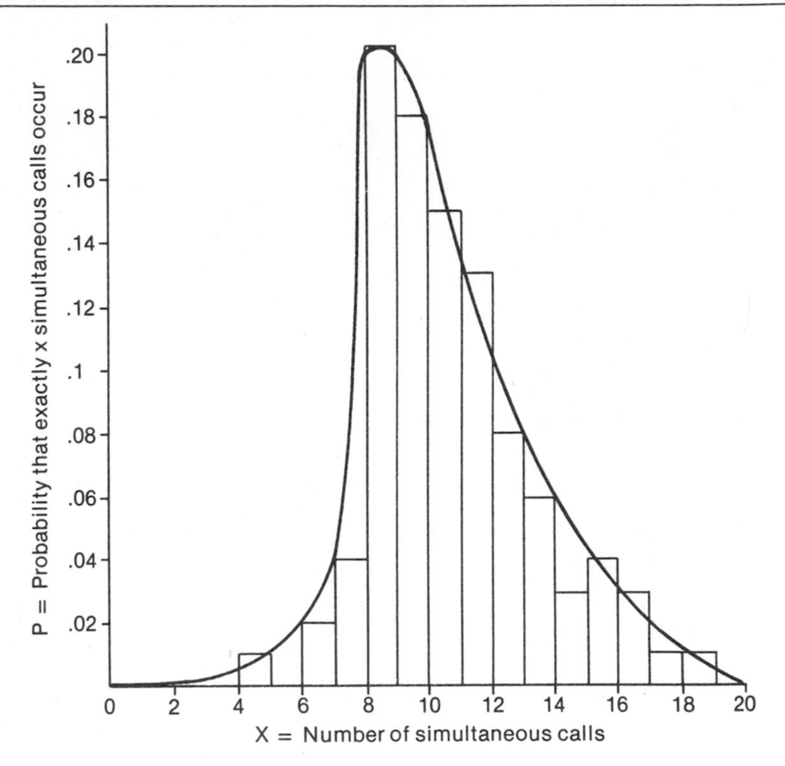

These distributions of attempts and holding time are modeled by tables or software that are selected according to the disposition of blocked calls. Assuming the arrival rate is random, the traffic formula is chosen according to the behavior of users in the queue. The Erlang B formula assumes blocked calls are cleared, that is, when a user encounters blockage, he does not immediately reenter the system. The Poisson formula assumes blocked calls are held. They are not actually held but instead the user immediately redials when blockage is encountered. For a given grade of service, the Poisson tables require slightly more trunks than Erlang B. Erlang C tables are used for the blocked calls delayed or queued assumption.

It is important to understand that traffic formulas are valid only if the attempts are random. Several things can affect randomness. A national or local emergency can drive the

FIGURE 13.4
Exponential Distribution of Service Times

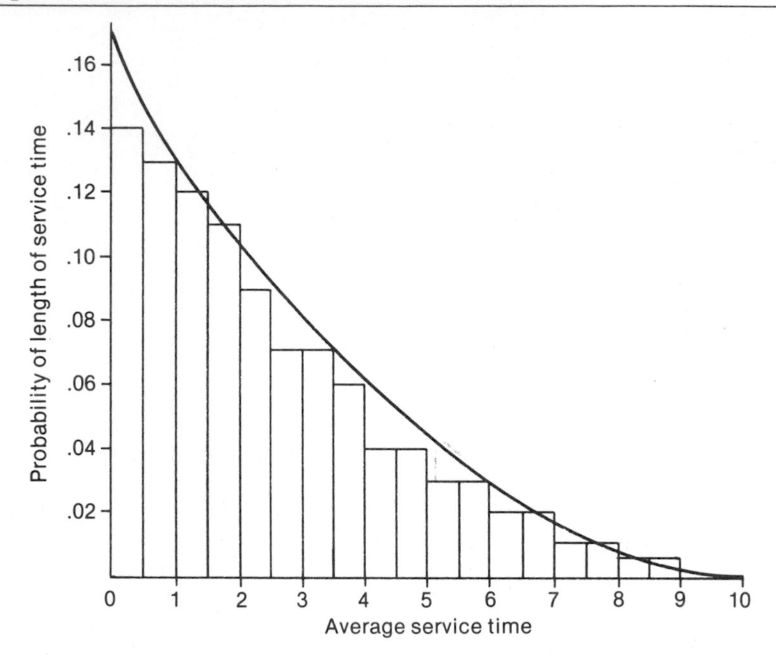

number of attempts far beyond the capacity of the network. This causes blockage and results in numerous retrials, which in turn generates more attempts. Most common control switching equipment exhibits a load-service response curve similar to that shown in Figure 13.5. With a gradually increasing load, service, as measured in amount of delay, follows a slope that is almost flat until a critical point is reached. At that point service degenerates rapidly, and the network, for all practical purposes, collapses. Networks are protected from overload by *flow control*, which is a series of procedures, described in a later section, to keep additional traffic off the network when excessive congestion will result.

It is also important to understand that in a common control network, overloading one critical element can cause the entire network to collapse. For example, if a network with ample trunks and switching capacity has too few DTMF receivers, the trunk and switching capacity will be unused because users will be unable to access them. Because equipment is generally sized on the basis of usage, *suppression* in one

FIGURE 13.5
Load/Service Curve of Typical Common Control Equipment

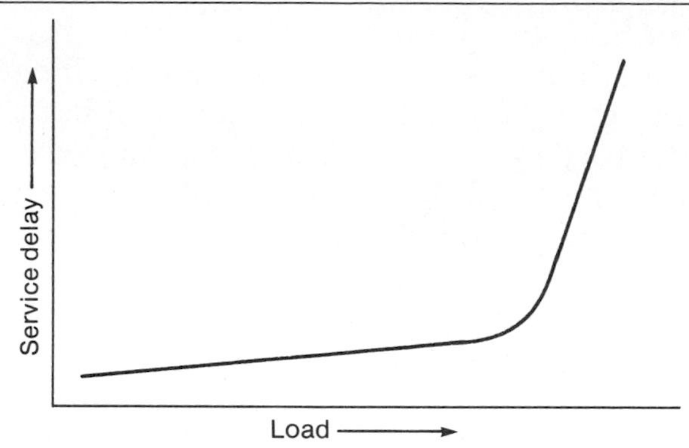

portion of the network makes it appear that capacity in other parts of the network has been provided too liberally. When the cause of suppression is removed, the true demand becomes apparent.

Common carriers, particularly those operating in a competitive environment, are well aware that blockage can result in loss of customers. Therefore, most common carriers design to avoid blockage in all but a small percentage of attempts. Blockage is often greater in private networks, however, because of the attempt to control communications costs by underproviding network capacity. Private networks can generally afford to provide a lower grade of service than public networks. If carried too far, however, the risk is great that the randomness of attempts will be lost, service will be poor, measurements will be invalid, and productivity will deteriorate as a result.

Busy Hour Determination

Since networks are designed to handle peak loads, knowledge of the peak-load periods, called the "busy hour," is needed to design the network. The busy hour is not a single hour but rather a composite network design point. Leveling is used to represent the design peak. A common factor for engineering switching machines is the 10 high-day (10HD) busy period. This factor is determined by averaging the amount of traffic in the busiest hour of the 10 days during the year when the highest traffic load is carried.

The busy hour for the network as a whole is not necessarily the busy hour for all circuit groups or all equipment. In fact, each group is likely to have its own busy hour because of the varying flow of traffic between nodes, particularly when they are in different time zones. Unless these load peaks are carefully observed and chosen as design criteria, congestion in some parts of the network will result.

Grade of Service

Networks are described as "loss" systems in which attempts are turned back when they encounter blockage or "delay" systems in which blocked attempts are placed in queue waiting for an idle server. The service grade in a loss system such as a trunk group is expressed as the probability of blockage. Trunk groups are sized by using tables or computer programs that express blockage as a decimal number. For example, a P.01 grade of service means that a call has a 99 percent chance of finding a vacant circuit and a 1 percent chance of being blocked. Traffic tables used for sizing circuit groups are indexed by grade of service; therefore, selecting a grade-of-service objective is the starting point in designing a network.

The grade of service for a delay system such as access to service circuits is expressed in terms of the percent of calls that encounter a delay in excess of the objective. For example, dial tone delay is often expressed as "no more than 1.5 percent of the attempts will be delayed more than three seconds."

It must be recognized that a network consists of several elements connected in tandem, and each with a possibility of blockage. For example, after a call has been dialed, it must contend for a path through the switching network, next for a trunk to the destination, then for a path through the terminating switching network. As each of these elements has its own probability of blockage, the probability of completing a call is the sum of the individual probabilities.

This additive nature of blockage probability can result in a network design that is wasteful of capacity if it is not carefully controlled. For example, if a circuit group is designed to a P.005 grade of service (0.5 percent blockage probability) but incoming calls encounter a terminating switching network with a high degree of blockage, a considerable amount of trunk capacity will be wasted in carrying calls that cannot be completed because they are blocked in the final link of the chain. This waste of circuit capacity is not detected by traffic load measurements because measures of attempts and holding time do not discriminate between successful and unsuccessful calls.

Traffic Measurements

The most difficult task in network design is obtaining accurate load data. With accurate data, a skilled designer can design a network that meets service objectives. In public networks with a clientele that remains fairly constant, the load can be predicted with reasonable accuracy. The number of call attempts and holding time can be predicted, and change occurs gradually enough that it can be accommodated with minor adjustments in capacity. However, when major fluctuations in load occur, it is sometimes impossible to adjust capacity quickly enough, and service deterioration is experienced in the form of slow dial tone or blocked calls.

The best predictor of traffic load is a historical analysis of past usage, but in many networks historical information is unavailable for predicting the traffic load. When an organization first establishes a new private network, the only source of information may be records of toll billed by the telephone company. If WATS lines, foreign exchange lines, and tie lines share the load and are unbilled prior to establishing the new network, insufficient data may exist to make a reliable prediction. The required capacity of circuits and equipment is predicted by observing past calling patterns. Historic information is a good predictor of the future if the conditions that generated the original demand remain unchanged. However, when changed conditions affect the demand, the original design may be rendered invalid and require modification.

Usage is measured by traffic usage recording (TUR) equipment that is either an external device attached to the circuit for measuring attempts and holding time or it consists of

TABLE 13.1
Partial Poisson Traffic Table

	Carried Traffic Load in CCS		
Trunks	*P.01*	*P.05*	*P.10*
1	0.4	1.9	3.8
2	5.4	12.9	19.1
3	16	29.4	39.6
4	30	49	63
5	46	71	88
6	64	94	113
7	84	118	140
8	105	143	168
9	126	169	195
10	149	195	224
20	399	477	523
30	675	778	836
40	964	1,088	1,157
50	1,261	1,403	1,482

Steps in using traffic tables:

1. Choose the appropriate queuing discipline.
2. Choose the objective grade of service.
3. Locate the load in CCS in the proper column.
4. Read the number of trunks required.

software registers that are assigned to groups of circuits and equipment items. Measurements are produced in raw number form and must be processed before the results can be applied to network design. For example, assume that a group of circuits is connected between two switching machines. From TUR readings the load is expressed in CCS. Traffic tables derived from the Poisson or Erlang formulas are consulted to determine the quantity of circuits needed to fill the demand for an objective grade of service. Table 13.1 is an example of a typical table and shows how it is used to determine the required number of circuits.

If the data in Table 13.1 is converted to call carrying capacity in CCS per circuit, the curve in Figure 13.6 is obtained. This curve shows an important factor in network design; large circuit groups are much more efficient in their traffic-carrying ability than small groups because large groups have less idle circuit time for a given grade of service.

FIGURE 13.6
Circuit Capacity as a Function of Size of Circuit Group

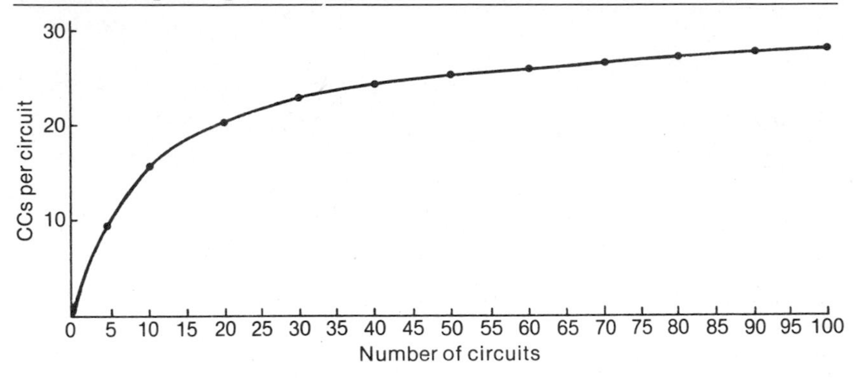

Often, usage data is not available to predict the demand accurately. For example, it may be known that an organizational change or a circuit routing change will affect demand on the trunk group between machines. In the absence of valid measurements, estimates must be made from other sources such as toll statements. Circuits are added on the basis of the estimates, but usage must be measured continually to evaluate service. Usage measurements indicate only the load that was actually carried. In addition to usage it is important to measure ineffective attempts, the number of calls queued, and length of time in queue (if queuing is used) to derive a more valid indication of the true demand.

A network can be visualized as a black box consisting of sources and dispositions (also called *sinks*) of calls as shown in Figure 13.7. Every attempt must be accounted for by its possible dispositions. The network administrator must determine whether the service provided by each disposition meets the organization's cost and service objectives. The combination of all calls—intramachine, those terminating on a tone trunk such as busy or reorder, off-net local calls, toll calls, WATS calls, etc.—comprise the total load on the switching network. The various ports must be sized to accommodate the offered load or blockage will result. The network design task becomes one of computing the quantities of circuits and equipment to accommodate each of the possible dispositions of a call.

FIGURE 13.7
Sources and Dispositions of Network Load

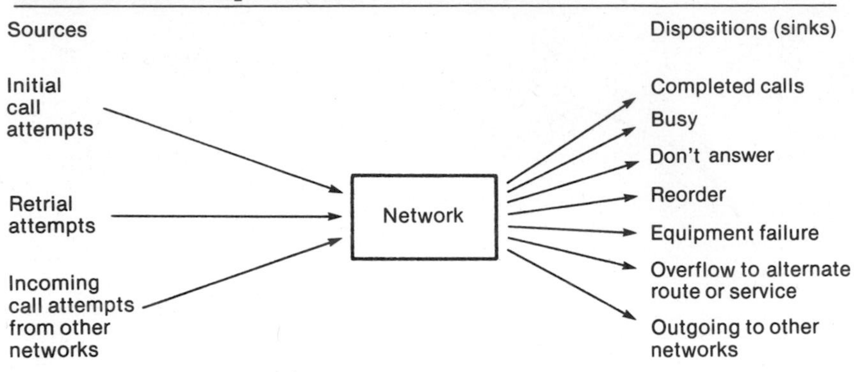

FIGURE 13.8
High-Usage and Final Routes in a Three-Node Network

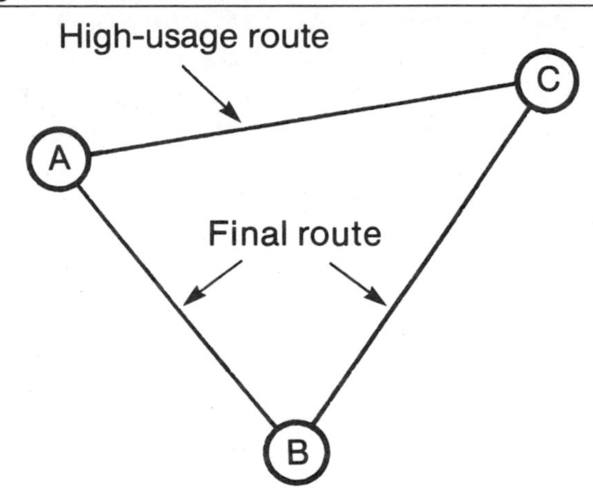

Alternate Routing

One of the possible dispositions of a call, as shown in Figure 13.7, is overflow to an alternate route. Alternate routing is one of several techniques used to maintain circuit occupancy at a high level. The decision whether to use alternate routing is based on cost. For example, assume that a network consists of three nodes as shown in Figure 13.8. The circuits between A and C are designed for a high level of occupancy and are called *high-usage* circuits. When the A–C trunk group is blocked, calls are routed through the tandem switch B over a *final route*. Obviously, different design criteria must be used on final and high-usage trunk groups. On high-usage groups an Erlang B table is used to size the group. On final groups the Poisson table is used at a higher grade of service.

Simulation

The techniques discussed so far for determining demand and capacity use mathematical modeling to calculate required capacity given an objective grade of service and estimated demand. Modeling is a valid way of designing a network if the demand is random and if it follows a known distribution of arrivals and holding time. Furthermore, modeling is a relatively inexpensive way of designing a network. Although computer design tools are useful in modeling, manual design using traffic tables is equally valid and can be conducted by a trained person with little apparatus beyond a book of tables and a hand calculator.

The modeling technique falls apart, however, when demand is not random and fails to follow predictable arrival and service rates. As shown previously, this occurs during blockage conditions. It is impossible to compute what happens to service when a network is blocked. Therefore, a design is valid only within the limits prescribed by the original assumptions. Besides overloading, the modeling technique is often unable to describe the load when the network handles a significant amount of data traffic. Data terminals and computers behave entirely differently than people using voice communications.

The holding time of data terminals tends to follow the probability distribution shown in Figure 13.9. Many computer calls are established only long enough to send one page or less of text, which at 1,200 b/s takes about 15 seconds, much less than the average telephone call. Other circuits used for bulk data transfer may have holding times much longer than average telephone calls—in the order of 30 minutes or more. Furthermore, the data busy hour may be different from the telephone busy hour. For example, workers may dial their electronic mail boxes immediately upon returning from lunch, putting a high momentary load on the system that is unlike that imposed by voice traffic.

FIGURE 13.9
Distribution of Holding Time Probability in a Data Network

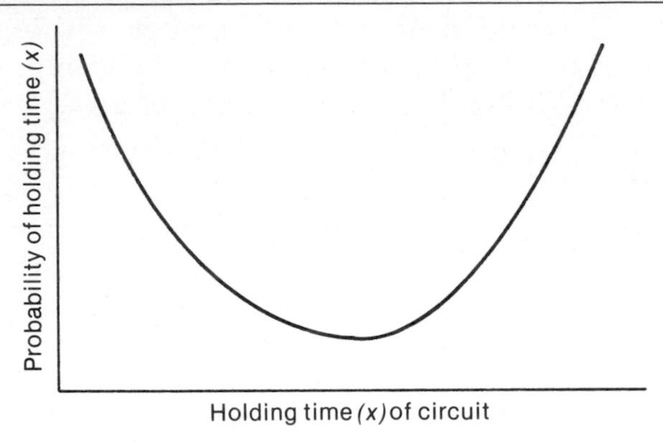

Where this lack of uniform distribution and randomness occurs, *simulation* can be employed to determine the effect on cost of service or capacity changes. A simulation program treats the network as a black box, and by varying the parameters, the costs and effects on service can be observed. Unlike modeling, simulation requires an elaborate program in addition to a trained operator to produce valid designs. Data from traffic measurements is still required to operate the program, but simulation differs from modeling in that service results can be observed outside the limits of the modeling formulas.

The chief limitation on any network design is the validity of observations about users. Furthermore, this behavior rarely remains constant. The smaller the network, the more difficult it is to predict behavior because small perturbations tend to have a large effect. For example, in a small private network a reorganization can change the calling patterns of the users and blockage may occur where previously capacity was ample.

Network Topology

Choice of topology, the pattern of interconnection of terminals and nodes in the network, is another element of network design. The principal topologies are described in Chapter 3. In both voice and data networks, all nodes can be

FIGURE 13.10
Multinode Concentrator Data Network

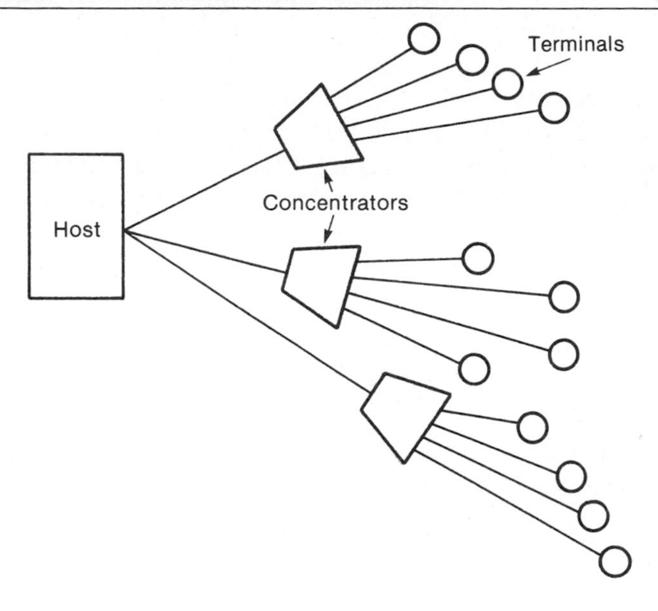

afforded equal access level, or they can be configured hierarchically as in AT&T Communications' network and the local telephone access networks. Private networks can also be hierarchically ordered with high-usage trunks between low-level nodes and final routing through a higher level node. The chief advantage to a hierarchical network is in the ability to conserve circuits by concentrating traffic in shared final circuit groups where insufficient traffic exists to justify high-usage groups.

Data circuits are also frequently configured in a hierarchical fashion. For example, Figure 13.10 shows a typical data network composed of terminals, concentrators, and a host computer. The design problem in such a network is where to place concentrators given capacity and cost of the concentrator, capacity and cost of interconnecting circuits, and the traffic volume between points. No formula has been developed for solving the node placement problem; and as the number of nodes increases, a trial-and-error approach becomes unmanageable.

To solve this problem, several heuristic algorithms have been developed to offer an orderly approach to the problem. Two such algorithms are:

- **Add algorithm.** All sites are initially assumed to be connected to the host. A concentrator location is chosen, and the cost of connecting each site to the concentrator is calculated. Additional concentrator locations are chosen and sites moved from the host or other concentrators until adding concentrators no longer improves cost.
- **Drop algorithm.** Several provisional concentrator locations are chosen, and terminals are initially connected. Concentrators are then dropped and terminals reassigned. If the total cost increases, the concentrator is restored; otherwise it is omitted.

Neither approach is necessarily superior. In complex networks with numerous nodes, different answers may be obtained from the different approaches. However, the final solution is likely to be dictated by other variables such as the location of administrative headquarters, availability of technical support, floor space, and other such factors.

DATA NETWORK DESIGN

Many of the same principles described for voice network design are applicable to data networks. Queuing theory is used to design data concentrator and statistical multiplexer networks. Other shared data equipment such as dial-up ports into a computer or pooled modems can be designed by measuring usage and comparing it to capacity determined from traffic tables.

Other types of data networks, however, have no counterparts in voice networks. Two examples are polled and packet switched data networks. Like voice networks, these can be sized by modeling or simulation, but with different techniques.

A typical application, terminal to CPU communication, is illustrated in Figure 13.11. In a multidrop network such

FIGURE 13.11
Multidrop Polled Network

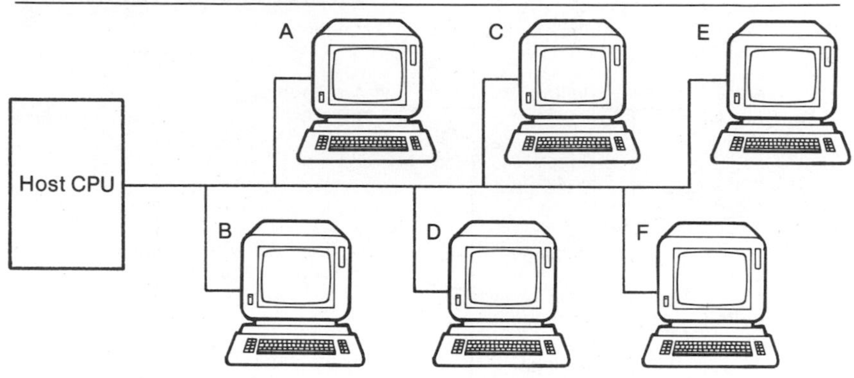

as this, response time is expressed as the interval between sending a request to the CPU and the time the first character of the response is received. A generalized response time model is shown in Figure 13.12. A terminal in a polled network sends either a block of data or returns a negative response to indicate that it has no data to send. The terminal response time of such a network depends on these factors:

- CPU response time.
- Transmission speed.
- Error rate.
- Data block size.
- Number of overhead bits (NAK messages, CRC, headers, etc.).
- Modem turnaround time (for a half-duplex circuit).
- Propagation speed of the transmission medium.
- Protocol error recovery method.

With these factors known, the network throughput and terminal response time can be calculated.

Packet switched networks are designed with similar criteria. The network throughput is related to transmission speed, block length, error rate, the number of overhead bits, propagation speed, and the network's packet routing algo-

FIGURE 13.12
Multidrop Polled Network Generalized Response Time Model

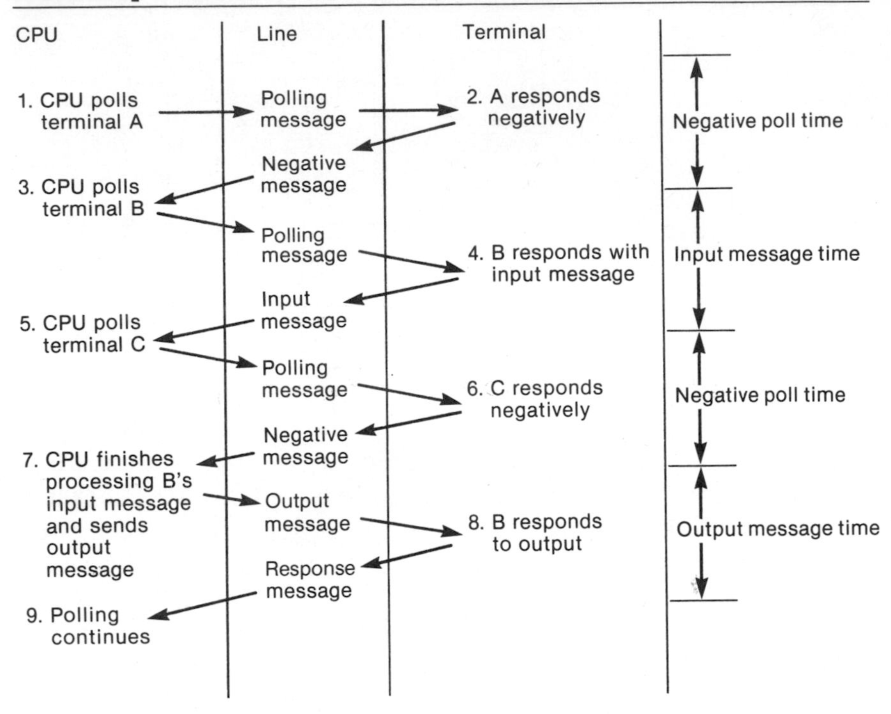

rithm. Throughput can be increased by using an efficient algorithm. The primary algorithms are:

- **Flooding.** Incoming packets are transmitted on every route except the one it arrived on.
- **Static.** Incoming packets are transmitted according to a fixed routing table.
- **Centralized.** A network control center optimizes the route based on load and congestion reports from the nodes.
- **Isolated.** Nodes adjust their routing tables from their knowledge of network load and congestion.
- **Distributed.** Nodes adjust their routing tables by exchanging information about network load and congestion.

Local area networks are a specialized form of packet switching network that use design criteria similar to long-haul packet networks. Contention networks present a unique design problem because not only are the other criteria of transmission speed, block length, error rate, and protocol important, but the frequency of collisions presents an additional factor that is difficult to model. Generalized design criteria are impossible to specify because of variations in the method of implementing a contention network.

Token passing networks with their deterministic access method are far easier to design. Given the characteristics of the input data and number, length, and frequency of messages, throughput for any terminal becomes a function of the network transmission speed, error rate, and the priority of the individual terminal. With all local networks, the manufacturer's instructions should be followed for design.

Like voice networks, data networks require flow control to prevent congestion. The network access nodes queue incoming messages and allow them on the network up to the capacity limits, holding the remainder in a buffer, or stopping them at the source. The physical layer protocol provides a method of flow control by actuating the "clear to send" (CTS) lead of the interface. To stop a terminal from overloading the network, its control node turns off the CTS signal until congestion has cleared, thereby holding the message in the terminal. Similarly, a receiving terminal is able to stop a transmitting terminal from overrunning its buffer capacity by sending an "X OFF" character. When it is capable of receiving data again, an "X ON" character is sent.

NETWORK MANAGEMENT

A network is a dynamic thing. The original premises on which it was designed never remain static, which means the network must be monitored continually and adjusted when change is indicated by the results. The process of monitoring and adjusting the network is called *network administration*.

Network Administration Tools

Network administration requires a number of skills and tools to keep the network in the proper cost/service balance. The first requirement is a current and valid report of the demand on all network elements. These reports, as previously discussed, must include attempts, usage, and ineffective attempts on every circuit and equipment group in the network.

Next, the administrator must have a full understanding of the network itself. Current records are required of all circuits and equipment, including identification, end points, and capacity. A network administrator must possess a comprehensive knowledge of how the network functions, what the service objectives are, and what the capacity is of each element. The administrator must also have information on common carrier services and rates. When capacity adjustments are warranted from the service results, changes in capacity must be made to retain the best cost/service balance. This requires knowledge of common carrier tariffs and the cost of other alternatives.

Network administration also requires knowledge of the current status of all circuits and apparatus. Although the administrator may not be directly responsible for maintenance, knowledge of trouble frequency and history are essential for diagnosing service results and timing adjustments of capacity. Status records should indicate, preferably on a real-time basis, all cases of equipment and circuits out of service. The length of outage should be monitored carefully so excessive clearing times can be brought to the attention of maintenance forces or the common carrier.

When blockage beyond planned levels is observed, the network administrator should have the ability to administer controls to relieve the blockage. The tools for relieving blockage include the following:

- Additional capacity can be provided. This may require temporarily removing restrictions on overflow to other networks.
- Some code groups can be blocked. This may require the ability to block codes in an originating machine to keep congestion caused by ineffective attempts off the network.
- Circuit groups can be rerouted temporarily. It may be necessary to cancel alternate routing or to revise the

rerouting tables in some machines temporarily to relieve congested machines or circuit groups.

Trunk directionalization can be applied. Two-way trunks can be temporarily converted to one-way trunks to allow a congested machine to place outgoing calls while insulating it from excessive incoming traffic.

Managing flow controls such as these requires that the administrator have the necessary skills and understand the network well enough to understand where and how to apply controls to relieve blockage. The feasibility of applying these controls is a function of the complexity of the network. In complex networks, mechanized assistance may be required.

Automatic Network Controls

Administration of a network composed of stored program control switching machines that are designed for mechanized network management can be simplified by applying automatic controls. Mechanized network management requires a central computer with a map of the total network and two-way data links to the network nodes. The network control system monitors load and transmits orders to the nodes to relieve congestion. Such systems are complex, expensive, and warranted only for networks composed of multiple nodes and large quantities of trunks.

STANDARDS

Network design and administration are not covered by standards. Instead, networks are designed and administered to cost and service objectives that are achieved by applying network design principles. The bibliography of this chapter lists publications that describe these principles more fully.

GLOSSARY

Blocked calls delayed (BCD): A variable used in queuing theory to describe the behavior of the input process when the user is held in queue when encountering blockage.

Blocked calls held (BCH): A variable used in queuing theory to describe the behavior of the input process when the user immediately redials when encountering blockage.

Blocked calls released (BCR): A variable used in queuing theory to describe the behavior of the input process when the user waits for a time before redialing when encountering blockage.

Busy hour: The composite of various peak load periods selected for the purpose of designing network capacity.

Erlang: A unit of network load. One erlang equals 36 CCS and represents 100 percent occupancy of a circuit or piece of equipment. Also used to define the input process under the BCC (Erlang B) and BCD (Erlang C) blockage conditions.

Exponential: A curve that describes the service time of users in a queue.

Flow control: The process of protecting network service by denying access to additional traffic that would add further to congestion.

Grade of service: The percentage of time or probability that a call will be blocked in a network.

Hundred call seconds (CCS): A measure of network load. Thirty six CCS represents 100 percent occupancy of a circuit or piece of equipment.

Modeling: The process of designing a network from a series of mathematical formulas that describe the behavior of network elements.

Network administration: The process of monitoring loads and service results in a network and making adjustments needed to maintain service and costs at the design objective level.

Network design: The process of determining quantities and architecture of circuit and equipment to achieve a cost/service balance.

Partial dial: A condition that exists when a user hangs up before dialing a complete telephone number.

Poisson: A formula that describes the distribution of arrival times at the input to a service queue.

Simulation: The process of designing a network by simulating the events and facilities that represent network load and capacity.

Suppression: A condition that exists when total service demand in a network is limited by the lack of capacity.

BIBLIOGRAPHY

Bellamy, John C. *Digital Telephony.* New York: John Wiley & Sons, 1982.

Cooper, Robert B. *Introduction to Queuing Theory.* New York: North Holland Publishing, 1981.

Doll, Dixon R. *Data Communications.* New York: John Wiley & Sons, 1978.

Freeman, Roger L. *Telecommunication System Engineering.* New York: John Wiley & Sons, 1980.

Lawson, Robert W. *A Practical Guide to Teletraffic Engineering and Administration.* Chicago: Telephony Publishing, 1983.

Martin, James. *Telecommunications and the Computer.* Englewood Cliffs, N.J.: Prentice-Hall, 1976.

———. *Computer Networks and Distributed Processing.* Englewood Cliffs, N.J.: Prentice-Hall, 1981.

Sharma, Roshan La.; Paulo T. deSousa; and Ashok D. Inglé. *Network Systems.* New York: Van Nostrand Reinhold, 1982.

Tannenbaum, Andrew S. *Computer Networks.* Englewood Cliffs N.J.: Prentice-Hall, 1981.

CHAPTER FOURTEEN

Power, Distributing Frames, and Other Common Equipment

Auxiliary equipment common to all types of central offices is called *common equipment* in telecommunications terminology. It includes:

- Interbay cabling and wiring.
- Central office alarm equipment.
- Distributing frames.
- Power equipment.
- Relay racks, cabinets, and supporting structures.

PBXs and tandem switches designed for private use are self-contained with common equipment included within switching cabinets. However, central office techniques are usually employed to borrow power for auxiliary equipment and to interconnect lines and trunks.

CABLE RACKING AND INTERBAY WIRING

Central office equipment is mounted either in cabinets or open relay racks, with the latter the most common for all systems except for PBXs and small tandem switches. In cabinetized equipment the interbay cabling is contained within the cabinet. In relay rack mounted equipment the cabling is external and is supported by overhead cable troughs or run through raceways in the floor. Because of the quantities of cables involved and the need for physical separation in some cables, overhead racking is the most common.

To control noise and crosstalk in central offices, the manufacturer's specifications must be followed for the type and layout of cabling. As with outside plant cable, the twist in central office cable is designed to control crosstalk and to prevent unwanted coupling between circuits. Furthermore, cables must often be run in separate troughs that are segregated by signal level and kept physically separated by enough distance so that signals from high-level cables cannot crosstalk into low-level cables. For some types of cable, shielding is required to further reduce the possibility of crosstalk.

DISTRIBUTING FRAMES

Temporary connections or those requiring periodic rearrangement are terminated on cross-connect blocks mounted in distributing frames. Distributing frames also provide an access point for testing cable and equipment. The size and structure of a distributing frame are dictated by the quantity of circuits to be connected. Figure 14.1 shows a typical frame layout. Cabling to the central office equipment is routed through openings at the top of the frame, is fastened to vertical members, and is turned under a metal shelf that supports the cross-connect blocks. The cross-connect blocks are composed of multiple metallic terminals mounted in an insulating material and fastened to the distributing frame. Equipment and lead identity are stenciled on the blocks.

Cross-connects are made by running "jumper" wire between blocks in a supporting wire trough. Connections are made at the block by one of three methods. In the oldest method, rarely used in modern equipment, the wire is stripped and soldered to the block. The second type uses wire wrapped connections in which the wire is stripped and

FIGURE 14.1
Central Office Distributing Frame

Courtesy Cook Electric Division of Northern Telecom Inc.

tightly wrapped around a post with a wire wrap tool. The third type of connection uses a split "quick clip" terminal that clamps the wire and pierces the insulation as it is inserted in the terminal with a special tool.

Some installations use modular distributing frames, which are carefully designed to keep the length of cross-connects to a minimum and are often administered by a computer. In small installations, distributing frame administration is usually not a problem. However, in large centers with thousands of subscriber lines and trunks terminated in the office, distributing frame congestion becomes a significant problem as quantities of cross-connect wires are piled in troughs. In large installations, it is important that the distributing frame be carefully designed and administered to keep the wire length to a minimum.

Protector Frames

Incoming circuits that are exposed to power or lightning are terminated on protector frames using ironwork similar to that in Figure 14.1. The protector module forms the connection between the cable pair and the attached equipment. As described in Chapter 7, if excessive current flows in the line the protector opens the circuit to the central office equipment. If excessive voltage strikes the line, carbon blocks inside the protector module arc across to ground the circuit. An exploded view of a protector module is shown in Figure 14.2. Modules are manufactured with gas tubes where these are needed to protect vulnerable central office equipment such as digital switches.

Combined and Miscellaneous Distributing Frames

Small installations use combined protector and distributing frames. The incoming cable pairs terminate on the protector frame where they are cross-connected to equipment terminated on the distributing frame. In large central offices the physical size of the frame may dictate separate protector and distributing frames. One or more distributing frames ter-

FIGURE 14.2
Protector Module

Courtesy Cook Electric Division of Northern Telecom Inc.

minate trunks, switching machine line terminations, and miscellaneous equipment such as repeaters, range extenders, and signaling equipment. Where multiple frames are required, tie cables are used to enable cross-connecting between cable and equipment terminated on different frames.

RINGING AND TONE SUPPLIES

Common equipment includes ringing, dial tone, call progress tone, and recorded announcement apparatus. In digital central offices, many of these tones are generated in software, so no external equipment is needed. In electromechanical and analog electronic central offices, external supplies are required.

Ringing machines are usually solid state supplies that generate 20 Hz ringing current at about 90 volts. Older ringing machines also generate dial tone. Older dial tone supplies are not compatible with DTMF dialing because they generate harmonics that fall inside the band pass of DTMF receivers. To prevent this, dial tone supplies are required with precisely generated tones. Busy tone and reorder supply tones are generated in the same manner as dial tones—either with solid state generators or with a branch circuit from the ringing machine.

Recorded announcements are stored in digital form in solid state memory or in analog form on magnetic tapes or drums. Most recorders contain multiple tracks for the several types of messages used in central offices.

ALARM AND CONTROL EQUIPMENT

All telecommunications equipment is designed with integral alarms in any circuit that can affect service. The extent and type of alarming varies with the manufacturer, but generally alarms are intended to draw attention to equipment that has failed or is about to fail, and to direct the technician to the defective equipment.

Equipment alarms light an alarm lamp on the equipment chassis and also operate external contacts that are used for remoting the alarm and for operating external audible and visual alarms. Most central offices contain an office alarm system to aid in locating failed equipment. Alarms are segregated into major and minor categories to indicate the seriousness of the trouble; different tones are sounded to alert maintenance personnel to the alarm class and location. In addition to audible alarms, alarm lamps are used to guide maintenance personnel to the room, equipment row, bay, and the specific equipment in trouble.

In offices designed for unattended operation, telemetering equipment is used to transmit the alarms to a distant center over ordinary telephone circuits. The alarm remote is generally a slave that reports only the identity of the alarm point. The central is typically equipped with a processor and data base that pinpoint the trouble and may also diagnose the cause. Some equipment, including most electronic switching systems, are designed to communicate with a remote that provides the equivalent of the local switching machine console. Other remote alarms report building status such as open door, temperature, smoke, and fire alarms.

Central offices designed for unattended operation are frequently equipped with control apparatus for sending orders from a distant location over a data circuit. For example, microwave and fiber optic equipment are equipped with control systems that enable transferring working equipment to a backup channel. Offices equipped with emergency generators are frequently arranged for engine start and shut down, and transfer to and from commercial power.

POWER EQUIPMENT

Central office equipment is designed to operate from direct current, usually −48 volts, which is the typical voltage supplied by central office charging and battery plants. Microwave equipment is usually designed to work on −24 volts DC in radio relay stations and −48 volts DC in central offices. Some central office equipment is designed to operate from alternating current (AC) and requires a DC to AC converter known as an *inverter* to provide an uninterrupted power source during power failures. AC operated equipment includes tape and disk drives, computers, and other such equipment that is not normally designed for DC operation. Commercial uninterruptible power source (UPS) equipment is an alternative device that contains a built-in battery supply.

Under normal conditions, battery charging equipment is driven by commercial AC power sources. Chargers, batter-

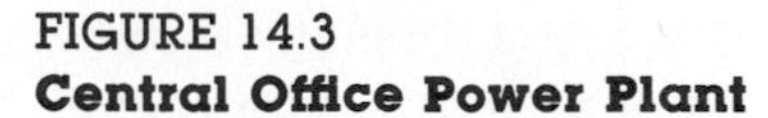

FIGURE 14.3
Central Office Power Plant

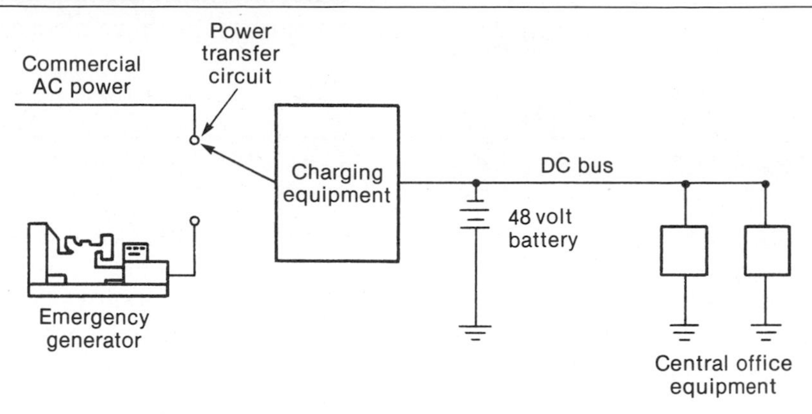

ies, and central office equipment are tied to a common bus bar as shown in Figure 14.3. When commercial power is on-line, the charging equipment carries the central office load; the batteries absorb only enough power to compensate for internal leakage and to filter noise on the power bus. When the power fails, the batteries carry the entire central office load up to the limits of their capacity. Most central offices are equipped with an emergency generator to carry the load and keep the batteries charged in case of prolonged power outage. The emergency generator is connected to the charging equipment through a power transfer circuit that cuts off commercial power while the generator is on line. Offices lacking emergency power equipment are usually equipped with circuitry for connecting an external generator.

Storage batteries use technology similar to automobile batteries. Lead acid and nickel cadmium cells are common, and some equipment uses batteries with solid electrolyte called *gel cells*. Power is distributed from the battery plant to the central office equipment over bus bars, which must be designed large enough to carry current for the total equipment load. To minimize the amount of voltage drop, batteries are installed as close to the equipment as possible.

Some types of central office equipment require voltage higher or lower than the nominal −48 or −24 volts used by most equipment. These voltages are supplied by either a separate charging and battery plant, or by using solid state power *converters.* Except for very high current loads, power converters are the preferred method of supplying other voltages.

STANDARDS

Common equipment is generally built to manufacturer's standards. The principal voltages used in central offices, −48 and −24 volts, are accepted by convention but are not regulated by telecommunications standards. The National Electrical Safety Codes and local codes apply to wiring commercial power to charging equipment, but the voltages used on central office equipment are too low to be considered hazardous.

The Bell New Equipment-Building System (NEBS) guidelines are followed by most manufacturers in their bay dimensions. Relay racks are standardized at 7, 9, and 11.5 feet in height and support equipment with 19- or 23-inch wide mounting panels. The mounting screw holes are also spaced at standard intervals.

Bell Communications Research Publications

PUB 49002 General Remote Surveillance Philosophy and Criteria for Interoffice Transmission Equipment.

PUB 51001 New Equipment-Building System (NEBS)—General Equipment Requirements.

EIA

RS-310-C Racks, panels, and associated equipment.

Institute of Electrical and Electronic Engineers

IEEE Standard 446-1980 IEEE recommended practice for emergency and standby power systems for industrial and commercial applications.

APPLICATIONS

Common equipment is separately engineered in central offices. In private networks, common equipment will usually be engineered and furnished by the switching equipment supplier. This section discusses the primary considerations in evaluating common equipment in a local private network environment.

Evaluating Common Equipment

Evaluation considerations discussed in previous chapters are equally applicable to common equipment. As with all telecommunications equipment, high reliability is imperative. Compatibility is important with alarm and control systems, but with power, distributing frame, and cabling, compatibility is generally not a problem if the equipment meets the specifications of the manufacturer of the interfacing equipment.

Environmental Considerations

An early consideration in establishing a system is to provide the floor space and environment required for its operation. The manufacturer's recommendations should be followed with respect to heating, air conditioning, air circulation, cabling, and mounting. The primary considerations are provision of:

- Sufficient floor space for expansion.
- Sufficient air conditioning and heating capacity.
- Ducts and raceways where required.
- Separate power equipment room where recommended by the manufacturer.

Work Space

Equipment areas should be designed to provide a physical working environment with adequate space and lighting for equipment maintenance. The manufacturer should install equipment to its standards and should specify aisle space between equipment lineups, lighting standards, and commercial AC outlets for powering test equipment.

Protection and Distributing Frames

Frame terminations should be provided for all equipment that requires rearrangement or reassignment. The primary considerations are the density of frame blocks and the amount of trough space provided for jumper wire. Block density is a trade-off between the amount of floor space consumed by the distributing frames and the difficulty of running multiple wires to small or congested blocks.

All cable pairs exposed to lightning strikes or power cross should be protected. This includes all pairs furnished by the local telephone company unless protection is included with the service. The manufacturer's recommendations should be followed with respect to gas tube protection.

Distributing frames should always be provided conforming to a plan designed to eliminate congestion and to facilitate productivity in placing and removing cross connects.

Power Equipment

The primary consideration in evaluating a power plant is its capacity. If a plant is designed to power a switching machine, it is important to ensure that enough spare capacity is included to power external transmission and signaling equipment. The amount of capacity required is a function of current consumed by the equipment and the amount of battery reserve if batteries are provided. In a private network the provision of batteries depends on whether the network is expected to remain in operation during power failures or whether it will be allowed to fail until the power is restored. The decision whether to provide an emergency generator is based on whether the battery reserve is enough to survive a long power outage.

Cabling

The number of leads to cable to distributing frames is an important consideration in installing equipment. Apparatus usually has numerous wiring options designed to accommodate special services. If not all leads are cabled to the distributing frame, use of certain options may be precluded in the future unless the equipment is recabled. However, if unneeded leads are terminated on to the distributing frame, extra costs will be incurred in cabling, frame blocks, and installation labor, and more frame space will be consumed by the extra terminations. Determining how many and what leads and options to wire to the frame is an important facet of central office equipment engineering.

Also of importance is proper segregation of cables. For example, cables carrying low-level carrier signals are usually separated from high-level cables, or the cables are shielded to prevent crosstalk. Data bus cables in many SPC switching systems must be isolated from other cables to prevent errors. The manufacturer's specifications must be rigidly adhered to in designing cable racks and troughs.

Maximum cable length is usually specified by the manufacturer. Most telecommunications equipment has critical lead lengths that must not be exceeded.

GLOSSARY

Converter: A device for changing central office voltage to another DC voltage for powering equipment.

Cable racking: Framework fastened to bays to support interbay cabling.

Control equipment: Equipment used to transmit orders from an alarm center to a remote site to perform operations by remote control.

Distributing frame: A framework holding terminal blocks that are used to interconnect cable and equipment and to provide test access.

Inverter: A device for generating AC from central office battery.

Jumper: Wire used to interconnect equipment and cable on a distributing frame.

Relay rack: Open ironwork designed to mount and support electronic equipment.

BIBLIOGRAPHY

AT&T Bell Laboratories. *Engineering and Operations in the Bell System*, 2d ed. Murray Hill, N.J.: AT&T Bell Laboratories, 1983.

Freeman, Roger L. *Reference Manual for Telecommunications Engineering*. New York: John Wiley & Sons, 1985.

MANUFACTURERS OF CENTRAL OFFICE COMMON EQUIPMENT

Cable and Central Office Structural Components

AT&T Network Systems
Amphenol
Anaconda-Ericsson Inc.
Brand-Rex Co.
General Cable Co.
3M Co. Telcomm Products Division
Newton Instrument Co. Inc.
Siecor Corporation
Standard Wire and Cable Co.

Distributing and Protector Frames

AT&T Network Systems
3M Co. Telcomm Products Division
Northern Telecom Inc., Cook Electric Division
Porta Systems Corp.
Reliance Comm/Tec

Alarm and Control Equipment

AT&T Network Systems
Badger/TTI
Communication Manufacturing Co.
NEC America Radio and Transmission Division
Rockwell International, Collins Transmission Systems Division

Power Systems and Storage Batteries

AT&T Network Systems
Exide Corporation
ITT Power Systems Corp.
Kohler Co.
LorTec Power Systems
Power Conversion Products, Inc.
Reliance Comm/Tec
Sola Electric

CHAPTER FIFTEEN

Microwave Radio

Microwave radio serves as the predominant long-haul transmission facility in the United States today. Since it was introduced shortly after World War II, microwave has undergone continual improvement, resulting in a low-cost, reliable broadband facility. Microwave's primary advantage is its ability to bridge obstructions where right-of-way is expensive to obtain or difficult to span. Furthermore, its capacity is significant. A single radio channel can carry as many as 6,000 voice channels in 30 MHz of bandwidth.

The primary drawback to microwave results from its free space radiation. Only a limited frequency spectrum is available, and coordination between microwave paths is required to prevent interference. Because of existing congestion, it is often impossible to obtain frequency assignments in metropolitan areas. Table 15.1 lists the microwave common carrier and industrial or operational-fixed frequency band assignments in the United States.

MICROWAVE CHARACTERISTICS

The general principles of microwave radio are the same as lower frequency radio. A radio frequency (rf) signal is gen-

TABLE 15.1
Common Carrier and Operational Fixed (Industrial) Microwave Frequency Allocations in the United States

Common Carrier	*Operational Fixed*
2.110–2.130 GHz	1.850–1.990 GHz
2.160–2.180 GHz	2.130–2.150 GHz
3.700–4.200 GHz	2.180–2.200 GHz
5.925–6.425 GHz	2.500–2.690 GHz (television)
10.7–11.7 GHz	6.575–6.875 GHz
	12.2–12.7 GHz*

*Based on noninterference with direct broadcast satellite service.

erated, modulated, amplified, and coupled to a transmitting antenna. It radiates in free space to a receiving antenna where a sample of the radiated energy is received, amplified, and demodulated. The primary differences between microwave and lower frequency radio are the length and behavior of the radio waves. For example, VHF television channel 2 has a wavelength of about 20 feet. To gain the maximum efficiency, a half-wave antenna receiving element is about 10 feet long. A 4 GHz microwave signal (1 GHz equals 1,000 MHz) has a wavelength of about 3 inches so that an effective antenna at microwave frequencies is small compared to those at lower frequencies.

At microwave frequencies, radio waves behave similarly to light waves. They can be focused with large parabolic or horn antennas similar to the ones shown in Figure 15.1, or they can be reflected with large flat passive reflectors to redirect the path around obstructions.

A second significant fact about microwave is the amount of bandwidth available in the microwave spectrum. To put this bandwidth in perspective, assume that the effective microwave frequencies extend from 2 GHz to 14 GHz (higher frequency bands are usable but at short repeater spacing). This 12 GHz of bandwidth is six times the total frequency spectrum ranging from very low frequency to the bottom of the microwave band. The essential point is that even though bandwidth is limited, a significant amount is available for point-to-point communication in the microwave bands. However, competition for it is intense.

FIGURE 15.1
Microwave Antennas

Courtesy Andrew Corporation.

a. Parabolic microwave antenna.

Courtesy Andrew Corporation.

b. Horn microwave antennas.

A third essential characteristic of microwave also relates to its lightwave-like behavior. Because microwaves follow line-of-sight paths, it is possible to operate stations on the same frequency in close proximity without mutual interfer-

FIGURE 15.2
Direct and Reflected Microwave Paths between Antennas

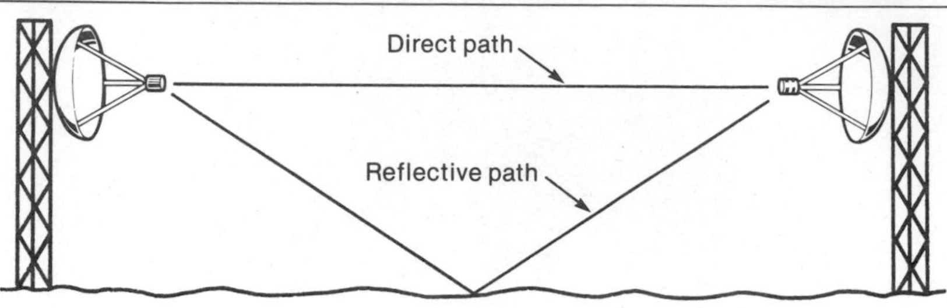

ence. On lower frequencies, radio waves cannot be narrowly focused to prevent them from radiating in all directions and interfering with other services on the same frequency.

On the minus side of the ledger, microwaves have some of the undesirable characteristics of light waves, particularly at the higher frequencies. The primary problem is fading. Microwave fading is caused by multipath reflections and attenuation by heavy rain. Multipath reflections occur when the main radio wave travels a straight path between antennas, but a portion of it is reflected over a second path as shown in Figure 15.2. The reflected path is caused by some changing condition such as a temperature inversion, a heavy cloud layer, or reflection off a layer of ground fog. The reflected wave, taking a longer path, arrives at the receiving antenna slightly out of phase with the transmitted wave. The two waves added out of phase cause a drop in the received signal level. A second cause of fading is heavy rain, which absorbs part of the transmitted power at frequencies higher than about 10 GHz. The two primary causes of microwave path disruption, fading and equipment failures, are partially alleviated by diversity as described in a later section.

MICROWAVE TECHNOLOGY

Microwave routes are established by connecting a series of independent radio paths with repeater stations. Repeater spacing varies with frequency, transmitter output power, antenna gain, antenna height, receiver sensitivity, number of voice frequency channels carried, and the free space loss of the radio path. At the high end of the band, around 18 to 24 GHz, repeaters are required at intervals as close as one mile. At the low end of the band, repeater spacings of up to 100 miles are possible.

Modulation Methods

Microwave systems are broadly classed as digital or analog, depending on the modulation method.

Analog Modulation

Analog radio is either amplitude or frequency modulated, with the latter comprising the bulk of equipment in service today. Recent developments have proven the suitability of single sideband (SSB) modulation for long-haul analog microwave. The primary advantage of SSB is its more efficient use of bandwidth than FM. In the 30 MHz bandwidth of the 6 GHz common carrier band, 2,400 voice channels is the theoretical limit that current FM technology can support while meeting channel noise objectives. This is an efficiency of 12.5 KHz of microwave bandwidth per voice frequency channel. With SSB AM, modern equipment is capable of carrying 6,000 channels for an overall efficiency of 5 KHz per voice frequency channel.

Digital Modulation

Direct modulation of an rf carrier with a digital signal has been commercially feasible since about 1975. The principal limit in the past has been the FCC requirement that digital radio use the spectrum as efficiently as analog radio. The problem with digital radio lies in the 64 kb/s coding of the voice channels. With a coding scheme that yields one bit per Hz of bandwidth, a digital channel would occupy 64 KHz, compared to the 12.5 KHz bandwidth efficiency of an FM analog radio. This excessive consumption of frequency spectrum impeded the use of digital radio until more efficient modulation methods could be developed. The earliest systems used phase shift modulation. Two radio channels were applied to the same frequency on cross-polarized antennas. Cross-polarization means that one radio operates with waves vertically polarized and the other radio with horizontally polarized waves. When the system works properly, the cross-polarization discrimination is enough to prevent interference between channels on the same frequency.

Later modulation systems employ more efficient phase shift keying (PSK) or quadrature amplitude modulation (QAM) systems. QAM, which is the system used in the latest microwave generation, is a simultaneous amplitude modulation-phase shift system. The earliest are 16 QAM systems, which are still being manufactured. This system supports 1,344 voice channels on a 30 MHz radio channel at 90 mb/s for an efficiency of 3 bits per Hz. The latest technology, 64 QAM, supports 2,014 channels at 135 mb/s for a spectral efficiency of 4.5 bits per Hz.

The major advantage of digital radio results from regeneration of the signal at each repeater point. If the incoming signal is sufficiently free of interference to allow the demodulator to distinguish between zeros and ones, digital radio provides the same high-quality, low-noise channel that T carrier provides. However, unlike analog radio that becomes progressively noisy during fades, digital radio remains quiet until a failure threshold is reached, at which point the bit error rate (BER) becomes excessive and the radio "crashes."

Although the signal is regenerated at each repeater, errors are cumulative from station to station and cannot be corrected. Therefore, the errors that occur in one section are repeated in the next section where additional errors may occur, until finally the signal becomes unsuitable for data transmission. For voice, however, the errors have relatively little effect. Research is currently underway into forward error correction techniques similar in concept to data forward error correction discussed in Chapter 3. This technique will extend the usable range of digital microwave.

In addition to the advantage of higher quality, digital microwave offers the advantage of directly interfacing T-1 carrier circuits without use of channel banks. This is particularly advantageous for transporting circuits between digital switching systems.

Data under Voice (DUV)

Until transcontinental fiber optic systems are completed, cross-country terrestrial digital circuits are available only by digitizing a portion of the capacity of analog microwave radios. AT&T's data under voice (DUV) is implemented on a system that provides one DS-1 channel per radio channel in the frequency range of 0 to 470 KHz. This frequency band is unused capacity below the 540 KHz lower range of the analog multiplex systems that ride on the radio.

DUV uses a seven-level partial response modulation method to encode the 1.544 mb/s digital signal into a 470 KHz-wide frequency band. Unlike digital radios, the signal is not regenerated at every repeater point. Instead, it is amplified as an analog signal and regenerated at approximately every third repeater station. It is primarily intended for implementing Dataphone® Digital Service (DDS) that is described in more detail in Chapter 17.

Bit Error Rate

The most important measure of digital microwave radio performance is the bit error rate (BER). BER is expressed as the number of error bits per transmitted bit and is usually abbreviated as an exponential fraction. For example, one error per billion transmitted bits is expressed as 10^{-9}. In a 90 mb/s system this would result in an average of one error in 11.1 seconds. A BER of 10^{-6} is generally accepted as the highest BER that can be tolerated for digital data transmission over microwave. At a BER of 10^{-3} a radio is considered failed, although voice transmission can still take place at this error rate.

Diversity

To guard against the effects of equipment and path failure, microwave systems use protection or diversity. *Space diversity* is achieved by spacing receiving antennas several feet apart on the same tower. This system protects against multipath fading because the wavelength of the signal is so short that the phase cancellation that occurs at one location will have little effect on an antenna located a few feet away.

Another protection system, permitted on common carrier bands, is *frequency diversity.* This system uses a separate radio channel operating at a different frequency to assume the load of a failed channel. When fades occur they tend to affect only one frequency at a time, so frequency diversity provides a high degree of path reliability. The primary disadvantages of this system are the use of the extra frequency spectrum and the cost of the additional radio equipment.

Frequency diversity is not permitted in most noncommon carrier frequency bands. Therefore, many microwave systems are established with *hot standby* diversity. In a hot standby system, two transmitter and receiver pairs are coupled to the antenna, but only one system is working at a time. When one system fails, the hot standby unit automatically assumes the circuit load. Hot standby protection is effective only against equipment failure. Fading and absorption, which affect the microwave path between stations, cannot be overcome by hot standby protection.

Transfer to a protection system is limited by the received noise level in an analog radio or by BER in a digital radio. When the noise or BER becomes excessive on one of the protected channels, the diversity switch initiator sends an order to the transmitting end to switch the entire input signal to the protection channel. Switches can also be initiated manually to clear a channel for maintenance. In any protection system, some loss of signal is experienced before the protection channel assumes the load. Many systems are capable of performing a "hitless" switch when a channel is manually transferred, but in case of equipment failure or fade, degradation will be experienced in the form of noise, excessive data errors, or both.

Protection systems are designed to protect working channels on a "one-by-one" or "one-by-N" basis with "N" being the number of working channels on the route. One-by-one protection is not permitted by the FCC where the application requires more than one radio channel because this method is wasteful of frequency spectrum. Figure 15.3 illustrates the three different applications of protection; frequency diversity, space diversity, and hot standby.

FIGURE 15.3
Microwave Protection System

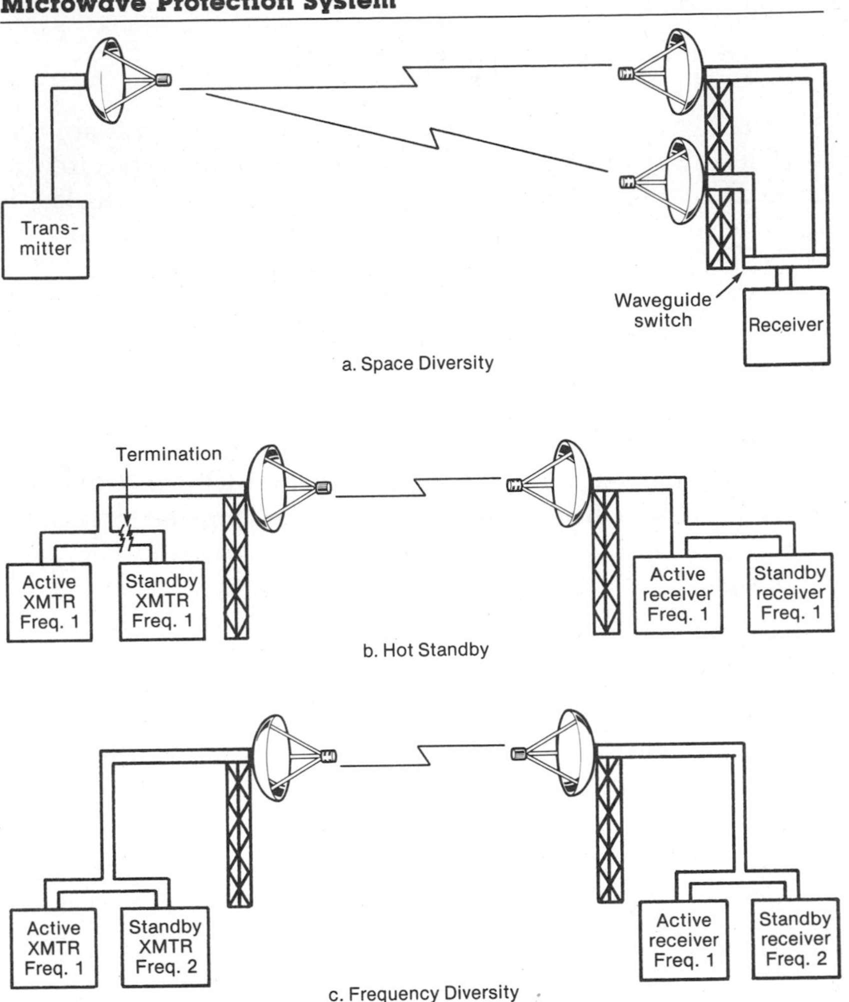

Microwave Impairments

Microwave signals are subject to impairments from these sources:

- Equipment, antenna, and waveguide failures.
- Fading and distortion from multipath reflections.

- Absorption from rain, fog, and other atmospheric conditions.
- Interference from other signals.

Microwave reliability is expressed as percent availability, which is the percentage of the time communications circuits on a channel are usable. Because of path uncertainties, a satisfactory reliability level is attainable only with highly reliable equipment. Fortunately, equipment reliability has progressed to the point that little down time is caused by equipment failures, and those failures that do occur can be protected by diversity.

Path Engineering

Microwave path reliability is less predictable and controllable than equipment performance. The first factor to consider is obtaining a properly engineered path. The path is designed by selecting repeater sites with availability of real estate, lack of interference with existing services, accessibility for maintenance, and sufficient elevation to overcome obstacles in the path.

In microwave path engineering it is not enough to have line-of-sight communication between stations; it is also necessary to have a minimum amount of clearance over obstacles. If insufficient clearance exists over buildings, terrain, or large bodies of water, the path will be unreliable because of reflection, or path bending. Reliable microwave paths must be engineered to minimize reflections as shown in Figure 15.4.

Rain Absorption

Rain absorption is also a common cause of signal fading at frequencies above about 10 GHz. At these frequencies the raindrop size is a significant fraction of the signal wavelength (wavelength is about 1 inch at 10 GHz). Absorption is a most significant impairment in areas of heavy rainfall with large drop size such as the Gulf Coast and southeastern United States. Conventional diversity is not effective against rain absorption because rain fading is not frequency selective. The most effective defenses are frequency diversity using a

FIGURE 15.4
Microwave Path Profile

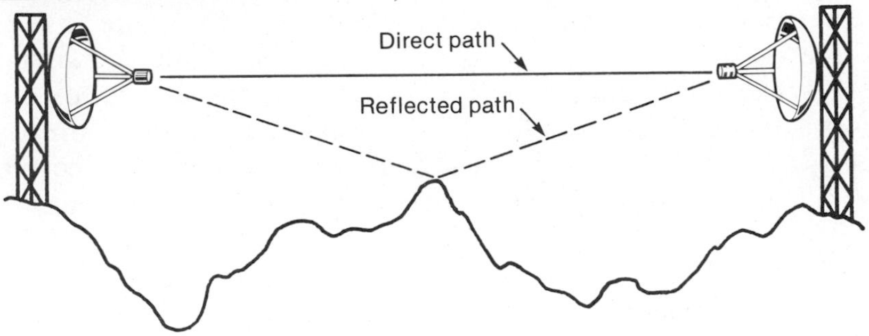

Adequate clearance is required over obstacles to minimize fading effects of reflected path.

lower band such as 6 GHz, if permitted by the FCC, and use of large antennas and closely spaced repeater stations.

Multipath Fading

Multipath fading is a source of impairment in both analog and digital microwave. It is caused by conditions that reflect a portion of the signal so both the main wave and the reflected wave arrive at the receiving antenna slightly out of phase. The phase differences between the two signals causes a reduction in the received signal level (RSL). Multipath reflections usually do not affect all frequencies within a band equally, which results in signal distortion within the received pass band. Distortion is of particular concern with digital microwave, which is susceptible to a higher BER under multipath fading conditions. One way of minimizing the effects of distortion is by use of an *adaptive equalizer,* a device inserted in the receiver to cancel the effects of distortion within the passband.

The *fade margin* is an important characteristic of a microwave system. It is defined as the depth of fade that can be tolerated before the channel noise level becomes excessive in an analog radio or before the BER becomes intolerable in a digital radio. Digital radio specifications usually include the *dispersive fade margin,* which states the tolerance of the radio to the frequency selective fades that cause received signal distortion.

Both frequency and space diversity are effective defenses against multipath distortion. With a second receiving antenna mounted a few feet below the first on the same tower, the main and reflected paths do not equally affect the signal received in both antennas. The system selects the best of the two signals. Frequency diversity is also an effective multipath distortion defense because of the frequency selective nature of signal reflections. However, frequency diversity is not permitted for all types of service.

Other defenses include an effective path profile study with proper site selection and tower height to provide adequate clearance over obstacles. Also, the use of large antennas focuses the transmitted signal more narrowly and increases the received signal level at the receiver.

Interference

Adjacent channel and overreach interference are other microwave impairments. Overreach is caused by a signal feeding past a repeater to the receiving antenna at the next station in the route. It is eliminated by selecting a zigzag path or by using alternate frequencies between adjacent stations.

Adjacent channel interference is also a potential source of trouble in a microwave system. Digital radios, particularly those using QAM modulation, are less susceptible to adjacent channel interference than PSK or FM analog radios because of the bandpass filtering used to keep the transmitter's emissions within narrow limits. In multichannel radio installations, cross-polarization is employed to prevent adjacent channel interference. In this technique, channel combining networks are used to cross polarize the waves of adjacent channels. Cross-polarization discrimination adds 20 to 30 dB of selectivity to adjacent channels.

Heterodyning versus Baseband Repeaters

Analog microwave repeaters use either of two techniques to amplify the received signal for retransmission: *heterodyning* or *baseband.* In a baseband repeater the signal is demodulated to the multiplex (or video) signal at every repeater point. In heterodyne repeaters the signal is demodulated to an intermediate frequency, usually 70 MHz, and modulated or

FIGURE 15.5
Channel Dropping at Main Microwave Repeater Stations

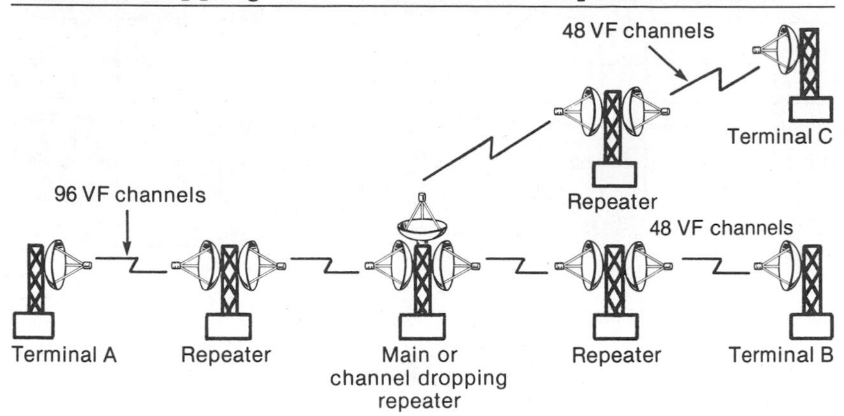

heterodyned to the transmitter output frequency. Heterodyne radio is reduced to baseband only at main repeater stations where the baseband signal is required to drop off voice channels.

The primary advantage of baseband radio is that some carrier channel groups can be dropped off at repeater stations. For example, in Figure 15.5 at the point where the microwave route branches, FM terminal equipment would be required in a heterodyne radio to split the baseband. Heterodyne radio has the advantage of avoiding the distortions caused by multiple modulation, demodulation, and amplification of a baseband signal. Therefore, heterodyne radio is employed for transcontinental use with drop-off points only at major junctions.

Multiplex Interface

Digital microwave is interfaced to multiplex equipment through either a standard or a special digital interface. Most systems marketed in the United States provide a standard DSX interface to one, two, or three DS-3 signals or to one DS-2 signal. Some radios designed for light route application have sub DS-3 interfaces consisting of 12 or 14 DS-1s and requiring special multiplexes to interface DS-1 signals to the radio.

FIGURE 15.6
General Model of an Analog Microwave System

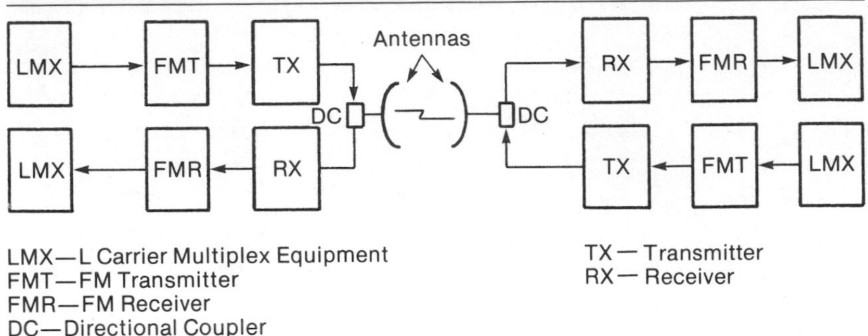

LMX—L Carrier Multiplex Equipment
FMT—FM Transmitter
FMR—FM Receiver
DC—Directional Coupler
TX — Transmitter
RX — Receiver

Analog microwave is connected to analog multiplex through frequency modulated transmitter (FMT) and frequency modulated receiver (FMR) equipment. The multiplex baseband signal is connected to the input of an FMT, which generates a frequency modulated intermediate frequency (if), usually 70 MHz. This signal is applied to the input of the radio as shown in Figure 15.6 and is modulated to the final rf output frequency. At the receiver the incoming signal is boosted by rf and if amplifiers and is connected to the input of the FMR. The output of the FMR is a baseband signal that is coupled to the multiplex equipment.

Microwave Antennas, Waveguides, and Towers

Microwave antennas are manufactured as either parabolic "dishes" or horns and range in size from 1 or 2 feet for short, high-frequency hops, to antennas 100 feet in diameter for earth station satellite service. Antenna efficiency is expressed in terms of "gain." Gain is a relative term referred to the performance of a free-space mounted dipole antenna. The antenna itself does not technically provide amplification, but it does provide a gain in efficiency over a dipole. This concept is illustrated in Figure 15.7, which shows the difference between a dipole radiating equally in all directions and a microwave antenna that focuses the signal to provide a narrower beam consisting of a major center lobe and side lobes of lesser intensity. The amount of gain is proportional to

FIGURE 15.7
Antenna Radiation Patterns

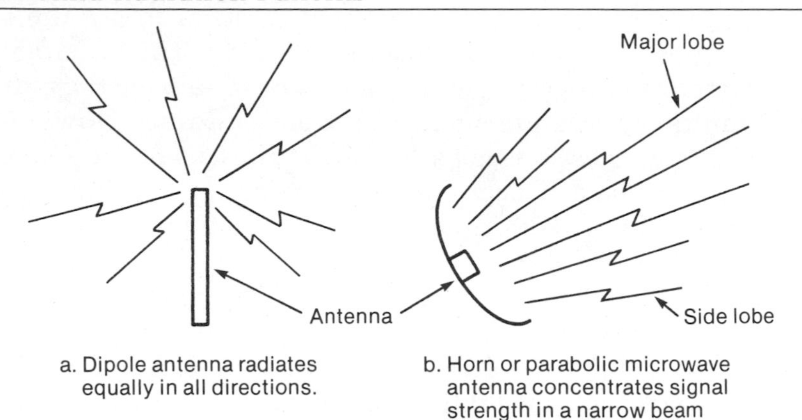

a. Dipole antenna radiates equally in all directions.

b. Horn or parabolic microwave antenna concentrates signal strength in a narrow beam with minor side lobes.

the physical characteristics of the antenna, most notably its diameter.

At lower frequencies, microwave antennas are fed with coaxial cable. Coaxial cable loss increases with frequency; therefore, most microwave systems use waveguide for the transmission line to the antenna. Waveguide is circular or rectangular, with dimensions designed for the frequency range.

Multiple transmitters and receivers can be coupled to the same waveguide and antenna system by the use of *branching filters. Directional couplers* are waveguide hybrids that allow coupling of a transmitter and receiver to the same antenna. This technique is often used in repeater stations to enable use of one antenna for each direction of transmission.

Antennas are mounted on rooftops if possible, or if more elevation is needed, on towers. Antennas must be precisely aligned. They are first oriented by eye or calculated azimuth and then are adjusted to maximum received signal level.

Microwave towers are supplied in guyed and self-supporting configurations. Self supporting towers require less space and must be designed more rigidly to support the antenna against the effects of weather. If enough land is available to accommodate down guys, a less expensive guyed tower can be used. Tower rigidity is important because excessive flexing can disorient the antennas.

Entrance Links

The facility to connect the final radio station in a route to the terminal equipment is the *entrance link*. The preferable way to terminate a microwave route is by mounting the radio and multiplex equipment in the same building. However, frequency congestion or path obstructions in metropolitan areas often make it necessary to terminate the radio some distance from the multiplex terminal. In this case, an entrance link is required.

Entrance links operate at baseband or, for short distances, at intermediate frequencies. Either coaxial cable or lightwave can be used for digital radio entrance links. With analog radio, coaxial entrance links are the rule because lightwave is not linear enough to support analog multiplex.

Digital Termination System

In 1981, the FCC opened the frequency band of 10.55 to 10.68 GHz for Digital Termination System (DTS), a short-range microwave service designed for metropolitan digital communications. DTS is implemented in sectors similar to the techniques used in cellular mobile radio. A central node broadcasts a digital signal through a sector of 90 to 120 degrees. Interference between sectors is prevented by a combination of different frequencies and cross-polarization of antennas. The central node broadcasts to one or more remotes in a sector at data speeds up to 1.544 mb/s. As many as 24 devices can be connected to the DTS equipment. The remote-to-central link consists of one or more transmitters, all operating on a single frequency. Remote units use a time division multiple access (TDMA) system that allows each unit to transmit in an allotted time slot to prevent mutual interference.

DTS has a range of 7 to 10 miles from the central node to the remotes. Range depends on the amount of rain absorption and on multipath reflections, which can be significant because of buildings and other obstructions. In the southwest, the longer range is possible, while in heavy rain areas of the southeast the 7-mile limit is generally the maximum achievable.

Short-Haul Digital Microwave

As microwave technology advances, frequencies that were once considered experimental are being opened for general use. Two bands of frequencies in the 18 and 23 GHz range are authorized for point-to-point use in both the common carrier and operational-fixed services. These systems are designed for operation over distances of 2 to 14 miles, with the range depending on terrain, polarization, and rainfall. Transmission over flat, reflective terrain with horizontal polarization in high-rainfall areas will yield the shorter ranges, while transmission over rugged terrain with vertical polarization in low-rainfall areas will allow greater range.

The radio is designed for easy installation. It can be mounted on rooftops, or, in some cases, the unit can be aimed out of windows. Units are available for DS-1 (24 voice channels) or DS-2 (96 voice channels) operation. The primary use is for linking local area networks and voice and data transmission between locations where broadband common carrier facilities are unavailable or prohibitively expensive. The equipment is relatively inexpensive, easy to install, and not difficult to license. It primary drawback is the occasional outages that occur when the received signal level fades.

STANDARDS

Microwave standards and specifications are set by the Federal Communications Commission in the United States and by the Consultative Committee on International Radio (CCIR) internationally. The FCC licenses transmitters only after the equipment has been type accepted. FCC rules and regulations list the operating rules for radio equipment within the United States. The EIA has established wind loading zones in the United States. for use in radio, tower, and antenna design, and several electrical and mechanical criteria for antenna and waveguide design. The FAA specifies tower lighting requirements.

Bell Communications Research

CB 119 Interconnection Specification for Digital Cross-connects.

PUB 43501 6 GHz Digital Radio Requirements and Objectives.

PUB 43502 11 GHz Digital Radio Requirements and Objectives.

CCIR

Volume IV/IX-2 of the 15th Plenary Assembly of the CCIR Frequency Sharing and Coordination Between Systems in the Fixed-Satellite Service and Radio Relay Systems.

Volume IX of the 15th Plenary Assembly of the CCIR Fixed Service Using Radio Relay Systems.

CCITT

G.411 Use of radio-relay systems for international telephone circuits.

G.412 Terminal equipments of radio-relay systems forming part of a general telecommunications network.

G.423 Interconnection at the baseband frequencies of frequency-division multiplex radio-relay systems.

G.441 Permissible circuit noise on frequency-division multiplex radio-relay systems.

EIA

RS-195-B Electrical and Mechanical Characteristics for Terrestrial Microwave Relay System Antennas and Passive Reflectors.

RS-200-A Circular Waveguides.

RS-222-C Structural Standards for Steel Antenna Towers and Antenna Supporting Structures.

RS-210 Terminating and Signaling Equipment for Microwave Communication Systems.

RS-250-B Electrical Performance Standards for Television Relay Facilities.

RS-252-A Standard Microwave Transmission Systems.

RS-261-B Rectangular Waveguides.

RS-271-A Waveguide Flanges—Pressurizable Contact Types for Waveguide Sizes WR90 to WR2300.

RS-285 Waveguide Flanges—Dual Contact Pressurizable and Miniature Type for Waveguide Sizes WR90 to WR975.

FCC

Part 21 Rules and Regulations Domestic Public Fixed Radio Service (Common Carrier).
Subpart I Point-to-Point Microwave Radio Service.
Part 94 Rules and Regulations Private Operational-Fixed Radio Service.
Subpart F Digital Termination Systems.

APPLICATIONS

Following World War II, when microwave was used on a limited basis, the technology has been transformed from an experimental medium with limited reliability to the point that it now is the predominant carrier of long-haul telecommunications circuits. Microwave also is finding application in short-haul services with the opening of bands above 15 GHz to commercial service. This section lists the most important factors to evaluate in applying microwave.

Evaluation Considerations

The factors of reliability, power consumption, availability, floor space, and the ability to operate in a variety of environmental conditions are important with microwave as with other telecommunications equipment. In addition to these considerations, which are covered in previous chapters, the following factors must also be evaluated.

System Gain

When a microwave signal is emitted into free space, it is attenuated by losses that are a function of the frequency, elevation, distance between terminals, and atmospheric conditions such as rain, fog, and temperature inversions. The amount of free space loss that a system can overcome is known as the *system gain*. System gain is expressed in decibels and is a function of the output power of the transmitter and the sensitivity of the receiver. Receiver sensitivity is a measure of how low the signal level into the receiver can be while still meeting noise objectives in an analog system or BER objectives in a digital system. For example, if a microwave transmitter has an output power of +30 dBm (1 watt) and a receiver sensitivity of −70 dBm, the system gain is 100 dB.

With other factors being equal, the greater the system gain, the more valuable the system because repeaters can be spaced farther apart. With the same repeater spacing a microwave radio with higher system gain has a greater fade margin than one with lower system gain. System gain can be improved in some microwave systems by the addition of optional higher power transmitters, low noise receiver amplifiers, or both.

Spectral Efficiency

Microwave radio can be evaluated by its efficiency in use of limited radio spectrum. The FCC prescribes minimum channel loadings for a microwave before it is type accepted. Within the frequency band, maximum bandwidth is limited by the license granted by the FCC. Where growth in voice frequency channels is planned, the ability to increase the channel loading is of considerable interest to avoid adding more radio channels. Spectral efficiency in both analog and digital radios is a function of the modulation method. The controlling factor is noise in analog radio and BER in digital radios.

Fade Margin

The *fade margin* refers to the amount of fading of the received signal level that can be tolerated before the system "crashes." A crash in an analog radio is defined as the maximum noise level that can be tolerated. In a digital microwave, fade margin is the difference between the signal level that yields a maximum permissible BER (usually 10^{-6}) and the crash level (usually 10^{-3}). Analog radios fade more gracefully than digital radios. As the received signal diminishes, the channel noise level increases in analog radio, but communication may still be usable over a margin of about 20 dB. The fade margin of a digital radio is quite narrow—in the order of 3 dB. Either a digital signal is very good or it is totally unusable with only a narrow margin between the two points.

Protection System

The need for protection in a microwave system is determined by the user's availability objectives. Availability is affected by equipment failures and fades. Equipment availability can be calculated from the formula:

$$\text{Percent availability} = \frac{\text{MTBF} - \text{MTTR}\ (100)}{\text{MTBF}}$$

Availability as affected by fades can be determined by a microwave path engineering study. It is possible to calculate percent availability within a reasonable degree of accuracy for both fades and failures, but it is impossible to predict when failures will occur. Therefore, protection may be necessary to guard against the unpredictability of failures even though the computed availability is satisfactory.

Another factor weighing in the decision to provide diversity is the accessibility of equipment for maintenance. In a service such as DTS, the equipment is normally mounted in an office building and accessible within a few minutes. On that basis it may be reasonable to provide spares, but to forego diversity to save cost. In a system with remote repeaters, diversity is usually needed because of difficulty in reaching the site in time to meet availability objectives.

Alarm and Control Systems

All microwave radios should be equipped with an alarm system that provides both local and remote failure indications. An alarm system is evaluated on the basis of how accurately the alarm is identified. Primitive systems indicate only that trouble exists but not what it is. Sophisticated systems provide a complete remote diagnosis of radio performance. Microwave systems equipped with protection and emergency power also require a control system to switch equipment and operate the emergency engine.

Standard Interfaces

Most digital microwave systems should be designed to connect to a standard digital signal interface such as DSX-1, DSX-2, or DSX-3. Systems designed for the operational-fixed band sometimes use nonstandard interfaces such as 12 or 14 DS-1 signals. Special multiplexers are required to implement these interfaces.

Frequency Band

The choice of microwave frequency band is often dictated by frequency availability. Where choices are available, the primary criteria are the number of voice frequency channels required, the availability of repeater locations, and the required path reliability. As stated earlier, path reliability decreases with increasing radio frequency because of rain absorption. Reliability can be improved by decreasing the repeater spacing or increasing the antenna size.

Path Engineering

A microwave path should not be attempted without an expert path survey. Several companies specialize in frequency coordination studies and path profile studies, and should be consulted about a proposed route. Sites should be chosen for accessibility and availability of real estate and a reliable power source. Tower heights are chosen to obtain the elevation dictated by the path survey. The antenna structure is chosen to support the size of antenna in a wind of predicted velocity. Wind velocities for various parts of the country are specified by EIA as indicated in the standards section of this chapter.

Environmental Factors

Frequency stability is a consideration in evaluating microwave equipment. FCC rules specify the stability required for a microwave system, but environmental treatment may be needed to keep the system within its specifications. Air conditioning is usually not required, but air circulation may be necessary. Heating may be required to keep the equipment above zero degrees Celsius. Battery plants lose their capacity with decreasing temperature. Therefore, in determining the need for heating it should be remembered that battery capacity is lowest during abnormal weather conditions when power failures are most apt to occur.

Test Equipment

All microwave systems require test equipment to measure frequency, bandwidth, output power, and receiver sensitivity. This equipment, which should be specified by the manufacturer, is required in addition to test equipment needed to maintain multiplex equipment.

GLOSSARY

Adaptive equalizer: A circuit installed in a microwave receiver to compensate for distortion caused by multipath fading.

Bit error rate (BER): The number of error bits in a signal expressed as a fraction of the number of transmitted bits.

Branching filter: A device inserted in a waveguide to separate or combine different microwave frequency bands.

Cross-polarization: The relationship between two radio waves when one is polarized vertically and the other horizontally.

Cross-polarization discrimination (XPD): The amount of decoupling between radio waves that exists when they are cross polarized.

Directional coupler: A device inserted in a waveguide to couple a transmitter and receiver to the same antenna.

Dispersive fade margin: A property of a digital microwave signal that expresses the amount fade margin under conditions of distortion caused by multipath fading.

Diversity: A method of protecting a radio signal from failure of equipment or the radio path by providing standby equipment.

Entrance link: A coaxial or fiber optic facility used to connect the last terminal in a microwave signal to multiplex or video terminating equipment.

Fade: A reduction in received signal level in a radio system caused by reflection, refraction, or absorption.

Fade margin: The depth of fade, expressed in dB, that a microwave receiver can accommodate while still maintaining an acceptable circuit quality.

Frequency diversity: Protection of a radio signal by providing a standby radio channel on a different frequency to assume the load when the regular channel fails.

Heterodyning: The process of shifting a radio frequency by mixing it with another frequency and selecting the desired frequency from the resulting modulation products.

Hot standby: A method of protecting a radio system by keeping a duplicate system tuned to the same frequency but decoupled from the antenna.

Multipath fading: A radio system fade caused by reflection of a portion of the transmitted signal so that it takes a longer path to the receive antenna and arrives slightly out of phase. The phase difference results in a reduced receive signal level.

Phase shift keying (PSK): A method of digitally modulating a radio signal by shifting the phase of the transmitted carrier.

Quadrature amplitude modulation: A method of digitally modulating a radio signal by combinations of phase shift and amplitude variations.

Received signal level (RSL): The strength of a radio signal received at the input to a radio receiver.

Receiver sensitivity: The magnitude of the received signal level necessary to produce objective BER or channel noise performance.

Space diversity: Protection of a radio signal by providing a separate antenna located a few feet below the regular antenna on the same tower to assume the load when the regular channel fades.

Spectral efficiency: The efficiency of a microwave system in its use of radio spectrum, usually expressed in bits per Hz for digital radios and KHz per voice channel in analog radios.

System gain: The amount of free space path loss that a radio can overcome by a combination of transmitted power and receiver sensitivity.

Waveguide: A rectangular or circular metallic tube capable of coupling a microwave signal from radio equipment to an antenna.

BIBLIOGRAPHY

American Telephone & Telegraph Co. *Telecommunications Transmission Engineering.* Bell System Center for Technical Education, vol. 1, 1974; vol. 2, 1977; and vol. 3, 1975.

Bellamy, John C. *Digital Telephony.* New York: John Wiley & Sons, 1982.

Freeman, Roger L. *Telecommunication System Engineering.* New York: John Wiley & Sons, 1980.

———. *Reference Manual for Telecommunications Engineering.* New York: John Wiley & Sons, 1985.

Feher, Kamilo. *Digital Communications Microwave Applications.* Englewood Cliffs, N.J.: Prentice-Hall, 1981.

MANUFACTURERS OF MICROWAVE RADIO EQUIPMENT

Microwave Transmitters and Receivers

AT&T Network Systems

Aydin Microwave Division

California Microwave

Digital Microwave Co.

Ericsson, Inc.

Farinon Division, Harris Corporation

NEC America, Inc.

Northern Telecom, Inc.

Rockwell International Collins Transmission Systems Div.

Terracom Division, Loral Corporation

Microwave Antennas, Waveguides, and Towers

Andrew Corporation

Gabriel Electronics Inc.

NEC America, Inc.

Rockwell International Collins Transmission Systems Division

Scientific-Atlanta, Inc.

CHAPTER SIXTEEN

Lightwave Communications

In the face of limited microwave frequency spectrum and radio's susceptibility to fading, fiber optic or lightwave technology is rapidly becoming the preferred method of digital transmission. Fiber optics overcomes the disadvantages of microwave radio. It is inexpensive to manufacture, has vast bandwidth, is not susceptible to interference and fading, and communications can be conducted over a fiber optic system with almost complete assurance of privacy.

Fiber optic cable is also an important replacement for twisted pair cable because of its greater capacity and smaller physical diameter. Diameter is important when conduits are congested and must be augmented to contain more voice frequency or T carrier cables. Replacing a single copper cable with fiber optics can usually gain enough capacity to forestall conduit additions for the foreseeable future. Furthermore, because the medium does not radiate into free space, no licensing from the FCC is required. Its primary disadvantage is the expense of right-of-way and of keeping it free of damage. Like copper cables, fiber optics can be interrupted by excavations, slides, vandalism, and accidents.

A fiber optic system is similar in concept to a microwave radio system in many ways. The primary exceptions are that the transmission medium for lightwave is a tiny glass waveguide rather than free space, and that transmission takes place at lightwave frequencies that have a much shorter wavelength than microwave. Where microwave is generally designated by its frequency band, lightwave is designated by wavelength. At light frequencies, the wavelength is so short that its unit of measurement is the nanometer (nm), one one-billionth of a meter. With present technology, the usable lightwave communications spectrum extends from approximately 800 to 1,600 nm.

The cost and difficulty of obtaining right-of-way precludes fiber optic ownership for many private networks. However, many local telephone companies offer lightwave capacity so private network users can lease fiber optic pairs and multiplex them in any way they choose. Fiber optic cable is also used in many local network applications. Low-cost optical transceivers are coupled to the ends of the fiber cable for high-speed transmission within a building or campus for either a local area network, as described in Chapter 19, or for point-to-point digital communications.

LIGHTWAVE TECHNOLOGY

Lightwave communication is an idea that has been around for more than a century, but it has become feasible only within the past few years. Alexander Graham Bell, in the first known lightwave application, received a patent for his "Photophone" in 1880. The Photophone was a device that modulated a light beam focused from the sun that radiated in free space to a nearby receiver. The system reportedly worked well, but free space radiation of light has several disadvantages that could not be overcome with the devices available at the time. Like many other ideas, this one was merely ahead of its time. Free space light communication is now technically feasible if the application can tolerate occasional outages caused by fog, dust, atmospheric turbulence, and other path disruptions. Free space light communication is covered in a later section.

Two developments raised lightwave communication from the theoretical to the practical. The first development was the laser in 1960. A laser produces an intense beam of highly collimated light; that is, its rays travel in parallel paths. The second event that advanced lightwave was the development of glass fiber of such purity that only a minute portion of a light signal emitted into the fiber is attenuated. With a laser source that is triggered on and off at high speed, the zeros and ones of a digital communication channel can be transmitted to a detector, usually an avalanche photo diode (APD) or PIN diode. The detector converts the received signal pulses from light back to electrical pulses and couples them to the multiplex equipment. Figure 16.1 shows the elements of a lightwave communication system. Repeaters or regenerators are spaced at regular intervals with the spacing dependent on the loss of the fiber and the system gain at the transmission wavelength. System gain in fiber optics is a concept similar to system gain in a microwave system and is discussed in a later section.

Like microwave, lightwave is normally protected by a standby channel that assumes the load when the regular channel fails. The two directions of transmission are normally protected separately between the digital signal input and output points. The cable, terminal equipment, and any repeaters are simultaneously switched to the protection channel.

The advantages of lightwave accrue from the protected transmission medium of the glass fiber. These tiny waveguides isolate the digital signal from the fading and interference characteristics of free space. The optical fiber attenuates the light signal, however, and unlike microwave, the transmission medium loss is not linear across the spectrum. The spectral loss typical of modern glass fiber is shown in Figure 16.2 on p. 420.

The earliest fiber optic systems were implemented at 850 nm because suitable lasers were first commercially available at that wavelength. As lasers became available at 1,300 nm, applications have shifted to this wavelength because of its lower loss in fiber. Glass fiber exhibits slightly lower losses in its third "window" at about 1,550 nm. The first commercial fiber optics system, installed in 1977, operated at 45 mb/s with repeaters required at 4-mile (6.4 km) intervals. Current systems operate at bit rates up to 560 mb/s with

FIGURE 16.1
Block Diagram of a Typical Fiber Optic System

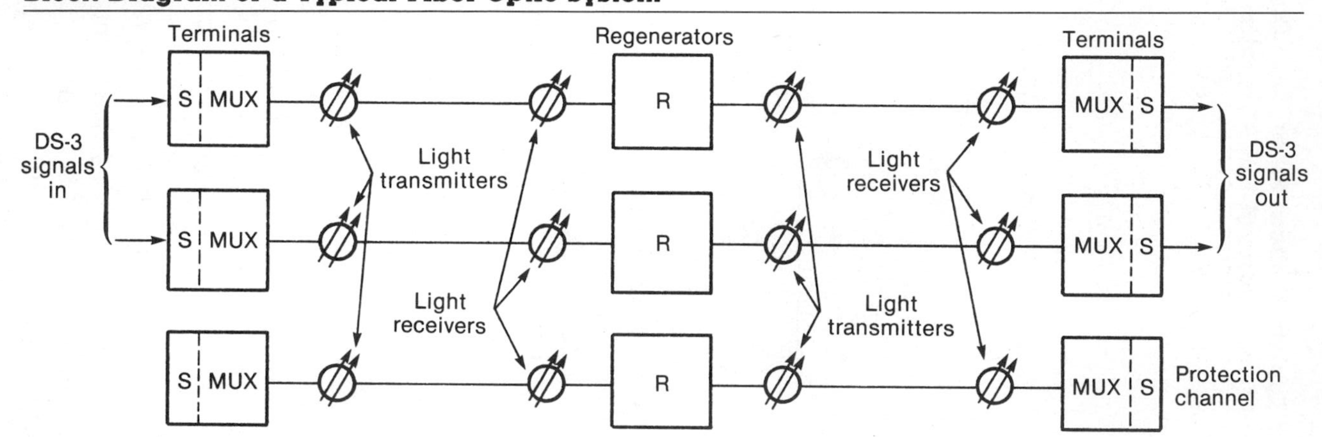

Note: Only one direction of transmission is shown.

FIGURE 16.2
Spectral Loss for a Typical Optical Fiber; Loss Disturbances Labeled OH^- Result from Hydroxyl Ion Absorption

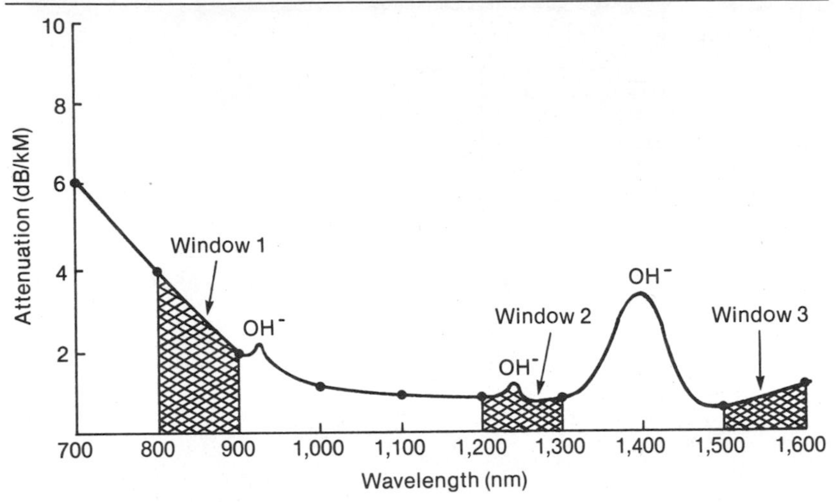

developmental systems operating at bit rates greater than two gb/s (2,000 mb/s). At two gb/s a pair of fibers, one for transmit and one for receive, can accommodate about 30,000 voice frequency channels. At 560 mb/s repeaters can be spaced as far apart as 30 miles (50 km) and can transport more than 8,000 voice channels.

Lightguide Cables

A digital signal is applied to a lightguide by pulsing the light source on and off at the bit rate of the modulating signal, and the signal is propagated to the receiver at slightly less than the speed of light. The lightguide consists of three parts: the inner core, the outer *cladding,* and a protective coating around the cladding. Both the core and the cladding are of glass composition; the cladding has a greater *refractive index* so that most of the incoming light waves are contained within the core. Light entering an optical fiber is propagated through the core in *modes,* which are the different possible paths a lightwave can follow. Optical fiber is roughly grouped into two categories: *single mode* and *multimode.* In single-mode fiber, light can take only a single path through a core that

FIGURE 16.3
Light Ray Paths through a Step Index Optical Fiber

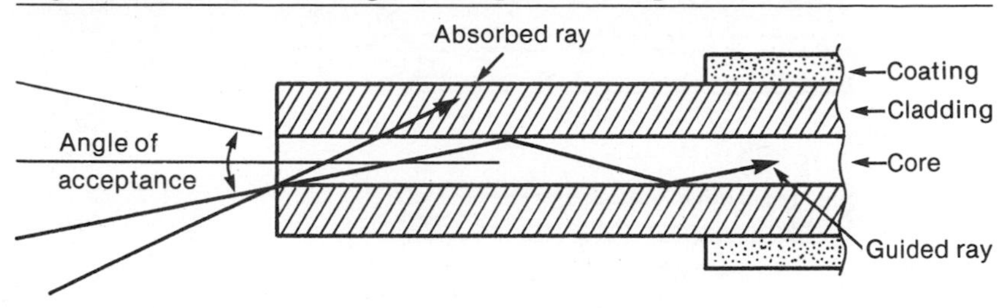

measures about 10 microns in diameter. (A micron is one one-millionth of a meter.) Multimode fibers have cores of 50 to 200 microns in diameter. Single-mode fiber is more efficient at long distances for reasons that are discussed below, but the small core diameter requires a high degree of precision in manufacturing, splicing, and terminating the fiber.

Lightwaves must enter the fiber at a critical angle known as the *angle of acceptance.* Any waves entering at a greater angle escape through the cladding as shown in Figure 16.3. The reflected waves take a longer path to the detector than those that propagate directly. The multipath reflections arriving out of phase with the main signal attenuate the signal, round, and broaden the shoulders of the light pulses. This pulse rounding is known as *dispersion.* It can be corrected only by regenerating the signal. The greater the core diameter, the greater the amount of dispersion. The small core diameter of single-mode fiber reduces the amount of dispersion and enables wider repeater spacing.

Fiber is also classified by its refractive index into two general types: *step index* and *graded index.* With step index fiber, the refractive index is uniform throughout the core diameter. In graded index fiber, the refractive index is lower near the cladding than at the core. Lightwaves are propagated at slightly lower speeds near the core than near the cladding. Consequently, dispersion is lower and the distance between regenerators can be lengthened. Wave propagation through the three types of fiber is shown in Figure 16.4. The effects of dispersion are minimized by using single-mode fiber. Although single-mode fiber is immune to the pulse rounding

FIGURE 16.4
Wave Propagation through Different Types of Optical Fiber

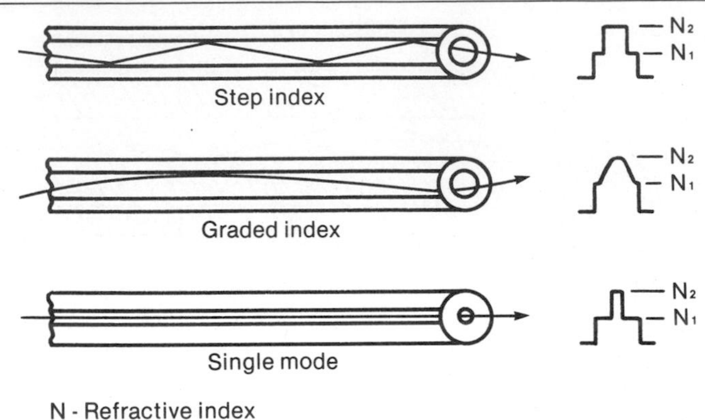

Note: Gradual reflection of graded index fiber is compared to abrupt reflections of step index fiber. Waves are not reflected in single mode fiber, resulting in minimal dispersion.

caused by *modal dispersion,* it is subject to another type known as *material dispersion.* Material dispersion results from the broad range of wavelengths contained in a pulse. Because the refractive index varies with wavelength, some wavelengths are attenuated more than others. A single mode pulse is rounded by material dispersion, but the effects are far less severe than modal dispersion, permitting greater regenerator spacing than with multimode fiber optics. Material dispersion is the primary factor inhibiting the use of bit rates in the 1 to 2 gb/s range. To achieve these high speeds a laser that emits a narrow band of frequencies is required.

In addition to the effects of dispersion, fiber optic regenerator spacing is limited by loss. Loss is caused by two factors: *absorption* and *scattering.* Absorption results from impurities in the glass core, imperfections in the core diameter, and the presence of hydroxyl ions or "water" in the core. The water losses occur most significantly at wavelengths of 1,400, 1,250, and 950 nm as shown in Figure 16.2. Scattering results from variations in the density and composition of the glass material. These variations are an inherent by-product of the manufacturing process.

Glass fibers are made with a process known as *modified chemical vapor deposition* (MCVD). The process starts with

FIGURE 16.5
Multiple Strand Fiber Optic Cable

Courtesy Siecor Corporation.

a pure glass tube about 6 feet long and 1½ inches in diameter. The tube is rotated over a flame of controlled temperature while a chemical vapor is introduced in one end. The vapor is a carrier for chemicals that are deposited on the interior of the glass by heat from the flame. The deposited chemicals form a tube composed of many layers of glass inside the original tube. When the deposition process is complete, the tube is collapsed under heat into a solid glass rod known as a *preform*. The preform is placed at the top of a drawing tower where the fiber is heated to the melting point and drawn into a hair-thin glass strand.

Multiple fiber strands are wound together around a strength member and enclosed in a sheath as shown in Figure 16.5. Like copper cable, fiber cable sheaths are made of polyethylene and can be enclosed in armor to protect against

FIGURE 16.6
Splicing Fiber Optic Cable

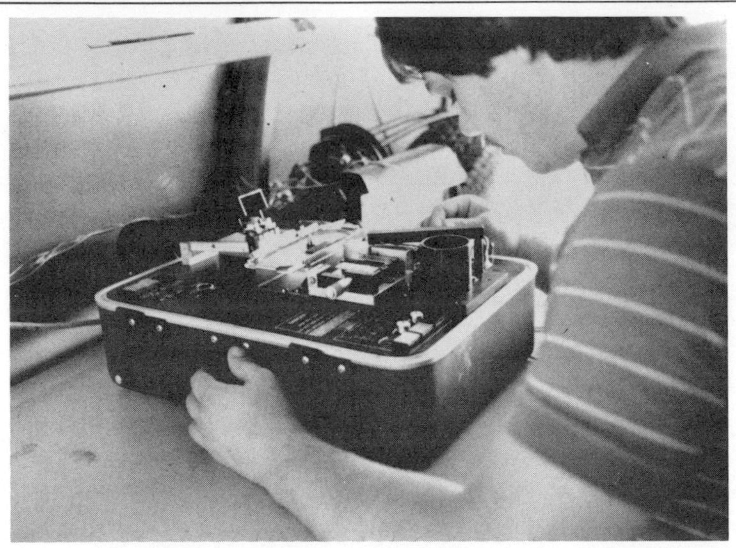

Courtesy Siecor Corporation.

Fibers are precisely aligned under magnification and fused electrically.

damage. Fiber optic cable is suitable for direct burial, pulling through conduit, suspension from an aerial strand, or submersion in water.

Fiber cables are spliced by adhesion or fusion. In the adhesion process, fibers are placed in an alignment fixture and joined with epoxy. The fusion method employs a splicing fixture of the type shown in Figure 16.6. The two ends of the fiber are aligned precisely under a microscope and fused with a short electric pulse. After splicing, the loss is measured to ensure that splice loss was acceptable. Splices are made with sufficient slack in the cable that they can be respliced, if necessary, until the objective loss is achieved.

Fiber Optic Terminal Equipment

Fiber optic systems are designed with separate transmit and receive fibers, the opposite ends of which are terminated in a light transmitter and receiver. The light transmitter, enclosed in the terminal shown in Figure 16.7, employs either a light emitting diode (LED) or a laser as its output element. Lasers have a greater system gain than LEDs because their output is higher and because a greater portion of the light signal can be coupled into the fiber without loss. The primary advantage of a LED transmitter is its lower cost. In applications where system gain is unimportant, the cost saving can easily justify the use of LED transmitters.

The multiplex equipment is wired to the input of the transmitter and the fiber optic cable is coupled to the output through a precision connector. Most fiber optic systems use digital modulation, but analog modulation is achieved by varying the intensity of the light signal, or by modulating the pulse rate or pulse width. Although analog modulation is normally not linear enough for transmitting analog multiplex, it is suitable for transmitting a video signal and is used in cable television systems.

The light receiver is an APD or PIN diode that is coupled to the optical fiber on the input end and to the multiplex equipment on the output end. The diode converts the light pulses to electrical pulses, which are reshaped into square wave pulses. A lightwave regenerator consists of back-to-back receiver and transmitter pairs that are coupled through a pulse reshaping circuit.

Fiber optic systems accept standard digital signals at the input, but each manufacturer develops its own output signal rate. Error checking and zero suppression bits are inserted to maintain synchronization and to monitor BER to determine when a switch to the protection channel is required. Because of differences in the line signals, lightwave systems are not end-to-end compatible between manufacturers.

Wavelength Division Multiplexing

The capacity of a fiber pair can be multiplied by the use of *wavelength division multiplexing* (WDM). WDM assigns services to different light wavelengths in much the same manner as frequency-division multiplexing is used to apply multiple carriers to a coaxial cable. Different frequencies are selected by using light sensitive filters to combine light frequencies at the sending end and separate them at the re-

FIGURE 16.7
Fiber Optic Terminal

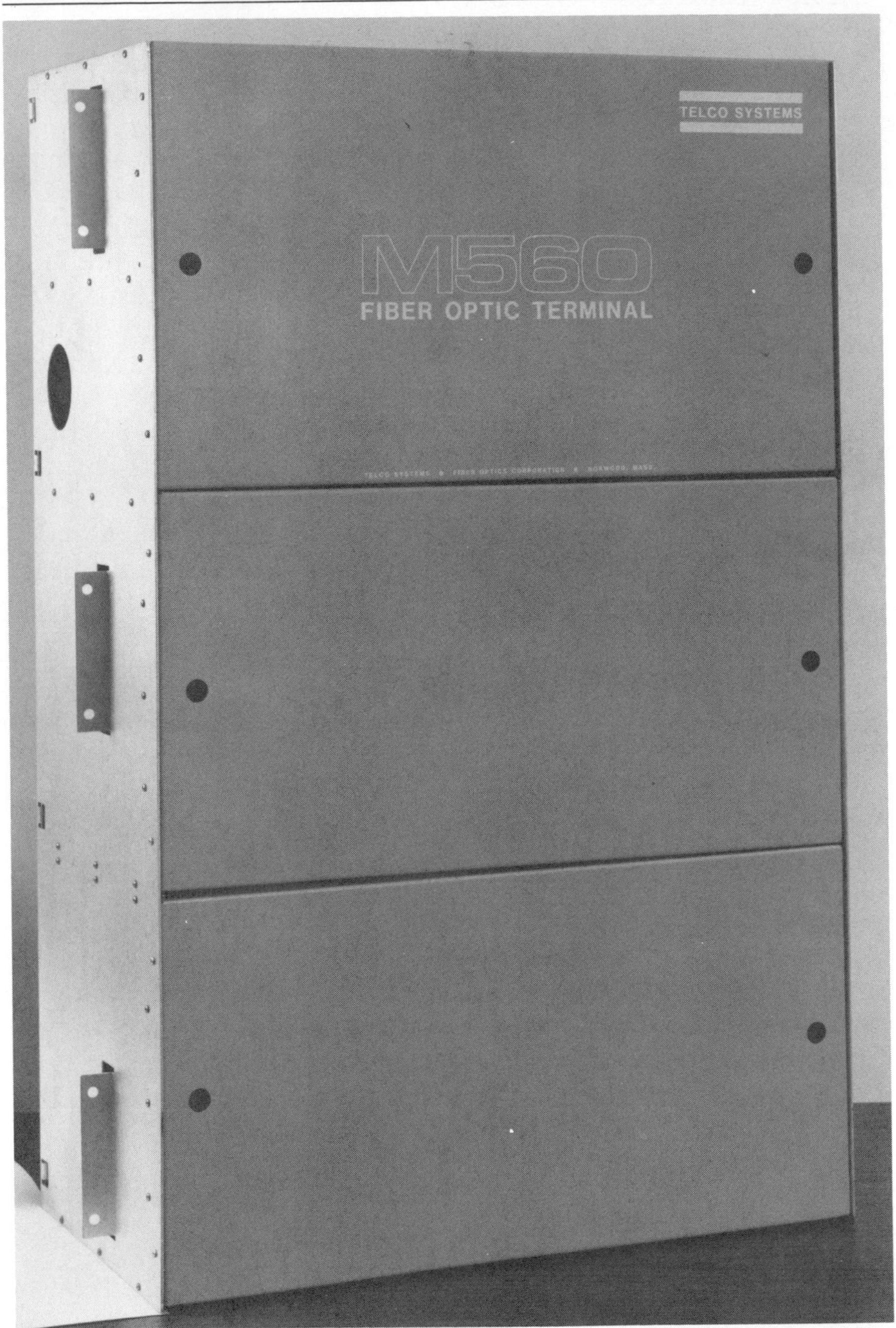

Courtesy Telco Systems Fiber Optics Corporation.

FIGURE 16.8
Wave Division Multiplexer

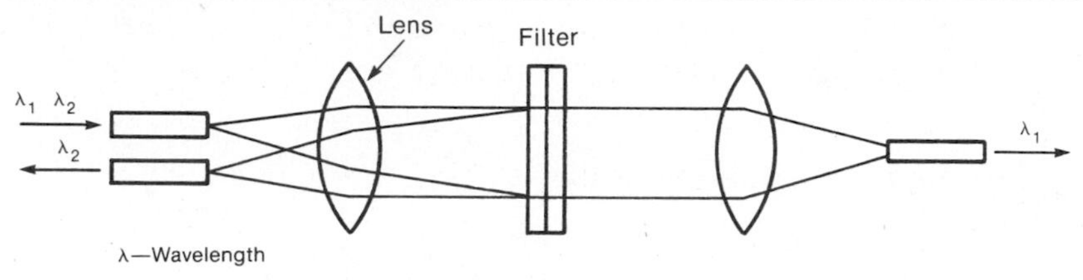

Filter is reflective to wavelength 2 and transparent to wavelength 1.

ceiving end as shown in Figure 16.8. Because the filter introduces loss, the distance between regenerators is reduced with WDM and the path length is limited by the wavelength with the highest loss.

Star Couplers

One limitation of fiber optics is that it is usable only for point-to-point service. It is difficult to tap optical fiber, which limits its use in multipoint applications. This disadvantage is of no concern in applications such as long haul and metropolitan trunks because these services are point-to-point by nature. However, in local networks characterized by multiple terminals needing rapid access to one another, multiple access is required. To accomplish this, *star couplers* are employed to bring multiple terminals into a central point. The star coupler radiates the light wave equally into all branches through a reflector or an active device to boost the signal before it is transmitted. The primary application for a star coupler is local area networks.

LIGHTWAVE SYSTEM DESIGN CRITERIA

Fiber optic systems are designed by balancing capacity requirements with costs comprising cable, cable placing, terminal equipment, and regenerators. In most systems a prime objective is to eliminate midspan regenerator points so regenerators are placed only in buildings housing other telecommunications equipment. This objective may require reducing bit rates, providing higher quality cable, and stretching the system design to preclude future WDM. The three primary criteria for evaluating a system are:

- Information transfer rate.
- System attenuation and losses.
- Cutoff wavelength.

Information Transfer Rate

The information transfer rate of a fiber optic system is a function of the bandwidth, which is a function of dispersion rate and of the distance between terminal or repeater points. Bandwidth in graded index fiber is quoted as a product of length and frequency. For example, a fiber specification of 1,500 MHz-km could be deployed as a 150 MHz system at 10 km or a 30 MHz system at 50 km. Fiber optic transmission systems are quoted according to the number of DS-3 systems they can support, with current commercial products ranging from fewer than 1 to 12 DS-3 systems per pair. Special-purpose fiber optic systems intended for short range private data transmission have much lower bit rates and typically use cables with considerably more bandwidth than is required.

System Attenuation and Losses

In any fiber system a key objective is to avoid placing repeaters between terminals, if possible, because of the expense of right-of-way and maintenance. Therefore the system loss and attenuation in conjunction with available bandwidth is a key factor in determining usable range. As with microwave systems, system gain in fiber optics is the difference between transmitter output power and receiver sensitivity. For example, a typical system with a transmitter output of −5dBm and a receiver sensitivity of −40 dBm has a system gain of 35 dB.

From the system gain, designers compute a *loss budget*, which is the amount of cable loss that can be tolerated within the available system gain. In addition to cable loss, allowances must be made for:

- Loss of initial splices plus an allowance for future maintenance splices.
- Loss of connectors used to couple fibers and terminal equipment.
- Temperature variations.
- Measurement inaccuracies.
- Future WDM.
- Safety margin.
- Aging of electronic components.

These additional losses typically subtract about 10 to 12 dB from the span between terminal points, which leaves a loss budget of about 25 dB for cable. Cable cost is a function of loss, so system designers choose a cable grade to match the loss budget.

Cutoff Wavelength

The cutoff wavelength is defined as the longest wavelength that can be transmitted within allowable loss limits. The cutoff wavelength is selected on the basis of future plans for WDM. With present technology, products under development are expected to support wavelengths of 1,500 nm or more.

FREE SPACE LIGHTWAVE TRANSMISSION

Free space lightwave communication devices operate on principles similar to fiber optics, except that the signal is radiated directly from a light transmitter to a receiver mounted a mile or less away. Free space lightwave is subject to the same effects that cause microwave fading, only to a more significant degree. Free space lightwave has the advantage of spanning short distances without the need for securing licenses or the cost of obtaining right-of-way. It is used for short-range communications such as between buildings, or over longer ranges when path reliability is not vital.

Because of the short wavelengths involved, infrared light transmission is attenuated by fog and dust particles. It is also subject to fading because of differences in the refractive index of the atmosphere, the effect that causes shimmering mirages to appear over flat surfaces on hot days. Scattering attenuates the signal when the optical wavelength is shorter than the size of particles suspended in the atmosphere. Free space light is subject to interference from other light sources including sunlight. Even with these drawbacks, however, for many applications the low cost of this transmission medium makes it a feasible alternative compared to other media. It is particularly adaptable to short spans such as between buildings across public right-of-ways such as streets.

STANDARDS

Except for special applications, the digital signal input of fiber optic systems is designed to interface Bell or CCITT digital signals. The output signal is determined by the manufacturer with no accepted standards. Therefore, fiber optic systems are not end-to-end compatible between manufacturers. EIA issues physical standards on fiber and connectors.

Bell Communications Research

CB 119 Interconnection Specification for Digital Cross-Connects.

CB 135 MX3/FT3 Digital Multiple/Lightwave Digital Transmission System Compatibility Specification.

PUB 43806 Generic Metropolitan Interoffice Digital Lightwave Systems Requirements and Objectives.

CCITT

G.651 Characteristics of 50/125 micrometer graded index optical fiber cables.

EIA

RS-440 Fiber Optic Connector Terminology.

RS-455 Standard Test Procedures for Fiber Optic Fibers, Cables, Transducers, Connecting and Terminating Devices (includes five addenda and 21 separate testing specifications).

RS-458 Standard Optical Waveguide Fiber Types.

RS-459 Standard Optical Waveguide Fiber Material Classes.

RS-475 Generic Specification for Fiber Optic Connectors.

RS-509 Generic Specifications for Fiber Optic Terminal Devices.

APPLICATIONS

The high cost of right-of-way often stands in the way of companies installing fiber optic systems, but the advantages of this medium will undoubtedly result in more private applications. The Martin Marietta Company is one user who has successfully applied fiber optics in a private voice and data network in a metropolitan area.

The Martin Marietta Fiber Optic Network

The Martin Marietta Corporation, whose primary businesses consist of a data processing service bureau, a systems integrator, and an aerospace manufacturer, has a private fiber optic network linking three plants in Orlando, Florida, its main facility; its computer facility located in the Orlando Central Park industrial complex; and its East facility located 20 miles northeast of the Central Park computer center. The first phase of the project, installed in 1980, is a 10 km (6.2 mile) nonrepeatered fiber optic cable between the Main plant and Central Park. This system operates at a wavelength of 830 nm and has a data rate of 45 mb/s.

The driving force behind the first phase was a requirement for high-speed facilities to implement a computer aided design and manufacturing (CADAM) project that initially required 12 T-1 carriers. The initial network, shown in Figure 16.9, consisted of a single cable containing four multimode fibers. By 1982, the initial 12 T-1 lines had grown to 28 T-1 lines, the maximum that one fiber optic pair can accommodate at the 45 mb/s data rate. To ensure 100 percent network availability, a second fiber optic cable was installed in 1982 to accommodate the growth of both voice and data circuits. Two fibers were activated in each cable with the second fiber pair reserved for future use and protection.

Both digital data and analog voice services are carried on this system. For voice, Martin Marietta uses ITT D448 channel banks, which provide 48 circuits over two T-1 carrier lines. Data circuits are carried over General Datacom T-1 Megamux multiplexers, which provide ports for multiple 9.6 and 56 kb/s circuits. Altogether, the network supports more than 2,000 terminals.

In 1983, Martin Marietta approved formal plans for an aerospace facility at a separate plant known within the company as the "East Facility." This location, 32 km (20 miles) northeast of the data systems center, had initial requirements for both voice and data circuits with future requirements for video. The distance between Central Park and the East plant is too great for nonrepeatered multimode fiber. Although repeaters presented no technical barriers to expanding the system, their additional cost, reduction in reliability, and the need for real estate and a small building for a repeater enclosure led the company to investigate the feasibility of single-mode fiber. In 1983, this technology was emerging from the laboratory but had not been proved in extensive field use.

A joint research and development team was formed between Martin Marietta, the cable manufacturer, Valtec, and the equipment manufacturer, Telco Systems. The purpose of the project was to study the feasibility of extending the network from Central Park to the East facility over single-mode fiber optic cable without a repeater. Technical demonstrations by the two vendors convinced Martin Marietta of the feasibility of single-mode fiber at 90 mb/s, and demonstrated that in the future, expansion to 560 mb/s would be feasible.

FIGURE 16.9
Martin Marietta Fiber Optic Network

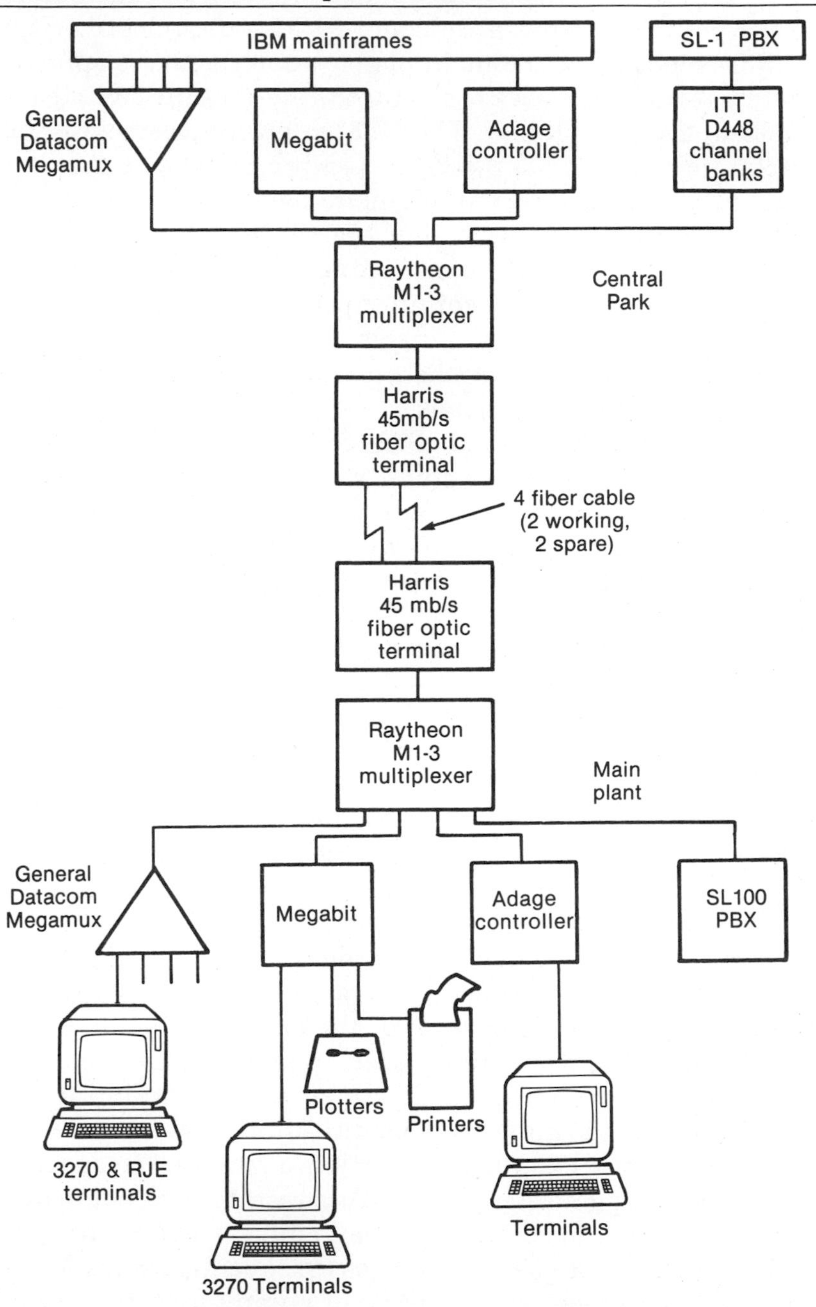

Courtesy Martin Marietta Data Systems.

During the feasibility testing phase, Martin Marietta's aerospace division decided to replace its Centrex telephone system with a Northern Telecom SL-100 digital PBX. The SL-100 is located at the main plant, where it serves approximately 4,000 users, with a remote line module (RLM) installed at the East facility. The RLM communicates with the SL-100 host over a DS-3 facility (45 mb/s). The remote-to-host link could have been implemented over common carrier facilities or over microwave, but Martin Marietta found that the fiber optic cable was more economical than other alternatives. Therefore, voice communication over the SL-100 is supported by the fiber optic link as well as data circuits. With the 90 mb/s data rate of the single-mode fiber, each pair supports two 45 mb/s signals.

Martin Marietta installed 70 percent of its fiber optic cable aerially on existing power poles, with attachment rights acquired from the power, telephone, railroad, and cable television companies. Of the remaining cable, half was installed in conduit and the remainder buried. Cable with an all-dielectric design (one that contains no metallic material in the cable) was required because so much of the cable is adjacent to high-voltage power lines and also because Orlando is in a high-lightning area. The cable is a loose-tube gel-filled design. The gel-filled tube allows the fibers to move freely within the sheath, which reduces microbending, a phenomenon that increases cable loss during temperature changes. The central strength member is epoxy-impregnated Kevlar; the outer sheath is polyethylene. The jelly filling prevents water from building up within the sheath.

As this system is constructed today, it can be expanded significantly by increasing the bit rate of the terminal equipment. The multimode portion of the cable can be increased to 135 mb/s. The single mode section between Central Park and the East facility can be expanded to at least 560 mb/s. Martin Marietta is implementing full-motion video in 1985, which consumes a significant portion of the spare capacity.

During the planning phases, Martin Marietta compared alternatives of both common carrier facilities and microwave radio, and elected to use fiber optics for several reasons. First, fiber optics was less expensive than common carrier facilities. The existing network, which supports 56 T-1 carriers, has a payback period of less than 14 months compared to using an equivalent number of T-1 facilities from the telephone company. The broadband T-1 facilities were significantly more cost effective than using individual 56 kb/s or 9.6 kb/s common carrier facilities.

Compared to microwave radio, fiber optics offered two significant advantages. Because Orlando is in a heavy rainfall area, the anticipated outage from digital radio would have reduced system reliability. Second, microwave radio is less secure than fiber optics. For a company that transmits confidential information, the greater security of fiber optics is an important reason for choosing it over microwave. Moreover, the capacity of fiber optics is increasing much more rapidly with changes in technology. As requirements expand with changes in services, the fiber optics technology can grow to provide increased capacity.

Evaluation Criteria

Fiber optic equipment is purchased either as an integrated package of terminal equipment and cable for specialized private applications or as separate components assembled into a system for trunking between switching nodes. For the former applications, which include local area, point-to-point voice, data, and video networks, the evaluation criteria discussed below are not critical. In such systems the main question is whether the total system fits the application. In all fiber optic systems the questions of reliability, technical support, cost, and compatibility are important. Fiber optic systems do not vary widely in their power consumption or space requirements, so these criteria may be safely ignored in most instances. In longer range trunking applications, the following criteria should be considered in evaluating a system.

System Gain

In selecting lightwave terminating equipment, the higher the system gain, the more gain that is available to overcome cable and other losses. The cost of a lightwave system is directly related to the amount of system gain. High-output lasers and high-sensitivity diodes are more expensive than devices producing less system gain. The least expensive transmitters use light emitting diodes for output and have less system gain than lasers. When the limits of lightwave range are being approached, obtaining equipment with maximum system gain is important. For applications with ample design margin, low system gain is acceptable.

Cable Characteristics

Cable is graded according to its loss and bandwidth. The cable grade should be selected to provide the loss and bandwidth needed to support the ultimate circuit requirement. If the cable can support ultimate requirements, there is little reason to spend extra money to purchase a higher grade. Unless some compelling reason for purchasing multimode cable exists, single mode cable should be purchased for all applications. The price of single mode fiber is approximately the same as multimode, and its greater bandwidth makes it considerably more valuable for future expansion. The cable composition should be selected with inner strength members sufficient to prevent damage when cable is pulled through conduits or plowed in the ground. Armoring should be considered where sheath damage hazards exist.

Wavelength

With present technology the most feasible wavelength to choose is 1,300 nm. Cable should be purchased with a 1,550 nm window if circuit requirements will ultimately justify the use of WDM. For most applications, 850 nm should be avoided because of its greater loss. Exceptions are in local networks and private networks implemented by using leased fibers. With leased fibers the 850 nm window can be used with WDM as a way of increasing capacity without leasing more fiber, providing the distance between terminals supports the use of 850 nm and WDM. In other applications such as local networks the wavelength may be predetermined by the manufacturer. If the total system has enough gain and bandwidth to support the application, the wavelength is of little or no concern to the user.

System Integration

In some applications such as local networks, fiber cable and terminal equipment are purchased as part of a total system furnished and designed by one vendor. In most applications, however, terminal equipment and cable are purchased from separate vendors. Because of the lack of standardization of fiber optic equipment it is advisable to purchase cable through the terminal equipment vendor or to the equipment vendor's specifications so the equipment vendor is able to assume responsibility for total system operation.

Wavelength Division Multiplexing

The question of whether to plan a fiber optics system with future WDM designed into the transmission plan is a balance between future capacity requirements and costs. WDM can double or triple the capacity of a fiber pair for little additional cost, or it can convert a single optical fiber into a full-duplex mode of operation by transmitting in both directions on the same fiber. However, it accomplishes this by reduced regenerator spacing, which is important in long systems but unimportant on systems that do not require an intermediate regenerator. On very short systems the cost of the WDM equipment may be greater than the cost of extra fibers.

GLOSSARY

Absorption: The attenuation of a lightwave signal by impurities or fiber core imperfections.

Angle of acceptance: The angle of light rays striking an optical fiber aperture, within which light is guided through the fiber. Light outside the angle of acceptance leaks through the cladding.

Avalanche photo diode (APD): A light detector that generates an output current many times greater than the light energy striking its face.

Cladding: The outer coating of glass surrounding the core in a lightguide.

Collimate: The condition of parallel light rays.

Core: The inner glass element that guides the light rays in an optical fiber.

Dispersion: The rounding and overlapping of a light pulse that occurs to different wavelengths because of reflected rays or the different refractive index of the core material.

Graded index fiber (multimode): Optical fiber that is made with a progressively lower refractive index toward the outer core region to reduce the effects of dispersion.

Loss budget: The sum total of all factors that introduce loss between the transmitter and receivers.

Material dispersion: Broadening of light pulses that occur because of differences in the refractive index at different wavelengths.

Mode: The different paths lightwaves can take through a transmission medium.

Modal dispersion: Pulse rounding that occurs because of the different paths taken by different light modes that arrive at the detector out of phase.

Modified chemical vapor deposit (MCVD): A process for manufacturing a fiber preform by progressively depositing glass chemicals on the inside of a tube under heat.

PIN diode: A photodiode manufactured with an intrinsic layer of undoped material between doped P and N layers and used as a lightwave detector.

Preform: A glass rod formed and used as the source material for drawing an optical fiber.

Refractive index: The ratio of the propagation speed of light in free space to its propagation in a given transmission medium.

Scattering: The loss that occurs in an optical fiber because of lightwaves striking molecules and imperfections in the core.

Single mode fiber: Optical fiber with a small core that propagates only one light mode.

Star coupler: A device that couples multiple fibers at a central point and distributes the signal from one fiber into all others simultaneously.

Step index fiber: A type of optical fiber with a uniform index of refraction throughout the core.

Wavelength division multiplexing (WDM): A method of multiplying the capacity of an optical fiber by simultaneously operating at more than one wavelength.

BIBLIOGRAPHY

Adams, M. J. *An Introduction to Optical Waveguides.* New York: John Wiley & Sons, 1981.

Boyd, O. T. *Fiber Optics Communications Experiments and Projects.* Indianapolis: Howard W. Sams, 1982.

Howes, M. J. and D. V. Morgan, eds. *Optical Fibre Communications.* New York: Wiley & Sons, 1980.

Kao, Charles K. *Optical Fiber Systems: Technology, Design and Applications.* New York: McGraw-Hill, 1982.

Lacy, Edward A. *Fiber Optics.* Englewood Cliffs, N.J.: Prentice-Hall, 1982.

Mims, Forrest M. *A Practical Introduction to Lightwave Communications.* Indianapolis: Howard W. Sams, 1982.

Palais, Joseph C. *Fiber Optic Communications.* Englewood Cliffs, N.J.: Prentice-Hall, 1984

Sandbank, C. P., ed. *Optical Fibre Communications Systems.* New York: Wiley & Sons, 1980.

MANUFACTURERS OF LIGHTWAVE PRODUCTS

Lightguide Cable

AT&T Network Systems

Corning Glass Works, Telecommunications Products Division

ITT Telecom, Network Systems Division

Northern Telecom Inc., Optical Systems Division

Siecor Corporation

Times Fiber Communications, Inc.

Valtec

Lightwave Transmission Products

AT&T Network Systems

GTE Communication Systems

Harris Corporation Fiber Optic Systems

ITT Telecom, Network Systems Division

NEC America Inc., Radio & Transmission Division

Northern Telecom Inc., Optical Systems Division

Stromberg Carlson Corp.

Telco Systems Fiber Optics Corp.

Infrared Transmission Equipment

American Laser Systems, Inc.

NEC America Inc., Radio & Transmission Division

Light Communications Corporation

CHAPTER **SEVENTEEN**

Satellite Communications

Like other telecommunications technologies, communications satellites have advanced dramatically in the last two decades, evolving from an experimental technique to the stage that they are now commonplace. Telstar 1, the first communications satellite launched in 1962, carried only 12 voice circuits. By contrast, INTELSAT VI, scheduled for 1986 launching, will carry 33,000 voice circuits plus four television channels. Telstar was launched in a low elliptical orbit, circling the earth in about two hours. It was tracked by a ground station for the short time it was visible, often less than one-half hour. The Telstar 1 low orbit was impractical because a chain of several satellites were needed to provide continuous service, and the satellite antennas required constant reaiming. Present-day communications satellites orbit the equator at a *geosynchronous* altitude of 22,300 miles. The equatorial orbit has the advantage of covering both the northern and southern hemispheres. Except for the extreme polar regions, about one third of the earth's longitudinal surface can be covered by a single equatorial satellite.

At geosynchronous orbit the satellite travels at the same speed as the earth's rate of spin and is held at a fixed position with relation to a point on the earth. From this distance three

TABLE 17.1
Communications Satellite Frequency Bands

Band	*Uplink*	*Downlink*
C	5,925–6,425 MHz	3,700–4,200 MHz
Ku	14.0–14.5 GHz	11.7–12.2 GHz
Ka	27.5–31.0 GHz	17.7–21.2 GHz

satellites can theoretically cover the entire earth's surface, with each satellite subtending a radio beam 17 degrees wide. The portion of the earth's surface that is illuminated by a satellite is called its *footprint.*

International satellite communications are controlled by the International Telecommunications Satellite Organization (INTELSAT), an international satellite monopoly operating under treaty among 109 member nations. Sixteen satellites are operated by INTELSAT. Communications Satellite Corporation (COMSAT) is the U. S. partner in this venture, and owns 24 percent of INTELSAT. Domestic satellites are owned and operated by COMSAT, AT&T, Satellite Business Services (SBS), Western Union, RCA, American Satellite, and GTE.

As shown in Table 17.1, the frequencies available for communication satellites are limited. The 4 and 6 GHz C band frequencies are the most desirable from a transmission standpoint because they are the least susceptible to rain absorption. The C band frequencies are shared with common carrier terrestrial microwave, requiring close coordination of spacing and antenna positioning to prevent interference. Interference between satellites and between terrestrial microwave and satellites is prevented by using highly directional antennas. Presently, satellites are spaced about the equator at three-degree intervals. The FCC has proposed two-degree spacing to allow more carriers to occupy the frequency spectrum. Figure 17.1 shows conceptually how satellites are positioned in equatorial orbit.

The Ku band of frequencies is coming into use as the C band becomes congested. The primary advantage to K band frequencies is that they are exclusive to satellites, allowing users to construct earth stations virtually anywhere, even in the metropolitan areas where frequency congestion pre-

FIGURE 17.1
Communication Satellites Are Spaced in an Equatorial Orbit 22,300 Miles above the Earth's Surface

Courtesy Satellite Business Systems.

cludes placing C band earth stations. The primary disadvantage of the Ku band is rain attenuation, which results in lower reliability. The Ka band frequencies are even more susceptible to attenuation. Even though considerable bandwidth is available, further development is needed before Ka frequencies come into general use.

The terms *uplink* and *downlink* used in Table 1 refer to the earth-to-satellite and the satellite-to-earth paths respectively. The lower frequency is always used from the satellite to the ground because earth station transmitting power can be used to overcome the greater path loss of the higher frequency, but satellite output power is limited by its solar battery capacity.

Satellites have several advantages over terrestrial communications.

- Costs of satellite circuits are independent of distance within the coverage range of a single satellite.

- Impairments that accumulate on a per hop basis on terrestrial microwave circuits are avoided with satellites because the earth-station-to-earth-station path is a single hop through a satellite repeater.
- Sparsely populated or inaccessible areas can be covered by a satellite signal, providing high-quality communications service to areas that are otherwise difficult or impossible to reach. The coverage is also independent of terrain and other obstacles that may block terrestrial communications.
- Earth stations can verify their own data transmission accuracy by listening to the return signal from the satellite.
- Because satellites broadcast a signal they can reach wide areas simultaneously.
- Large amounts of bandwidth are available over satellite circuits, making high-speed voice, data, and video circuits available without using an expensive link to a telephone central office.
- The satellite signal can be brought directly to the end user, bypassing the local telephone facilities that are expensive and limit bandwidth.
- The multipath reflections that impair terrestrial microwave communications have little effect on satellite radio paths.

Satellites are not without their limitations, however. The greatest drawback in the long run is the lack of frequencies. If higher frequencies can be developed with reliable paths, plenty of frequency spectrum is available, but atmospheric limitations may prevent their use for commercial grade telecommunications service. Other limitations include these:

- The round-trip delay between earth stations is nearly one-half second. This delay is tolerable for voice when echo cancelers are used but is detrimental to block-mode data protocols and polled data circuits.
- Multihop satellite connections impose delay that is detrimental to voice communications and is generally

avoided. Because direct satellite-to-satellite transmission is not yet feasible, multiple hops are required when the distance between earth stations exceeds the satellite's footprint.

- Path loss is high (about 200 dB) from earth to satellite.
- Rain absorption affects path loss, particularly at higher microwave frequencies.
- Frequency crowding in the C band is high with potential for interference between satellites and terrestrial microwave operating on the same frequency.

Some observers predict that terrestrial fiber optics will cut deeply into satellite business because fiber optics eliminates satellite impairments without introducing comparable impairments of its own. There can be no doubt that transcontinental fiber optics will have an important place in future communications, but satellites have unique properties that assure them of a role in the future because no other service can blanket the earth's surface with signals at a cost that is independent of whether the user is in New York City or rural Kansas.

SATELLITE TECHNOLOGY

A satellite circuit consists of five elements as shown in Figure 17.2: two terrestrial links, an uplink, a downlink, and a satellite repeater. If the earth station is mounted on the user's premises, the terrestrial links are eliminated. The satellite itself consists of six subsystems described below:

- Physical structure.
- Transponder.
- Attitude control apparatus.
- Power supply.
- Telemetry equipment.
- Station-keeping apparatus.

FIGURE 17.2
Satellite Circuit

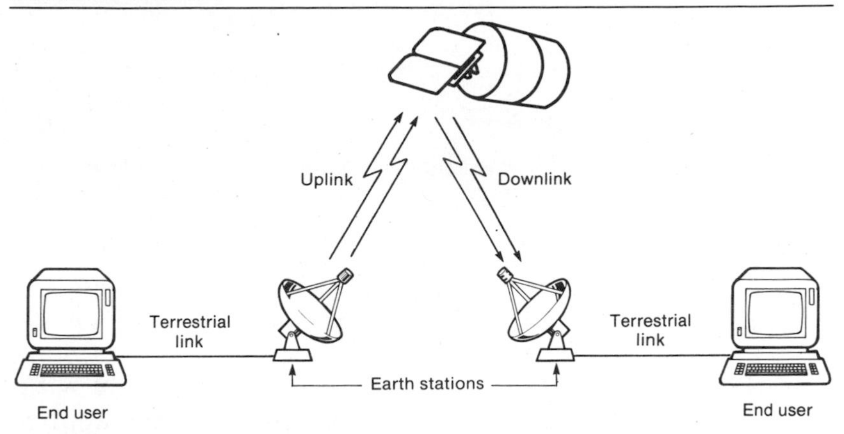

Earth stations may be mounted directly on user's premises, eliminating the terrestrial links.

Physical Structure

The size of communications satellites has been steadily increasing since the launch of Early Bird, the first commercial satellite, in 1965. Size is limited by the capacity of launch vehicles and by the requirement to carry enough solar batteries and fuel to keep the system alive for its design life of 5 to 10 years. Advances in space science are making larger satellites technically feasible. Launch vehicles are capable of carrying greater payloads, and the demonstrated ability of the space shuttle to service a satellite in flight or return it to earth for maintenance is changing design considerations that previously limited satellite size.

A large physical size is desirable. Not only must the satellite carry the radio and support equipment, but it must also provide a platform for large antennas to achieve the high gain needed to overcome the path loss between the earth station and the satellite.

Transponders

A *transponder* is a radio relay station on board the satellite. Transponders are technically complex, but their functions are identical to those of a terrestrial microwave radio relay

FIGURE 17.3
Satellite Transponder

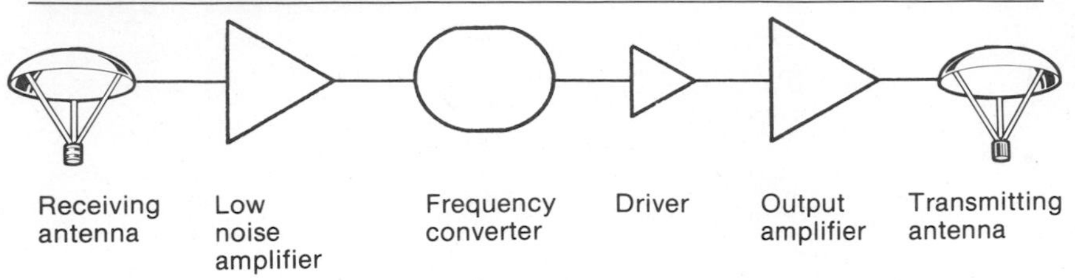

station. The diagram in Figure 17.3 shows the major elements. The incoming signal from the earth station is picked up by the antenna and amplified by a low noise amplifier (LNA), which is designed to boost the received signal without adding noise. The LNA output is amplified and applied to a mixer that reduces the incoming signal to the downlink frequency. The downlink signal is applied to a high-power amplifier (usually between 5 and 15 watts) using a traveling wave tube (TWT) as the output amplifier. The output signal is coupled to the downlink antenna. Most satellites carry multiple transponders, each with a bandwidth of 30 to 70 MHz. For example, AT&T's Telstar 3 contains 30 transponders, 24 working and 6 spare, with bandwidths of 36 MHz.

Cross-polarization is used on both satellite and earth station antennas to double the capacity of a single frequency. In some satellites, beam focusing techniques are used to concentrate rf energy on a relatively small spot of earth. For example, some satellites illuminate the continental United States with spots focused on Alaska, Hawaii, and Puerto Rico. Future satellites are expected to use on-board switching of spot beams to minimize interference. This will become important as two-degree positioning comes into being.

Attitude Control Apparatus

Satellites must be stabilized to prevent tumbling through space and to keep antennas precisely aligned toward earth. With present-day equipment, alignment can be maintained within 0.1 degrees. Satellite stabilization is achieved by two methods. A *spin stabilized* satellite rotates on its axis at about 100 RPM. The antenna is despun at the same speed to provide constant positioning and polarization toward earth. The second method is *three-axis stabilization,* which consists of a gyroscopic stabilizer inside the vehicle. Accelerometers sense any change in position in all axes, and fire positioning rockets to keep the satellite at a constant attitude.

Power Supply

Satellites are powered by solar batteries. Power is conserved by turning off unused equipment by signals from the earth. On spin stabilized satellites the cells are mounted on the outside of the unit so that one third of the cells always face the sun. Three-axis stabilized satellites have cells mounted on solar panels that are extended like wings from the satellite body. Solar cell life is one of the limiting factors in the working life of a satellite. Solar bombardment gradually weakens the cell output until the power supply is no longer capable of powering the on-board equipment.

A nickel-cadmium battery supply is also kept on board most satellites to power the equipment during solar eclipses, which occur during two 45-day periods for about an hour per day. The eclipses also cause wide temperature changes that the on-board equipment must withstand.

Telemetry Equipment

A satellite contains telemetry equipment to monitor its position and attitude, and to initiate corrections of any deviation from its assigned station. Through telemetry equipment, the earth control station initiates changes to keep the satellite at its assigned longitude and inclination toward earth. Telemetry also monitors the received signal strength and adjusts the receiver gain to keep the uplink and downlink paths balanced.

Station Keeping Equipment

Small rockets are installed on board the vehicle to keep it on station. When the satellite drifts from position, rockets fire to return it. The activities that keep the satellite on position are called *station keeping.* The fuel required for station keeping is the other factor, together with solar cell life, that limits the design life of the satellite. With future satellites, refueling from the space shuttle may become feasible, extending the design life accordingly.

EARTH STATION TECHNOLOGY

Earth stations vary from a simple, inexpensive, receive-only earth station that can be purchased by an individual consumer, to elaborate two-way communications stations that offer commercial access to the satellite's capacity. An earth station consists of microwave radio relay equipment, terminating multiplex equipment, and a satellite communications controller.

Radio Relay Equipment

The radio relay equipment used in an earth station is similar to the terrestrial microwave equipment described in Chapter 15 except that the transmitter output power is considerably higher than terrestrial microwave. Also, antennas up to 30 meters in diameter are used to provide the narrow beam width required to concentrate power on the targeted satellite. Figure 17.4 is a photograph of an earth station.

Because the earth station's characteristics are more easily controllable than the satellite's and because power is not the problem on earth that it is in space, the earth station is assigned the major role in overcoming the path loss between the satellite and earth. Path loss ranges from about 197 dB at 4 GHz to about 210 dB at 12 GHz. Also, the higher the frequency the greater the loss from rainfall absorption. Therefore the uplink is always operated at the higher frequency where higher transmitter output power can be used to overcome absorption, while the lower frequency is reserved for the downlink where large transmitting antennas and high-power amplifiers are not feasible.

Antennas are adjustable to compensate for slight deviations in satellite positioning. Antennas at commercial stations are normally automatically adjusted by motor drives, while inexpensive antennas are manually adjusted as needed. Thirty-meter antennas provide an extremely narrow beamwidth, with half-power points 0.1 degree wide.

FIGURE 17.4
K Band Satellite Earth Station Mounted on Rooftop

Courtesy Satellite Business Systems.

Satellite Communications Control

A satellite communications controller (SCC) is employed to apportion the satellite's bandwidth, to process signals for satellite transmission, and to interconnect the earth station microwave equipment to terrestrial circuits. The SCC formats the received signals into a single integrated bit stream in a digital satellite system or combines FDM signals into a frequency modulated analog signal in an analog system.

Multiplexing

The multiplex interface of an earth station is conventional. Satellite circuits use either analog or digital modulation, with interfaces to terrestrial circuits of the types described in Chapters 4 and 5.

Access Control

Satellites employ several different techniques to increase the traffic carrying capacity and to provide access to the capacity. *Frequency Division Multiple Access* (FDMA) divides the transponder capacity into multiple frequency segments between end points. One disadvantage of this system is that users are assigned a fixed amount of bandwidth that cannot be adjusted rapidly or easily assigned to other users when it is idle. Also, the guard bands between channels use up part of the capacity.

Time Division Multiple Access (TDMA) uses the same concept of time sharing the total transponder capacity described under Digital Termination Systems in Chapter 15. Earth stations transmit only when permitted to do so by the access protocol. When the earth station is given permission to transmit, it is allotted the total bandwidth of the transponder, typically 50 mb/s, for the duration of the station's assigned time slot. Access is controlled by a master station or by the earth stations' listening to which station transmitted last and sending their burst in preassigned sequence. Each earth station receives all transmissions but decodes only those addressed to it. TDMA provides priority to stations with more traffic to transmit by assigning those stations more time slots than low priority stations. Therefore, a station with a growing amount of traffic can be allotted a greater share of total transmission time.

An alternative to preassigned multiple access is *Demand Assigned Multiple Access* (DAMA). DAMA equipment keeps a record of idle radio channels or time slots. Channel assignments are made on demand by one of three methods: polling, random access with central control, and random access with distributed control. Control messages are sent over a separate terrestrial channel or are contained in a control field in the transmitted frame from a TDMA station.

A further method of assigning capacity is *Spacecraft Switched Time Division Multiple Access* (SSTDMA). Under this system the satellite aims its power at a single station based on instructions from the earth. SSTDMA is a technique that can be used in the future to reduce satellite spacing.

Signal Processing

The SCC conditions signals between the terrestrial and satellite links for transmission. The type of signal conditioning depends on the vendor and may include compression of digital voice signals, echo cancellation, forward error correction, and digital speech interpolation to avoid transmitting the silent periods of a voice signal.

Satellite Transmission

Much of the previous discussion is of only academic interest to those who use satellite services. However, satellite circuits and terrestrial circuits are not identical in their transmission characteristics. Users should be aware of the differences so satellite circuits are applied where their use is both technically and economically feasible.

Satellite Delay

The half-second round-trip delay between two earth stations is noticeable in voice communications circuits, but most people become accustomed to it and accept it as normal if the circuit is confined to one satellite hop. Data communications circuits are another matter. Throughput on circuits using a block transmission protocol such as IBM's binary synchronous is reduced to an unacceptably low level through a satellite because blocks are transmitted only after the preceding block has been acknowledged by the receiver. At this rate only one block per second could be transmitted. Throughput on polling circuits is likewise drastically reduced because a poll from a host computer consists of an inquiry and a response, requiring two earth-station-to-earth-station links, and nearly a full second of propagation delay.

The effects of delay in data circuits are mitigated with a satellite delay compensator as shown in Figure 17.5. Delay compensation cannot, however, resolve the deficiencies of a satellite in a polling network because acknowledgements must come from the terminals themselves. In a delay compensator, the DTE communicates in its native protocol, but communication is with the delay compensator rather than with the DTE at the other end of the circuit. The delay compensator buffers the transmitted block awaiting acknowledgement from the distant end. If a negative acknowledge message indicating an error block is received, the delay compensator retransmits either the error block and all succeeding

FIGURE 17.5
Data Transmission through Satellite Delay Compensators

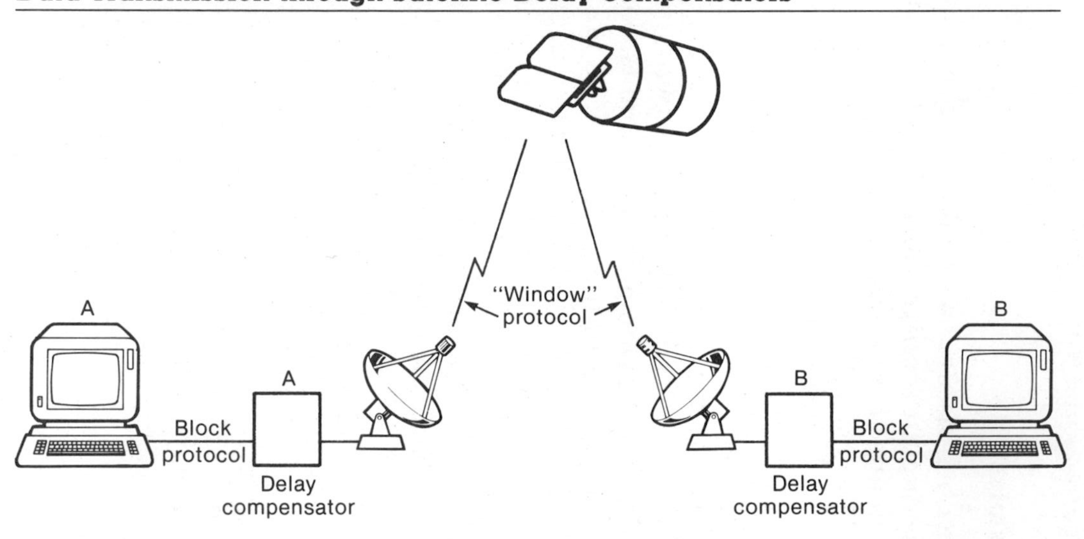

DTE A sends block to DC A
DC A stores block and acknowledges receipt to DTE A
DC A forwards block to DC B
DC B acknowledges receipt to DC A
DC B forwards block to DTE B
DTE B acknowledges receipt to DC B
DC A receives acknowledgment from DC B and removes block from buffer

blocks (go back N), or only the error block (selective retransmission). Figure 17.5 lists the steps the DTEs and the delay compensator use. Throughput is reduced somewhat compared to a terrestrial circuit because the delay compensator interrupts transmission until an error has been corrected. Throughput is a function of error rate as it is on terrestrial circuits, although satellite circuits react more severely to a high error rate because of delays during error correction. The alternative to using a delay compensator is to change to a protocol such as HDLC or SDLC that permits multiple unacknowledged blocks. However, this is often not economically feasible.

Rain Absorption

Rain absorption has a dual effect on satellite communications. Heavy rains increase the path loss significantly and may change the signal polarization enough to impair the cross-polarization discrimination ability of the receiving antennas. (See Chapter 15 for an explanation of cross-polarization.) Unfortunately, the greatest impairment exists at the higher frequencies where interference is less and greater bandwidths are available. Rain absorption can be countered by these methods:

- Choosing earth station locations where heavy rain is less likely to occur.
- Designing sufficient received signal margin into the path to enable the circuits to tolerate the effects of rain.
- Locating a diversity earth station far enough from the main station on the expectation that heavy rain storms will be localized.

Technical considerations limit the first two alternatives. Transmit power and antenna gain from the satellite can be increased only within limits dictated by the physical size of the satellite and the transmit power available. Locations with low precipitation cannot always be selected and still deliver service where it is needed. These considerations mandate the use of earth station diversity at higher frequencies, which suffers the disadvantage of being costly.

Sun Transit Outage

During the spring and fall equinoxes for periods of about 10 minutes per day for 6 days, the sun is positioned directly behind the satellite and focuses a considerable amount of high-energy radiation directly on the earth station antenna. This solar radiation causes a high noise level that renders the circuits unusable during this time. Solutions are to route traffic through a backup satellite or to tolerate the outage.

Interference

Interference from other satellites and to and from terrestrial microwave stations is always a potential problem with satellite circuits. The FCC requires interference studies of all proposed licensees before either satellite or terrestrial licenses are granted. Some licenses in the 12 GHz operational-fixed microwave band are allowed to exist only on the basis of noninterference with direct broadcast satellite (DBS) services.

Carrier to Noise Ratio

Satellite transmission is measured on the basis of carrier-to-noise ratio, which is analogous to signal-to-noise ratio measurements on terrestrial circuits. The ratio is relatively easy to improve on the uplink portion of the satellite circuit because transmitter output power and antenna gain can be increased to offset noise. On the downlink portion of a circuit the effective isotropic radiated power (EIRP), which is a measurement of the transmitter output power that is concentrated into the downlink footprint, can be increased only within the size and power limits of the satellite.

REPRESENTATIVE SATELLITE SERVICES

In this section, three different types of satellite services are discussed to illustrate the versatility of communications satellites. Two of the services, maritime and direct broadcast television, are not feasible except through communications satellites. The third service replaces conventional terrestrial communications and is advantageous in that signals are brought directly to the user without requiring the last link in a communications path—the local telephone loop—that is often expensive and bandwidth limiting.

Satellite Business Services

The SBS network is illustrative of the kind of services that are available through a domestic satellite system. SBS operates three satellites for regular and backup communications. Only large users will be able to meet SBS's minimum capacity requirements, however. At least three serving nodes are required to meet the minimums. The basic node is a Network Access Center (NAC), which is a complete earth station. The other node type is a Serving Point (SP), which accesses an earth station that is shared between multiple users. Service is measured in Transmission Units (TUs), which are channels with a bandwidth of 224 kb/s. The basic Communications Network Service (CNS) requires a minimum usage of three TUs.

The NAC is mounted directly on the user's premises. For nodes too small to justify a full earth station, access is over terrestrial facilities to an SP. The primary users of SBS service are large, geographically dispersed organizations that require combinations of voice, video, facsimile, and data communications.

SBS's network is controlled through a Network Control Center (NCC) that monitors, tests, and switches redundant elements to ensure service continuity. The NCC monitors network usage, allocates capacity to customers, and routes and changes traffic on demand. The satellites are tracked and controlled by earth stations at Castle Rock, Colorado, and Clarksburg, Maryland.

The satellites each have 500 MHz of bandwidth divided into eight transponders of 48 mb/s each. Earth station access is controlled by TDMA. The basic frame is 15 milliseconds long. The earth stations broadcast their capacity needs in a control field so capacity can be assigned on demand. If capacity beyond the contracted number of TUs is needed, it is assigned from spare TUs that are obtained from a pool.

Figure 17.6 shows a block diagram of a NAC. This unit is mounted on the user's premises operating in the Ku band.

FIGURE 17.6
Satellite Business Systems Network Access Centers

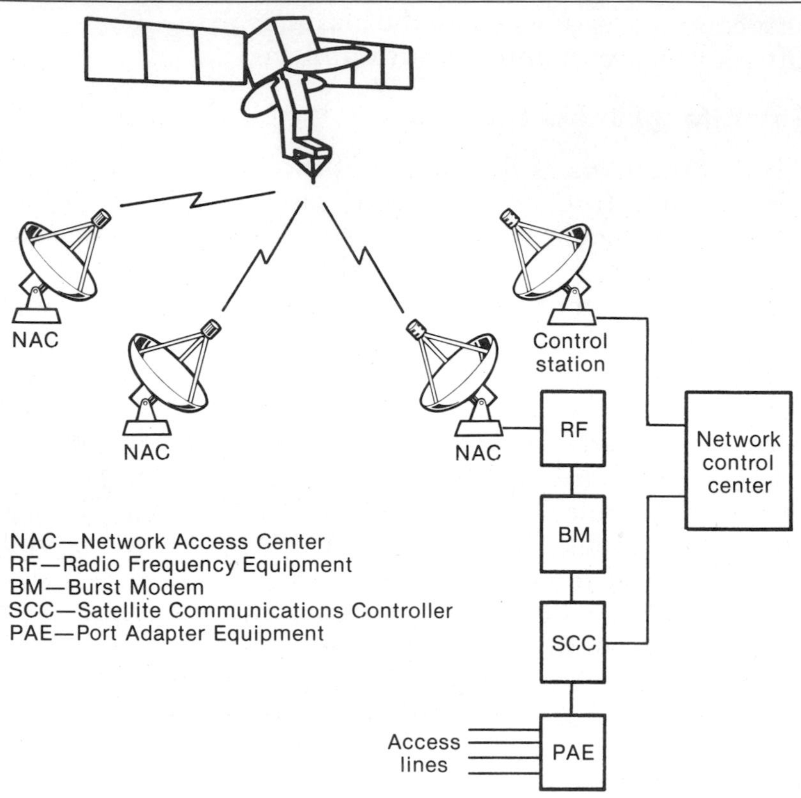

Signals from the user are connected over coaxial cable or twisted pair wire to the port adapter equipment. The port adapter converts analog signals to continuous variable-slope delta modulation. The signals are combined in the SCC into a single bit stream and forwarded to the burst modem where they are framed into 48 mb/s bursts and applied to the transmitter. Received signals are demodulated by the reverse of this process.

International Maritime Satellite Service (INMARSAT)

INMARSAT is an international maritime satellite service under the auspices of the International Maritime Organization (IMO), a United Nations agency. The INMARSAT system consists of a network of 17 existing and planned coastal earth stations. These stations form one terminal of a circuit; the other terminal is the ship earth station. The ship earth station is mounted above decks and automatically kept on position with satellite tracking equipment. Shipboard equipment is type accepted and regulated by INMARSAT.

INMARSAT provides the same kinds of communications services for ships at sea that land stations can access through satellite or terrestrial circuits. In the past the principal method of communication from ships was telex and Morse code over high-frequency radio. Now data circuits are replacing those modes of communication. Voice circuits replace the high-frequency ship-to-shore radio that often suffered from poor signal propagation reliability. In addition, services such as video and facsimile can be carried over INMARSAT. Other services that do not generally apply to land stations can also be accessed through INMARSAT. Ship locations can be monitored precisely through polling equipment. Distress calls can be received and rebroadcast to ships in the vicinity but out of radio range. Broadcasts such as storm warnings can be made to all ships in an area.

Direct Broadcast Satellite (DBS)

The services discussed to this point have been two-way communications between earth stations. A substantial demand exists for receive-only satellite services. Such services have existed for several years to transmit television signals to cable TV services such as Home Box Office, Movie Channel, and Cinemax. Many of these services are picked up by privately owned earth stations for personal use. These services are not intended for personal usage, but a late 1984 court ruling has declared such reception to be legal, leading to a decision by many services to scramble the signal.

DBS service is commercial television intended for individual reception. The first DBS stations are expected to go on the air in 1985 to 1986, broadcasting with a transmitting power of up to 200 watts. The power, considerably higher than that used in communications satellites, is needed to limit the size of receiving antennas to about 1 meter in di-

FIGURE 17.7
Direct Broadcast Television

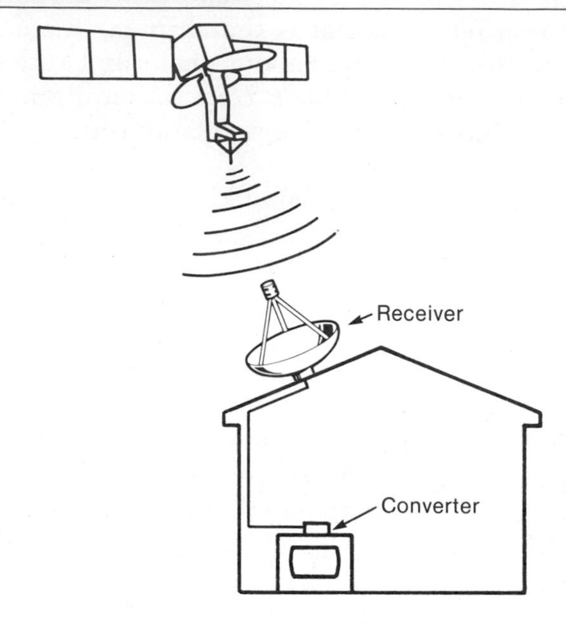

ameter. Viewers receive TV channels through a personal earth station consisting of the components shown in Figure 17.7. A frequency down-converter is mounted on the antenna to reduce the 12 GHz downlink signal to approximately 1 GHz. This signal is down converted to a TV channel with a multiple channel selector at the TV set. Signals will be encoded in most cases, requiring a descrambler in the TV receiver. DBS is expected to have a wide impact on TV broadcasting in the United States because of the ability of broadcasters to reach a nationwide market from a single source. DBS is also intended for reception by broadcast, master antenna, and cable television operators as well as individual users. In many cases, DBS will supplement rather than replace existing services.

STANDARDS

Satellite communication is regulated by the FCC in the United States and by CCITT and CCIR internationally. Satellite carriers are free to design systems to their own standards and objectives, but the use of the radio frequency spectrum and satellite positioning must conform to standards set by the FCC. The majority of users obtain their services from a satellite carrier and are therefore not concerned with the performance of the satellite and earth station equipment but are concerned with circuit performance. Circuit performance criteria are established by the carrier. CCITT recommends circuit performance objectives, but compliance is voluntary in the United States.

CCIR

Volume IV-1 of the 15th Plenary Assembly of the CCIR Fixed-Satellite Service.

Volume IV/IX-2 of the 15th Plenary Assembly of the CCIR Frequency Sharing and Coordination between Systems in the Fixed-Satellite Service and Radio-Relay Systems.

Volume X/XI-2 of the 15th Plenary Assembly of the CCIR Broadcasting Satellite Service (Sound and Television).

CCITT

E.210, F.120, Q.11 *ter* Ship station identification for VHF/UHF and maritime mobile-satellite services.

E.211, Q.11 *quater* Numbering and dialing procedures for VHF/UHF and maritime mobile-satellite telephone services.

G.434 Hypothetical reference circuit for communication-satellite systems.

G.445 Noise objectives for communication-satellite system design.

G.473 Interconnection of a maritime mobile satellite system with the international automatic switched telephone service; transmission aspects.

M.1100 General maintenance aspects of maritime satellite systems.

Q.14 Means to control the number of satellite links on an international telephone connection.

Q.60 General requirements for the interworking of the terrestrial telephone network and the Maritime Mobile-Satellite Service.

U.60 General requirements to be met in interfacing the international telex network with maritime satellite systems.

U.61 Detailed requirements to be met in interfacing the international telex network with maritime satellite systems.

EIA

RS-411 Electrical and mechanical characteristics of antennas for satellite earth stations.

FCC

Part 21 Rules and Regulations Domestic Public Fixed Radio Services.

Part 25 Satellite Communications.

Part 100 Direct Broadcast Satellite Service.

APPLICATIONS

Many organizations use satellite services in their private networks, but few are as extensive as the Hercules Corporation's network described in this section. Because Hercules is widely dispersed geographically, it is particularly adaptable to satellite communications.

The Hercules Corporation Telecommunications Network

Hercules Incorporated produces a broad line of natural and synthetic materials and products including cellulose and natural gum thickeners, flavors and fragrances, natural and hydrocarbon rosins and resins, polypropylene fibers and films, explosive products, graphite fibers and products for aero-

FIGURE 17.8
The Hercules Network Includes Nine Earth Stations in the United States.

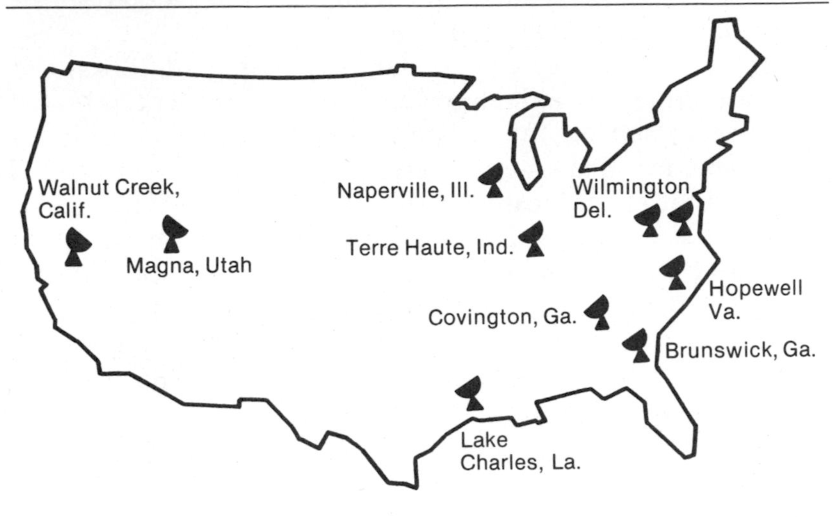

Courtesy Hercules, Incorporated.

space, including propellants and structures for military and aerospace use.

Hercules employs over 26,000 people at its more than 80 domestic and international locations. To improve productivity and reduce travel and communications costs, Hercules launched an extensive program in 1979 to provide automated office services and to tie the company together with a private terrestrial and satellite network that is one of the largest in the country.

The network, known as Hercules Integrated Telecommunications System (HITS), became operational in 1982. HITS long-haul facilities are largely composed of satellite services obtained through SBS's Communications Network Service. Ten earth stations are located throughout the country as shown in Figure 17.8. The network is used for voice, voice messaging, video teleconferencing, and data transmission, including electronic mail. Satellite services are particularly adaptable to this form of network because the broadcast nature of satellites is well suited to video conferencing. Also, many of Hercules' offices are located in smaller metropolitan areas where wideband digital facilities are unavailable. The 10 earth stations in the network are located on Hercules' premises with outlying locations linked to the network through DDS facilities.

Hercules' terminal equipment, consisting of PBXs, numerous computers, and teleconferencing equipment, is linked to the earth stations by twisted pair wire, coaxial cable, or DDS facilities. End-to-end digital connections are provided to users that are colocated with a NAC or linked to a NAC with digital facilities. The network's performance is monitored by SBS's network control center (NCC) in McLean, Virginia. SBS field engineers are located at each NAC. Trouble reports from users are received by a Hercules network services operator, logged, and referred to the SBS field engineer.

Least-cost routing (LCR) for off-network calls is provided to all users by an SBS satellite communication controller at each earth station. With centralized LCR, the service is available to all users, including those served by PBXs that lack this feature.

SBS-HITS Interface

The HITS network is implemented over the SBS-1 satellite, which consists of 10 transponders of 48 mb/s each. The TDMA access method is used to allot Hercules a specified portion of the traffic field of each frame. During peak load periods, Hercules can draw on an unallocated capacity pool. Figure 17.9 is a diagram of a typical earth station showing the interface between Hercules' equipment and the NAC.

Voice Services

More than 8,000 users are served by 50 PBXs furnished by various vendors. Access to the network is obtained by dialing "8" from a PBX and then dialing seven digits to reach any terminating location on the network. Off-network voice communications are handled over WATS lines and local and toll telephone services as well as satellite services. Digital voice channels over the satellite portion of the network are provided over 32 kb/s delta modulated channel banks using a digital speech interpolation system that provides 2:1 compression. Hercules has a voice messaging system located

FIGURE 17.9
Block Diagram of Hercules Earth Station

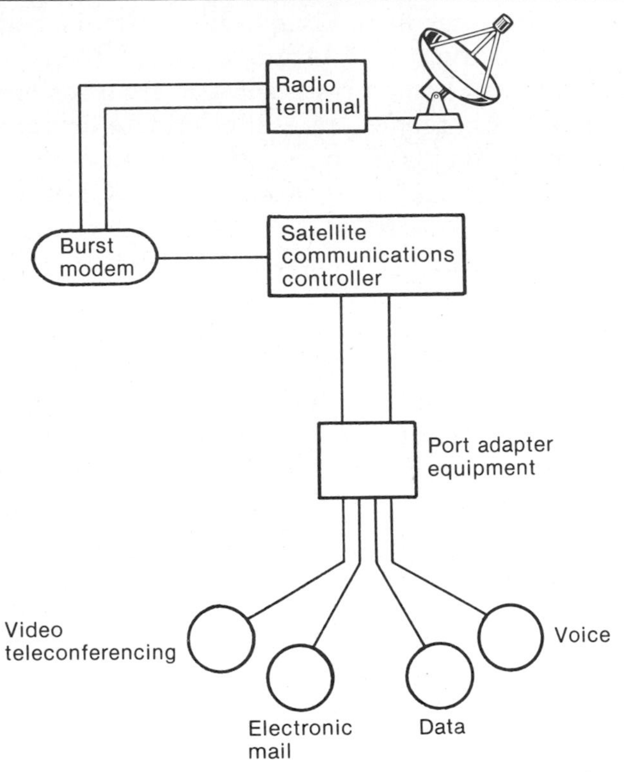

Courtesy Hercules, Incorporated.

Earth station on Hercules Network accepts voice, data, document, or electronic mail, and video teleconferencing transmissions.

in Wilmington that serves 2,900 users with about 50,000 messages per month. The system can be accessed over the public telephone network as well as over the HITS network.

Video Conferencing

Twenty-four video conference rooms, including one in London, have access to the HITS network. The bulk of the conference facilities use freeze-frame video. In locations with access to 56 kb/s trunks, the pictures are refreshed as often as every 12 seconds. In locations lacking 56 kb/s circuits, conferencing is implemented over four telephone circuits providing a 19.2 kb/s channel and a picture refresh rate of 35 to 40 seconds.

Between Wilmington and Naperville, Illinois, full-motion video is planned over a 1.544 mb/s circuit. The system will use NEC codecs to transmit signals that are interleaved from two cameras. This technique will eliminate camera switching by enabling signals from two separate cameras to be displayed simultaneously on two monitors. Graphics equipment will also provide for slides and charts. When a graphic image is transmitted, the video images will be frozen for two seconds.

Data Transmission

Both interactive and batch data are transmitted between the corporate data center in Wilmington and data centers in other locations. A typical configuration between Wilmington and Oxford, Georgia, using IBM 3705 Communications Controllers is shown in Figure 17.10. A 1.544 mb/s satellite-based data link is being planned between the Wilmington and Magna, Utah, data centers. The data link will be used for backup and load sharing between computers located there. The primary protocols used on the Hercules data network are SDLC, BSC, X.25, and asynchronous at speeds of 1.2 to 56 kb/s.

Experience Summary

Before HITS was established, Hercules had no formal network, which eased the transition into the HITS network but meant that few statistics were available for sizing. The initial network was designed with sufficient capacity to ensure a high degree of call completions, with the ability to reduce capacity wherever service was overprovided.

SBS has assumed responsibility for the bulk of network administration. Hercules is able to monitor usage and to control cost and service through the information provided by the NCC, but they rely on SBS's telecommunications expertise for day-to-day administration and have avoided building a large internal staff of their own. Hercules users report a high degree of satisfaction with network quality and reliability. Error performance of the satellite circuits has been superior to terrestrial data communications services. The

FIGURE 17.10
Hercules Telecommunications Network Diagram

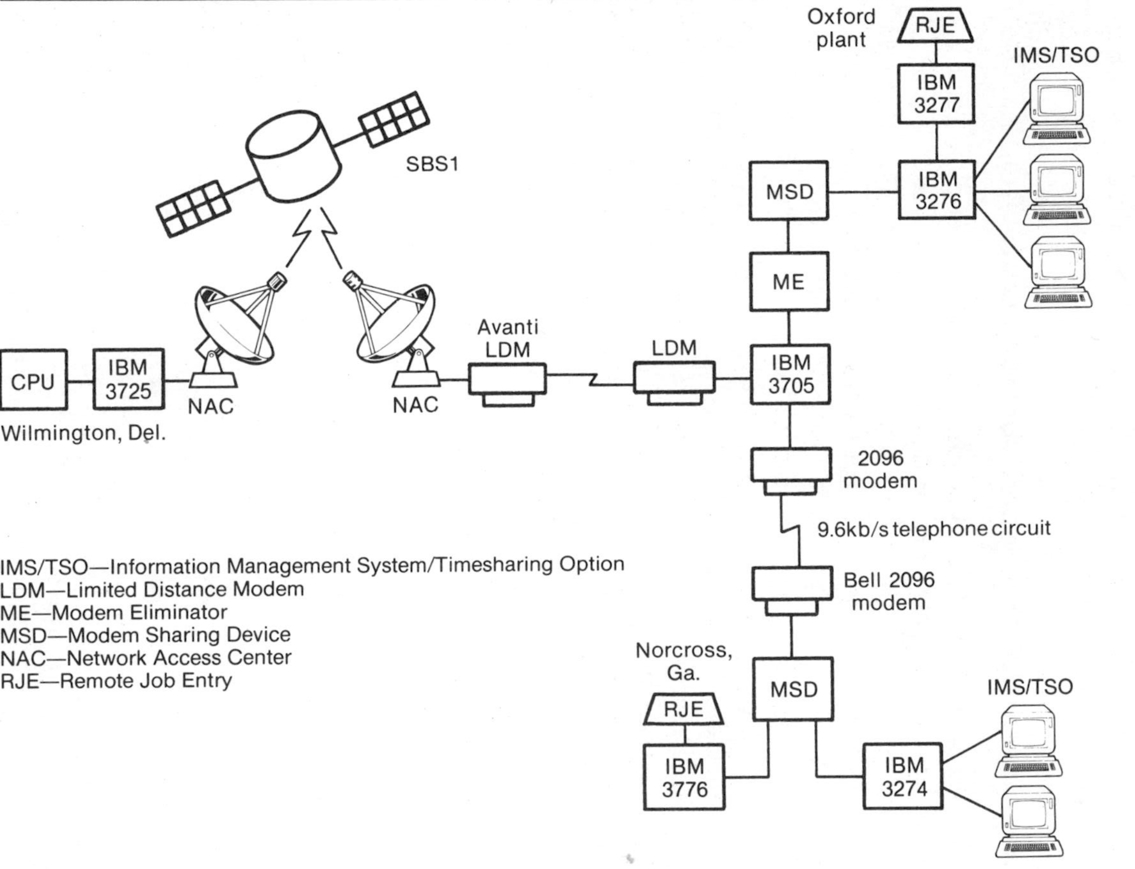

IMS/TSO—Information Management System/Timesharing Option
LDM—Limited Distance Modem
ME—Modem Eliminator
MSD—Modem Sharing Device
NAC—Network Access Center
RJE—Remote Job Entry

Courtesy Hercules, Incorporated.

Note: Configuration that is typical on the network. Batch RJE and interactive IMS/TSO applications at Oxford, Ga., plant are mixed with similar data coming from nearby Norcross facility and transmitted to earth station at Covington, a few miles away. The satellite links that station with Wilmington.

video conferencing facilities have achieved their objectives of reducing travel and improving meeting quality as well as developing closer relationships between the headquarters and field locations.

Satellite Service Evaluation Considerations

Satellite space vehicle evaluation criteria are complex, technical, and of interest only to designers, owners, and manufacturers of satellites and on-board equipment. Therefore, these criteria are omitted from this discussion. Likewise, common carrier earth station equipment evaluations are not included in this discussion. Evaluation criteria discussed in Chapter 15 on microwave equipment generally apply to satellite services except that multipath fading is not a significant problem in satellite services. Also, alarm and control systems in terrestrial microwave are different from those used in satellite systems.

The following factors should be considered in evaluating satellite services and privately owned earth station equipment:

Availability

Circuit availability is a function of path and equipment reliability. To the user of capacity over a carrier-owned earth station, equipment reliability is a secondary consideration. The important issue is circuit reliability measured as percent error-free seconds in digital services and percent availability within specified noise limits for analog services.

These same availability criteria apply with privately owned earth stations, but the carrier can quote availability only on the basis of path reliability. In this case, equipment availability is a function of MTBF and MTTR, as discussed in Chapter 15, and must be included in the overall reliability calculation. The frequency and duration of any expected outages because of solar radiation or solar eclipse should be evaluated.

Access Method

Satellite carriers employ a variety of techniques to increase the information-carrying capacity of the space vehicle. Techniques such as DAMA can potentially result in congestion during peak load periods with the potential that earth station buffer capacity can be exceeded or access to the system blocked. Some carriers employ delta modulation or adaptive differential pulse code modulation to increase the voice circuit carrying capacity of the satellite and may therefore limit data transmission speeds. Users should determine what methods the carrier uses to apportion access, whether blockage is possible, and whether transmission performance will meet objectives.

Transmission Performance

The carrier's loss, noise, echo, envelope delay, and absolute delay objectives should be evaluated. Except for absolute delay, which cannot be reduced except by using terrestrial facilities to limit the number of satellite hops, satellite transmission evaluation should be similar to the criteria discussed in Chapter 2.

Earth Station Equipment

Earth station equipment is evaluated against the following criteria:

- Equipment reliability.
- Technical criteria, such as antenna gain, transmitter power, and receiver sensitivity, that provides a sufficiently reliable path to meet availability objectives.
- Antenna positioning and tracking equipment that is automatically or manually adjustable to compensate for positional variation in the satellite.
- Physical structure that can withstand the wind velocity and ice loading effects for the locale.
- The availability of radome or deicing equipment to ensure operation during snow or icing conditions.

GLOSSARY

Carrier-to-noise ratio: The ratio of the received carrier to the noise level in a satellite link.

Demand Assigned Multiple Access (DAMA): A method of sharing the capacity of a communications satellite by assigning capacity on demand to an idle channel or time slot from a pool.

Direct broadcast satellite (DBS): A television broadcast service that provides television programming services throughout the United States from a single source through a satellite.

Downlink: The radio path from a satellite to an earth station.

Effective Isotropic Radiated Power (EIRP): Power radiated by a transmitter compared to the power of an isotropic antenna, which is one that radiates equally in all directions.

Earth station: The assembly of radio equipment, antenna, and satellite communication control circuitry that is used to provide access from terrestrial circuits to satellite capacity.

Footprint: The earth's area that is illuminated by the rf output signal of a satellite.

Frequency Division Multiple Access (FDMA): A method of sharing the capacity of a communications satellite by frequency division of the transponder.

Geosynchronous: A orbit that positions a satellite at a constant point with respect to a point on the earth's surface.

INMARSAT: The International Maritime Satellite Service that provides communications services to ships at sea.

Satellite Communications Control (SCC): The earth station equipment that controls such communications functions as access, echo suppression, forward error correction, and signaling.

Satellite delay compensator: A device that compensates for absolute delay in a satellite circuit by communicating with data terminal equipment with the DTE's own protocol.

Spacecraft Switched Time Division Multiple Access (SSTDMA): A method of sharing the capacity of a communications satellite by on-board switching of signals aimed at earth stations.

Spin stabilization: A method of preventing a satellite from tumbling by spinning it about its axis.

Spot beam antenna: A satellite antenna that is capable of illuminating a narrow portion of the earth's surface.

Station keeping: The process on board a satellite for keeping it at its assigned longitude and inclination.

Sun transit outage: Satellite circuit outage caused by direct radiation of the sun's rays on an earth station receiving antenna.

Three-axis stabilization: A method of preventing a satellite from tumbling by use of a gyroscope inside the satellite.

Time Division Multiple Access (TDMA): A method of sharing the capacity of a communications satellite by allotting access to the satellite to earth stations transmitting on the same frequency.

Transponder: A satellite-mounted radio repeater that amplifies and converts the uplink frequency to the downlink frequency.

Uplink: The radio path from an earth station to a satellite.

BIBLIOGRAPHY

Bhargava, V. K. et. al. *Digital Communications by Satellite.* New York: John Wiley & Sons, 1981.

Chorafas, Dimitris N. *Telephony Today and Tomorrow.* Englewood Cliffs, N.J.: Prentice-Hall, 1984.

Freeman, Roger L. *Telecommunication System Engineering.* New York: John Wiley & Sons, 1980.

———. *Reference Manual for Telecommunications Engineering.* New York: John Wiley & Sons, 1985.

Fthenakis, Emanuel. *Manual of Satellite Communications.* New York: McGraw-Hill, 1984.

Martin, James. *Telecommunications and the Computer.* Englewood Cliffs, N.J.: Prentice-Hall, 1976.

Prentiss, Stan. *Satellite Communications.* Blue Ridge Summit, Pa: TAB Books, 1983

Rosner, Roy D., ed. *Satellites, Packets, and Distributed Telecommunications.* Belmont, Calif.: Lifetime Learning Publications, 1984.

Singleton, Loy A. *Telecommunications in the Information Age.* Cambridge, Mass.: Ballinger Publishing, 1983.

Van Trees, H. L., ed. *Satellite Communications.* New York: IEEE Press, 1979.

VENDORS OF SATELLITE EQUIPMENT AND SERVICES

Earth Station Equipment

Aydin Microwave Division

Equatorial Communications Co.

M/A-COM DCC Inc.

NEC America Inc., Radio & Transmission Division

Rockwell International Collins Transmission Systems Division

Satellite Transmission Systems, Inc.

Scientific-Atlanta, Inc.

Satellite Transmission Services

AT&T Communications

American Satellite Corporation

GTE Satellite Corporation

RCA Americom

Satellite Business Systems

Western Union Telegraph Co.

CHAPTER **EIGHTEEN**

Data Communications Networks

A substantial portion of the circuits in U. S. telephone networks are occupied with data transmission. Just what the ratio between voice and data is, no one knows because so much data communication takes place over switched voice facilities. The use of the telephone network for data has advantages, the primary advantages being its accessibility and worldwide coverage. However, it has several disadvantages that generate demand for separate data communications networks. The primary drawback to telephone circuits is the fact that they are analog and require modems at each end of the connection. The modems are not only expensive, but they also must be matched in speed and modulation method or they are incompatible. The voice telephone channel also limits data transmission speed. A substantial number of calls are carried over digital facilities that are capable of 64 kb/s transmission, but data must be converted to analog to pass through local loops and local central offices.

The shortcomings in the telephone network result from its being designed for voice communication at a time when neither the technology nor the motivation existed to develop a digital network. Now that both the technology and the motivation exist, networks are changing. Chapter 25 discusses the integrated services digital network (ISDN) that will evolve gradually over the next few decades. In the ISDN, communications will be digital with the analog conversion taking place in the telephone. With the magnitude of investment in analog facilities, however, the change to ISDN will occur gradually. In the meantime, separate data networks, both publicly and privately owned, carry data traffic that cannot economically be carried on the telephone network.

The explosive growth of distributed data processing and of personal computers and public data bases has resulted in a corresponding growth in data communications. By the mid 1980s, businesses and governments remain the primary users of data communications. However, we are on the threshold of several developments that may expand personal and residential data communications. For one, demand for videotex, which is interactive computer-generated information, is gradually developing. Videotex services include catalog shopping, news broadcasts, stock quotations, reference material, entertainment—the list is limited only by human imagination and the willingness of consumers to pay, which at the present is still in doubt.

To fulfill all applications, the ideal data communications network will require several criteria that cannot be met in total by the present telephone network:

- The information channel will be readily available to every user.
- The channel will be usable for all types of communications: voice, data, and video.
- The channel will be affordable.
- The channel will be capable of being linked to a multiplicity of networks including metropolitan, long haul, and local area.
- The channel will provide error-free communications between terminals.

At present, cable television is an alternative that may come into general use. Its broad bandwidth offers capacity that cannot be matched with twisted pair cable. However, like the telephone network, CATV is inherently an analog medium.

Circuit switching, which is used for many data services today, has drawbacks for data communications in its present form. The greatest advantage of circuit switching is that a channel is dedicated to users for the duration of the session, but with certain types of data a great deal of the capacity is wasted because data is not flowing continuously in both directions of transmission. Data signals often carry the destination address as part of the message and can readily share communication channels under conditions that would be detrimental to voice. The most economical type of network for data depends greatly on the nature of the application. Data traffic is characterized by the length of message, the response time that can be tolerated, and the degree of balance in usage between the two directions of transmission. Usage can be grouped into categories such as these:

- Inquiry/response. This is typical of information services where a short inquiry is followed by a lengthy response from the host.
- Conversational. This mode, typical of terminal-to-terminal communication, consists of short messages that are of approximately equal length in both directions.
- Bulk data transfer. This is typical of applications such as mainframe-to-mainframe communications where large files are passed, often at high speed in only one direction.
- Remote job entry. This is typical of applications in which terminals send information to a host. The bulk of the transmission is from the terminal with a short response from the host.

Short, bursty messages such as inquiry/response and conversational modes can be handled efficiently on a packet switched network, but waste circuit capacity over a circuit switched network because so much of the communication is in only one direction at a time. These modes usually require a full-duplex circuit. Remote job entry applications can make effective use of message switching and usually work over reversible half-duplex lines. Polling can work efficiently in this type of application. Bulk data transfer is most efficiently conducted over point-to-point circuits. If enough traffic exists to warrant a full-time circuit, private line service is appropriate. If not, circuit switching is generally most efficient. As with other types of communications service, data networks are chosen by balancing cost and service.

Data networks are classed as long haul, metropolitan, or local. Local networks are covered in Chapter 19. The other two types are explained in this chapter. Networks are also classified as point to point or switched, referring to the methods stations use to access one another. In point-to-point networks, stations are directly wired to the network. Messages are routed from the station to the host by addressing or by polling. In switched networks, stations are connected by circuit switching, packet switching, or message switching. With some exceptions, data network protocols and equipment are not standardized but use proprietary methods unique to the manufacturer or vendor. Even where networks use the same protocol, the options in the particular application may still prevent compatibility. Any communication between noncompatible networks requires a gateway circuit or protocol converter to translate between protocols.

POINT-TO-POINT NETWORK TECHNOLOGY

The first data communications networks consisted of teletypewriters wired to a common circuit. All machines printed all the traffic on the network; the operator removed messages from the machine, discarded those intended for other stations, and delivered messages to the addressee. Security in such a system obviously left a great deal to be desired, and the operation was labor intensive.

Addressing

To improve security and to reduce the time of screening unwanted messages, *addressing* was introduced to teletypewriter networks. Using an electromechanical selector, the sender preceded messages with a call directing code that blinded all machines but the addressee. Addressing is used on modern data networks by preceding the message with the address of the stations it is intended for. This system is common in local area networks where all stations except the addressee electronically discard messages. Addressing has several advantages. First, it is simple to design and administer. Also, it is inexpensive because all machines share the transmission medium. However, as the network has all the characteristics of a large telephone party line, interference between stations may be a problem.

LANs use various systems that are described in Chapter 19 to minimize interference. Although stations are programmed to respond only to their own addresses, it is not difficult to arrange a terminal to respond to all messages. Therefore, if absolute security is needed, encryption is required. A further drawback of asynchronous point-to-point networks is their lack of error detecting and correcting capability.

Polling

A multipoint or multidrop network, as shown in Figure 18.1, can use *polling* to apportion access. A host computer is connected to a circuit, usually through a front end processor, which is a computer equipped for telecommunications that is intended to relieve the host computer of teleprocessing chores. Each station is assigned a code. The host polls the stations by sending each a short message. If the station has no traffic, it responds with a negative acknowledgement message. If it has traffic, it responds with a block of data.

Polling networks are designed as full duplex or half duplex. In half-duplex networks, the modem is reversed at the end of a poll or response. Polling is an efficient way of sharing a common data circuit. However, it has drawbacks that limit its applications. First, a substantial portion of the circuit time is consumed in the overhead of sending polling messages, returning negative acknowledgements, and in reversing the modems. To some degree throughput can be improved by using hub polling. In hub polling, when a station receives a poll it passes its traffic to the host and passes the

FIGURE 18.1
Polled Multidrop Data Network

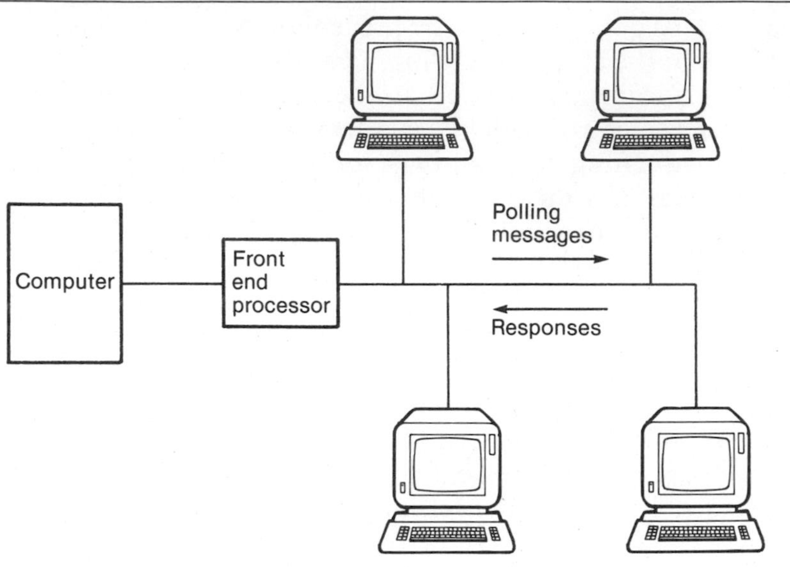

polling message to the next station in line. Hub polling is more complex than "roll call" polling and is not as widely used. Security is also a problem in polling networks. Any time stations share a common circuit, messages can be intercepted.

Point-to-Point Digital Facilities

Most point-to-point and multipoint data networks use analog facilities and modems. Several alternatives are available for end-to-end digital facilities. Very large users can justify communication over a digital satellite with earth stations mounted on their premises. Satellite communication has certain drawbacks, primarily absolute delay, for some types of data protocols. These problems are discussed in more detail in Chapter 17.

Private microwave and lightwave systems offer another alternative for digital communication. Facilities are obtained by private ownership or by leasing capacity from common carriers. Digital microwave is covered in Chapter 15 and lightwave in Chapter 16.

Dataphone® Digital Service (DDS), offered by telephone companies and AT&T Communications, is another method of obtaining end-to-end digital channels. With the breakup of AT&T, DDS is known by various names in the operating telephone companies and is called Accunet® DDS by AT&T. Its structure, however, remains essentially the same as pre-divestiture DDS.

Circuits are available at speeds of 2.4, 4.8, 9.6, and 56 kb/s. The user can multiplex signals to lower speeds if desired. DDS service objectives are 99.5 percent error-free seconds and 99.96 percent availability. DDS employs dataport channel equipment on T-1 carrier as discussed in Chapter 4. Long-haul circuits are transported on analog microwave using data under voice (DUV) equipment as discussed in Chapter 15. Service to the end user is over four-wire nonloaded cable facilities. DDS uses a bipolar signaling format requiring that the user's data signal be converted from the usual unipolar output of terminal equipment. This is accomplished with a data service unit (DSU) located on the user's premises. If the user's equipment is capable of providing bipolar output and timing recovery, the data signal is coupled to a channel service unit (CSU). Both units provide loop-back facilities so the local cable can be tested by looping the transmit and receive pairs together. The signal is fully synchronized from end to end. Data signals from multiple users are concentrated in a DDS hub office where they are connected to the long-haul network. The hub is also a testing point.

Bulk digital terrestrial transmission facilities operating at 1.544 mb/s have recently been offered by common carriers. For example, AT&T's Accunet® T1.5 is a premise-to-premise service offered at prices that are competitive with DDS when several 56 kb/s services are required between two points. Local telephone companies are also offering T-1 lines at prices that are more attractive than groups of individual special service lines. The disadvantages of these services compared to analog facilities is that they are not available in many locations and are more difficult for the end user to test at a circuit level than circuits with modems.

PACKET NETWORK TECHNOLOGY

Unlike circuit switched networks which provide a circuit between end points for the exclusive use of two or more stations, packet networks provide *virtual circuits,* which have many of the characteristics of a switched circuit. The difference is that the circuits are time shared rather than dedicated to the connection. As shown in Figure 18.2, a packet network consists of multiple nodes that are accessed through dedicated or dialed connections from the end user, using one of several access options that are described later. The switching nodes control access to the network and route packets to the destination over a backbone of high-speed data circuits.

Packet Switching Nodes

The packet switching nodes consist of processors that are interconnected by backbone circuits. High-speed analog or digital facilities operating at 9.6 or 56 kb/s are usually employed between nodes.

Packet Assembly and Disassembly

Data flows through the network in packets consisting of an information field sandwiched between header and trailer records. The packet used by CCITT's X.25 protocol is shown in Figure 18.3, p. 480. Note in the figure that *octet* is the CCITT word for byte. The X.25 packet is enclosed in a frame consisting of a one octet flag having a distinctive pattern that is not repeated in the data field. The second octet is an address code that permits up to 2,555 addresses on a data link. The third field is a control octet that sequences messages and sends supervisory commands. The X.25 packet consists of three format and control octets and an information field that is filled by the user. The final three octets of the trailer are a 16-bit cyclical redundancy check field (see Chapter 3) and an end-of-frame flag.

A message is sliced into packets by a packet assembler/disassembler (PAD). The PAD communicates with the packet network using a packet network protocol. X.25, which is described in a later section, is a CCITT protocol recom-

FIGURE 18.2
Packet Network Showing Access Options

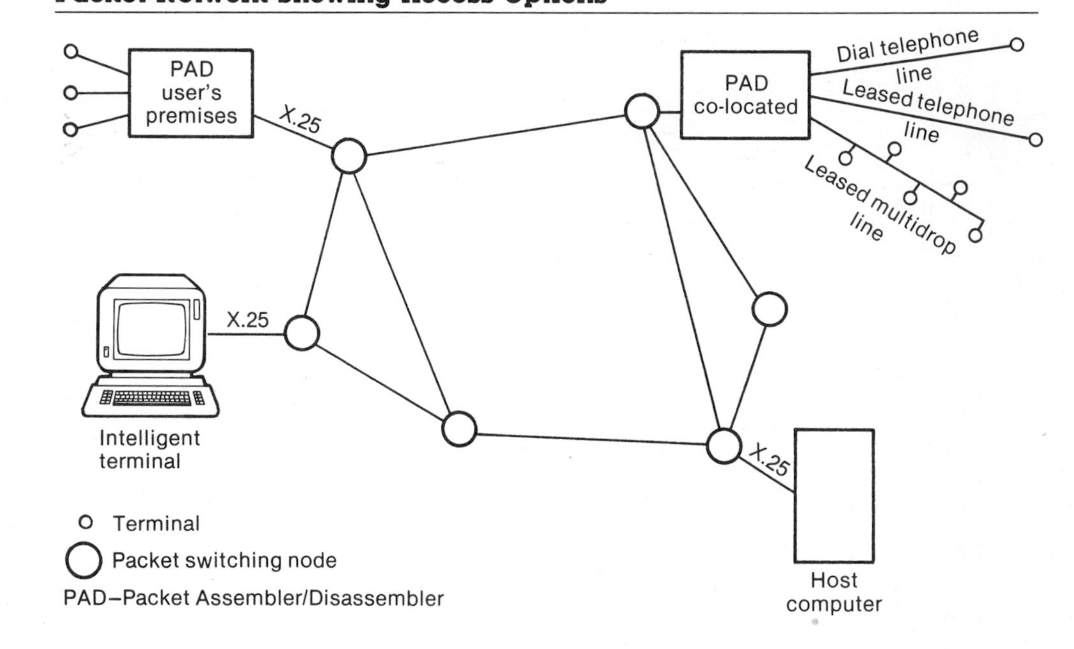

FIGURE 18.3
Level 2 Frame Enclosing X.25 Packet

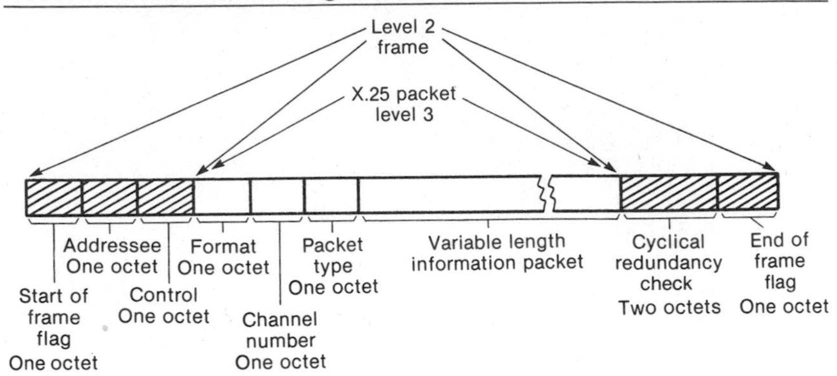

mendation that is used by most public and many private data networks. The packet network uses the address field of the packet to route it to the next node on the way to the destination. Each node hands off the packet following the network's routing algorithm until it reaches the final node. At the terminating node the switch sequences the packets and passes them off to the PAD. The PAD strips the data from the header and trailer records and reassembles it into a completed message.

Error checking takes place in each link in the network. If a block is received in error, the node rejects it and receives a replacement block. Because errors require resending blocks and because blocks can take different paths to the destination, it is possible for blocks to arrive at the destination out of sequence. The receiving node contains buffers to store the message so the blocks are presented to the PAD in the proper sequence.

In many ways packet networks are similar to message switching or store-and-forward networks discussed in Chapter 3; however, there are several differences:

- Packet switching networks are more effective for real-time operation. Message switching networks store messages for later delivery. Although the time is often short, it may be a substantial fraction of a minute under load.

- Message networks typically retain a file copy of the message for a given period. Packet networks clear the message from their buffers as soon as it is delivered.
- Messages are transported as a unit in message networks. In packet networks they are sliced into shorter blocks that are reassembled into a message at the receiving end of the circuit.

A special type of message known as a *datagram* is available in some packet networks. A datagram is a single packet that is routed through the network to the destination without acknowledgement. Although some requirement for this type of message exists, it is not supported by major packet networks. An alternative to the datagram is the *fast select* message, which is a single packet and response. A typical application for fast select messages are credit-checking terminals that transmit the details of a transaction from a point-of-sale terminal and receive an acknowledgment from a credit agency's data base.

The packet size is established by the network designer to optimize throughput. Since each packet consists of a fixed-length header and trailer record, short packets reduce throughput because of the time spent in transmitting overhead bits in the header and trailer. On the other hand, long packets reduce throughput because the switching node cannot forward the packet until all bits have been received, which increases buffer requirements at the node. Also, the time spent in retransmitting error packets is greater if the packet length is longer. Most packet networks operate with a packet length of 128 to 256 bytes.

Virtual Circuits

Packet networks establish two kinds of virtual circuits: permanent and switched. Permanent virtual circuits are the packet network equivalent of a dedicated voice circuit. A path between users is established, and all packets take the same route through the network. With a switched virtual circuit the network path is established with each session.

Packets consist of two types: control and data. Control packets, which contain information to indicate the status of the session, are analogous to signaling in a circuit switched network. For example, a call setup packet would be used to establish the initial connection to the terminating machine. An answer packet would be returned by the terminating machine. Control packets are used to interrupt calls in progress, disconnect, indicate acceptance of reversed charges, and other functions that are controlled by operators and supervisory signals on telephone networks. With switched virtual circuit operations, these control packets are used to establish a session. In permanent virtual circuit operation, the path is preestablished and no separate packets are needed to connect and disconnect the circuit.

Network Access Methods

Access to public data networks is provided by one of three methods as shown in Figure 18.4:

- A dedicated X.25 link between the user's host computer and the data network.
- A dedicated link between the data network and a PAD on the user's premises.
- Dedicated or dial-up access over the telephone network into a PAD provided by the network vendor.

In the first option the user's host computer performs the PAD functions. The second option involves a PAD provided either by the vendor on the user's premises or a user-owned PAD certified for network compatibility by the network.

The third option is the least complex and is the only method economically feasible for small users. However, dial access has several disadvantages. The first problem is the loss of end-to-end error checking. A few asynchronous error checking protocols have been proposed, but none have been accepted as a universal standard. Those that are supported by personal computer software reduce the effective throughput of the data circuit to a low value. In addition to the error-checking problem, the telephone network has other deficiencies for data communications. Data calls may progress through several switching machines, some of which may be electromechanical machines that introduce impulse noise.

FIGURE 18.4
Alternatives for Access to Packet Switched Networks

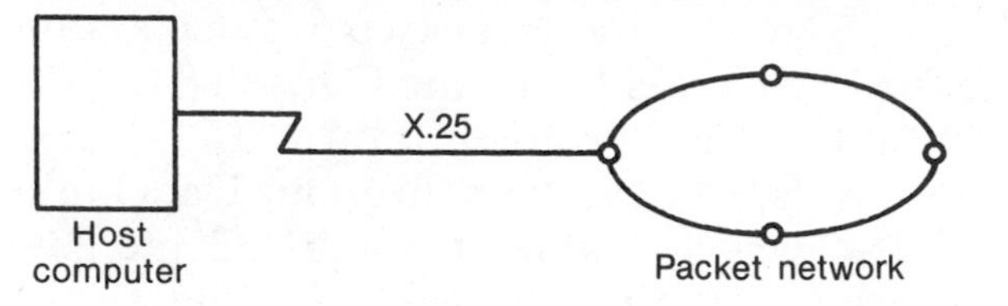

a. User Supplied Computer and Software on User's Premises with Certified Interface

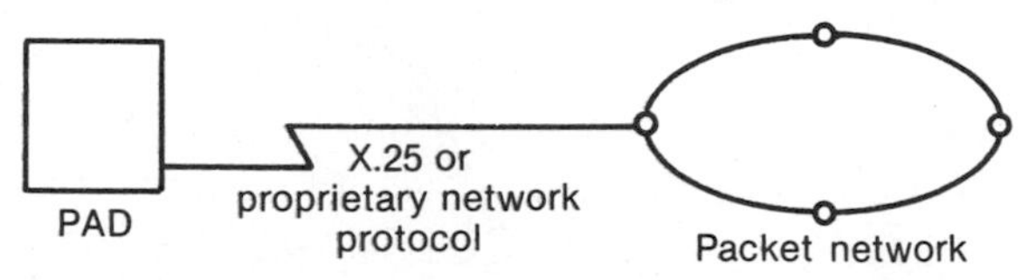

b. User or Network Supplied PAD on User's Premises

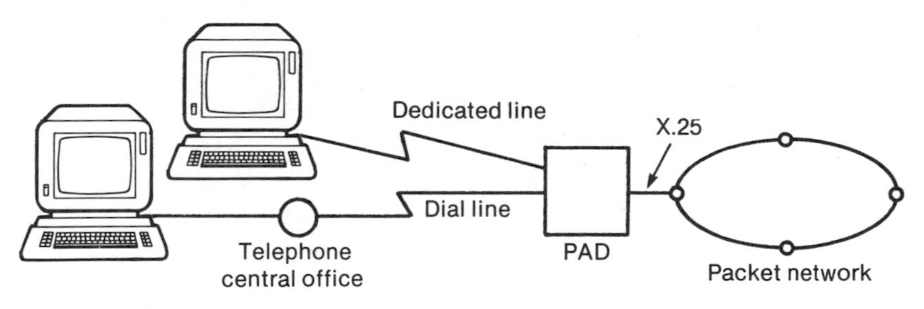

c. Network Supplied PAD on Network Vendor's Premises with Dedicated or Dial Access, Typically from Asynchronous Terminals

Furthermore, local telephone service costs may not be effective for data communications as service usage increasingly is measured. Also, intermachine communication requires code and speed compatibility. Unless the two devices are using identical protocols, they will be unable to communicate.

CCITT X.25 Protocol

The X.25 protocol is used as the interface between the PAD or user's host and the packet network. X.25 implements layers one through three of the ISO Open Systems Interconnection model that is described in Chapter 3. The level one or physical interface is contained in CCITT X.21 recommendations, or the alternative X.21 *bis,* which is virtually identical to RS-232-C. The data link layer-two protocols supported by X.25 are the High-Level Data Link Control (HDLC) in a version called Link Access Protocol (LAP). HDLC is a full-duplex protocol similar to IBM's Synchronous Data Link Control (SDLC). HDLC was developed from SDLC and has only minor differences, but the protocols are not identical.

When X.25 was first announced in 1974, it supported the LAP protocol which was a half-duplex protocol that was not widely accepted. In the second version of X.25, announced in 1980, a revised LAP-B protocol supporting full-duplex operation was recommended. This protocol has been adopted in some applications such as Local Area Data Transport that is described in a later section. In 1984, CCITT issued a third version of X.25 that deletes the datagram option and adds several new features.

The network or third level protocol is designed to establish logical channels and virtual circuits between the PAD and the network. The X.75 protocol recommends the interface for gateway circuits between packet networks.

Although X.25 is accepted by most data networks as a standard interface, the fact that a network supports X.25 does not assure universal compatibility. The CCITT recommendations include numerous options, of which about two thirds are discretionary. Therefore, public data networks accept X.25 connection from users only after their equipment has been certified compatible. Equipment vendors are responsible for obtaining this certification.

It is important to recognize that while X.25 presents an error-free network, it contains only the first three protocol layers. It cannot, therefore, ensure delivery of all packets. Higher level protocols above the network layer are needed to detect any packets lost by the network and to arrange for retransmission.

VALUE ADDED NETWORKS

Public data networks, also known as value added networks, provide an alternative to the telephone network for long-haul data communications. The term *value added* derives from processing functions that are added to the usual network functions of data transport and switching. The value-added services include such features as error checking and correction, code conversion, speed conversion, and storage.

Public Packet Switched Data Networks

Packet switching services have been provided in the United States over two nationwide networks since the 1970s: Tymshare Corporation's Tymnet and GTE Telenet. Other networks include AT&T's Accunet and Net 1000 service, United Telecom's Uninet, ADP's Autonet, Computer Science Corporation's Infonet, and several others. Public packet switching networks have developed more rapidly in Europe where it is more difficult to obtain private network facilities from the Postal Telephone and Telegraph agencies.

Public data networks in the United States are in many cases able to transport data overseas by interconnecting through a gateway to a data network in another country. The structure and services offered by some of these networks are discussed to demonstrate the structure of the networks and how users interface them to obtain data transport and value added services.

GTE Telenet

Telenet is a three-level hierarchical packet network. Nine Class 1 central offices are linked by 56 kb/s backbone circuits. Class 2 Telenet COs are linked by 9.6 kb/s circuits and serve Class 3 Telenet access controllers that provide PAD functions to end users. Telenet's access services are divided into Class A, B, and C cities with lower rates to the larger Class A locations. Class A cities offer dedicated connections up to 9.6 kb/s. Class A, B, and C locations offer dedicated and dial access at speeds up to 1.2 kb/s.

Telenet charges are based on access time and the number of packets transmitted. The packet length is 128 bytes, but partially filled packets are forwarded on receipt of a control character or after a user-specified time interval. The amount of data transmitted in an average packet depends on the character of the user's traffic, which must be carefully analyzed because the charge for a partially filled packet is the same as for a completely filled packet.

Access to Telenet is via a user-supplied certified X.25 host or by a Telenet processor that can be mounted on the user's premises or in a Telenet central office. Telenet processors support asynchronous and binary synchronous devices, and devices using HDLC protocol.

Telenet provides a high degree of reliability by using adaptive routing and dynamic alternate routing. Adaptive routing changes the flow of each packet based on the network load at the instant the packet was forwarded. Dynamic alternate routing allows the network to bypass a failed or congested node.

Tymnet

Access to Tymnet is through a user-supplied X.25 host or through a Tymnet engine, which uses Tymnet's own protocol to the packet net. Access is by dial-up or dedicated link to the engine on Tymnet or user's premises. Data is transported to the distant terminal on a bit oriented basis. Packets are shared among users so that a user need not fill a packet before it is forwarded. Consequently, billing is independent of packet size. Instead, it is based on connect time and the number of characters transmitted. Access charges are also based on the service area with the lowest charges to areas with the highest density.

Unlike Telenet, which uses dynamic routing, Tymnet uses a fixed routing plan. At the start of the session the route is dynamically chosen by the network, but it remains fixed for the duration of the session, unless the route becomes inoperative, in which case an alternate route is selected.

AT&T Information Systems Net 1000[1]

AT&T's Net 1000 is a value added network using AT&T Communications' Accunet® service as the transport medium. Accunet is available to users for data communication, but without VAN services. Although the analogy is not entirely accurate, Net 1000 is related to Accunet as some public processing and database services are related to their carriers, Tymnet and Telenet.

---

[1]As this book was going to press, AT&T announced it was discontinuing Net 1000.

Net 1000 provides processing services to its users. Users can store and execute programs in COBOL code and can access shared storage. The network also provides conversion between a limited number of protocols. Eventually, most popular protocols are expected to be supported.

Billing is based on the type and character of usage. Basic usage is based on volume. In addition, users are billed for computer resources used, storage used, delivery options, and a user-designated message priority.

LOCAL AREA DATA TRANSPORT

Data communications networks are designed for large business users. Many end-user services, such as videotex, requiring connection to public data networks or connection to other terminals and host computers now have little alternative but to use the local telephone network for data transport. Recognizing the need for a data communications service that is as ubiquitous and easy to use as telephone service, telephone companies are beginning to introduce a service called Local Area Data Transport (LADT). LADT uses the local loop for a specialized data communications service. As shown in Figure 18.5, LADT offers switched or direct access to a local packet switching system. In most cases, the packet switching network will cover a Local Access Transport Area (LATA). (See Chapter 9 for an explanation of LATAs.) A synchronous communications service using the data link layer of the X.25 protocol is provided between the end user and a Local Data Concentrator (LDC). The LDC is located in a local central office and is accessed on a dial-up basis from the resident switching machine or directly through data over voice (DOV) carrier superimposed on the user's local loop.

The user is equipped with network channel terminating equipment (NCTE) that interfaces the DTE and implements the X.25 protocol toward the LDC. The LDC and NCTE communicate using a LAP-B protocol. In the direct access method the NCTE and the LDC contain DOV carrier units that provide a full-duplex 4,800 b/s channel. The DOV carrier enables simultaneous voice and data operation over a

FIGURE 18.5
Local Area Data Transport

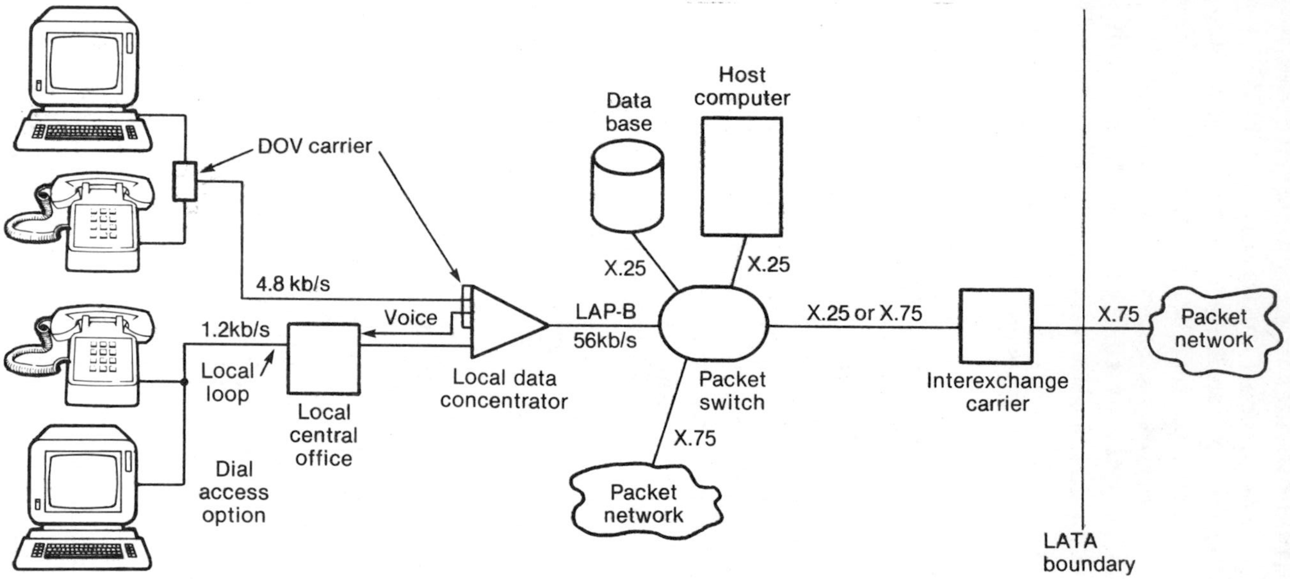

single local loop. However, the user's premises must be within 18,000 feet of the telephone central office because load coils used on longer cable pairs block the DOV signal.

The DTE interfaces the NCTE through an eight-pin connector, of which four leads are used: two for a balanced transmit pair and two for a balanced receive pair. The voice interface is a standard two-wire telephone circuit. The line code between the NCTE and the LDC is a bipolar format in which the polarity of every other "one" pulse is reversed.

Dial access service provides the same features as direct access except that DOV carrier is not used. Transmission is at 1,200 b/s seconds over the voice frequency local loop. Simultaneous voice and data operation are not possible with dial access. Although dial access still uses an analog central office as the access vehicle, the number of switches in a connection is reduced. Depending on how the service is configured by the telephone company, access may involve only one local switch.

The LDC, besides demultiplexing direct access signals from the user, provides a full error detecting and correcting protocol in conjunction with the NCTE. It also registers call billing details and traffic statistics. The LDC is interfaced to the packet switching network over 56 kb/s digital channels. The packet network provides direct access to LDCs, public data networks, and host computers.

LADT terminals are numbered according to the CCITT X.121 numbering plan. A four-digit Data Network Identification Code (DNIC) is assigned to LADT in the United States. The network terminal number (NTN) is a 10-digit code. The first three digits are the standard telephone area code plan. The terminal number is a seven-digit number that carries the same format as telephone numbers. Because the DNIC is separate from the telephone network, seven-digit NTNs can be assigned without interference with a local telephone number.

LADT is designed to provide a high degree of availability and reliability. The network is designed for 99.6 percent availability with an error rate objective of 10^{-8}. Users are able to select their own packet size from 128 to 256 bytes. The system is equipped with buffers to store incoming data pending delivery. When the buffer capacity is exceeded, the user's terminal is prevented from sending more data. The system is capable of providing hunting service for incoming calls. The user subscribes to the desired number of lines. Calls are placed to the main number and automatically hunt to an idle port.

Calls over LADT are placed in a manner similar to making telephone calls. Over a dial-up terminal a call is dialed to the telephone number of the LDC. From a direct access terminal the user goes off hook. The LDC returns a tone and with a message prompts the user for the called number. Upon receiving the tone the user dials the NTN of the called station. The LDC verifies the validity of the called number and places the call, meanwhile returning a message to the user to indicate the call is going through.

SWITCHED 56 KB/S SERVICE

In February 1985, AT&T filed with the FCC for authorization to offer a service called Accunet® Switched 56. This service would allow a user to set up switched 56 kb/s connections between locations that are equipped for the service. At the time of this writing, the FCC had not authorized the service, but if clearance is granted, AT&T plans to offer it in 78 cities by the end of 1987. The service is intended for applications that need a digital connection for a time too short to justify the use of a dedicated channel, generally three or fewer hours per day. Examples are video conferences, high-speed facsimile, graphics, and part-time extensions to existing digital networks. As with other switched services, Accunet Switched 56 is charged on the basis of usage time and carries a rate slightly higher than an ordinary voice connection.

Before the breakup of the Bell System, AT&T had proposed a similar service called Circuit Switched Digital Capability (CSDC). Under CSDC, a connection was to be established through the Class 5 switching machine. The connection would be established on a voice frequency basis and transferred to a digital service. The service required modifications to the Class 5 switching machines to accommodate it.

After divestiture, the service was feasible only as a joint project of the local operating telephone company and AT&T. A few local Bell companies offer a similar service called Public Switched Data Service (PSDS) within their serving areas, but no local companies had signed an agreement with AT&T to offer CSDC. Therefore, AT&T elected to make the service an interLATA only offering. Instead of access through the local central office, access is directly to one of AT&T's No. 4 ESS tandem switching machines via a dedicated 56 kb/s DDS line or over a channel of a T-1 line. The service will be initially available only where AT&T has No. 4 ESS machines and calls can be placed only to locations that have the necessary dedicated digital end link. Calls are placed by dialing 700-56X-XXXX. (The "X" is any digit.)

The service is priced reasonably, but price of the dedicated access channel is the main limiting feature in this service. Until arrangements are made with the local operating companies to switch the end link, the use of Accunet Switched 56 service will be limited.

NETWORK CONTROL CENTERS

Large data networks are usually equipped with a network control center (NCC) to monitor and control access, maintenance, and billing. The format of the NCC is related to the network architecture and equipment used. Only the general functions of the NCC are discussed here.

The NCC is linked by data circuits to the major network nodes. It monitors traffic loads and nodal performance. In packet networks with centralized dynamic routing, the NCC alters routing based on traffic volumes and congestion. The NCC collects performance statistics from the nodes for both short-term and long-term load evaluation. The NCC also serves as a central network testing location. Alarm and status messages are received from the nodes and are used for trouble isolation and referral.

STANDARDS

Data network standards are established internationally by CCITT with supporting U.S. standards set by ANSI, EIA, and by major equipment vendors whose products take on the character of *defacto* standards. For example, in data communications, IBM's Systems Network Architecture (SNA) is so widely used as to constitute a *defacto* standard. The list of protocol standards is so extensive that no attempt is made to reproduce it here. Instead, the primary CCITT and EIA standards that affect data communications networks are included. Complete lists of proprietary protocols can be obtained from the organizations listed in Appendix B. Refer to standards section of Chapter 3 for physical interface standards.

Bell Communications Research

TR 880-22135-84-01 Item 360 Circuit Switched Digital Capability Network Access Interface Specification.

CCITT

Series V Recommendations: Data Communications over the Telephone Network

V.5 Standardization of data signaling rates for synchronous data transmission in the general switched telephone network.

V.6 Standardization of data signaling rates for synchronous data transmission on leased telephone-type circuits.

V.20 Parallel data transmission modems standardized for universal use in the general switched telephone network.

V.50 Standard limits for transmission quality of data transmission.

V.52 Characteristics of distortion and error-rate measuring apparatus for data transmission.

Series X Recommendations: Data Communication Networks Services and Facilities, Terminal Equipment, and Interfaces

X.2 International user services and facilities in public data networks.

X.3 Packet assembly/disassembly facility (PAD) in a public data network.

X.20 Interface between data terminal equipment (DTE) and data circuit-terminating equipment (DCE) for start-stop transmission services on public data networks.

X.21 Interface between data terminal equipment (DTE) and data circuit-terminating equipment (DCE) for synchronous operation on public data networks.

X.25 Interface between data terminal equipment (DTE) and data circuit-terminating equipment (DCE) for terminals operating in the packet mode on public data networks.

X.28 DTE/DCE interface for start-stop mode data terminal equipment accessing the packet assembly/ disassembly facility (PAD) in a public data network situated in the same country.

X.29 Procedures for the exchange of control information and user data between a packet assembly/disassembly facility (PAD) and a packet mode DTE or another PAD.

X.75 Terminal and transit call control procedures and data transfer system on international circuits between packet switched data networks.

X.121 International numbering plan for public data networks.

X.150 DTE and DCE test loops for public data networks.

EIA

RS-334-A Signal quality at interface between data terminal equipment and synchronous data circuit-terminating equipment for serial data transmission.

RS-366-A Interface between data terminal equipment and automatic calling equipment for data communication.

RS-404 Standard for start-stop signal quality between data terminal equipment and nonsynchronous data communication equipment.

RS-484 Electrical and mechanical interface characteristics and line control protocol using communication control characters for serial data link between a direct numerical control system and numerical control equipment employing asynchronous full-duplex transmission.

RS-491 Interface between a numerical control unit and peripheral equipment employing asynchronous binary data interchange over circuits having RS-423-A electrical characteristics.

APPLICATIONS

In Chapter 12, the Rockwell data communications network is described in detail. Also, in Chapter 17, data communications over the Hercules Corporation's network is described. These two networks are more complex than the average but offer an interesting insight into the alternatives for nationwide data communications. Also, see Chapter 3 for more detailed information on IBM's Systems Network Architecture (SNA).

Evaluation Considerations

Data networks are evaluated by criteria that differ significantly in many cases from circuit switched networks. For example, the short length of many data messages makes setup time, which is of little concern in voice networks, an important factor. Also, error considerations are important in data networks. Circuit noise that is merely annoying in a voice network may render the channel unusable for data. Also, because many data networks' billing is not distance sensitive and because billing is based on volume rather than connect time, cost evaluations are significantly different between the two types of network.

X.25 Compatibility

Connection to a public data network requires certification of protocol compatibility by the network. Users planning to interface a PAD or host computer to a public data network must obtain assurance from the equipment vendor that their implementation of the X.25 protocol has been certified compatible by the public network.

Costs

Cost comparisons between public data networks are difficult to make because of differences in the way data is handled. Cost is a function of how the network handles messages and bills for usage. Considering networks such as Tymnet and Telenet that bill on a character and packet basis respectively, the costs should be evaluated to determine whether the costs of partially filled packets tend to outweigh the slightly higher per-character transmission costs. The geographic area to be covered has a significant effect on the cost of public data network services. Costs are higher in low-density areas. With dial access the cost of measured local telephone service or long-distance charges to the node when none is available in the locality must be considered.

Network Setup Time

Setup time is critical to most data applications. When using dial access over a public data network, setup time is significant, often considerably longer than the message transmission time. Whether data communications are conducted over a public data network or the telephone network, however, call setup time is apt to be equivalent. With dedicated access to a public data network, the dialing, answer, and authentication routines are eliminated, greatly reducing call setup time. The time of establishing the call through the network to the terminating station remains a significant factor and is comparable whether dial or dedicated access is used. This time can be reduced by using a permanent virtual circuit, if offered by the network, which establishes a preassigned and dedicated path between terminals.

Response Time

Response time in a data communications network is evaluated by analyzing the time required for each step of the

FIGURE 18.6
Generalized Data Network Response Time Model

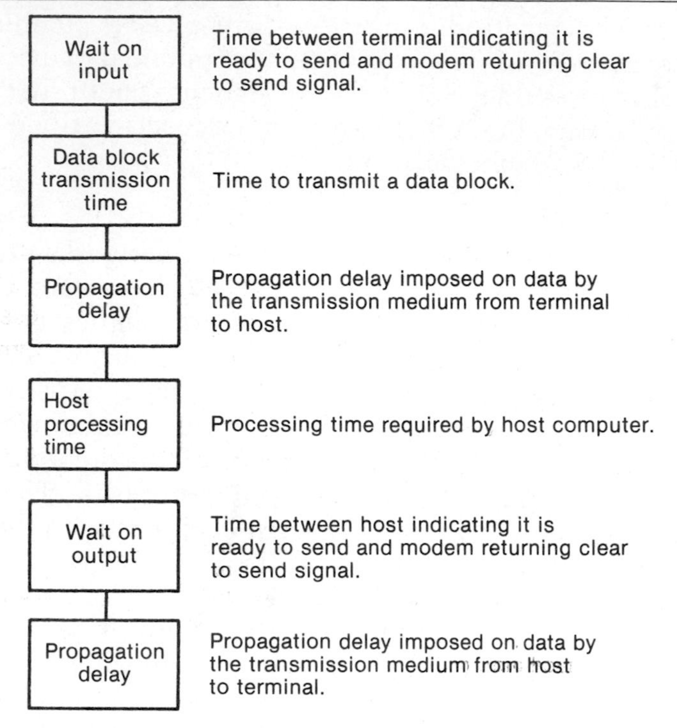

process shown in Figure 18.6. Response time definitions vary, but the most generally accepted definition, and the one used here, is the interval between the terminal operator's sending the last character of a message and the time the first character of the response from the host arrives at the terminal. This time is independent of message length and consists of the steps outlined in Figure 18.6.

Note that modem reversal time applies only to half-duplex circuits. In public data networks the model shown can be simplified to eliminate the waiting times on input and output. The transit time of a public data network can be substituted for the medium propagation delay. Response time should be evaluated on the basis of the characteristics of the user's messages. Response time is affected by packet size and the process the network uses in handling packets and assembling messages. The considerations in evaluating response time are:

- Type of protocol. A full-duplex protocol eliminates modem turnaround time. Block protocols are more sensitive to absolute delay. Error recovery methods affect response time.
- Modem reversal time varies with modem design in circuits using half-duplex protocol. The shorter the reversal time, the more rapid the response time.
- CPU response time is independent of the telecommunications network and has no effect on the choice of data communications facilities.
- Absolute delay varies with the type of transmission medium and the choice of network alternative. Satellite services interpose a minimum delay that cannot be reduced except by operating one direction of transmission over terrestrial facilities. Packet networks introduce delays that may vary with load.
- Error rate affects the number of rejected messages that must be retransmitted, and thus increases response time.

Type of Interface

Value-added networks provide three general types of interface as shown in Figure 18.4. Small users have little option except for dial access. Large users requiring end-to-end error checking will select an interface using a PAD or host computer on the user's premises. The primary considerations are:

- Certification of compatibility with the network.
- Lease versus ownership cost comparisons for interface equipment.
- Transmission speed supported by the interface.
- Type of terminals and protocols supported by the PAD.
- Number of devices supported.

Features

Data communications features required by the user should be examined and compared to the services available from the different data communications network alternatives. Features to be considered include:

- Virtual circuit or datagram service. Users with very short messages may require a datagram or fast select service.
- Type of message delivery. Consider whether message delivery is to be automatic or delayed awaiting a busy or unattended terminal.
- Message storage or electronic mail services may be offered by the network or a value-added service on the network.
- Multidestination message service. Consider whether it is important to broadcast messages to numerous stations simultaneously.
- Billing service. Consider whether detailed call accounting is required or whether message charges are bulked to a user number.
- Security. Networks should offer password security and should also offer encryption and a private storage facility that can be unlocked only with an additional private code.
- Closed user groups. These are private networks within the network that are designated for the exclusive use of users who gain access only with proper authentication. They are used either for terminal-to-terminal communication or for privately accessed store-and-forward service.
- Protocol conversion. Communication between terminals using unlike protocols may be supported by the network.
- Disconnect of stations that are connected for a given interval without sending traffic should be provided.
- Abbreviated or mnemonic addressing. These may be

provided with the system generating the data network number from a simplified address.

Line Conditioning

Analog private line data transmission facilities that are used at high speeds may require conditioning, a concept that is explained in Chapter 3. Type C conditioning improves the attenuation distortion and envelope delay distortion in the facility. Type D conditioning also improves noise performance and harmonic distortion of the circuit. Some types of modems equipped with adaptive equalization may not require conditioning, but other modems will require conditioned lines at 9.6 kb/s and may require it at 4.8 kb/s.

Availability

Data networks are evaluated by their percent availability, which measures the percentage of time the network is available for service and the number of error-free seconds provided by the network. Availability should run as close to 100 percent as possible, typically at least 99.9 percent. It is affected by the failure rate and redundancy in network paths and switching equipment. The error-free second criterion applies only to facilities. If the network provides an end-to-end error checking process, transmission can be presumed to be error free.

GLOSSARY

Conditioning: Special treatment given to a transmission facility to make it acceptable for high-speed data communication.

Data Network Identification Code (DNIC): A 14-digit number used for worldwide numbering of data networks.

Datagram: A single packet that is routed through a network without acknowledgement.

Fast select: A network service that transmits one packet and a return acknowledgement.

Hub polling: A polling system in which a polled station sends its traffic and passes the polling message to the next station.

Local Area Data Transport (LADT): A telephone network-based service to provide a synchronous data channel with error correction to the end user over telephone loops.

Modem turnaround time: The time required for a half-duplex modem to reverse the direction of transmission.

Network terminal number (NTN): The number assigned to a data terminal under the Data Network Identification Code system.

Packet assembler/disassembler (PAD): A device that accepts messages from terminals using the terminals' protocols, slices the message into packets, and delivers them to a network using the network's protocol. Incoming packets are reassembled into messages and delivered to terminals.

Permanent virtual circuit: A virtual circuit that provides the equivalent of dedicated private line service over a packet switching network. The path between users is fixed.

Polling: A network sharing method in which remote terminals send traffic upon receipt of a polling message from the host.

Response time: The interval between the terminal operator's sending the last character of a message and the time the first character of the response from the host arrives at the terminal.

Switched virtual circuit: A virtual circuit composed of a path that is established with each session.

Value-added network: A data communication network that adds processing services such as error correction and storage to the basic function of transporting data.

Videotex: An interactive information retrieval service that usually employs the telephone network as the transmission medium.

Virtual circuit: A circuit between users that is derived over a logical path shared with other users.

BIBLIOGRAPHY

Doll, Dixon R. *Data Communications.* New York: John Wiley & Sons, 1978.

Deasington, R. J. *A Practical Guide to Computer Communications and Networking.* New York: John Wiley & Sons, 1985.

Glascal, Ralph. *Techniques in Data Communications.* Dedham, Mass.: Artech House, 1983.

Held, Gilbert, and Ray Sarch. *Data Communications: A Comprehensive Approach.* New York: McGraw-Hill, 1983.

Held, Gilbert. *Data Compression: Techniques and Applications.* New York: John Wiley & Sons, 1983.

Martin, James. *Computer Networks and Distributed Processing.* Englewood Cliffs, N.J.: Prentice-Hall, 1981.

Rosner, Roy D. *Packet Switching: Tomorrow's Communications Today.* Belmont, Calif.: Lifetime Learning, 1981.

Sherman, Kenneth. *Data Communications: A Users Guide.* Reston, Va.: Reston Publishing, 1981.

Sharma, Roshan La.; Paulo T. deSousa; and Ashok D. Inglé. *Network Systems.* New York: Van Nostrand Reinhold, 1982.

Stuck, B. W., and E. Arthurs. *A Computer and Communications Network Performance Analysis Primer.* Englewood Cliffs, N.J.: Prentice-Hall, 1985.

Tanenbaum, Andrew S. *Computer Networks.* Englewood Cliffs, N.J.: Prentice-Hall, 1981.

VENDORS OF DATA NETWORK SERVICES

Packet Switched Network and Value-Added Services

ADP Autonet

AT&T Communications

AT&T Information Systems

CompuServe, Inc.

Computer Sciences Corporation

GTE Telenet Communications Corporation

Grafnet, Inc.

MCI Telecommunications Corporation

RCA Cylix Communications Network

Western Union Telegraph Co.

Other Data Communications Network Services

AT&T Communications

MCI Telecommunications Corporation

ITT Worldcom

RCA Americom

RCA Global Communications

Uninet, Inc.

CHAPTER NINETEEN

Local Area Networks

Studies of organizational communications show that on the average about 80 percent of all documents travel a short distance—typically within an office, a building, or a campus. Even with the declining costs of global communications, in the automated office an increasing portion of messages will be transmitted between machines over these narrow ranges because it is less costly and more convenient than verbal or handwritten transmittal.

Communication has become part of most computing systems; in some functions such as electronic mail, it is the essence of the application. Declining computer costs have made it economical to distribute many operations that were previously centralized, but related costs have remained high. The price of printers, high-speed disk drives, and, most important, program development costs have not followed the rapid decline of processor costs. Consequently, the need arises for a low-cost, high-speed network to allow many small computers and terminals to share resources and to communicate with one another.

Limited-range messages require a different form of network than traditional networks such as the telephone system. Local networks must meet the demand for high-speed, short-range communication at a cost low enough to justify appearances at nearly every desk. Although it is possible to use an individual telephone line for data terminals, the common carrier network is prohibitively expensive for short-range communications, particularly as metered service becomes common.

The most economical solution for many offices and industrial applications is the *local area network* (LAN). A LAN is a network dedicated to a single organization, limited in range, and connected by a common communication technology. This definition has important implications. Because the network is private, it can be specialized for a function. Security may be less critical than in multiuser networks; and because range is limited, a LAN can operate at high speeds.

LANs are accessed by multiple terminals through circuitry that goes by several different names, but which we shall call a *network access unit* (NAU). The form of the NAU is determined by the designer. Although various manufacturers call the NAU by other names, the following functions are common to all LANS:

- Provides a physical interface to the transmission medium.
- Monitors the busy/idle status of the network.
- Buffers the speed of the terminal, typically 300 b/s to 9,600 b/s, to the speed of the network, typically 1 mb/s to 10 mb/s.
- Converts terminal protocol to the network protocol.
- Monitors for and corrects errors.
- Assembles the transmitted data stream into packets for transmission on the network and restores the data stream at the receiving end.
- Recovers from *collisions* resulting from simultaneous transmissions.

LANs share many characteristics with long-haul and metropolitan networks. They differ, however, in many subtle ways that must be considered in order to understand their application. The principal issues facing the user of LANs are discussed below:

- Network topology.
- Access method.
- Modulation methods.
- Transmission media.
- Throughput.
- Size of network.
- Interconnecting LANs with other networks.
- Local area network standards.

NETWORK TOPOLOGY

Network topology or architecture is the pattern of interconnection of the terminals. LANs use the same topologies as global and metropolitan networks: the star, bus, ring, and branching tree, but rarely the mesh. The most common topology is the bus, in which terminals are connected to a single circuit, as shown in Figure 19.1a. Messages or packets are broadcast simultaneously to all terminals on a bus. Access is either allocated by a control node or the nodes contend for access.

Although star networks, shown in Figure 19.1b, are widely used with switched local networks such as PBXs and data switches, the star is not used in LANs except where the physical wiring of the transmission medium bridges all conductors together at a common point creating the electrical equivalent of a bus. Switched local networks were covered in more detail in Chapter 11.

Branching tree networks, illustrated in Figure 19.1c, are often used in LANs that employ cable television technology. The branching tree is electrically identical to the bus except that its branches must be connected only through properly designed impedance-matching devices. Otherwise, data signal reflections will cause the network to malfunction.

The star, branching tree, and bus topologies function identically in a LAN—a station with a message to send gains

FIGURE 19.1
LAN Topologies

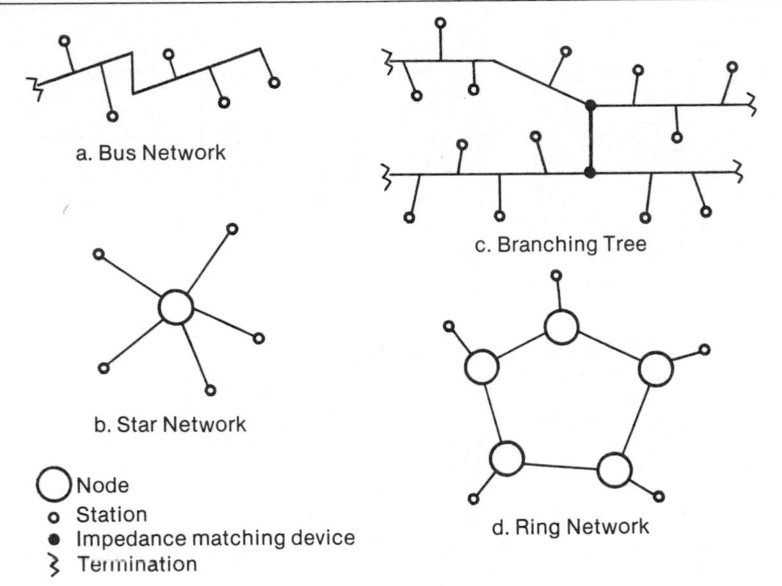

access to or *acquires* the network and broadcasts a signal that is received by all stations. The addressee retains the message; all other stations discard it.

In a ring topology, illustrated in Figure 19.1d, all stations are connected in series, and the signal is transmitted one bit at a time. Each station receives the signal, regenerates it, and transmits it to the next station in the ring. Bits always flow in only one direction. The addressee retains the message, but it continues to circulate until it returns to the sending station, which is the only station entitled to remove it from the ring.

ACCESS METHOD

A key distinguishing feature of LANs is the method of providing the stations access to the network. In a switched network, stations gain access by signaling over the network and transmitting an address to the controller. The controller determines the destination and sets up a path or circuit be-

FIGURE 19.2
Collisions in a Contention Network

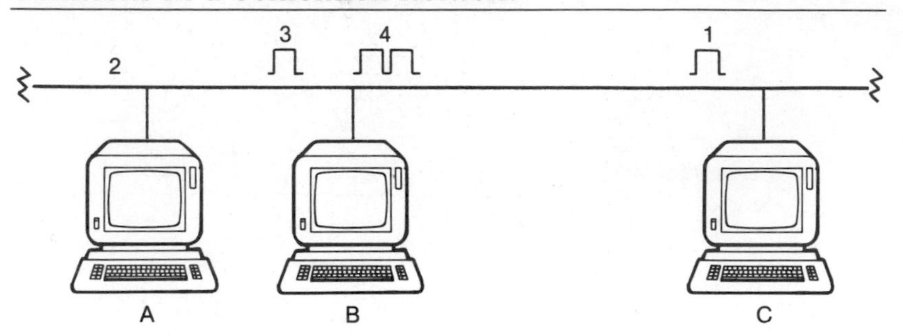

1. Station C transmits acquisition pulse.
2. Station A listens to network but pulse from C has not arrived.
3. Station A transmits pulse, which collides with pulse from C.
4. Station B detects collision and transmits jamming signal.
5. Both A and C stop transmitting and wait random time before retransmitting.

tween the sender and receiver. Because the network has multiple paths, stations cannot interfere with one another as long as the capacity of the network is not exceeded.

By contrast, a LAN has only one path to handle high-speed data. The total capacity of the path, however, far exceeds the transmission speed of any station. Stations are given exclusive access to the entire network for bursts of time long enough to send a packet of data or a message. LAN access methods are classified as contention or noncontention.

Contention Access

A contention network can be visualized as a large party line with all stations vying for access. In contention networks control is distributed among all stations. When a station has a message to send, it listens to the network and if the network is idle, the station sends a packet of data.

It is not always possible, however, for a station to determine when the network is idle. As shown in Figure 19.2, two stations may begin to transmit simultaneously. Because of the delay between the time a data pulse is transmitted and it is received at the distant end, neither station is aware that the other is sending. When simultaneous transmissions occur, the two signals *collide.*

During the time it takes a pulse to travel from the sending terminal to the furthest terminals on the network, known as the *collision window,* stations are blinded to potential collisions. In 1 kilometer of coaxial cable, the collision window is approximately 5 microseconds wide. Because of potential collisions, contention networks are restricted in diameter. The wider the network, the longer the collision window. The practical limitation in contention LANs operating at a data rate of 10 mb/s is in the order of 1 mile or 1.6 km.

If an entire packet was transmitted before a collision was detected, the collision would mutilate both signals, and valuable network time would be wasted in retransmission. The most common system for managing access and collisions in a contention network is known as *Carrier Sense Multiple Access with Collision Detection* (CSMA/CD). With CSMA/CD, a station acquires the network by transmitting a single pulse. It then monitors to determine if the pulse was mutilated by the acquisition pulse of another station. If not, it has acquired the network and is free to transmit a packet. If the pulse is mutilated, however, any station detecting a collision transmits a jamming signal, and both stations cease to transmit.

The procedures that stations follow when colliding is called their *backoff algorithm.* If stations attempted to acquire the network immediately following collision, repeated collisions between the same two stations would occur. To prevent this, stations must wait a random time before the next attempt.

A variation of CSMA/CD is CSMA/CA (collision avoidance). In the most common variation of this method, each station is assigned a time slot in which to transmit data. If the station has no traffic to send, the time slot remains unused. Therefore, this system is capable of less throughput than CSMA/CD.

Noncontention Access

CSMA/CD is known as a "statistical" access method, relying on the probability that its stations will get enough share of the network to send their traffic. Although unlikely, it is possible for a station to be excluded from network access during periods of heavy load.

A noncontention system called *token passing* introduces a form of control that overcomes the drawbacks of the free-for-all system used by CSMA/CD. A token is a unique combination of bits that circulate through the network following a predetermined route. When a station has data to send, it captures the token, transmits its message, and replaces the token on the network. Token passing is known as a "deterministic" system. If a station has traffic equal to or higher in priority than other traffic on the network, the control mechanism will allocate it a portion of the network's capacity.

The advantages of control are purchased at the price of greater complexity. One of the stations in a token network must be equipped to initiate recovery action if the token is lost or mutilated, which can occur if a station fails or loses power at the time it possesses the token. Other functions required of the control station include the removal of persistently circulating packets, removal of duplicate tokens, control of priority, and addition and removal of stations. Further complicating the process is the need for a recovery routine if the control station fails. All stations are equipped with the logic to assume control if necessary. Because of this greater complexity, token passing systems are generally more expensive than contention systems. Moreover, additions and deletions are more complex because all stations must be programmed with the route of the token.

Ring and bus topologies predominate in token networks. In a token ring, which is illustrated in Figure 19.3, each station receives each message and repeats it to the next station in turn. Sequencing is automatic; it always follows the same route in the same direction around the ring.

In a token bus network, messages are broadcast to all stations simultaneously. However, control follows a logical ring sequence as illustrated in Figure 19.4. When a station acquires the token, it is permitted to broadcast a message, but the token can be passed to any other station without regard to its physical position on the network.

FIGURE 19.3
Token Ring Network

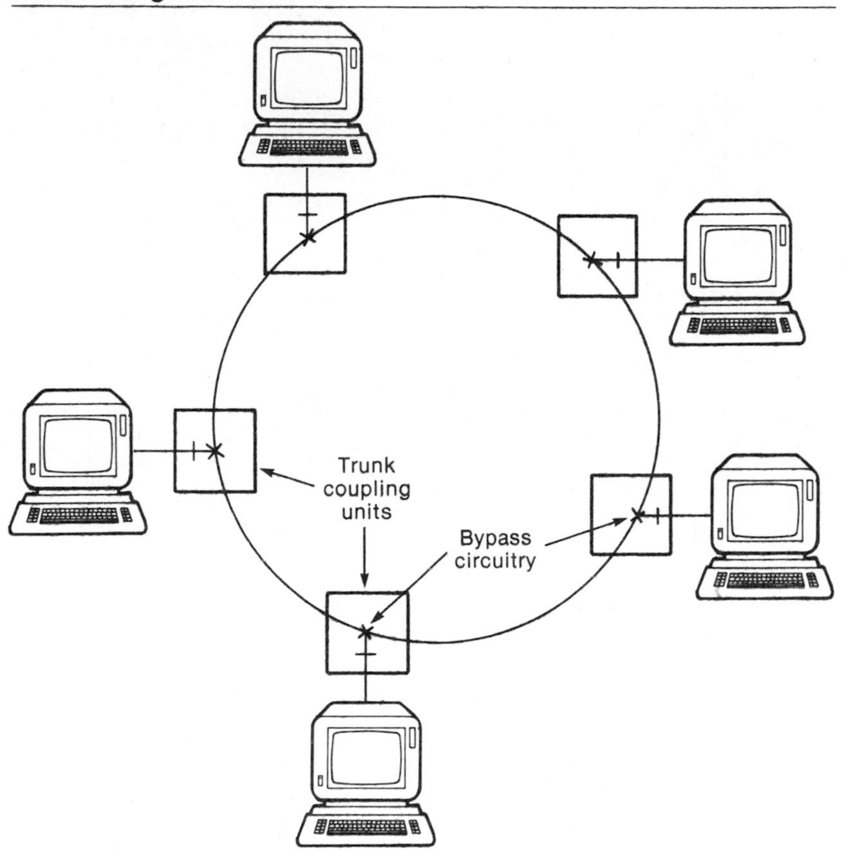

MODULATION METHODS

LANs use one of two methods of pulsing a data signal on the transmission medium. In the first, known as *baseband,* the signal is transmitted directly in the form of high-speed, square wave pulses of DC voltage. *Broadband* systems use cable television (CATV) technology to divide the transmission medium into frequency bands or channels. Each channel can be multiplexed to carry data, voice, or video.

Both baseband and broadband networks accept identical data streams from the terminal, but they differ in the network access unit. In a baseband system the NAU contains

FIGURE 19.4
Token Bus Network

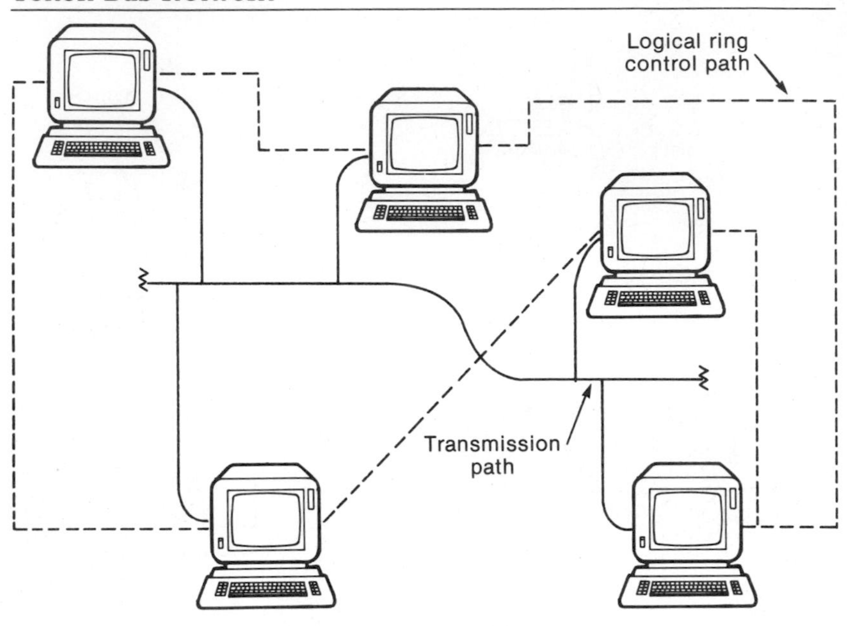

a *transceiver,* which is a simple cable driver that matches the impedance of the cable and transmits pulses at the data transfer rate. In a broadband network the NAU contains a *radio frequency modem* to modulate the data to an assigned channel.

Baseband

The primary advantage of baseband is its simplicity. No tuned circuits or radio frequency apparatus are required. A baseband system has no active components aside from the NAUs. Only the cable is common to the network, making baseband less vulnerable to failure than broadband, which contains amplifiers and other active components.

A baseband network is composed of terminals, NAU, and the transmission medium. The terminals send a data stream to the NAU, where it is formed into packets that are pulsed directly on the transmission medium. The active components of the NAU are frequently mounted directly on the cable to increase bandwidth and range. The transmission me-

FIGURE 19.5
CATV Frequency Allocations

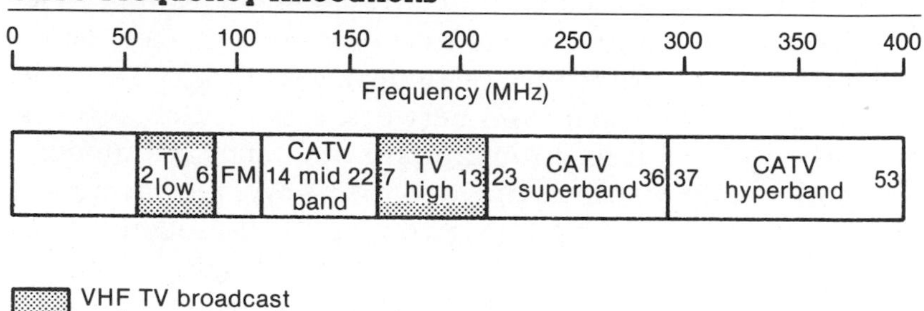

dium, discussed in a later section, can be ribbon or paired copper wire, coaxial cable, or fiber optic cable. In the latter, data pulses drive a light transmitter, which turns a laser or light emitting diode on and off corresponding to the zeros and ones of the data signal.

Broadband

Broadband networks use a coaxial cable and amplifier system capable of passing frequencies from about 5 mHz to 400 mHz as shown in Figure 19.5. Television channels each occupy 6 mHz, providing capacity for more than 60 one-way channels, each of which can be used for a LAN, video, or voice. The primary advantage of broadband LANs is their greater capacity compared to baseband. The equivalent of several baseband networks is derived by using multiple subcarriers for either increased LAN channels, video, or in some cases, voice.

Unlike baseband where signals are broadcast in both directions simultaneously, broadband is inherently a one-way system because of its amplifiers. Bi-directional amplifiers are available, but the transmitting and receiving signals must be separated to obtain bi-directional transmission. The reverse direction is handled either by sending on one cable and receiving on another or splitting the sending and receiving signals into two different frequencies. The first method is called a *dual cable* system, and the second a *single cable* system. *Head-end equipment* is used to couple the transmit cable to the receive cable in a dual cable system or to shift the transmit frequency to the receive frequency in a single cable system. Head-end equipment in a single cable system and amplifiers in all broadband systems are active elements; a failure can interrupt the entire network.

Terminals in a broadband network are interfaced to the transmission medium through radio frequency (rf) modems that contain a transceiver tuned to the network transmit and receive frequencies. There are two types of rf modems: *fixed frequency modems,* which are tuned to a single frequency; and *frequency agile modems,* which can be shifted. Frequency shifts are directed by a central controller that connects stations by selecting an idle channel, directing the two modems to the channel and dropping out of the connection.

Baseband and broadband LANs are similar in most respects except for the frequency separation in a broadband network and differing methods of collision detection. With some products it is possible to start at baseband and later convert to broadband by replacing the transceiver in the NAU with an rf modem. There are some differences in operation between the two systems. Collisions are detected on a direct current basis in a baseband network, but because broadband networks are incapable of passing DC, they employ a different method of collision detection. A common technique is for the station to listen for transmissions mutilated by collision. This reduces network throughput compared with baseband. Another system of collision detection is the bit comparison method in which a series of bits is transmitted to acquire the network. If a collision occurs, the first section detecting the collision transmits a jamming signal on a separate channel. Throughput is also reduced by the added length of the doubled cable. These two factors plus higher cost are the primary drawbacks of broadband.

TRANSMISSION MEDIA

All of the transmission media used in other networks—twisted pair wire, coaxial cable, fiber optics, radio, and light—are employed in LANs. Radio and light are usable for special applications; the other three are usable for general LAN application. The choice of transmission medium will usually be dictated by the method used by the vendor. However, because of differing characteristics, the vendor choice may be driven by the transmission medium required. The principal factors to consider are:

- Presence of electromagnetic interference (EMI).
- Network throughput.
- Bandwidth required.
- Network diameter.
- Multiple terminal access.
- Cost.
- Security.

Twisted Pair Wire or Ribbon Cable

The primary advantage of wire is cost. Wire is readily available from many vendors and can be installed with simple and inexpensive hand tools without a great deal of skill required. In addition, wire has enough bandwidth to handle data speeds of 1 mb/s for distances of about 1 mile; it is available in multiple pair cable or in flat ribbons that can be installed under carpet; and it is durable and capable of withstanding considerable abuse without damage (with its sheath intact, wire is impervious to weather).

Despite its advantages, wire has limitations that prevent its use in some applications. Its bandwidth is too narrow for the speeds required by many systems. Moreover, unless cable is shielded, EMI from such sources as elevators and industrial equipment can cause errors. Where security is important, wire is a poor choice, for it is easily tapped. For further information on the characteristics of wire, refer to Chapter 7.

Coaxial Cable

Coaxial cable or "coax" is the most widely used transmission medium for LANs. It is inexpensive, has wide bandwidth, and can be installed by moderately skilled workers. Coax can support both high speed data and video, and because it is widely used for cable television (CATV), coax is readily available at moderate cost.

Coaxial cable consists of one or more center conductors surrounded by a shield of flexible braid or semirigid copper or aluminum tube, with an outer jacket of PVC or Teflon®. When properly installed, the conductor is shielded from EMI and is reasonably impervious to weather. Special precautions are required to avoid unwanted effects in installing coaxial cables. In baseband networks, cables must be grounded only in one place; precautions are required to insulate connectors from ground. Branching points in broadband coax must be equipped with splitters and directional couplers to avoid impedance irregularities.

Coax can be tapped with little difficulty, which is advantageous when adding stations to a LAN without interrupting service. This, however, means that communications on the network are not entirely secure from unauthorized access, although compared to wire, coax is less susceptible to unauthorized taps. Refer to Chapter 20 for further discussion on CATV components.

Fiber Optics

Fiber optic cable with its wide bandwidth is capable of supporting data speeds far in excess of that needed by most LANs. It can be used in either baseband or broadband systems when equipped with lightwave transceivers. Perhaps more important in LANs, lightwave is immune to EMI, which is advantageous in many industrial applications.

Offsetting its advantages, lightwave has disadvantages that limit its applications in LANs. First, a lightwave system employs a combination of electronic and light technology that is more costly than coaxial cable drivers. Optical cable requires special tools and techniques for installation. The fibers must be carefully aligned into fixtures on the light transceivers, and they can be spliced only with special apparatus. Furthermore, glass fibers are difficult to tap. Although this is advantageous for security, it renders fiber optics almost unusable in a bus topology except when fibers are linked at a central point through a star adapter, creating the equivalent of a bus. Fiber optics can also be used in a ring configuration, with each node regenerating the light signal.

Although fiber optic cable and its associated light drivers are more expensive than coax, in terms of capacity, lightwave may be less expensive. Furthermore, the technology is advancing rapidly, resulting in both technical and cost improvements. Refer to Chapter 16 for additional information on lightwave.

Microwave Radio

Microwave radio is inherently a point-to-point medium. Microwaves travel in a straight line, so intermediate stations can be linked only if they are on the path of the radio beam. Therefore, radio has limited application in a LAN. However, radio is useful where right-of-way is a problem such as crossing obstructions and spanning moderate distances. It is also useful in connecting LANs or for linking a LAN with distant terminals.

Among its limitations, radio is not easily secured. It is impossible to prevent unauthorized detection of data signals over a microwave path so when security is important, encryption is required. Frequency allocations are coordinated by the Federal Communications Commission and may be difficult to obtain. Furthermore, microwave is expensive to purchase, requires trained technicians to install and maintain, and is susceptible to interference from outside sources. Refer to Chapter 15 for additional information on microwave.

Light

Optical transceivers are available for the same kinds of applications as described above for radio. These systems use light transceivers operating over short line-of-sight distances such as crossing a street between two buildings. Distances are limited, and transmission is not completely reliable because light beams can be interrupted by influences such as fog and dust. Its application is limited to short distances where other alternatives are prohibitively expensive.

THROUGHPUT

Throughput, or rate of transmission of information between stations, is an important issue with LANs as with any other network. LANs have high data transmission speeds, generally ranging between 1 and 20 mb/s. However, transmission speed should not be confused with throughput, which may be a fraction of the transmission speed. Throughput is limited by several factors in a LAN:

- Data transmission speed of the network.
- Overhead bits used by the network because of the coding system and the protocol.
- Time spent in collision and error recovery.
- Bandwidth of the transmission medium.
- Diameter of the network.
- Load imposed on the network.

Throughput is predictable in token ring networks, but it is difficult to predict in contention LANs. In contention networks, throughput can be determined experimentally by loading the network to see how response time is affected or by a computer simulation.

SIZE OF THE NETWORK

Many small offices have the same requirements of sharing high-cost peripherals and files, and communicating between terminals, yet are too small to justify the cost of a LAN. To meet that demand, a different class of network known as a *personal computer network* (PCN) has been developed. PCNs and LANs share many common characteristics and share similar technologies, but PCNs differ from LANs in the following ways:

- PCNs operate at lower speeds and therefore have lower throughput than LANs and can support fewer stations.
- PCNs have a narrower range or diameter than LANs.
- With few exceptions, PCNs operate only in the baseband mode.
- There are no existing standards for PCNs; each design is proprietary.

- PCNs use a less expensive transmission medium than most LANs and are therefore less expensive and easier to install.
- In many PCNs the NAU plugs directly into the bus or expansion slot in the PC, so the PCN is compatible only with PCs for which it is designed.
- Many PCNs lack gateways to public networks. To use the PC in off-net applications may require an external modem attached to the PC rather than to the network.
- Many PCNs lack security provisions so that confidential files must be kept off the network.

Most personal computer operating systems such as Microsoft's MS-DOS® are not designed for multiuser application, so most networks add a control layer to the operating system to allow users to share common resources without interfering with one another. These modifications may generate incompatibilities with some application software. Multiuser operating systems for PCs such as AT&T's Unix are beginning to enter the market, but they may be incompatible with many existing PCNs.

Although standards for PCNs do not exist, this problem will undoubtedly be solved in the future. At the time of this writing, several manufacturers of PCNs are petitioning IEEE to begin development of a low-speed version of the 802 standards described later in this chapter.

INTERCONNECTING LANS WITH OTHER NETWORKS

With the exception of closed loop manufacturing processes, many LANs require interconnection to other LANs, to public telephone or data networks, or to private networks. There are two methods of interconnecting networks. A *gateway* communicates between two networks, using each network's protocol. A gateway is needed for each pair of networks. The second method is by use of a *bridge.* A bridge interconnects networks through higher level protocols. Contrasted to a gateway, which communicates by using each network's own protocol, the bridge allows the protocols to coordinate, accommodating any difference between the two and making the protocols appear as if they were compatible.

LOCAL AREA NETWORK STANDARDS

Local area network standards originated in much the same way that other communications standards have been set. Manufacturers experimented with communications and access techniques, developed proprietary techniques, and gradually demonstrated their feasibility. Most protocols were proprietary, limiting the compatibility between the network and existing equipment and that of other manufacturers. Ethernet® is a case in point. Developed in the early 1970s by Xerox Corporation in its Palo Alto Research Center, Ethernet was offered by Xerox for licensing at a nominal cost. However, rarely do proprietary systems become adopted as standards without modification. Ethernet was no exception.

The IEEE 802 Committee

In 1980, the IEEE Computer Society appointed a committee to work on project 802, the development of LAN standards. The 802 committee's objectives were to establish standards for the physical and data link connections between devices. The following requirements were established:

- Existing data communications standards were to be incorporated into the IEEE standard as much as possible.
- The network was intended for light industrial and commercial use.
- The maximum network diameter was set at 2 kilometers.
- The data speed on the network was to be between 1 mb/s and 20 mb/s.
- The network standard was to be independent of the transmission medium.
- The failure of any device on the network was to not disrupt the entire network.

- There was to be no more than one undetected error per year on the network.

It was apparent that Ethernet would not suffice as a single standard because of blockage under heavy load conditions. Therefore, the committee set two incompatible standards. For light duty, a bus contention network similar but not identical to Ethernet was selected. For applications where assurance of network access is needed, token passing bus and ring standards were selected.

The 802 standards are published in five parts:

802.1 Overview Document (Containing the reference model, tutorial, and glossary.)

802.2 Link Layer Protocol Standard.

802.3 Contention Bus Standard.

802.4 Token Bus Standard.

802.5 Token Ring Standard.

The 802 standards are developed around layers 1 and 2 of the OSI protocols (see Chapter 3) and do not include all the functions of a complete network. Each manufacturer applies its own design to the network; as a result, even though a network complies with an 802 standard, apparatus is not necessarily interchangeable between networks. Furthermore, many products on the market do not conform to the IEEE standards. Therefore, it is important to understand that the 802 standard does not imply universal compatibility.

The CSMA/CD LAN IEEE 802.3

The 802.3 standard is a CSMA/CD network intended for commercial or light industrial use. The specification supports Media Access Units (MAUs) for baseband and broadband coaxial cable and baseband fiber cable. The standard is not a complete network; for example, although the protocol detects errors, it does not include an error recovery process.

The elements of 802.3 are illustrated in Figure 19.6. The link layer is divided into two sublayers: the Logical Link Control (LLC) and the Media Access Control (MAC), which

FIGURE 19.6
IEEE 802 Standard

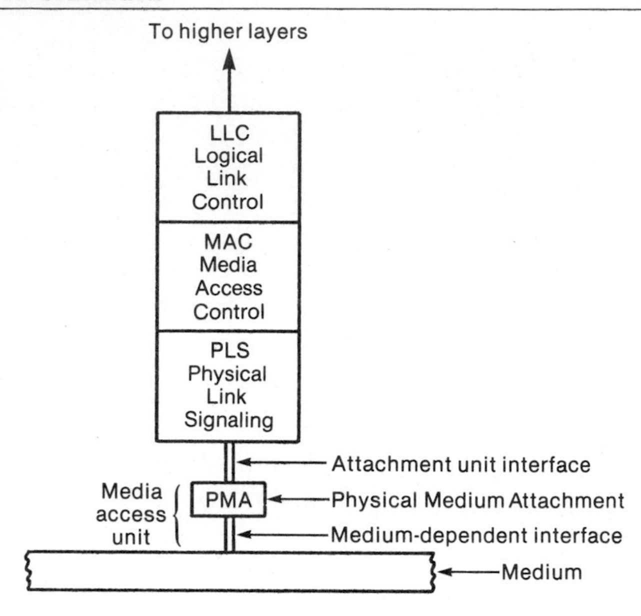

together correspond to the data link layer in the OSI model. The 802.3 standard interacts with higher layers for error recovery and network control. The interface between the media access sublayer and the LLC transmits and receives frames and provides status information to forward to higher levels for error recovery. The interface between the media access and the physical layers includes signals for framing, detecting, and recovering from collisions, and for passing serial bit streams between the two layers.

The network is composed of cable segments of a maximum of 500 meters (1,640 feet) long—the greatest distance that can be spanned at the maximum signaling rate of 20 mb/s. As many as five segments can be interconnected through a maximum of four repeaters as shown in Figure 19.7. The repeaters sense carriers from both cable segments to which they are connected, and also detect and avoid collisions. When a repeater detects a collision in one segment, it transmits a jamming signal to both segments. The design of the MAU is simple, consisting of a high impedance bridge on the transmission medium.

FIGURE 19.7
Maximum Configuration of IEEE 802.3 Network

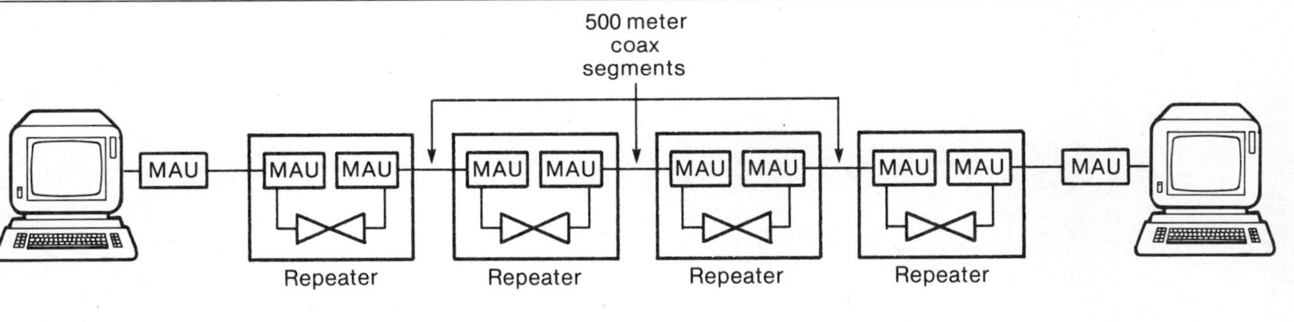

MAU—Media Access Unit

The Token Bus LAN IEEE 802.4

A token bus LAN, illustrated in Figure 19.4, uses the same topology as CSMA/CD but control flows in a logical ring. Although messages are broadcast as with CSMA/CD, control passes in sequence from station to station. Each station is programmed with the address of its preceding and succeeding stations.

Token passing allows a wider physical diameter than CSMA/CD. The diameter can range from 1,280 meters (4,200 feet) to 7,600 meters (25,000 feet) depending on cable grade. Although the network diameter can be increased by repeaters, for most LANs the range will be sufficient to make their use unnecessary.

The MAC in a token bus performs many of the same functions that it does in CSMA/CD, such as address recognition and frame encapsulation, but several functions are added to accomplish the more complex control. The primary functions of the MAC are to determine when its station has the right to access the medium, to recognize the presence of a token, and to determine when and how to pass the token to the next station. The MAC must be capable of initializing or resetting the network. MAC must be able to recognize when a token has been lost and to regenerate it when necessary; MAC must also be able to control the addition of a new station to the network or to recognize when a succeeding station has failed.

On the surface, token passing does not appear complicated, and if all goes well, the protocol has little work to do. Each MAC is programmed with the address of its successor and predecessor stations. When the MAC passes the token to its successor, it listens for the successor's transmission. If the successor fails to transmit, the MAC sends a second token and again listens for a response. If no response is heard, MAC assumes the next station has failed and transmits a message asking which station follows. Each MAC on the network compares its predecessor station number with the number contained in the "who follows" message. The station that follows the failed station responds to the message, and the failed station is bypassed.

The Token Ring LAN IEEE 802.5

The token ring, illustrated in Figure 19.3, is both a logical and a topological ring. Each station is a repeater, enabling greater diameter than with bus networks. Because each node repeats the data stream, failure of a node could disrupt the network. Therefore, nodes are equipped with trunk coupling units to bypass a failed station automatically.

Tokens contain a priority indicator. When a token circulates through a station that has traffic of a priority equal to or greater than the priority designator, that station can seize the token and send priority traffic until a timer within its MAC, known as the "token holding timer," expires. In this manner, the network ensures that traffic is always transmitted up to the capacity of the system and that low priority traffic is deferred.

One station on the ring is always designated as the active monitor (AM) to supervise the network. The AM controls error recovery, detects the absence of a token or valid frames of data, and detects a persistently circulating token or frame. Although only one station is designated as the AM, all stations contain its logic. If the AM fails, the station first detecting the failure assumes the role.

STANDARDS

The standards that follow are the principal standards that apply to LANs.

ANSI

X3T9.5 Local distributed data interface (LDDI), 70 mb/s coaxial cable network and fiber distributed data interface (FDDI), token passing ring architecture for local networks using optical fiber cable.

ANSI/IEEE 802.2 Logical Link Control.

ANSI/IEEE 802.3 Local Area Networks: Carrier Sense Multiple Access with Collision Detection (CSMA/

CD) Access Method and Physical Layer Specifications.

ANSI/IEEE 802.4 Token-Passing Bus Access Method.

CCITT

X.21 General-purpose interface between data terminal equipment and data circuit-terminating equipment for synchronous operation on public data networks.

X.25 Interface between data terminal equipment and data circuit-terminating equipment for terminals operating in the packet mode on public data networks.

V.24 Functions and electrical characteristics of circuits at the interface between data terminal equipment and data circuit-terminating equipment.

EIA

RS-232-C Interface between data terminal equipment and data communication equipment employing serial binary data interchange.

RS-449 General-purpose 37-position and 9-position interface for data terminal equipment and data communication equipment employing serial binary data interchange.

APPLICATIONS

The data communications industry is at the low end of the learning curve when it comes to applying LANs, but that is expected to change rapidly. An organization requiring a LAN will have one or more of these characteristics:

- Widespread use of terminals or personal computers.
- Use of automated office features such as electronic mail, computer messaging, computer graphics, and electronic filing.
- Sharing of common files and data bases.

- Sharing of high-cost peripherals such as files and printers.
- Use of computer-controlled manufacturing processes.
- Use of alternate forms of communication such as video for security and training.

As the use of PCs and automated office features grows, nearly every business that exceeds more than a handful of employees will be a potential user of LANs.

The San Diego State University Network

San Diego State University is a suburban university located about 8 miles from the heart of San Diego, California. The campus covers an area of approximately 300 acres. The university serves a student population of about 33,000 full-time, part-time, and graduate students majoring in such diverse subjects as art, business, engineering, music, and education.

In 1983, SDSU completed the first phase of a broadband LAN that is capable of serving as many as 24,000 stations in 28 buildings. The initial phase covered 400 stations in 9 buildings. Before deciding to install a LAN, SDSU considered several other options to serve its faculty and students. Among these were a microwave network, an optical fiber telecommunications system, and a point-to-point twisted pair network. A coaxial cable-based LAN was chosen because it had the necessary capacity and could be implemented with off-the-shelf components.

The network consists of a backbone of broadband coax using a tree topology, with branches into the buildings. It includes 14,000 feet of cable, 15 amplifiers, head-end equipment, splitters, and couplers. (See Chapter 20 for a description of cable television equipment used in this type of network.) The cable was placed in existing conduits by university personnel. The other components were placed, spliced, and aligned by a CATV contractor. The total installed cost of the system, exclusive of planning, was about $40,000. Connection to the network is made through Sytek interface devices known as TBOXs, which handle two stations, and TMUXs, which support eight stations.

The cable is a midsplit (see Chapter 20), with an upstream frequency band of 5 to 110 MHz and a downstream band of 160 to 300 MHz. (Upstream is the direction from the stations to the head end, and downstream is head end to the stations.) The cable has a 300 MHz bandwidth. The bandwidth is divided into six groups of 20 channels each. Each group occupies the 6 MHz bandwidth of a television channel. Each channel is capable of supporting up to 200 devices operating at 300 to 19,200 b/s. The channels operate on a packet switched basis at 128 kb/s.

The boxes used for interface to the user devices contain all the communications parameters. These parameters can be altered inadvertently by power surges, by changes made by the users, or can be accidentally changed when users fail to terminate sessions properly. This was a frequent cause of service calls and was largely cured by installing an Apple microcomputer on the network. At night the computer dials all the interface devices, verifies the parameters, and resets them.

The network links a combination of mainframe computers, minicomputers, microcomputers, terminals, word processors, and miscellaneous devices. It is used for data transmission, interoffice communications, security and alarm, energy management, and for various remote control functions. Applications include a library circulation system, a data acquisition system used for accounting data, and a student records system.

SDSU has had excellent results with the network and feels that a broadband LAN is an appropriate solution for an application of this type. The tree-structured arrangement of a broadband LAN is flexible. Additional buildings and devices can be added easily by bridging them on the network. Only one of the six groups is activated currently, but additional groups can be added for a small incremental cost.

Evaluating LANs

Because local area network technology is relatively new and the experience base is limited, selecting equipment requires a thorough understanding of an organization's needs. Each local network application is unique enough that the list of requirements should be custom designed to fit the organization. This section briefly discusses the primary factors to consider in evaluating LANs and selecting the best product.

Vendor Support

The choice of LAN will frequently be dictated by the degree of support offered by the vendor. The products are complicated enough that it is rarely advisable to purchase a LAN without assurance of continued local support from the vendor. The following support features should be evaluated:

Implementation and Installation Assistance. The design and engineering of LANs is specialized enough that vendor support will usually be required to implement a system. Design aids such as computerized modeling and simulation tools are valuable. For some types of transmission medium, it will be feasible to obtain installation labor from the open market, but in CATV and fiber optics systems, the vendor will normally be required to install the network. In the majority of cases, the vendor should have the resources to turn the network up and debug the software.

Maintenance Support. Even though the reliability of modern electronic equipment is high and the predicted mean time between failures is several months or even years, when the system fails, immediate and competent support is needed. The vendor should employ enough technicians to cover absences and resignations. The vendor should also have the necessary test equipment to diagnose both the transmission medium and the network electronics. The vendor should be capable of offering warranty support on both the hardware and software.

Integration Capabilities. Many networks are not complete in themselves. It may be necessary to apply cable from one vendor, media access units from another vendor, and terminals from a third. Also, most LANs require integration with existing applications programs and interfacing with hardware of diverse manufacture. Unless an organization has internal capabilities for integrating the network, it will be necessary to obtain the service from the vendor or a contractor.

Costs

LAN costs vary widely and are changing frequently enough that it is risky to generalize. Costs ranging from $200 to more than $1,000 per terminal are quoted, but it is essential to determine total costs before acquiring a network. The following cost factors should be considered:

- Design and engineering of the network including the cost of collecting usage data for sizing it.
- Purchase price of the equipment including spare parts, delivery, and taxes.
- Purchase price of new terminal equipment required because of incompatibilities.
- Installation costs including costs of labor, building and conduit rearrangements, special permits, and licenses.
- Software right-to-use fees.
- Cost of growth when the system exceeds its capacity.
- Transition costs including cutover from the old network, training of users and operators, preparation of new forms and passwords.
- Purchase of special test equipment such as sweep generators to maintain the network (unless maintenance is provided by the vendor).
- Documentation of the network so repairs will not be delayed by lack of information.
- Maintenance costs including the cost of finding and repairing trouble and lost production time during outages.
- Costs of periodic hardware and software upgrades.
- Administration costs including usage monitoring, service monitoring, and interpretation of network statistics.

Traffic Characteristics

The volume, character, and growth rate of data traffic are the first factors to consider in selecting a network. Short, bursty messages are most efficiently handled on a contention network. Bulk data transfer can usually be handled most economically on a point-to-point basis. When terminals have longer messages to send and priority is important, token passing will likely be the best solution. The ultimate size of the network must also be considered to avoid the disruption of replacing a network when its capacity is reached.

The most difficult part of designing any network is obtaining reliable traffic data. The network must be based upon some study of traffic characteristics to determine the peak, mean, and variance of message size and volumes. Both reliable data and a valid growth forecast are required to select a LAN with confidence in the result.

For predicting future requirements, it is valuable to have a network that includes the ability to gather statistics on current usage. Some networks may have the ability to gather statistics in a higher order protocol and will provide information needed to assess the health of the network exclusive of user reports. If the data is available in machine readable form and the vendor offers an interpretive service, this may be invaluable in determining when and by how much to expand the network.

Reliability

The initial costs of a network often pale in significance compared to the costs of outages. The cost of lost production time can mount rapidly when people are depending on the network for their productivity. Networks should be invulnerable to the failure of a single element. Where common equipment such as amplifiers, repeaters, and head-end equipment exist, spares and duplicates should be retained. Most important, qualified repair forces must be available within a short time of failure.

When a network element or the total network fails, the more rapidly it can be restored to service, the more valuable the network. Some systems are equipped with internal diagnostics that aid in rapid trouble isolation and restoral. The network should also be inherently designed for fault isolation. For example, a ring network should be designed to identify which node has failed and bypass a failed node, with the media access controller or network control center providing alarms indicating a loss of received signal to aid in rapid fault isolation.

Off-Net Communications

LANs differ widely in their ability to handle off-net communications. Gateways or bridges are required to interconnect networks using different protocols. Where the protocols are identical, repeaters will normally be required. For access to public data networks, local networks require gateways, usually using X.25 protocols toward the public network.

It is important to anticipate what other networks will be communicated with, what their protocols are, and what the speed and volume of traffic will be. A network under consideration should be capable of these external interfaces.

Compatibility with Existing Equipment

A company having a large investment in existing terminals and computers will find it essential to select a network that is compatible with the existing equipment. This may dictate a standard interface such as RS-232-C or RS-449 between the network and its terminals. Some networks can interface only through controllers that plug into specific apparatus and will for that reason be unsatisfactory for some applications.

Compatibility with Other Vendors' Equipment

When proprietary networks are purchased, it may be difficult to obtain equipment from other than the network supplier, even though existing equipment is initially compatible. Network integration, or converting equipment to be compatible with the network, is likely to be costly, and in the worst case it may be impossible.

Need for Voice and Video Communications

In selecting between baseband and broadband equipment, the choice often will depend upon the need for other forms of communication. Where video is required, a baseband system is feasible only by using a paralleling CATV system. An extensive voice communication requirement may mandate the use of a PBX in addition to or instead of a LAN.

Security

Both contention and token networks allow access to all messages by all stations on the network. Although the stations are programmed to reject messages where they are not the addressee, the potential of unauthorized reception exists. Where security is important, it may be necessary to select a network with additional security provisions such as encryption. The transmission medium is also an important element of security. Twisted pair wire is easy to tap. Radio transmissions can be easily intercepted. Fiber optic cable is difficult to tap, and coaxial cable is somewhere between fiber optics and twisted pair wire.

Throughput and Response Time

Ideally, a LAN should not impose response delays on the attached terminals. Throughput and response time will greatly affect the choice of access methods. Contention networks are the most likely to introduce delays during peak load periods. With token passing networks, the delays can be distributed so some terminals receive priority access.

The throughput of the network will be affected by the data transfer rate, the choice of transmission medium, the access method, and the protocol. Quotations on throughput should be obtained from the vendor, given the traffic characteristics of the application.

Network Diameter

Distance limitations may dictate the choice of a LAN. When the network diameter exceeds the design limitations of a contention network, token passing or circuit switching will be required, the network will have to be segregated through a repeater, the network speed will have to be reduced, or a transmission medium with higher propagation speed such as fiber optics must be used.

Accessibility

The most effective LANs are designed so all terminals on the network can be interconnected with simple addressing. The easiest systems offer complete flexibility to the operator; for example, the initials of the addressee are used with the LAN converting to the address needed by the network. The system must, however, be secure enough that the sender and receiver are authenticated to prevent transmissions from reaching unintended hands.

Standardization

The interfaces between terminals and the networks are the most likely interfaces to conform to some standard. Many networks conform to IEEE 802 standards, but as these specifications cover only the first two layers of the OSI model, LAN standards are still far from universal. Although standards are lacking, it is important in many applications that the network conform to 802 standards because of the greater likelihood of obtaining equipment from more than one manufacturer.

GLOSSARY

Access: The capability of terminals to be interconnected with one another for the purpose of exchanging traffic.

Backoff algorithm: The formula built into a contention network used after collision by the media access controller to determine when to reattempt to acquire the network.

Baseband: A form of modulation in which data signals are pulsed directly on the transmission medium without frequency division.

Bridge: Circuitry used to interconnect networks with a common set of higher level protocols.

Broadband: A form of modulation in which multiple channels are formed by dividing the transmission medium into discrete frequency segments.

Carrier Sense Multiple Access with Collision Detection (CSMA/CD): A system used in contention networks where the network interface unit listens for the presence of a carrier before attempting to send and detects the presence of a collision by monitoring for a distorted pulse.

Collision: A condition that occurs when two or more terminals on a contention network attempt to acquire access to the network simultaneously.

Collision window: The time it takes for a data pulse to travel the length of the network. During this interval the network is vulnerable to collision.

Contention: A form of multiple access to a network in which the network capacity is allocated on a "first-come, first-served" basis.

Ethernet: A proprietary contention bus network developed by Xerox, Digital Equipment Corporation, and Intel. Ethernet formed the basis for the IEEE 802.3 standard.

Gateway: Circuitry used to interconnect networks by converting the protocols of each network to that used by the other.

Local area network (LAN): A form of local network using one of the nonswitched multiple access technologies.

Media access controller: The control circuitry in a LAN that converts the protocols of the DTE to those required by the LAN.

Network access unit (NAU): The circuitry and connectors used in local area networks to enable DTE to access the transmission medium.

Multiple access: The capability of multiple terminals connected to the same network to access one another by means of a common addressing scheme and protocol.

Token: A software mark or packet that circulates among network nodes.

Token passing: A method of allocating network access wherein a terminal can send traffic only after it has acquired the network's token.

BIBLIOGRAPHY

Byers, T. J. *Guide to Local Area Networks.* Englewood Cliffs, N.J.: Prentice-Hall/Micro Text, 1984.

Chorafas, Dimitris N. *Designing and Implementing Local Area Networks.* New York: McGraw-Hill, 1984.

Davis, George R. *The Local Network Handbook.* New York: McGraw-Hill, 1982.

Derfler, Frank Jr., and William Stallings. *A Manager's Guide to Local Networks.* Englewood Cliffs, N.J.: Prentice-Hall, 1983.

Gee, K. C. E. *Local Area Networks.* Manchester, England: The National Computing Centre, Ltd. 1982.

Green, James H. *Local Area Networks.* Glenview, Ill.: Scott, Foresman, 1985.

Institute of Electrical and Electronic Engineers, Inc. IEEE Project 802 Local Area Network Standards *Draft IEEE Standard 802.5 Token Ring Access Method and Physical Layer Specifications.* Working Draft, December 1983.

Madron, Thomas W. *Local Area Networks in Large Organizations.* Hasbrouck Heights, N.J.: Hayden Book, 1984.

Stallings, William. *Local Networks: An Introduction.* New York: Macmillan, 1984.

MANUFACTURERS OF LOCAL AREA NETWORK EQUIPMENT

Coaxial, Twisted Pair, and Fiber Optic Cables

Alpha Wire Corp.

Anaconda Ericsson Inc. Wire and Cable Division

Belden Corp.

Brand-Rex Co.

General Cable Co.

Siecor Corporation

Standard Wire and Cable Co.

LAN Equipment

AT&T Information Systems

Bridge Communications

Concord Data Systems

Contel Information Systems

Complexx Systems, Inc.

Corvus Systems, Inc.

Datapoint Corporation

IBM Corp.

Interlan

Novell, Inc.

Siecor Fiberlan

Sytek, Inc.

Ungermann Bass Inc.

Wang Laboratories, Inc.

Xerox Corporation Information Products Division

CHAPTER TWENTY

Video Systems

Since the end of World War II, television has grown steadily as an entertainment medium. Many businesses have employed television as a security surveillance system and have used closed circuit television (CCTV) for internal communications and training. The use of television to enhance the effectiveness of the classroom and business meetings has long been recognized, but until recently, the cost of long-haul transmission facilities has been prohibitive for all but a few large organizations. Recent developments are changing television into an economical medium for enhanced communications at a time when travel costs are continuing to climb.

The first factor inducing the change is the rapid growth of community antenna television (CATV). CATV has been around since the 1950s, but its use for services other than translating signals into areas without television service is relatively recent. At first CATV was unregulated. As recently as 1959 the FCC ruled that it had no jurisdiction over CATV systems, but in 1966 it asserted jurisdiction. The first rules regulating CATV were issued in 1972, requiring FCC Certificates of Compliance from CATV companies and establishing minimum technical standards for CATV transmissions.

CATV systems have grown rapidly in major market areas that were once served exclusively by broadcasters. Companies seeking franchises have agreed to provide "institutional" channels for use by municipalities and school districts at little or no cost as part of the franchise agreement. Not only is the medium touted for television alone but the enormous amounts of bandwidth available in CATV systems are used for data communications in some metropolitan areas. In most cities today, a broadband communications channel is routed to a substantial portion of businesses and residences. Approximately 40 million households are served by CATV, representing nearly half the television-owning households in the United States. Of this number, about 25 million are on CATV systems with premium entertainment channels.

CATV has had several effects on the growth of video communication systems in business. First, CATV hardware is used for CCTV and increasingly for local area networks in many businesses. Second, CATV is used as a transmission medium to connect teleports, which are communications centers that concentrate broadband facilities in urban areas for connection to satellite communication systems. Third, CATV offers an alternative to common carrier point-to-point communication facilities.

The second significant development is the growth of communications satellites. Although direct broadcast satellite service is not yet in operation, many CATV programs are carried by satellite, which offers the potential of delivering service to the entire continent. Satellite broadcasts have the enormous social impact of bringing video accounts of worldwide events into millions of homes almost at the instant they occur. The effects of this are only beginning to be felt. Satellites also provide nationwide transmission facilities for multipoint video conferences and one-way broadcasts for a fraction of the cost of terrestrial facilities. The broadcast nature of satellites and the fact that rates are independent of distance makes video conferencing feasible over satellites where it would be prohibitively expensive over terrestrial facilities.

A third development with impact on video communications is the use of video compression techniques. As discussed in a later section, video compression takes video signals that occupy 6 MHz of bandwidth on an analog basis and compresses them into narrow digital channels. The Picturephone, which was introduced by AT&T in 1964, failed to attract enough interest to make it a commercially successful product at the time. However, like so many other product developments, it was merely ahead of its time. Advancements in video technology now make it technically (although not economically) feasible to compress a signal of reasonably good quality into a passband that can be accommodated by the majority of the subscriber loops in the United States. Although costs of compression technology are high, the limits have not been reached.

A fourth development that is not nearly as far reaching as the first three is the provision of video meeting services as a utility. Conference rooms equipped with cameras, monitors, and access to transmission facilities are available in many cities to any organization wanting to conduct a multipoint meeting. Portable equipment can bring conference facilities to the user's premises or to hotels, many of which have their own satellite receiving equipment. In larger organizations, these facilities are duplicated within the organization itself, but for the vast majority of businesses the cost is prohibitive and is likely to remain so.

VIDEO TECHNOLOGY

At its present state of development, video is inherently an analog transmission medium because of the method of picture generation. Video signals in the United States are generated under the National Television Systems Committee (NTSC) system. A video signal is formed by scanning an image with a video camera. As the image is scanned, an analog signal is created that varies in voltage with variations in the degree of blackness of the image. In the NTSC system, the television "raster" consists of 525 horizontal scans. The raster is composed of two fields of 262.5 lines each. The two fields are interlaced to form a *frame*, as shown in Figure 20.1. The frame is repeated 30 times each second; the persistence of the human eye eliminates flicker.

FIGURE 20.1
Interlaced Scanning

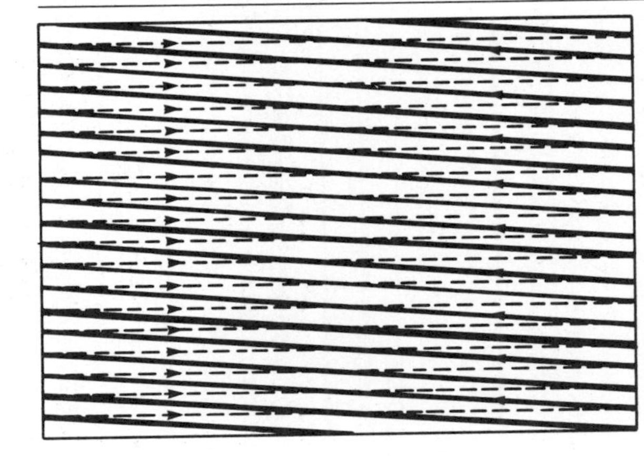

Active Fields
Heavy lines represent first field.
Light lines represent second field.

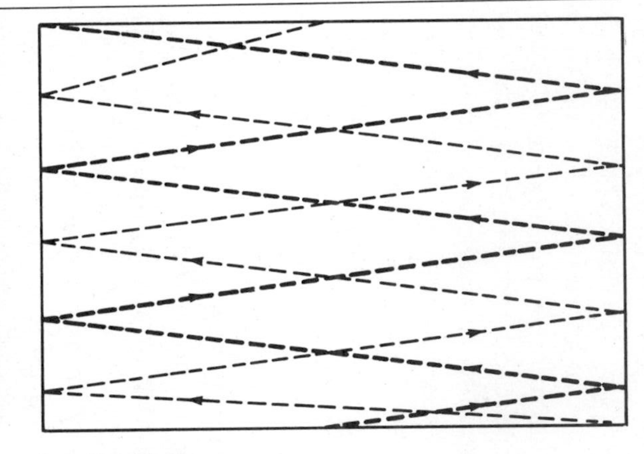

Inactive Fields
Heavy lines show retrace pattern from bottom to top of screen during vertical blanking interval. Heavy lines represent first field and light lines the second.

On close inspection, a video screen can be seen to consist of a matrix of tiny dots. Each dot is called a *picture element,* abbreviated *pixel.* The resolution of a television picture is a function of the number of pixels per frame. Also, the amount of bandwidth required to transmit a television signal is a function of the number of pixels. The NTSC system requires about 4 MHz of bandwidth for satisfactory resolution. Because of the modulation system used and the need for guard bands between channels, 6 MHz of bandwidth is assigned to U. S. television channels.

The signal resulting from each scan line varies between a "black" and a "white" voltage level as shown in Figure 20.2a. A horizontal synchronizing pulse is inserted at the beginning of each line. Frames are synchronized with vertical pulses as shown in Figure 20.2b. Between frames the signal is blanked during a vertical synchronizing interval to allow the scanning trace to return to the upper left corner of the screen. In teletext services, information is transmitted during this interval, known as the *vertical blanking interval.*

A color television signal consists of two parts, the *luminance* signal and the *chrominance* signal. A black-and-white picture is transmitted with only the luminance signal, which controls the brightness of the screen in step with the sweep of the horizontal trace. The chrominance signal modulates subcarriers that are transmitted along with the video signal. The color demodulator in the receiver is synchronized by a "color burst" consisting of eight cycles of a 3.58 MHz signal that is applied to the horizontal synchronizing pulse as shown in Figure 20.2a.

When no picture is being transmitted, the scanning voltage rests at the black level, during which the television receiver's screen is black. Because the signal is amplitude modulated analog, any noise pulses that are higher in level than the black signal level appear on the screen as "snow." A high-quality transmission medium must keep the signal level above the noise to preserve satisfactory picture quality. The degree of resolution in a television picture is a function of bandwidth. Signals sent through a narrow bandwidth are fuzzy with washed out color. The channel must also be sufficiently linear. Lack of linearity results in high-level signals being amplified at a different rate than low-level signals,

FIGURE 20.2
Synchronizing and Blanking in a Television Signal

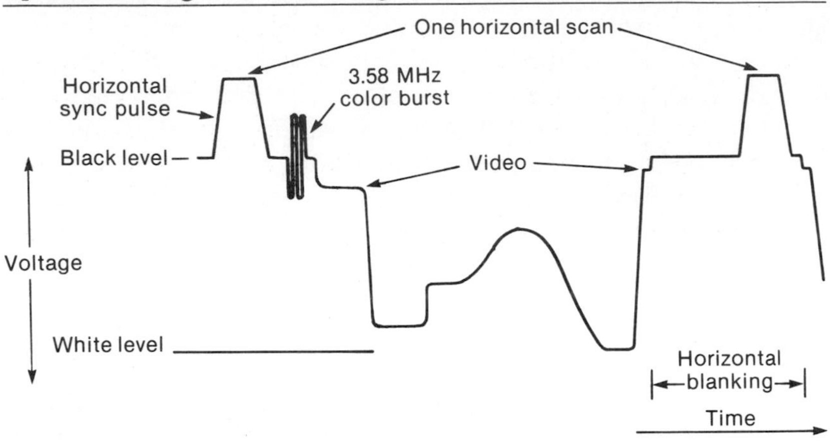

a. Voltage levels during one horizontal scan. As image is scanned, voltage varies from reference black to reference white level.

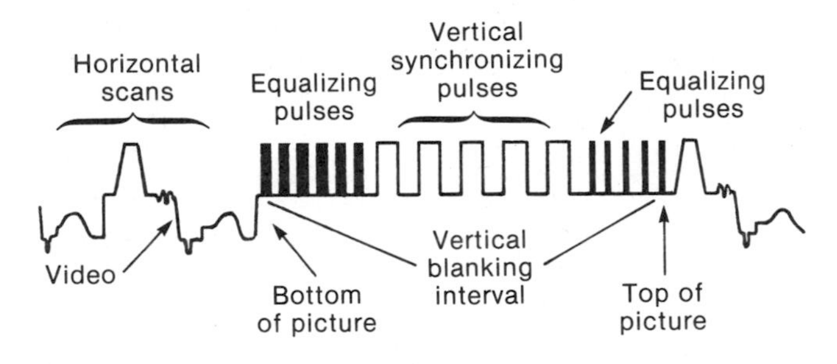

b. Vertical synchronization. Vertical blanking occurs during retrace of scanner to top of screen.

which affects picture contrast. Another critical requirement of the transmission medium is its envelope delay characteristic. If envelope delay is excessive, the chrominance signal arrives at the receiver out of phase with the luminance signal and color distortion results.

The four primary criteria for assessing a video transmission medium, therefore, are noise, bandwidth, amplitude linearity, and phase linearity. The primary media used are analog microwave radio, analog coaxial cable, and for broadcasting, free space. As video signals increasingly are being

FIGURE 20.3
Block Diagram of a CATV System

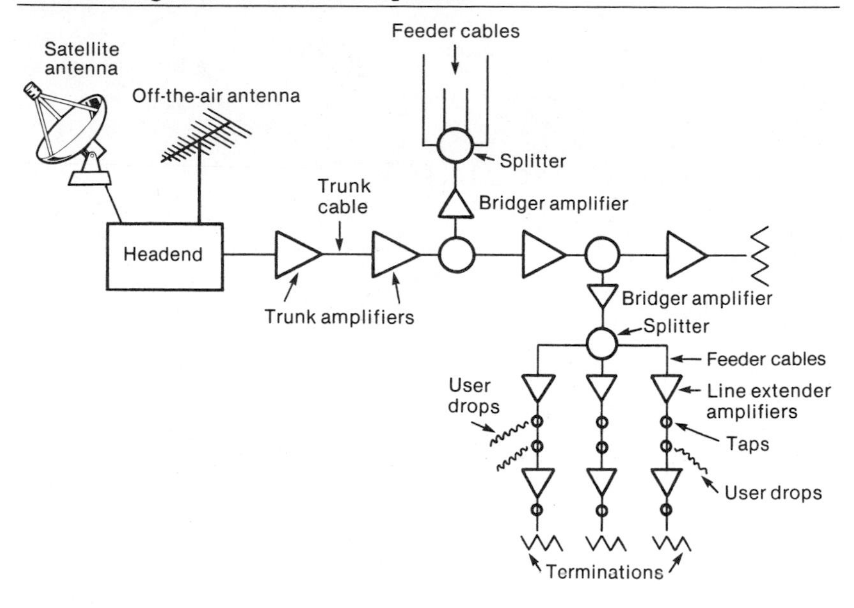

digitized, digital radio and fiber optics also are being used as transmission media. Analog signals are also transmitted over fiber optics using intensity modulation or frequency modulation of a pulse stream.

Cable Television Systems

Cable television systems are composed of three major components: head-end equipment, trunk cable, and feeder and user drop equipment. A block diagram of a CATV system is shown in Figure 20.3. A principal limitation of CATV systems is a lack of selectivity. All channels that originate at the head end are broadcast to all stations. Except for various devices to prevent unauthorized reception of pay television signals, any receiver can receive all services on the network. This is in contrast to the telephone network, which radiates lines in a star configuration to each user individually.

Head-End Equipment

Head-end equipment, as shown in Figure 20.4, is used to receive and generate video signals; and in the case of two-

FIGURE 20.4
Head End Equipment

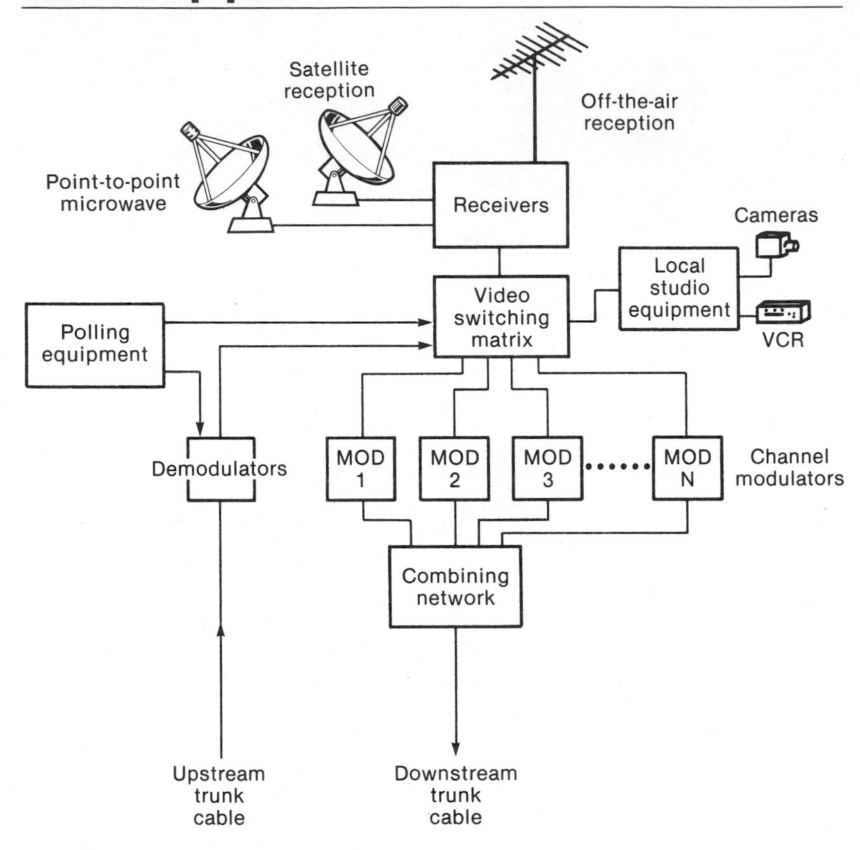

way systems, it is used to repeat the signal from the user on the upstream channels to terminals on the downstream channels. Signals are inserted at the head end from the following sources:

- Off-the-air pickup of broadcast signals including both television and FM radio.
- Signals received over communications satellites.
- Signals received from distant locations over microwave radio relay systems.
- Locally originated signals.

FIGURE 20.5
Cable Television Frequency Bands

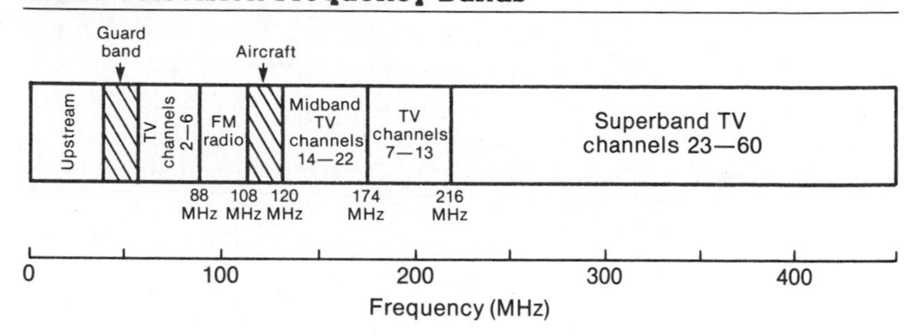

The head-end equipment modulates each television signal to a separate channel in the range of 50 to 450 MHz. Narrower bandwidth systems are used by some companies, but the present state of the art can accommodate up to 60 channels over this frequency range as shown in Figure 20.5. UHF television, which occupies frequencies from 470 to 890 MHz, cannot be transported over a CATV system, so off-the-air UHF channels are converted into another channel within the passband. Most CATV systems use a single coaxial cable to carry signals, but some systems double their capacity by using two cables. All amplifiers and cable facilities are duplicated in a two cable system, but the cost of placing the cables, which is often a significant part of the total investment, is little more than the cost of one.

Trunk Cable Systems

Head-end equipment applies the signal to a trunk to carry the signal to local distribution systems. In some systems, *hub head ends* are used to distribute the signal. Hub head ends are satellites of a master head-end and have the ability to add or distribute services before they are sent through the local feeder area. The cable between the master head end and hubs is called a *super trunk*. Hubs may be fed by point-to-point microwave radio operating in a band designated by the FCC as Cable Television Relay Service.

Early CATV systems used separate amplifiers for each television channel. This technique simplified the amplifier design but restricted the practical number of channels that could be transported. Current systems use broadband amplifiers that are equalized to carry the entire bandwidth. Amplifiers have about 20 dB of gain and are placed at intervals of approximately one-third mile. Amplifiers known as *bridgers* are used to split the signal to feeder cables.

Automatic gain control is applied to adjust amplifier gain as cable loss changes with temperature variations. Power is applied over the coaxial center conductor to amplifiers, with main power feed points approximately every mile. Many CATV systems are used for home security and alarm systems so that continued operation during power outage is essential. To continue essential services during power outages and amplifier failures, redundant amplifiers and backup battery supplies are provided.

Trunk cable uses a high grade of coaxial cable with diameters of ¾ inch to 1 inch. The cable often shares pole lines and trenches with power and telephone. The cable and amplifiers must be constructed to be free of signal leakage. Because the CATV signals operate on the same frequencies as many radio services, a leaking cable can interfere with another service or vice versa. FCC regulations curtail the use of CATV in frequency bands of 108 to 136 and 225 to 400 MHz in many localities because of the possibility of interference with aeronautical navigation and communication equipment.

As with other analog transmission media, noise and distortion are cumulative through successive amplifier stages. The serving radius of 60-channel CATV is limited by noise and distortion to about 5 miles from the head end.

Feeder and Drop System

Feeder or distribution cable is split from the trunk cable with bridger amplifiers. Multiple feeders are coupled with *splitters* or *directional couplers* which match the impedance of the cables. The feeder cable is smaller, less expensive, and has higher loss than trunk cable. Subscriber drops are bridged to the feeder cable through *taps,* which are passive devices that isolate the feeder cable from the drop. The tap must have enough isolation so that shorts and opens at the television set do not affect other signals on the cable.

Pay Television Equipment

As shown in Figure 20.5, only 12 of the 60 channels in a CATV system are within the normal tuning range of a television set. Converters at the set shift the midband and superband channels to a channel that is unused in the local broadcast area. Some newer types of converters are addressable; that is, they can be accessed by a signal at the head end to allow reception of only authorized channels.

Because of the high cost of addressable converters, other systems are often employed to prevent reception of unauthorized channels. This is particularly necessary as "cable-ready" television sets are capable of receiving all channels on a cable. The simplest system uses sync suppression, which removes synchronizing pulses from incoming channels, making it impossible to synchronize the television set without an adapter. A second system places a jamming carrier on the signal from the head end. The carrier is filtered out by a trap at the television set. Another system uses a trap at the tap to "suck out" channels the user is not paying for. The trap is surrounded by a sleeve to make it difficult to access without a special tool. Increasingly, scrambling is used to render unauthorized channels unusable without a descrambling device.

Two-Way CATV Systems

Two-way communication is available on relatively few CATV systems, but it is this that changes CATV from a medium reserved for entertainment to one that can accommodate two-way communications. The key to a two-way system is bi-directional amplifiers. Filters split the signal into the high band for downstream transmission and the low band for upstream transmission. Figure 20.6 shows three splitting methods. The subsplit is the most common in CATV systems with the midsplit and high-split methods often used in broadband local area networks. The two directions of transmission are separated by a guard band 15 to 20 MHz wide. The upstream direction shares frequencies with high-powered short wave transmitters operating on 5 to 30 MHz frequencies, so interference from these sources is a potential problem. Cable and amplifiers must be adequately shielded to prevent interference.

Head-end equipment is considerably more complex in a two-way than in a one-way CATV system. User terminals

FIGURE 20.6
Two-Way Cable Frequency Splitting

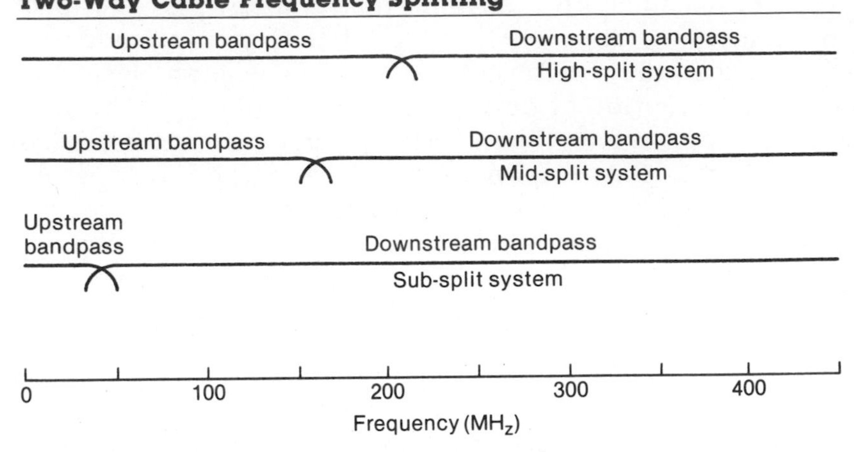

access the upstream cable by contention, described in Chapter 19, or by being polled from the head end. Polling is described in Chapter 18. The head end of a two-way system contains a computer to poll the terminals. Some systems use a transponder at the user end to receive and execute orders from the head end. For example, a polling message might instruct the transponder to read utility meters and forward the reading over the upstream channel.

Video Compression

When transmitted on an analog basis, a video signal with full color and motion uses 6 MHz of bandwidth. Digitizing a video signal results in a 96 mb/s data signal. This would use the capacity of two transponders on a satellite or most of the capacity of a digital microwave radio with 30 MHz of bandwidth. Digital television in uncompressed form is unable to compete economically with analog television. Video compression equipment has entered the market in the past few years to reduce video signals to narrower bandwidths. Current equipment can compress a video signal with full color and motion into a single 1.544 mb/s T1 line. It is expected that before the end of the decade a video signal will be compressed into 512 kb/s, or half of a T1 channel, without sacrificing quality. Equipment currently on the market can compress a video signal into much narrower bandwidths with reduced quality.

Video signals contain a great deal of redundancy from frame to frame. Often large portions of background do not change between frames, but in conventional transmission systems these are transmitted anyway. Video compression systems use techniques known as *interframe encoding* and *intraframe encoding* to transmit only changes between frames. In an interframe coding device, sample elements of succeeding frames are compared, and only pixels that have changed from the previous frame are transmitted. Approximately every two seconds the entire frame is refreshed, but in intervening frames, only the changed pixels are transmitted.

Intraframe encoding provides another element of compression. The picture is broken into blocks of 16 × 16 pixels. Blocks are transmitted only when pixels within the block have changed. Otherwise, the decoding equipment forwards the previous block to the receiver. By using combinations of inter- and intraframe encoding it is expected that it will be possible to transmit video with full color and motion over a 512 kb/s channel. The equipment that encodes and decodes the video signal is called a codec.

At the present state of development, video compression equipment costs in the order of $100,000 per codec for a full color and motion system. When development costs are recovered, the cost of codecs should come down, although it will be some time before they are economical for any but the largest organizations. At some sacrifice in quality, signals can be sent over narrower channels. At present it is possible to send a conference quality picture over a 512 kb/s channel. The trade-off with bandwidth is in the lack of ability of the equipment to convey motion. Abrupt motions smear across the screen until the codec is able to catch up. This characteristic is unsatisfactory for commercial broadcast use but is acceptable for video conferencing in exchange for the lower cost of transmission facilities.

In a video conference, the camera position largely determines the adverse effects of smearing. When the camera is positioned to encompass a passive group, smearing is generally unobjectionable. It is most objectionable when there is considerable action or when the camera shows a close up of the face of an individual who is talking. Equipment now on the market can compress a color picture into a 56 kb/s channel. The equipment is expensive—about $35,000 per codec—and the signal is degraded by motion, but it is suitable for many types of services.

Standards for video compression are virtually nonexistent. Codecs are made to manufacturer's specifications and are end-to-end compatible only with identical equipment. Standards for measuring picture quality also do not exist. Systems are evaluated by the subjective opinion of the users.

Freeze-Frame Video

Another form of video transmission known as *freeze-frame* can be used to transmit video signals over much narrower channels than compressed video. The video signal is digitized, and a sample frame is transmitted over a digital line or over a modem-equipped analog line. Unlike compressed video, which results in a continuously varying picture, freeze-frame video presents a still picture that is refreshed periodically. The refresh rate is a function of the transmission speed. For example, over a 56 kb/s line the signal is refreshed about every 15 seconds. The picture is scanned, digitized, and transmitted over the channel. At the receiving end, the picture gradually fills the screen one line at a time from top to bottom. The "waterfall" effect can be eliminated by using a frame storage unit to accept the frame and relay it to the monitor after the entire frame has been received. Freeze-frame video is particularly adaptable to conferences that make extensive use of graphics.

High-Definition Television

The present NTSC television standard was defined in 1941 when 525 line resolution was considered to yield excellent quality. The width-to-height ratio of the screen, called the *aspect ratio,* is approximately that of motion-picture screens at the time, 4:3. With wide screen movies and a growing trend to large screen television, the current scanning system provides definition that is considerably less than the original image. High-definition television (HDTV) has been successfully demonstrated using 1150 lines and a wide-screen aspect ratio of 5:3 or 2:1. The technology and desirability of HDTV are unquestioned, but before it can be introduced, several practical problems must be solved.

The first problem is the amount of bandwidth HDTV requires. HDTV transmits five times the number of pixels of an NTSC signal, and therefore requires five times the bandwidth, or 30 MHz. In the present television bands, this would mean sacrificing five standard channels to obtain one HDTV channel. An equally important question is how compatibility can be maintained with existing television sets. No system for HDTV has been proposed that would not obsolete millions of television sets.

Part of the bandwidth solution may lie in using video compression techniques. A 5:1 compression is not difficult to obtain with modern techniques, but this requires digitizing the signal, and thus makes it difficult to provide a system that is compatible with existing sets. HDTV may eventually become practical on direct broadcast satellite, cable, and video tape systems rather than broadcast. Even this solution has imitations because these systems are also analog.

VIDEO SERVICES AND APPLICATIONS

Entertainment is likely to remain as the primary use of video into the foreseeable future because it is a service used virtually everywhere in the world. A growing use of television for other services is predictable. These services are briefly examined here.

Cable Television Services

As CATV provides a broadband information pipeline into a substantial portion of U. S. households, the growth of non-entertainment services is expected. The services described below are largely for the use of residences, but business applications will follow the availability of the services.

Teletext

Teletext is the transmission of information services during the vertical blanking interval of a television signal. As discussed earlier, the vertical blanking interval is the equivalent of 21 horizontal scan lines of a television signal, during which no picture information is transmitted. Teletext services transmit information such as magazines and catalogs over a television channel. An adapter decodes the information and presents it on the screen.

The British Prestel system is the first widespread application of teletext. The service has not been used widely in the United States, but if the demand develops, teletext can become an inexpensive way of transmitting information for little more cost than the price of a converter.

Security

Television security applications take two forms: alarm systems and the use of closed circuit television (CCTV) for monitoring unattended areas from a central location. CCTV is used by many businesses for intrusion monitoring and is also widely used for intraorganizational information telecasts. Alarm services have principally used telephone lines to relay alarms to a center. To do so requires a separate line or automatic dialer. The expense of these devices can be saved by remoting the alarms over a CATV upstream channel, but to do so requires a terminal to interface the alarm unit. As described earlier, a computer in the head end scans the alarm terminals and forwards alarm information to a security agency as instructed by the user.

Data Communications

Many CATV companies offer two-way data communications over their systems. A modem converts the digital data signal to analog for application to the cable. Speeds up to 1.544 mb/s are available over systems equipped for two-way operation. Broadband local area networks, as described in Chapter 19, are a growing application that uses the technology but not necessarily the CATV facility.

Control Systems

Two-way CATV systems offer the potential of controlling numerous functions in households and businesses. For example, utilities can use the system to poll remote gas, electric, and water meters to save the cost of manual meter reading. Power companies can use the system for load control. During periods of high demand, electric water heaters can be turned off and restored when reduced demand permits. A variety of household services such as appliances and environmental equipment can be remotely controlled by a computer located at the head end. CATV companies themselves can use the system to register channels that viewers are watching and bill on the basis of service consumed. They can also use the equipment to control addressable converters to unscramble a premium channel at the viewers' request.

Opinion Polling

Experiments have been conducted with opinion polling over CATV. For example, CATV has been used to enable viewers to evaluate the television program they have just finished watching. The potential of this system for allowing viewers to watch a political body in action and immediately express their opinion has great potential in a democracy, although it has not yet been used to any extent.

Video Conferencing

Video conference facilities are offered as a service with conference rooms and full video facilities available for lease, or facilities are owned by large corporations that want to reduce travel and meeting costs. Conferencing equipment is categorized as audio only, audio with graphics, freeze-frame video, and full-motion video. Audio conferencing is discussed in Chapter 11, and the other systems are discussed here.

A conference facility consists of the following subsystems that are integrated into a unit:

- Audio equipment.
- Video and control equipment.
- Graphics equipment.
- Document hard-copy equipment.
- Communications.

A full description of these systems is beyond the scope of this book. They are discussed here briefly to describe the composition of a video conferencing facility.

Audio Equipment

Some analysts contend that audio is the most important part of a video conference facility. In large video conferences it is often impossible to show all participants, but it is important that everyone be able to hear and to be heard clearly. Audio equipment consists of microphones, speakers, and amplifiers placed strategically around the room, preferably coupled to a four-wire voice private line. Speaker telephones are sometimes used, but with generally less satisfactory results than a dedicated circuit. If the signal is transmitted by satellite, it is essential that the audio be transmitted over the same circuit to keep the sound and speakers' lips in synchronism.

Video and Control Equipment

Video equipment consists of two or three cameras and associated control equipment. The main camera is usually mounted at the front of the room and is sometimes controlled to follow the voice of the speaker. Zoom and azimuth controls are mounted on a console and are generally operated by conference participants. A second camera is mounted overhead and is used for graphic displays. The facility sometimes includes a third mobile camera that can be operated independently. A switch at the console operator's position selects the camera. Monitors are placed around the room so they can easily be seen by the participants. Usually one monitor set shows the picture from the distant end and another shows the picture at the near end.

Digitizing and encoding equipment are used to compress full motion video or to create freeze-frames. In addition, encryption equipment may be included for security. Digital storage equipment is sometimes used to enable transmitting presentation material ahead of time so conference time is not taken up with transmission of graphics. Equipment is available to freeze a full motion display for a few seconds to enable sending a graphic image over the circuit.

Graphics Equipment

Video conference facilities may include graphics generating equipment to construct diagrams with arrows, circles, lines, and other symbols. This type of equipment is similar to facsimile, which is covered in more detail in Chapter 21.

Document Hard Copy Equipment

Used in conjunction with graphics equipment, facsimile devices can be attached to a video conference circuit to produce hard copies of documents for the participants.

Communications

Communications facilities are composed of audio and video circuits with the type of facility depending on the bandwidth required. It ranges from voice circuits to full DS-1 video facilities and satellite earth stations. When satellite communication facilities are used, the audio channel must be diplexed over the video channel to keep the audio in synchronism with the picture.

STANDARDS

U. S. television signals are generated according to standards developed by the National Television System Committee (NTSC). Television broadcasts and CATV are regulated by rules and standards of the FCC. EIA specifies electrical performance standards of television signals and equipment. CCITT establishes standards for international television transmission and recommends signal quality standards. No standards aside from manufacturers' specifications exist for video compression equipment.

CCITT

Series J Recommendations: Sound-Program and Television Transmissions

J.61 Transmission performance of television circuits designed for use in international connections.

J.62 Single value of signal-to-noise ratio for all television systems.

J.63 Insertion of test signals in the field-blanking interval of monochrome and color television signals.

J.64 Definitions of parameters for automatic measurement of television insertion test signals.

J.65 Standard test signal for conventional loading of a television channel.

J.66 Transmission of one sound program associated with analog television signal by means of time division multiplex in the line synchronizing pulse.

J.73 Use of a 12 MHz system for the simultaneous transmission of telephony and television.

J.74 Methods for measuring the transmission characteristics of translating equipments.

J.75 Interconnection of systems for television transmission on coaxial pairs and on radio-relay links.

J.77 Characteristics of the television signals transmitted over 18 MHz and 60 MHz systems.

EIA

RS-170 Electrical Performance Standards—Monochrome Television Studio Facilities.

RS-189-A Encoded Color Bar Signal.

RS-215 Basic Requirements for Broadcast Microphone Cables.

RS-240 Electrical Performance Standards for Television Broadcast Transmitters.

RS-250-B Electrical Performance Standards for Television Relay Facilities.

RS-266-A Registered Screen Dimensions for Monochrome Picture Tubes.

RS-312-A Engineering Specifications Outline for Monochrome CCTV Camera Equipment.

RS-324-A Registered Screen Dimensions for Color Picture Tubes.

RS-330 Electrical Performance Standards for Closed Circuit Television Camera 525/60 Interlaced 2:1.

RS-343-A Electrical Performance Standards for High-Resolution Monochrome Closed Circuit Television Camera.

RS-375-A Electrical Performance Standards for Direct View Monochrome Closed Circuit Television Camera 525/60 Interlaced 2:1.

RS-378 Measurement of Spurious Radiation from FM and TV Broadcast Receivers in the Frequency Range of 100 to 1,000 MHz Using the EIA Laurel Broadband Antenna.

RS-412-A Electrical Performance Standards for Direct View High Resolution Monochrome Closed Circuit Television Monitors.

RS-420 Electrical Performance Standards for Monochrome Closed Circuit Television Cameras 525/60 Random Interlace.

RS-439 Engineering Specifications for Color CCTV Camera Equipment.

FCC Rules and Regulations

Part 76 Cable Television Service.

APPLICATIONS

Video conferencing equipment is used by many large organizations, among them Rockwell International and Hercules Corporation. Refer to the case studies presented in Chapters 12 and 17 for information on the application of video conferencing to a corporate network.

Evaluation Considerations

This section discusses evaluation considerations of video conferencing and transmission equipment. Other video services such as CATV are provided only as and where available, and they are not usually subject to choice. The exception to this is data communication over CATV facilities, which is an alternative to using the telephone network. This alternative is evaluated on the basis of criteria discussed in Chapter 18. Examples of the criteria are speed of transmission, data error rate, and cost.

Type of Transmission Facilities

Video conferences are classed as point-to-point or multipoint. With terrestrial facilities the distance and number of points served has a significant effect on transmission costs. Satellite facility costs are independent of distance, and except for earth station costs, they are independent of the number of points served if the points are within the satellite's footprint. If earth stations exist for other communications services, such as a companywide voice and data network, multipoint video conferences can be obtained for costs equivalent to point-to-point conferences.

The type of transmission facility is also a function of the bandwidth required. Full-motion, full-color video requires a 6 MHz analog channel or a 1.544 mb/s digital channel. With a sacrifice in clarity during motion, digital bandwidths can be reduced to as little as 512 kb/s. Analog bandwidth reductions result in a loss of color and picture resolution. Freeze-frame video can be accommodated over narrower bandwidths. Digital transmission is usually less economical for multipoint or one-way broadcasts because of the cost of video compression equipment.

Type of Video Compression

The fundamental decision in establishing a video conference facility is whether to use full-color, full-motion, or freeze-frame video. Freeze-frame video is more expensive than a straight audio conference, but it offers a feeling of presence and the ability to transmit pictures and graphic information. However, the lack of full motion is disconcerting to some participants. The decision is primarily one of cost. Video compression equipment and the broadband transmission channels they require are expensive, but many organizations can justify their cost by reduced travel expense. A secondary benefit is reduction in meeting time. If the facilities are scheduled for a set time, meetings are often completed in less time than those without a limit.

System Integration

Video conference equipment is usually an assembly of units made by different manufacturers. Except for the television signal itself, little standardization exists. To ensure compatibility, it is usually advisable to obtain equipment from a vendor who can assemble it into a complete system.

Security

The type of information being transmitted over the channel must be considered. Often, highly proprietary information is discussed during conferences. Both terrestrial microwave and satellite services are subject to interception; signals over fiber optics are less vulnerable. Scrambling of both the video and audio signals is warranted in many cases.

Public or Private Facilities

Private video conference facilities have a significant advantage over public access systems. Public facilities are unavailable in many localities, which may preclude holding many video conferences. The travel time to a public facility offsets some of the advantages of video conference.

Future Expansion

Video conference facilities should be acquired with a view toward future expansion plans. For example, a conference facility may start with freeze-frame with plans to convert to full-motion video at a later date. The facility should be expandable to other points if growth is foreseen.

Fixed or Portable Equipment

Portable video conference equipment is available and is required in some applications. It is most adaptable to freeze-frame television where equipment can be taken to a classroom or meeting without the need to set up elaborate apparatus or to take conferees to a central meeting room. Portable satellite equipment is also used in some video services, particularly one-way broadcasts to hotels or other meeting facilities.

GLOSSARY

Aspect ratio: The ratio between the width and height of a television screen.

Bridger amplifier: An amplifier installed on a CATV trunk cable to feed branching cables.

Chrominance: The portion of a television signal that carries color encoding information to the receiver.

Codec: Abbreviation for coder/decoder, a device in television transmission that compresses a video signal into a narrow digital channel.

Community antenna television (CATV): A network for distributing television signals over coaxial cable throughout a community. Also called cable television.

Closed circuit television (CCTV): A privately operated television system not joined to a public distribution network.

Directional coupler: A passive device installed on a CATV cable to isolate the feeder cable from another branch.

Downstream channel: The frequency band in a CATV system that distributes signals from the headend to the users.

Frame: A complete television picture consisting of two fields of interlaced scanning lines.

Freeze-frame video: A method of transmitting a video signal over a narrow channel by sending one frame at a time without motion.

Head end: The equipment in a CATV system that receives television signals from various sources and modulates and applies them to downstream channels.

High-definition television (HDTV): A system for transmitting a television signal with greater resolution than the standard NTSC signal.

Hub: A system of multiple head-end equipments in a CATV system consisting of master and satellite headends that relay signals to a local distribution area.

Interframe encoding: A method of video compression that transmits only changed information between successive frames.

Intraframe encoding: A method of video compression that divides the picture into blocks and transmits only changed blocks between successive frames.

Luminance: The portion of a video signal that carries brightness information.

Picture element (pixel): A single element of video information.

Raster: The illuminated area of a television picture tube.

Splitter: A passive device that divides a CATV signal into multiple legs.

Super trunk: A trunk between master and hub headends in a hub CATV system.

Tap: A passive device mounted on a feeder cable to distribute a television signal to a user.

Teletext: A videotext service that receives information during the vertical blanking interval of a television signal.

Upstream channel: A band of frequencies on a CATV cable reserved for transmission from the user's terminal to the headend.

Vertical blanking interval: The interval between television frames in which the picture is blanked to enable the trace to return to the upper left corner of the screen.

Video compression: A method of transmitting analog television over a narrow digital channel by processing the signal.

BIBLIOGRAPHY

Baldwin, Thomas I., and D. Stevens McVoy. *Cable Communication.* Englewood Cliffs, N.J.: Prentice-Hall, 1983.

Cunningham, John E. *Cable Television,* 2d ed. Indianapolis, Ind.: Howard W. Sams, 1980.

Hollowell, Mary Louise, ed. *The Cable/Broadband Communications Book,* vol. 3, 1982–1983. Washington, D.C.: Communications Press, 1983.

ITT. *Reference Data for Radio Engineers,* 6th ed. Indianapolis, Ind.: Howard W. Sams, 1984.

Singleton, Loy A. *Telecommunications in the Information Age.* Cambridge, Mass.: Ballinger Publishing, 1983.

MANUFACTURERS OF VIDEO EQUIPMENT

CATV Equipment

Reliance Comm/Tec

C-COR Electronics

Jerrold Division of General Instrument Corp.

Teleconferencing Equipment

Colorado Video

Compression Labs, Inc.

NEC America Inc., Radio & Transmission Division

Satellite Transmission Systems, Inc.

CHAPTER TWENTY-ONE

Facsimile Transmission

To transmit a page of information over a data circuit, two alternatives can be employed. The characters can be encoded in an eight-bit alphanumeric code and transmitted to a terminal that displays the characters on a screen or prints them on paper. The second alternative is to scan the printed page, encode it, and transmit the image in shades of black and white without identifying individual characters. The latter technique, known as *facsimile,* was invented in 1843, six decades before invention of the teletypewriter. Despite this early start, facsimile found little application for several decades. It first came into general use in the 1920s for transmitting wirephotos and weather maps and has been used extensively by law enforcement agencies for sending mug shots and fingerprints, but it has only recently gained acceptance as a business machine.

Facsimile has several significant advantages over alphanumeric data communication; it conveys graphic as well as textual information, source documents can be transmitted without rekeying, and facsimile is less affected by transmission errors than data communications. A noise burst that would obliterate several characters in ordinary data transmission may damage a facsimile scan line, but unless perfect reproduction is required, the document will be quite readable. Facsimile is also fast. Graphic information can be transmitted anywhere there is a telephone in from one to six minutes. Compared to the alternative of express mail services, facsimile is markedly faster. However, the comparison is not entirely valid because facsimile costs are directly proportionate to the quantity of information transmitted, whereas express mail services usually transmit several pages for a fixed fee.

The primary drawback to data terminals is their inability to transmit graphics. Signatures, logos, and illustrations cannot be sent from a data terminal. Furthermore, if the source document is not machine readable, before information can be transmitted by data communication it must be rekeyed from the source or converted with an optical character reader. Either of these alternatives is costly. Offsetting facsimile's advantages are two drawbacks. First, the terminal equipment is costly. Facsimile transceiver costs range from less than $1,000 for a basic machine to more than $20,000 for a full-featured system. Second, facsimile requires about 10 times the transmission time to send a standard 8½-by-11-inch page of text compared to sending it from a data terminal.

Facsimile can be used to implement one form of electronic mail. The primary differences between computer-based electronic mail and facsimile are the type of output and storage methods. Facsimile is inherently a hard-copy system contrasted to electronic mail, which can be implemented entirely in a soft-copy state (that is, information is displayed on a screen, but not printed). Although facsimile messages can be stored on electronic media, computer-based electronic mail messages can be stored in one tenth or less of the storage space and can be retrieved with any asynchronous terminal. Facsimile and electronic mail can be used interchangeably with low message volumes, but if the volume is high, the inherent advantages of each system tend to dictate which is the most feasible for a given application.

Facsimile has other applications that will become more important as the technology matures. Presently, many facsimile machines are capable of accepting ASCII or EBCDIC input and printing alphanumeric characters. In this mode they are used in lieu of printers. From this application it is a short but complex leap to linking a facsimile machine with a communicating word processor for printing documents remotely. To make this application practical, a set of compatible high-level protocols are needed. Such protocols are currently under development by CCITT in connection with their proposed Group 4 facsimile standards that are described later. As these standards develop, communication between word processors and facsimile machines will become more prevalent, and machines will be able to merge, transmit, and store combined textual and graphic information. A photograph of a modem facsimile machine is shown in Figure 21.1.

Although facsimile got off to a slow start compared to other telecommunications technologies, its use is expanding to the point that it is now indispensable for many forms of record communication. This has come about because machine costs have been dropping and standards are developed to enable machines of different manufacture to communicate.

FACSIMILE TECHNOLOGY

A facsimile machine consists of four major elements as diagrammed in Figure 21.2: scanner, printer, control, and communications. The scanner sweeps across a page, segmenting it into multiple lines in much the same way that a television camera scans an image. The scanner output is a continuous analog signal that varies between white and black level, or it is a digital output that converts the image to a binary code. Analog scanner output modulates an analog carrier. Digital scanner output is compressed and applied directly to a digital circuit, or through a modem to an analog circuit. Control equipment directs the scanning rate, and in the case of digital facsimile, it processes the signal to compress solid expanses of black or white. At the receiving end the incoming signal is demodulated and drives a print mechanism that reproduces the incoming image on paper.

Facsimile Machine Characteristics

Facsimile equipment is categorized by its modulation methods, its speed, its resolution, and its transmission rate. Both

FIGURE 21.1
Desktop Facsimile Transceiver

Courtesy Ricoh Corporation.

frequency and amplitude modulation are applied directly to telephone circuits. Digital machines produce a binary signal that is applied to the circuit through a modem. The digital transmission speed varies from 1,200 b/s to 56 kb/s; the speed of sending a page varies from six minutes down to less than one minute. Resolution is measured in lines per inch (lpi) and varies from slightly less than 100 lpi to 400 lpi.

Facsimile is divided by CCITT standards into four groups:

- Group 1. Frequency modulated analog, six minutes per

FIGURE 21.2
Block Diagram of a Facsimile System

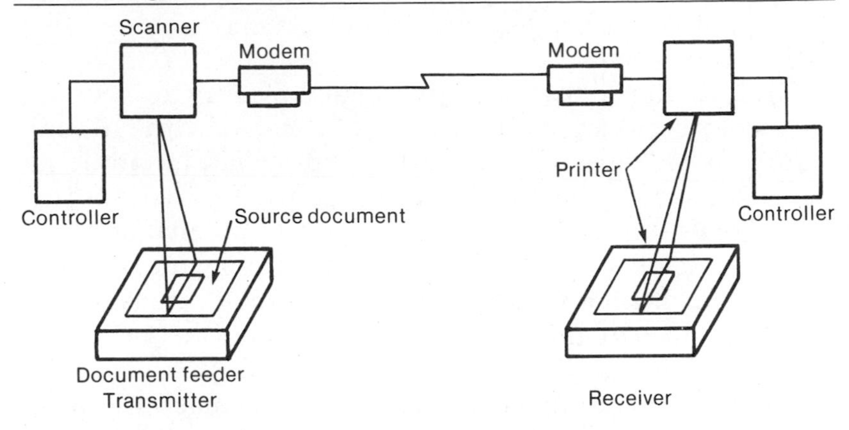

page transmission time, 100 lpi resolution, and 1,500 Hz white frequency and 2,400 Hz black frequency. (These are U. S. conventions. The CCITT standard is 1,300 Hz white and 3,100 Hz black.)

- Group 2. Amplitude modulated analog, two to three minutes per page transmission time, 100 lpi resolution, and 2,100 Hz carrier frequency using amplitude, phase, or vestigial sideband modulation.
- Group 3. Compressed digital, one minute or less per page transmission time, 200 lpi resolution, and 4,800 to 9,600 b/s data rate.
- Group 4 (under development). Compressed digital, subminute per page transmission speed, 200 to 400 lpi resolution, and data rates up to 56 kb/s.

CCITT sets standards for protocol, scanning rate, phasing, scans per millimeter, synchronization, and modulation method. Facsimile machines must be both phased and synchronized. *Phasing* is the process of starting the printer and the scanner at the same position on the page at the beginning of a transmission. Synchronization keeps the scanner and printer aligned for the duration of the transmission. Many machines are designed for compatibility with more than one CCITT group; for example, machines compatible with both

Groups 2 and 3 are common. Compatibility should always be verified between machines of unlike manufacture because subtle differences between manufacturers may yield less than satisfactory results.

Group 1 machines transmit a signal by shifting a carrier frequency between white and black levels. As with FM radio transmission, the constant amplitude of an FM facsimile signal is less susceptible to noise and amplitude variations than AM signals. Group 2 machines modulate an analog carrier with a signal that varies in level or phase from white to black. This more efficient modulation system enables transmission of a page in half the time or less compared to a Group 1 machine. These analog machines are capable of reproducing a full range of gray signal levels between the black and white limits.

Digital machines break a signal into dots or picture elements, which are analogous to the pixels in a video signal. Unlike a video signal, which varies in intensity, digital facsimile produces a binary signal that is either on (black) or off (white) for each picture element. The number of picture elements per square inch determines the resolution of a digital facsimile and also determines the transmission time of a page.

Transmission times of all digital and some analog machines vary with the density of the information on a page. It is customary, therefore, to specify transmission time using a standard ISO test page called "A4," which contains images to assist in making a subjective evaluation of quality as well as speed. Compared to Groups 1 and 2 machines, transmission time is reduced in Group 3 facsimile by increasing the speed of the modem and by compressing the data signal. Modem speeds of 4,800 b/s are standard with most digital machines, with 9,600 b/s optional. As with other data signals, 9,600 b/s cannot be transmitted reliably over the switched telephone network and often requires data circuit conditioning for transmission over leased facilities (see Chapter 3).

Data compression is employed in Group 3 facsimile to reduce transmission time. Many documents have expanses of white or black that are compressed by a process called *run-length encoding.* Instead of transmitting a string of zeros or ones corresponding to an long stretch of white or black, the length of the run is encoded to mark its limits. Run-length encoding is either limited to horizontal runs or encodes both horizontal and vertical runs. By using run-length encoding a digital facsimile machine compresses data into approximately one eighth of the number of nonencoded bits.

Even with data compression, facsimile is a less efficient way to send text than ASCII encoding. For example, the standard pica type pitch is 10 characters per inch horizontally and 6 lines per inch vertically. To encode a character over facsimile with 200 picture elements per inch resolution requires 660 bits as shown in Figure 21.3. Assuming an 8:1 compression ratio, a character can be transmitted with about 80 bits. Compared to the eight bits required to send a character in ASCII, facsimile requires 10 times the transmission time. Because of this difference, it is more economical of circuit time to transmit data in alphanumeric coded form than by facsimile.

The lower resolution of Groups 1 and 2 facsimile is acceptable for most forms of text where there is little ambiguity between characters. However, analog modulation methods require considerably more transmission time than digital. Offsetting the transmission cost advantage of digital is the cost of a 4,800 b/s modem, which is considerably more expensive than the simple modems used in Groups 1 and 2 facsimile. The economic trade-offs between Groups 1, 2, and 3 facsimile are determined by volume. At low speeds, communication costs control the overall per copy costs; and at high speed, machine costs are controlling. Labor costs enter into the decision whether to buy features for unattended operation such as automatic feed, document storage, and document routing.

Scanners

Document scanning produces either an analog or a digital output. Although numerous scanning techniques are used, a description of two representative techniques will illustrate how documents are converted to a signal for transmission over telephone lines. In the analog system illustrated in Figure 21.4a, a point source of light is swept across the docu-

FIGURE 21.3
Encoding a Character with Facsimile

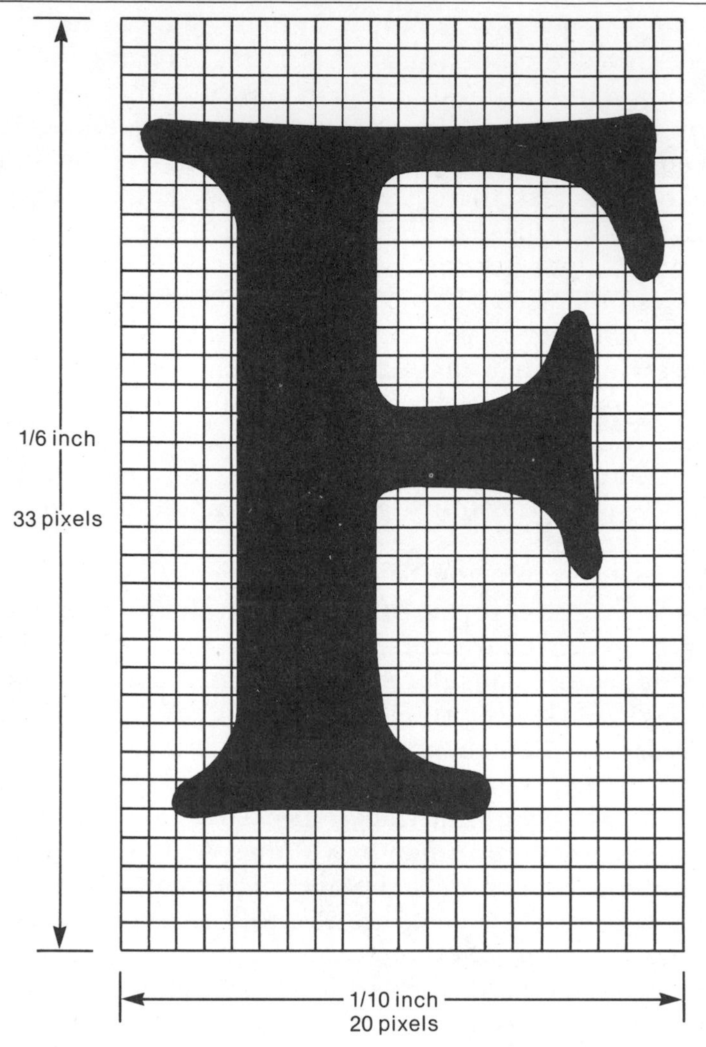

ment. The reflection, which varies in intensity with shading on the source document, is detected by a photocell. The photocell output produces a continuously varying analog signal that modulates the amplitude or frequency of a carrier within an audio passband. In the digital modulation system illustrated in Figure 21.4b, a row of photocells is used, consisting of one cell per picture element and corresponding to one

FIGURE 21.4
Analog and Digital Facsimile Scanning Process

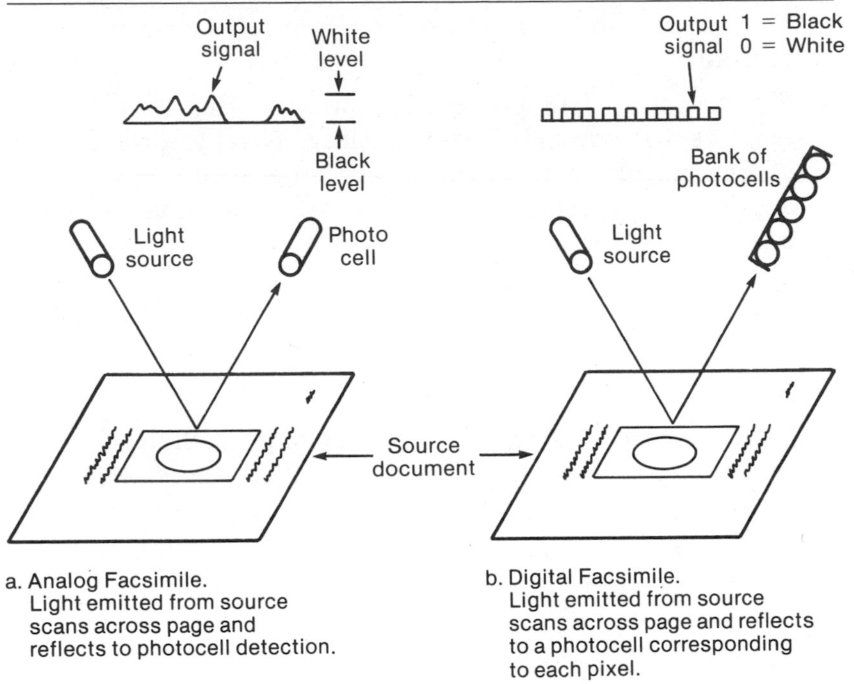

a. Analog Facsimile.
Light emitted from source scans across page and reflects to photocell detection.

b. Digital Facsimile.
Light emitted from source scans across page and reflects to a photocell corresponding to each pixel.

scan line. The photocells detect light from the source and emit a one or zero impulse, depending on whether the reflected light is above or below the threshold between black and white.

Printing

Numerous techniques are used for reproducing or printing documents encoded into a facsimile signal. The first two methods described below have stationery and movable styluses mounted above and below the paper. An electric current passing between the styluses through the paper changes the contrast of the paper in synchronism with the original image. The following printing methods are in general use:

- *Electrosensitive.* This method uses paper with a special coating that is burned off as current varies in proportion to image contrast.

- *Electrolytic.* Similar to the electrosensitive process, a paper with a special electrolytic coating is employed. The paper changes contrast in response to changes in current.
- *Electrostatic.* This process is similar to that used by electrostatic copiers. The incoming signal impresses a charge proportional to the image on a drum. Toner is applied to the drum and then transferred under heat to untreated paper.
- *Dielectric.* This process is similar to electrostatic, except that a charge proportional to the image is applied to treated paper. Toner is applied to the paper and fixed with heat.
- *Photographic.* The received image exposes light-sensitive paper, which is then chemically processed through a photographic stabilization processor.
- *Percussive.* Several techniques are used to apply an image directly to paper. In the ink jet method, the received signal drives a mechanism that applies droplets of ink directly to the paper in proportion to the signal. Other methods drive a hammer that strikes a ribbon to impress dots on the paper.

Facsimile printers vary in their ability to reproduce gray shades or halftones. Analog facsimile machines can reproduce a full range of gray tones, but digital facsimile machines are capable of sending only a black or white signal in each picture element. The degree of black intensity is regulated by the amount of resolution in the picture. Machines with greater resolution are more capable of reproducing halftones than machines with less resolution. The photographic process is capable of the best halftone quality.

Special Telecommunications Features

Some facsimile machines are equipped with features to make them a complete document communications center, designed for attended or unattended operation. For example, stations can be equipped with polling features so a master can interrogate slave stations and retrieve messages from queue. In some machines the master polls the remote, after which the remote redials the master to send the document. Other machines have a feature called reverse polling which enables the receiving machine to transmit a document on the initial poll. Some machines have automatic digital terminal identification capability and apply a time and date stamp to transmitted and received documents. Machines can be equipped with document feeders and stackers to enable them to send and receive documents while unattended. Some digital facsimile devices are also capable of storing digitized messages and routing them to designated addressees on either a selective or a broadcast basis.

Another feature of many machines is the capability of accepting ASCII or EBCDIC input. With this feature they can be used as printers. At 200 lpi resolution, however, these machines are slower than most printers and are economical only as a substitute for a printer in light-duty applications.

Group 4 Facsimile

CCITT is currently developing standards for Group 4 facsimile, which will be a high-speed, high-resolution digital system operating at speeds up to 56 kb/s. Resolution will be from 200 to 400 lpi using high compression and probably X.25 protocol. The draft resolution for Group 4 separates facsimile machines into three classes. All three classes are capable of 100 picture element per inch resolution. Classes 2 and 3 have 300 picture elements per inch resolution with options of 240 and 400 picture elements per inch. Table 21.1 lists the characteristics of the three proposed classes of Group 4 machines.

Besides the advantages of high-speed and improved resolution, Group 4 standards will facilitate communications between word processors. Currently, memory-to-memory data transfers are a feature of communicating word processors. This feature is known as *teletex* (not to be confused with teletext, which is the transmission of information during the vertical blanking interval of a video signal). Most manufacturers now use proprietary protocols that are incompatible with protocols of other equipment. Moreover, like other data terminals, word processors cannot transmit

TABLE 21.1
Proposed CCITT Group 4 Facsimile Terminal Characteristics

	Class		
Service	1	2	3
Facsimile	Transmit/ receive	Transmit/ receive	Transmit/ receive
Teletex		Receive	Create/ transmit/ receive
Mixed mode		Receive	Create/ transmit/ receive
Resolution:			
Standard	200	200/300	200/300
Optional	240/300/400	240/400	240/400
Page memory	No	Yes	Yes

graphic information. Group 4 facsimile standards are being developed to make it possible to integrate facsimile and communicating word processors. To overcome the inherent disadvantages of each type of system, Classes 2 and 3 machines will send textual information in alphanumeric form and graphic information in facsimile form. Terminals capable of this form of communication are called *mixed mode.* Page memory is required for memory-to-memory transmission.

Group 4 facsimile standards are being developed around the ISO Open Systems Interconnection model for data communications. See Chapter 3 for a description of the OSI model.

STANDARDS

The principal source of facsimile standards is CCITT, which sets the standards to which most but not all facsimile machines adhere. As with many other telecommunications technologies, equipment was produced and marketed well in advance of standards, resulting in considerable equipment that does not meet all CCITT specifications. Moreover, CCITT standards deal only with the protocols and line signals exchanged between machines. Manufacturer's specifications control other features intended to integrate facsimile into a network. Differences in storage, polling, machine identification, and other such features may result in incompatibility with some machine functions, even among machines conforming to the standards of one CCITT group.

ANSI

X3.117 Printable/Image Areas for Text and Facsimile Communication Equipment.

X3.98 Text Information Interchange in Page Image Format (PIF).

CCITT

F.160 General operational provisions for the international public facsimile services.

F.170 Operational provisions for the international public facsimile service between public bureaux (bureaufax).

F.180 Operational provisions for the international public facsimile service between subscribers' stations.

F.200 Teletex service.

H.41 Phototelegraph transmissions on telephone-type circuits.

H.42 Range of phototelegraph transmissions on a telephone-type circuit.

H.43 Document facsimile transmissions on leased telephone-type circuits.

X.400 Series message-handling facilities.

Series T Recommendations: Terminal Equipment and Transmission for Facsimile Services

T.0 Classification of facsimile apparatus for document transmission over the public networks.

T.1 Standardization of phototelegraph apparatus.

T.2 Standardization of Group 1 facsimile apparatus for document transmission.

T.3 Standardization of Group 2 facsimile apparatus for document transmission.

T.4 Standardization of Group 3 facsimile apparatus for document transmission.

T.10 Document facsimile transmissions on leased telephone-type circuits.

T.10 *bis* Document facsimile transmissions in the general switched telephone network.

T.11 Phototelegraph transmissions on telephone-type circuits.

T.12 Range of phototelegraph transmissions on a telephone-type circuit.

T.13 Phototelegraph transmission over combined radio and metallic circuits.

T.20 Standardized test chart for facsimile transmissions.

T.21 Standardized test chart for document facsimile transmissions.

T.30 Procedures for document facsimile transmission in the general switched telephone network.

EIA

RS-465 Group 3 Facsimile Apparatus for Document Transmission.

RS-466 Procedures for Document Facsimile Transmission.

RS-497 Facsimile Glossary.

APPLICATIONS

The business application for facsimile equipment is in organizations that need rapid transmission of combined text and graphic information. Facsimile can be used for a form of electronic mail, but until Group 4 standards are implemented in a new generation of equipment, the output can be handled only in hard-copy form.

Evaluation Criteria

The criteria discussed below apply primarily to the telecommunications aspects of facsimile. A discussion of the technical requirements and features of facsimile machines is beyond the scope of this book. Refer to the bibliography for further information on terminal equipment and operational considerations.

Transmission Facilities

Facsimile transmission facilities should be evaluated and selected on the same basis as voice and data circuits. Groups 1 and 2 facsimile use voice frequency facilities that should conform to loss, noise, and echo criteria of a telephone circuit. Digital facsimile can be transmitted over either analog or digital circuits. These circuits should meet the same requirements as for any other data communication service applied to the telephone network or to a public data network.

Many facsimile machines are capable of either half- or full-duplex operation. However, four-wire leased circuits are required for full duplex. Most facsimile applications do not require full-duplex transmission, although if a large volume of documents is being sent in both directions, savings in circuit costs may be possible by using full duplex. If transmission is in only one direction, it is possible to save money by acquiring receive-only and transmit-only machines instead of transceivers. In this case, of course, a half-duplex circuit is all that is necessary.

Compatibility

CCITT Group 3 standards were finalized in 1980, so most digital machines manufactured prior to 1981 are incompatible with Group 3 unless they have been converted. Many machines are capable of operation on two or three groups; if communication is planned with machines controlled by other organizations, this feature should be investigated. However, many machines of different manufacture are compatible with one another only at reduced speed. If possible, test transmissions should be made to determine the degree of compatibility with dissimilar machines.

Document Characteristics

The primary factors in determining which group of facsimile equipment to acquire are the document volume and resolution required. The least costly machines use FM modulation and have a maximum speed of six minutes per page. Digital machines are more expensive than analog machines, although they are generally capable of higher resolution and greater speed, thus saving both labor and circuit costs.

If the documents being transmitted are primarily text, the lower resolution of analog machines may be acceptable because even the poorest resolution of a properly adjusted machine is still readable. However, if copy quality is important, a high-resolution machine should be selected. Some AM machines have dual speeds, with the slower speed intended for higher resolution. The need for halftones in documents should also be considered. Facsimile machines vary widely in their ability to handle halftones.

Laborsaving features such as automatic document feed, automatic dialing and answer, document storage, and document routing should be considered. If confidential information is being transmitted, encryption should be considered.

Network Interface

Like data terminals, facsimile machines may be directly connected to the telephone network if they are registered with the FCC. Unregistered machines can be connected to the network through a Data Access Arrangement (DAA). Most newer machines are registered. A second consideration is whether a built-in modem is included in the machine. If the machine will be used on analog facilities, a modem is required. However, it may be more economical to acquire the modem separately from the facsimile machine or to use modems from a PBX controlled pool. If the machine will be used on digital transmission facilities, no modem is required.

GLOSSARY

Communicating word processor: A word processor that includes protocols for enabling memory-to-memory transfer over a telecommunication circuit.

Facsimile: A system for scanning a document, encoding it, transmitting it over a telecommunication circuit, and reproducing it in its original form at the receiving end.

Mixed mode: A system that is capable of encoding data in both alphanumeric and facsimile form for integrating text and graphics.

Printer: The portion of a facsimile machine that converts a received signal to page copy output.

Phasing: The process of ensuring that both sending and receiving facsimile machines start at the same position on a page.

Resolution: The number of scan lines or picture elements per unit of vertical and horizontal dimension on a document.

Run-length encoding: A facsimile encoding process that converts an expanse of white or black to a code corresponding to the length of the run.

Scanner: The portion of a facsimile machine that converts a source document to an electrical signal.

Teletex: A service enabling users to exchange correspondence automatically between machine memories over telecommunication networks.

BIBLIOGRAPHY

Costigan, Daniel M. *Electronic Delivery of Documents and Graphics.* New York: Van Nostrand Reinhold, 1978.

Datapro. *Datapro Reports on Data Communications.* Delran, N.J.: Datapro Research Corporation, 1984.

Freeman, Roger L. *Telecommunication System Engineering.* New York: John Wiley & Sons, 1980.

MANUFACTURERS OF FACSIMILE EQUIPMENT

AT&T Information Systems

Burroughs Corporation

Canon USA

IBM Corporation

3M Company, Business Communication Products Division

NEC America Inc.

Panafax Corporation

Pitney Bowes Facsimile Systems

Rapicom, Inc.

Ricoh Corp.

Xerox Corporation Information Products Division

CHAPTER TWENTY-TWO

Mobile Radio

Compared to most telecommunications technologies, mobile telephone has been only moderately applied, and until recently, it has undergone relatively little advancement. Mobile telephone was introduced in 1946, about 25 years after the first mobile radio system went into operation. The term *mobile radio* is often used synonymously with mobile telephone. Although the two services use technology and equipment that are essentially the same, they differ in these ways:

- Mobile telephone uses separate transmit and receive frequencies, making full-duplex operation possible. Mobile radios operate either on the same frequency in a simplex mode or on different frequencies in a half-duplex mode.
- Mobile telephones are connected directly to the telephone network and can be used to originate and terminate telephone calls with billing rendered directly to the mobile telephone number. Mobile radio, if connected to the telephone network, is connected through a coupler to a telephone line. Billing, if any, is to the wireline telephone.

- Mobile telephone signaling is based on a 10-digit dialing plan. Mobile radios use loudspeaker paging or selective signaling that does not fit into the nationwide dialing plan.

Conventional mobile telephone service suffers from several drawbacks as a communications medium. First, the frequency spectrum is limited, and demand greatly outstrips capacity, resulting in long waiting lists for service in many parts of the country. Furthermore, a mobile telephone channel is a large party line with the disadvantages of limited access and lack of privacy. Also, coverage is limited in some parts of most serving areas. When a vehicle leaves a coverage area, transmission is poor and the conversation must often be terminated and reestablished on a different channel, or deferred until signal strength improves. Within the coverage area, communication is apt to be sporadic, or at times impossible.

In 1974, the FCC designated part of the UHF television spectrum between 800 and 900 MHz for a new cellular radio service. The concept of cellular radio had been studied for more than two decades, but it was impeded by the lack of FCC approval, a sufficiently large block of clear frequencies, and lack of a suitable control technology. Although the frequencies were allocated in 1974, the FCC delayed approval of the service pending a lengthy hearing process, which included a solicitation of proposals for demonstration systems. The first cellular radio demonstration system in the United States was installed by AMPS, an AT&T subsidiary, in Chicago in 1978.

In 1981, the FCC authorized 666 cellular radio channels in two bands of frequencies, 825 to 845 MHz and 870 to 890 MHz. The lower half of each band is designated for "wireline" carriers, which are defined roughly as operating telephone companies. The upper half of the band is designated for "nonwireline" carriers, which are any nontelephone company common carriers. Licenses are granted in both bands to serve a Cellular Geographic Serving Area (CGSA). A CGSA corresponds to a Standard Metropolitan Statistical Area, which is a major metropolitan area defined by the Office of Management and Budget. The FCC requires that license applicants cover 75 percent of the CGSA's surface area within two years of the date of operation. Applications have been filed and licenses granted in most of the major SMSAs in the United States, with service expected to be available in all the top 30 areas by 1986.

Cellular mobile is not highly developed in the United States at the time of this writing. Although many of the large markets in the United States are currently provided with cellular radio service, most of the smaller metropolitan areas have only conventional mobile telephone service. By contrast, the Scandinavian countries of Sweden, Norway, Finland, and Denmark are the most fully developed nations in the world for cellular radio. Service is provided by the Nordic Mobile Telephone Agency, and more than 100,000 units are now in service.

CONVENTIONAL MOBILE TELEPHONE TECHNOLOGY

As an aid to understanding cellular radio, it is instructive to review the operation of conventional mobile telephone service. Forty-four channels are assigned to Public Mobile Service in three ranges: 35 to 44 MHz, 152 to 158 MHz, and 454 to 512 MHz. Coverage in all three bands is essentially line-of-sight with the lower frequencies providing the widest coverage. Under some propagation conditions in the 35 MHz band, coverage is so broad that mobile units frequently communicate with unintended base stations. To prevent interference, channels can be reused only with a geographical buffer of about 50 to 100 miles between base stations.

A metropolitan mobile telephone service area consists of transmitters centrally located and operating with 100 to 250 watts of output. Because of the difference between mobile and base station transmitter output power, common carriers often install receivers in more than one site, as shown in Figure 22.1, to improve coverage. These receivers are called *voting receivers* in that a central unit measures the relative signal-to-noise ratio of each receiver and selects the one with the best signal. This helps to improve the power imbalance between the mobile and the base station unit that may cause the mobile to be able to hear the base station while the base

FIGURE 22.1
Diagram of Conventional Mobile Telephone Service

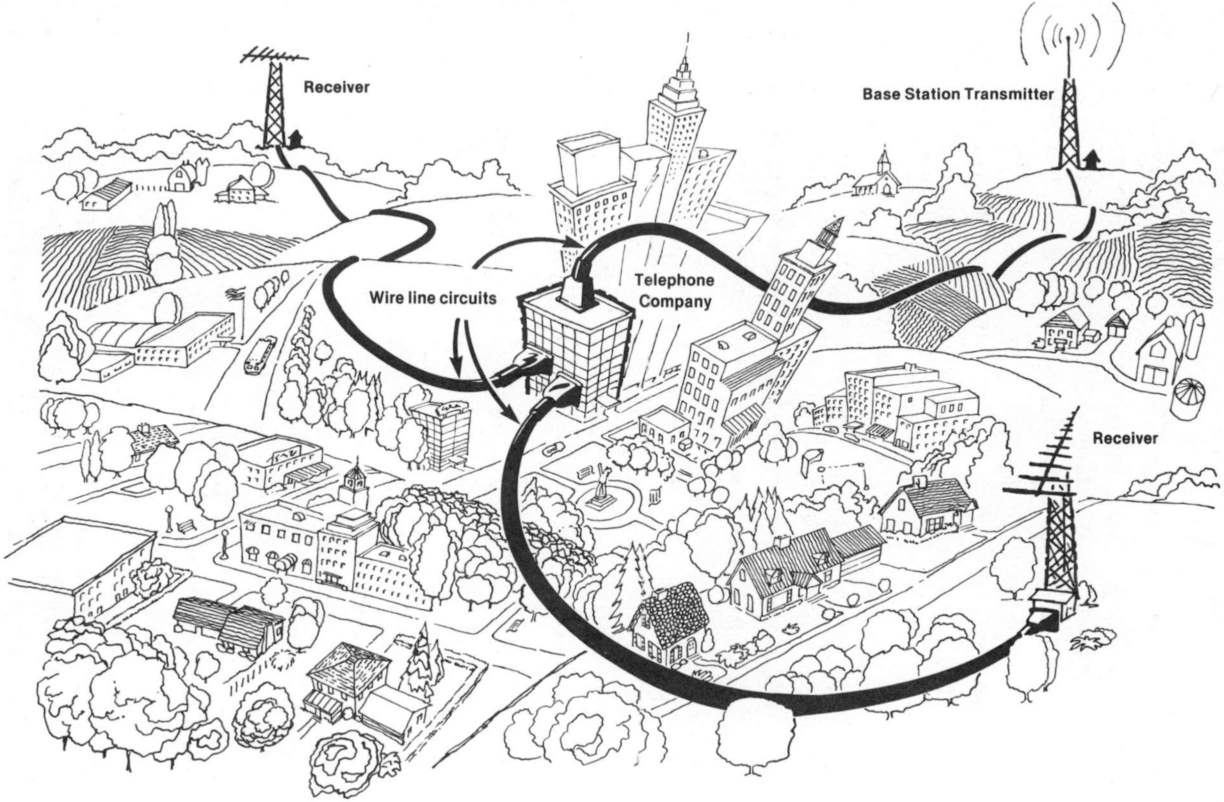

is unable to hear it, or vice versa. Most coverage areas are equipped with several radio channels. Transceivers can shift between channels within the same band, but not between bands. However, modern mobile telephone sets are available that employ a single control head that can switch between a cellular and conventional mobile telephone.

For about the first 20 years of mobile telephone, the service was manually operated. Users placed calls by lifting the handset and keying on the mobile transmitter momentarily to signal the operator. The operator placed and timed the call to a wireline telephone or other mobile unit. With this system the operator supervised only the wireline telephone. Mobile-to-mobile calls were manually monitored to detect the start and end of the conversation. The operator signaled the mobile telephone by multifrequency dialing; a selector inside the mobile transceiver was coded to respond to a five-digit number. Because of frequency congestion, many users purchased multiple-channel sets. The greater number of channels improved the chances of finding an idle channel for outgoing service, but incoming service to the mobile unit was not improved because users could be called only on the channel they monitored.

In 1964, AT&T introduced "Improved Mobile Telephone Service" (IMTS) in an effort to align mobile telephone service more closely with ordinary telephone service. The IMTS mobile receiver automatically seeks an idle channel and tunes the transceiver to that channel. When the user lifts the handset, the system returns dial tone, and the call is dialed like a conventional telephone call. Calls from wireline to mobile units are dialed directly without operator intervention. The base station automatically selects the first idle channel and signals the mobile unit over that channel. IMTS, with its idle channel-seeking capability, improved service for users by eliminating the need for manual channel changes and by making more channels available to reduce congestion.

With both manual and IMTS systems, the base station configuration presents several disadvantages. The coverage area is more or less circular; the actual coverage area depends on directivity of the antenna system and on terrain. Obstructions are a problem with ordinary mobile telephone service. Hills some distance from the base station typically create a

FIGURE 22.2
Frequency Reuse in a Cellular Mobile Serving Area

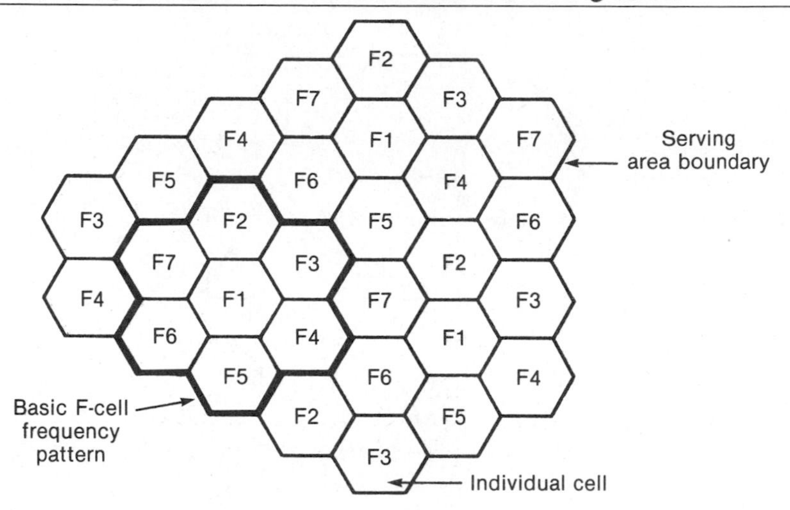

radio signal shadow and impair transmission, even within the range of the primary coverage area. *Roamers,* who are users that travel between serving areas, present a problem with mobile telephone service. Mobile users have a designated "home" channel and can be called only while they are tuned to that channel. When users leave their home areas, they must manually inform potential callers of their location or they cannot be called.

CELLULAR MOBILE RADIO TECHNOLOGY

Cellular mobile radio overcomes most of the disadvantages of conventional mobile telephone. A coverage area is divided into hexagonal *cells* as shown in Figure 22.2. Frequencies are not duplicated in adjacent cells, which reduces interference between base stations. It also allows the carrier to reuse frequencies within the coverage area with a buffer between cells operating on the same band of frequencies. This technique greatly increases the number of radio channels available compared to a conventional mobile telephone system in which a frequency is used only once in a coverage area. Of the 666 frequencies allocated by the FCC, 333 are des-

FIGURE 22.3
Cellular Radio Serving Plan

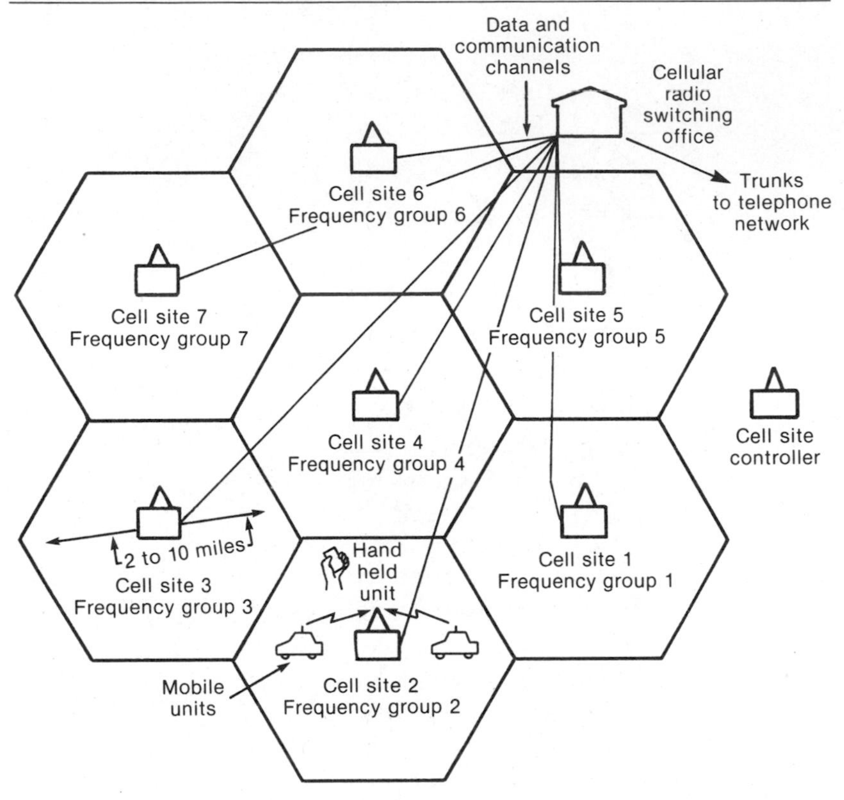

ignated for wireline and 333 for nonwireline carriers. Of these, 312 frequencies are designated for voice and data channels and 21 for control in each band. The FCC has recently proposed 12 MHz of additional spectrum to be divided equally between the wireline and nonwireline systems.

The general plan of cellular radio is shown in Figure 22.3. The number and size of cells are selected by the carrier to optimize coverage, cost, and total capacity within the serving area. These factors are not specified by FCC rules and regulations. The mobile units are *frequency agile,* that is, they can be shifted to any of the 312 voice channels. Mobile units are equipped with processor-driven logic units that respond to incoming calls and shift to radio channels under control of the base station. Each cell is equipped with transmitters, receivers, and control apparatus. One or more frequencies in each cell are designated for calling and control. For incoming and outgoing calls the *cell-site controller* assigns the channel and directs a *frequency synthesizer* inside the mobile unit to shift to the appropriate frequency.

An electronic central office serves as a *cellular radio switching office* (CRSO) and is used to control mobile operation within the cells. The cell-site controllers are connected to the CRSO over data links. The CRSO switches calls to other mobile units and to the local telephone system, processes data from the cell-site controllers, and records billing details. It also controls *handoff* so a mobile leaving one cell is automatically switched to a channel in the next cell.

A major drawback of conventional mobile telephone service, the lack of supervision from the mobile unit, is overcome with cellular radio. Cellular radio uses the control channel to supervise the mobile station. Unlike conventional mobile telephone service in which call timing is based on supervisory signals from the wireline telephone, in cellular radio timing is controlled by either the wireline or the mobile unit. This greatly facilitates mobile-to-mobile calling, which closely approximates ordinary telephone service.

Cell-Site Operation

A cell site consists of one radio transmitter and two receivers per channel, the cell-site controller, an antenna system, and data links to the CRSO. The hexagon was chosen for the cell shape because this shape is a practical way of covering all of an area without the gaps and overlaps of circular cells. As a practical matter, cell boundaries are not precise. Directional antennas are used to approximate the shape, but the decision when to switch a user from the coverage of one cell into the adjacent cell is made by the CRSO on the basis of signal strength reports from the cell-site controllers. The handoff between cells is nearly instantaneous, and users are generally unaware that it has occurred. The handoff, which takes about 0.2 seconds, has little effect on voice transmission aside from an audible click. However, data errors will result from the momentary interruption. As many as 128 channels can be provided per cell, with the number of channels based on demand; most cells will operate with about 70 or fewer channels.

Cell sites are selected to provide coverage with the relatively low power of the cell-site transmitters. FCC rules limit cellular transmitters to 100 watts output, but higher powers are used only if necessary to cover large cells. At the UHF frequencies of cellular radio, transmission is line-of-sight, so considerable planning is needed to define the coverage area of the individual cell while minimizing the need to realign cells in the future.

A minimum of one channel is provided per call for control of the mobile units from the cell-site controller. The cell-site controller directs channel assignments, receives outgoing call data from the mobile unit, and pages mobile units over the control channel. When the load exceeds the capacity of one channel, separate paging and access channels are used.

The cell-site controller, which is shown in Figure 22.4, manages the radio channels within the cell. It receives instructions from the CRSO to turn transmitters and receivers on and off, and it supervises the calls, diagnoses trouble, and relays data messages to the CRSO and mobile units. The cell-site controller also monitors the mobile units' signal strength and reports it to the CRSO. It scans all active mobile units operating in adjacent cells and reports their signal strengths to the CRSO, which maps all working mobile units. This map is used to determine which cell should receive a mobile unit when handoff is required.

The number of users that can be supported in a single cell is a function of traffic characteristics. As cellular radio is introduced to an area, usage is expected to be low. As the prices of portable units and monthly service charges drop, the demand is expected to grow, necessitating increases in cell capacity. Cell capacity can be expanded by adding radio channels up to the maximum. When channel capacity is reached, cells can be *sectored,* that is, subdivided into two to six sections with frequencies reused within the cells. Interference is avoided by providing directional antennas as shown in Figure 22.5. Sectored patterns are also used near mountains, water, and other terrain obstructions to direct radio frequency energy away from areas where it is unneeded.

FIGURE 22.4
Cell-Site Controller

Courtesy General Electric and Northern Telecom.

Capacity can also be increased by splitting cells. One strategy for introducing cellular radio is to begin with large cells as shown in Figure 22.6. As demand increases, a larger cell can be subdivided into smaller cells by reducing power and

FIGURE 22.5
Sectored Cell

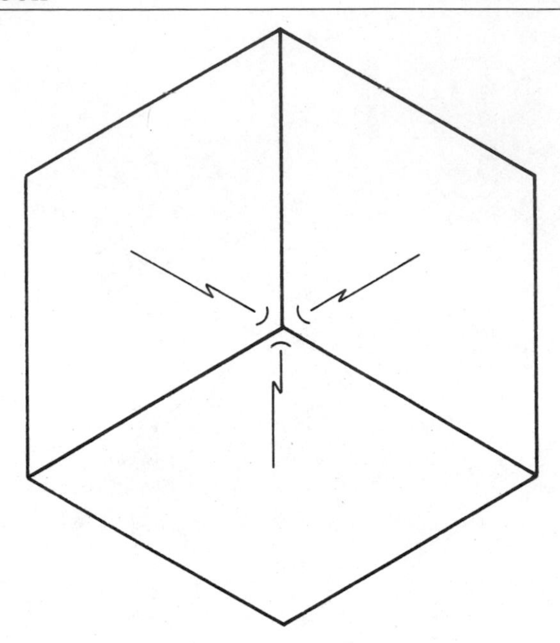

FIGURE 22.6
Increasing Capacity by Splitting Cells

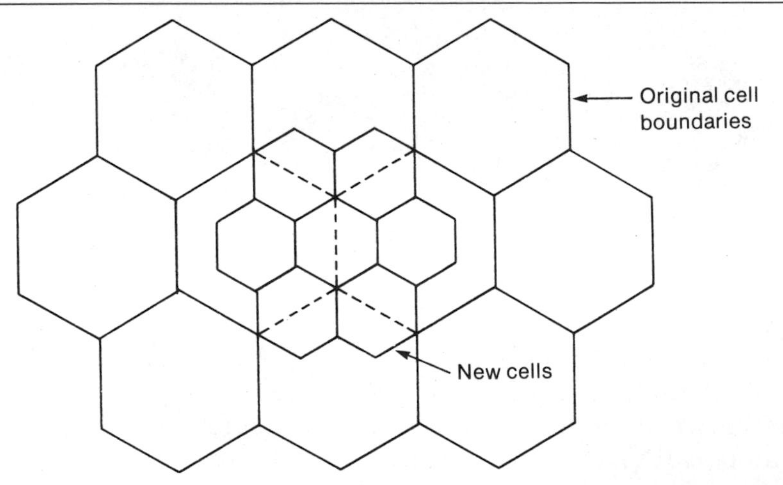

changing the antenna patterns. A fourth method of increasing capacity is to borrow unused channels from an adjacent cell.

A technique using *supervisory audio tones* (SAT tones) is used to prevent a mobile unit from talking on the same frequency at separate cell sites. One of three SAT tones is sent by the base station and transponded by the mobile. If the SAT tone returned by a mobile unit is different from the one sent by the cell site, the CRSO will not accept the call.

Mobile Telephone Serving Office

The CRSO is essentially a Class 5 switching machine of the type described in Chapter 10 with a special-purpose generic program for cellular radio operation. A typical CRSO is shown in Figure 22.7. Not all CRSOs are Class 5 switching machines; some products are designed specifically for cellular radio. With the exception of AT&T Network Systems' No. 1 ESS, CRSOs are digital switching machines. The objective of most service providers is to offer cellular radio features that are essentially identical to wireline telephone features.

The CRSO is linked to the cell-site controller with data circuits for control purposes and with four-wire voice circuits for communication channels. When a call is received from the mobile unit, the cell-site controller registers the dialed digits and passes them over the data link. The concept is similar to common channel signaling as described in Chapter 8, but X.25 protocol is used. The CRSO registers the dialed digits and switches the call to the telephone network over an intermachine trunk or to another cellular mobile unit within the system. When mobile-to-mobile calls or calls from the local telephone system are placed, the CRSO pages the mobile unit by sending messages to all cell-site controllers.

The CRSO receives reports from the cell-site controller on the signal strength of each of the mobile units transmitting within the coverage area. Data is relayed to the CRSO to enable it to decide which cell is the appropriate serving cell for each active unit. The CRSO also collects statistical information about traffic volumes for allowing the system administrator to determine when additional channels are

FIGURE 22.7
DMS Mobile Telephone Exchange

Courtesy Northern Telecom.

needed. Records of usage are stored as the source for generating bills.

Mobile Units

In the mid-1980s, cellular mobile units are costly with the most expensive units selling for about $3,000. Less expensive units are marketed for about $1,000, with the price expected to decline farther in the future. The transceiver is

FIGURE 22.8
Block Diagram of a Cellular Mobile Radio Transceiver

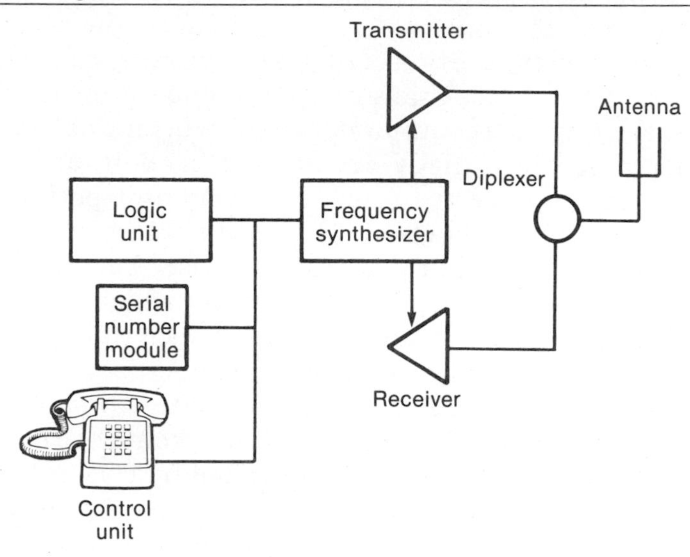

a sophisticated device that is capable of tuning all 666 channels within an area. Unlike conventional mobile transceivers that use individual crystals for setting the frequency of each channel, cellular transceivers are frequency agile. That is, they use frequency synthesizers, which are circuits that generate the end frequency by multiplying from a reference frequency. When cellular theory was first examined in 1947, the science of solid state electronics was undeveloped and control circuitry was electromechanical, bulky, and consumed considerable power. Transceivers of today are small enough to carry in a briefcase and are easily powered by rechargeable batteries.

The major components of a typical mobile unit are shown in the block diagram in Figure 22.8. The transmitter and receiver are coupled to the antenna through a *diplexer,* which is a device that isolates the two directions of transmission so the transmitter does not feed power into its own receiver. The transmit and receive frequencies are generated by the frequency synthesizer under control of the logic board. The synthesizer generates a reference frequency from a highly stable oscillator and divides, filters, and multiplies it to generate the required frequency.

The most complex part of the mobile unit is the logic equipment. This system communicates with the cell-site controller over the control channel and directs the other systems in the functions of receiving and initiating calls. These functions include recognizing and responding to incoming signals, shifting the rf equipment to the working channel for establishing a call initially and during handoff, and interpreting users' service requests. The logic unit periodically scans the control channels and tunes the transceiver to the channel with the strongest signal. The user communicates with the logic unit through a control unit, which consists of the handset, dial, display unit, and other such functions that approximate a conventional telephone set.

The unit also includes a power converter to supply the logic and rf equipment with the proper voltages from the battery source. The transmitter is limited by FCC rules to 7 watts output, but many mobile units use 3 watts or less to reduce battery drain. Battery drain is particularly important in handheld or portable units. The FCC segregates mobile units into three power classes ranging from 0.6 watts to the maximum power permitted.

Another module generates the unit's 32-bit binary serial number electronically to prevent fraudulent or unauthorized use of a mobile unit. The serial number is communicated during each call to the cell-site controller for comparison with a data base maintained by the CRSO. The unit's 10-digit calling number and a station class mark are also built into memory and transmitted to the cell-site controller with each outgoing call. The class mark identifies the station type and power rating.

Mobile Telephone Features

A full description of mobile telephone features is beyond the scope of this book. The following is a brief description of the most popular cellular radio mobile features. Among the most important features are those that improve safety by enabling the user to operate the system while driving.

A/B System Selection. The A/B System selection offers a choice between wireline and nonwireline carrier bands.

Call Timer. The call timer displays elapsed time of calls, displays accumulated time to aid in estimating billing costs, and provides a preset interval timer during a call.

Dialed Number Display. The dialed number is displayed in a readout on the handset. Misdialed digits can be corrected before the number is outpulsed.

Last Number Dialed. Like a conventional telephone, this feature stores the last number dialed so it can be recalled with a touch of a button.

Muting. The handset is cut off with a button so the distant party cannot hear a private conversation within the car.

On-Hook Dialing. This feature allows the user to dial a number while the unit is on hook. In conjunction with a dialed number display, the number can be pulsed into a handset and reviewed before it is sent.

Repertory Dialing. A list of telephone numbers is stored in the unit and selected by dialing one or two digits.

Scratch Pad. This feature allows storing numbers in a temporary memory location by entering them from the keypad.

Security. Because of the high price of the units, theft is a potential problem. Units are internally equipped to report their serial number to the cell-site controller each time a call is placed. This identifies the specific transmitter a call is placed from and makes it difficult to change the telephone number for fraudulent reporting. Other security features include both key locks and digital locks that must be activated before the unit can be operated. Some units are designed for easy removal so they can be stored in the car's trunk or glove compartment.

Self-Diagnostics. Internal diagnostics indicate trouble in transceiver and control unit to aid in rapid trouble shooting.

Speaker. Some units include a speaker and remote microphone so conversations can be monitored by others in the car and to enable hands-free operation of the unit.

Special Signaling. A "call in absence" indicator turns on a light when a call is received but not answered. Auxiliary signals are optionally provided to honk the horn or turn on the lights when a call arrives while the unit is unattended.

Cellular Radio Services

Cellular radio is intended to duplicate the services of wire line carriers as nearly as possible. Some carriers offer operator services to allow users to place collect, third number, and credit card calls. An attendant service is also required so roaming users can check in when they leave one carrier's service area and enter another. Although the mobile unit automatically identifies the user, the different jurisdictions require identification and registration of the user for billing purposes.

RADIO PAGING

Paging is a radio application that is considerably less sophisticated and costly than cellular radio. Developed under the centralized transmitter plan of conventional mobile telephone service, radio paging offers dial access from a wireline telephone to actuate a signal in a pocket receiver. Portable paging units sell for $100 to $450 plus a monthly fee. Two-way communication is not provided over paging systems. The user receives a page and dials a predetermined telephone number to receive the message. Some units are equipped with a dual number system so a user can respond to one wireline telephone for one signal and a second number for the other signal. The most sophisticated systems have calling number display capability so the user can return the call to the specific calling telephone number.

MOBILE DATA TRANSMISSION

Data can be transmitted over cellular mobile radio. The interruption of about 200 milliseconds when a mobile unit is handed off from one cell to another can be detrimental to data transmission, but otherwise the service is equivalent to wireline data transmission. The mobile unit is equipped with a special modem, sometimes built into the radio equipment, and a data terminal intended for vehicular operation. Mobile terminals are also used on private radio systems for such

FIGURE 22.9
Metropolitan Radio Data Network

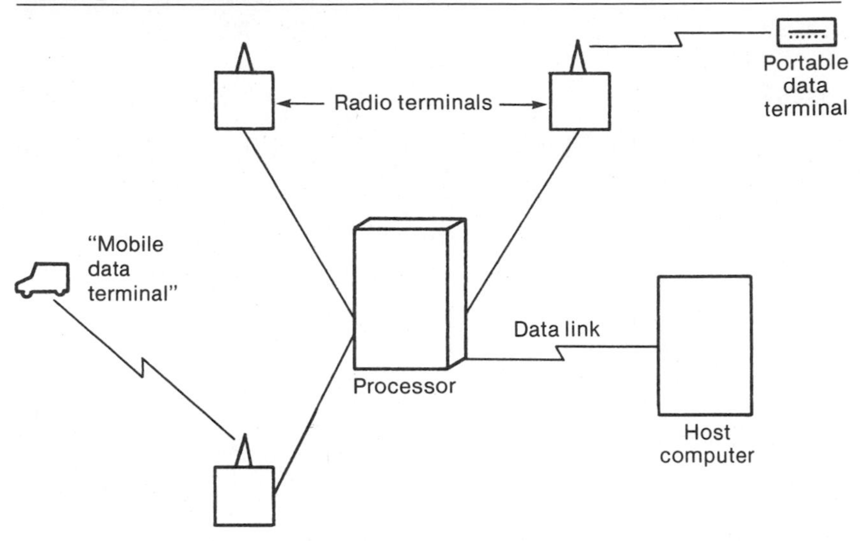

purposes as linkage between a law enforcement agency and a vehicular or law enforcement data base.

Handheld mobile data equipment is also available for use in a metropolitan data network as shown in Figure 22.9. The mobile data terminal is a small handheld terminal that operates on a single full-duplex channel in the 800 MHz frequency range. Data is transmitted by keying into a small compact keyboard. Return messages are received in a portable display. A processor controls multiple radio channels, which are operated on the same frequency within the metropolitan area. In the Motorola KDT™ system, data is transmitted at 4.8 kb/s. The system is designed for two-way applications such as dispatching.

STANDARDS

Mobile radio standards are issued by the FCC in their rules and regulations in the United States. FCC rules establish the authorized frequencies, power levels, bandwidth, frequency stability, signaling formats, and other such variables in the public mobile service. Cellular radio standards are outlined in the Cellular Mobile/Land Station Compatibility Specification issued by the FCC and EIA in 1981. Internationally, CCIR issues mobile radio recommendations.

ANSI

ANSI/IEEE 145 Definition of Terms for Antennas.

ANSI/IEEE 149 Test Procedure for Antennas.

ANSI/IEEE 377 Measurement of Spurious Emission from Land-Mobile Communication Transmitters.

Bell Communications Research

Compatibility Bulletin 146 Specifications for Channels Provided to Cellular Carriers.

Technical Reference 43301 Bell System Domestic Public Land Mobile Radio Service Interface Specifications for Customer-Provided Dial Mobile Terminals—Preliminary.

Technical Reference 43303 Bell System Public Switched Telephone Service Interconnection Criteria for Domestic Public Land Mobile Radio Service, Domestic Public Cellular Telecommunications Service and Maritime Radio Service—Preliminary.

CCIR

Volume I of 15th Plenary Assembly of the CCIR Spectrum Utilization and Monitoring.

Volume VIII of 15th Plenary Assembly of the CCIR Mobile Services.

EIA

RS-220-A Minimum Standards for Land-Mobile Communication Continuous Tone-Controlled Squelch Systems.

RS-316-B Minimum Standards for Portable/Personal Radio Transmitters, Receivers and Transmitter/Receiver Combination Land Mobile Communications FM or PM Equipment.

RS-329-A Minimum Standards for Land-Mobile Communication Antennas Part I Base or Fixed Station Antennas.

RS-329-I Minimum Standards for Land-Mobile Communication Antennas Part II Vehicular Antennas.

RS-374A Land Mobile Signaling Standard.

RS-450 Standard Form for Reporting Measurements of Land Mobile, Base Station, and Portable/Personal Radio Receivers in Compliance with FCC Part 15 Rules.

FCC Rules and Regulations

Part 22 Public Mobile Services

Sub Part G Public Land Mobile Service.

Sub Part K Domestic Public Cellular Radio Telephone Service.

Sub Part K Offshore Radio Telephone Service.

Part 90 Private Land Mobile Radio Services

Sub Part B Public Safety Radio Services.

Sub Part C Special Emergency Radio Services.

Sub Part D Industrial Radio Services.

Sub Part E Land Transportation Radio Services.

APPLICATIONS

Cellular radio is rapidly becoming a commodity for those who can afford it, much like ordinary telephone service. This section includes considerations users of cellular radio service should evaluate, both in terms of the service itself and the mobile radio equipment.

Evaluation Criteria

Cellular mobile telephone equipment is evaluated on much the same basis as other telecommunications equipment. Reliability and the ability to obtain fast and efficient service is a paramount concern. Cost is also an important consideration. If cellular radio follows the pattern of similar services and equipment, the price can be expected to remain high until the demand develops. Costs will likely decrease with a second generation of equipment, resulting in significant price reductions in first generation equipment, which will likely lack features provided in the second generation. A cost-related decision is whether to buy or lease the mobile unit. Such considerations are beyond the scope of this book but should be closely evaluated based on the expectation of future cost reductions and obsolescence.

Service Quality

The FCC established the frequency allocation plan for cellular radio with the intention of encouraging competition. In areas where competition exists, differences in price may be reflected in differences in service quality. If the carriers offer coverage demonstrations, it is advisable to try the service before making a commitment to ensure that the coverage area is satisfactory and that channels are not overloaded. If demonstrations are not provided, it may be possible to rent a mobile-equipped car to evaluate service quality. If extensive use will be made of portable units, coverage checks should be made to ensure that the unit will deliver the needed performance.

Usage Charges

Cellular radio charges are based on duration of both originating and terminating calls. Charges are made for total air time, and other message charges such as long distance are added on. Therefore, both originating and terminating calls are charged to the terminating mobile number; outgoing long-distance usage is billed twice—once for the air time and once for the wireline long-distance service. Extra charges may be imposed for roaming between service areas.

GLOSSARY

Cell: A hexagonal-shaped subdivision of a mobile telephone service area containing a cell-site controller and radio frequency transceivers.

Cell-site controller: The cellular radio unit that manages radio channels within a cell.

Cellular Geographic Serving Area (CGSA): A metropolitan area in which cellular radio licenses are granted by the FCC.

Cellular Radio Switching Office (CRSO): The electronic switching machine that switches calls between mobile and wireline telephones, controls handoff between cells, and monitors usage. This equipment is known by various trade names among different manufacturers.

Diplexer: A device that couples a radio transmitter and receiver to the same antenna.

Frequency agility: The ability of a cellular mobile telephone to shift automatically between frequencies.

Handoff: The process of changing radio channels when a mobile unit moves from the coverage area of one cell to another.

Improved Mobile Telephone Service (IMTS): A type of mobile telephone service that allows direct dialing between wireline and mobile units without operator intervention.

Nonwireline: A term describing cellular radio service providers who are not operating telephone companies.

Roaming: A mobile telephone user who uses service in more than one basic service area.

Sectoring: The process of dividing a cell into 180-, 120-, or 60-degree segments to increase capacity of the radio frequencies assigned to the cell.

Standard Metropolitan Statistical Area (SMSA): A metropolitan area consisting of one or more cities as defined by the Office of Management and Budget and used by the FCC to allocate the cellular radio market.

Voting receivers: A group of mobile base station receivers operating on the same frequency with a control unit to pick the best signal from among them.

Wireline: A term describing cellular radio service providers who are operating telephone companies. Also used to distinguish between mobile radio users and conventional telephone users in a standard telephone-to-mobile call.

BIBLIOGRAPHY

Datapro. *Datapro Communications Solutions.* Delran, N.J.: Datapro Research Corporation, 1983.

Prentiss, Stan. *Introducing Cellular Communications.* Blue Ridge Summit, Pa.: TAB Books, 1984.

MANUFACTURERS OF MOBILE RADIO EQUIPMENT

Cellular Central Office Equipment

AT&T Network Systems

Ericsson Inc., Communications Division

NEC America Inc. Switching Systems Division

Motorola Communications & Electronics, Inc.

Northern Telecom Inc., Digital Switching Systems

Cellular and IMTS Mobile Equipment

BBL Industries

CIT-Alcatel, Inc.

General Electric Mobile Communications Division

Glenayre Electronics

Harris Corporation, RF Communications Group

E.F. Johnson Company

NEC America Inc. Switching Systems Division

Motorola Communications & Electronics, Inc.

M/A-COM Land Mobile Communications, Inc.

NEC America Inc., Radio & Transmission Division

CHAPTER TWENTY-THREE

Testing Principles

Telecommunications network testing has two objectives:

- To measure operating variables that have been designed into the network and to confirm that design objectives have been achieved.
- To locate faulty network elements by sectionalizing and isolating defective items of equipment.

The first category of tests is usually called acceptance or proof-of-performance tests. These are conducted to establish a data base that can later be the basis for fault-locating tests. The second category of tests usually does not require the depth and sophistication of the former. It is normally sufficient to measure a single variable such as circuit loss when searching for a defective element. When the cause of excessive loss is determined, the tester will likely also locate the source of poor frequency response or excess noise. Therefore, the tests used for routine circuit verification can be reduced to single-frequency measurements, which in turn can be automated to survey a complex network rapidly and reliably.

Network reliability has increased dramatically over the years in spite of the ever-increasing complexity of the network. Nevertheless, every element of a telecommunications system is subject to failure, and when a circuit or service is in trouble, it is often difficult to determine what portion of the network is at fault. Service irregularities and end-to-end tests can reveal the presence of a fault, but the defective element can be identified only by testing to sectionalize the trouble.

As we evolve into an integrated digital network, testing will be simplified in comparison with today's networks. A bit stream will originate on the user's premises and be carried over the network to the distant end without crossing many of the physically wired interfaces that circuits cross today. Circuit operation will either be near-perfect or catastrophically failed, although with properly designed diversity, total failures should be rare. Furthermore, the advent of packet switching networks that automatically confirm the error-free reception of each packet and retransmit flawed packets until correctly received improves data transmission reliability. With today's networks, testing is complicated by several factors:

- Network elements are obtained from multiple vendors.
- Trouble must be sectionalized by testing to interfaces between vendors.
- The responsibility for impairments such as high noise and data errors is often unclear, and vendors are not quick to claim responsibility.
- Incompatibility at interfaces may arise under some conditions, and lacking interface standards, the user may be left to negotiate the solution between vendors.

Until the advent of low-cost processors, network testing was almost entirely manual. Some electromechanical equipment was used (and in some cases still is) to test switching systems and circuits. Most of this equipment has been replaced by automatic equipment for bulk testing, but many individual tests are still made manually. Tests are designed specifically for a particular class of equipment; the testing principles are essentially the same whether the test is manual or automatic. This chapter describes the principles of testing the major components of a telecommunications network, and where appropriate, the chapter discusses the results that should be expected.

Transmission and switching systems require specialized test equipment to perform general tests and some tests that are unique to the manufacturer. For example, two types of tests are required on microwave radio: system performance and service. FCC-required system performance tests include frequency and bandwidth measurements and are similar for both analog and digital microwave. However, service criteria for the two types of microwave is different. Analog microwave service is closely related to its channel noise performance; the primary service indicator of digital microwave is its error rate.

Switching systems, nearly all of which contain microprocessors, require test systems that are similar to those used for maintaining computers. Because of the specialized nature of these systems and their complexity, they are excluded from this discussion.

TEST ACCESS METHODS

A common problem for all types of circuit tests is how to obtain test access to the circuit. Access is obtained by one of three methods:

- Manual access through jacks, test points, or distributing frames.
- Switched access.
- Permanently wired test equipment.

Manual access methods include jacks and terminal strips that allow direct connection of test equipment to the circuit conductors. The primary drawbacks to manual access methods are that they are labor intensive and often result in delays before a technician can be dispatched to a remote location. Switched access methods provide circuitry for connecting remote test equipment to a circuit from a distant location. Switched circuits are often accessible through the switching machine. Dedicated circuits may be connected through a separate switching matrix that connects the transmission path to colocated test equipment. Test equipment may also be permanently wired into a circuit or designed into the terminating equipment. For example, some modems provide a low speed channel for performing tests while the circuit is in operation.

A fundamental principle in designing a network is that it must include a plan for testing, sectionalizing, and clearing trouble. It is sometimes possible to obtain network services from a vendor who assumes full end-to-end testing and maintenance responsibility, but this may be more expensive for many organizations than the alternative of providing that service for itself. In any event, it is rare that all of an organization's communications needs can be supplied by a single vendor who assumes full responsibility for performance.

ANALOG CIRCUIT TESTING PRINCIPLES

Certain tests are common to all elements of the network. The testing equipment may vary widely, but the principles are identical. The tests discussed in this section are applied to both voice frequency and data circuits, and where indicated, they may also be applied to broadband transmission facilities. Analog data circuits require all the tests described in this section, but many impairments that cause trouble on data circuits are undetectable when the circuit is used for voice. The differences between data and voice tests are pointed out in the following discussion.

Loss Measurements

Loss (or gain) measurements are made by injecting a signal source into the input of a circuit and reading the result at the output. Voice frequency measurements are made at a nominal frequency of 1,000 Hz; 1,004 Hz is actually used to prevent interference with digital transmission equipment. The frequency source or *oscillator* generates a pure audio tone. The signal is applied across the input of a circuit and measured at the output as shown in Figure 23.1. The difference between the two readings is the loss or gain of the circuit. This same principle is applied in measuring loss on radio and carrier systems. High-frequency signal sources and level measuring sets are used to measure loss at baseband or at carrier or radio frequencies.

In addition to single-frequency loss measurements, the gain-frequency or attenuation distortion characteristics of most apparatus is also of interest. Attenuation distortion is a variation in loss at the audio frequencies within a circuit passband. It is measured by sending and receiving tones across the audio passband or by sweeping the circuit with a signal that varies continually in frequency. A "swept" channel can be displayed on a cathode ray tube to present a visual display of the circuit passband. Similar tests are made at carrier and radio frequencies to examine the linearity of a broadband facility. Spectrum occupancy and the presence of extraneous or high-level frequencies is readily determined with a spectrum analyzer, a test instrument that displays the frequency domain of a signal.

Transmission level measurements are typically expressed as dB related to 1 milliwatt (dBm) as explained in Chapter 2. To ensure accurate results, test equipment must be calibrated by comparing the test set to a reference frequency and level standard.

Noise Measurements

As explained in Chapter 2, noise is expressed in dB compared to a reference noise level of −90 dBm. Noise measurements are made with a measuring set that reads the noise power through a weighting filter and registers it on a digital or meter readout. Voice frequency circuits are normally measured through a C message filter, which approximates the human perception of the interfering effect of noise on a telephone conversation. Special service and program lines are measured through flat weighting filters, which equally weight all frequencies within the passband.

When power is applied to a circuit, noise may increase because the greater loading reacts with nonlinearities in the

FIGURE 23.1
Transmission Measurement on a Circuit between TLPs

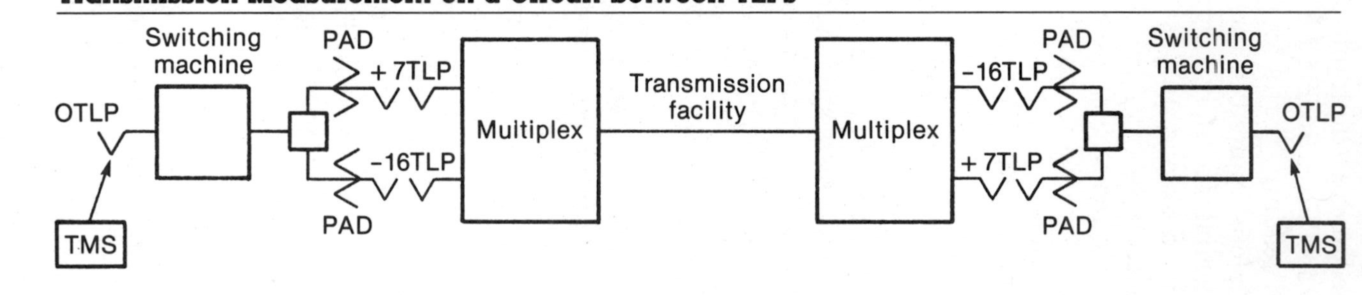

circuit to increase the intermodulation noise. This effect is measured by injecting a single frequency into the circuit at the sending end and removing it with a filter at the receiving end so that only the intermodulation products remain. Circuits that are equipped with companders exhibit very low noise in the idle state because the receiving end expander has no signal and consequently reduces the volume level of its output. To obtain a realistic estimate of the noise that will be interfering with a live conversation, it is necessary to place a test tone of appropriate level on the circuit to activate the compressor and expander combination and in turn permit a useful assessment of the circuit's noise performance. The test tone, of course, must be removed at the receiving end by a narrow band filter so that only the residual noise under approximately active conditions is measured. When the residual noise is measured through a C weighted filter by this technique, it is designated as *C notched noise.*

Impulse noise measurements are of particular interest in evaluating a circuit for data transmission. A common source of impulse noise is electromechanical relay operations that induce a sharp spike of noise into a circuit. Impulse noise has little effect on voice communication, but it can be devastating to data transmission. An impulse noise measuring set establishes a threshold level and counts the number of impulses that exceed the threshold. The set also includes a timer so impulses above reference level can be measured over a fixed period of time.

Envelope Delay

Envelope delay expresses the difference in propagation speed of the various frequencies within the audio pass band. It is made with an envelope delay measuring set that applies a pair of frequencies at the originating end of the circuit and registers the relative delay of each frequency at the receiver as the test signals are moved through the transmission band. Envelope delay affects high-speed data but has no discernible effect on transmission over circuits used for voice transmission.

Return Loss

Return loss measurements determine the amount of energy returned from a distant impedance irregularity such as the

FIGURE 23.2
Phase Jitter

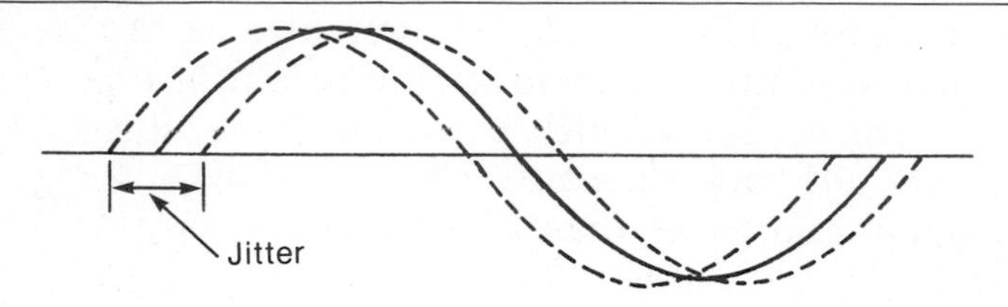

mismatch between a two-wire circuit and the network on the balancing port of a hybrid. Return loss measurements are made over a band of frequencies between 500 and 2,500 Hz by transmitting a white noise signal source on the circuit as described in Chapter 2.

Phase Jitter

Phase jitter is any variation in the phase of a signal as shown in Figure 23.2. Jitter is measured with a special test set that detects phase variations in a steady state tone injected at one end of a circuit and measured at the other.

Peak-to-Average Ratio

A peak-to-average ratio (P/AR) test gives an effective index of the overall quality of an analog circuit for data transmission. A P/AR transmitter sends a repetitive pulse consisting of a complex combination of signals. The P/AR receiver measures the envelope of the received signal and indexes the circuit quality on a scale of zero to 100. The P/AR measurement is an attempt to provide a single composite indicator of circuit quality.

Harmonic Distortion

As described in Chapter 2, a harmonic is a multiple of the fundamental frequency applied to a circuit. For reliable data communication, the second harmonic must be at least 25 dB lower than the fundamental frequency, and the third harmonic must be 28 dB lower than the fundamental. Harmonic distortion is measured with a special test set that reads directly the amount of second and third harmonic distortion.

SUBSCRIBER LOOP MEASUREMENTS

As discussed in Chapter 7, the subscriber loop is the part of a telecommunications circuit most susceptible to transmission impairment. Tests on the subscriber loop can be grouped into three categories: manual tests from a local test desk attached to a switching machine, automatic tests through a switching machine, and manual tests from the user's premises or intermediate points on the cable to the central office. Subscriber loop tests differ from trunk tests in that full loss and noise measurements on a loop require manual or remotely controlled tests from the user's end of the circuit. Equipment to perform remote measurements can be justified only in locations with large concentrations of high-cost circuits. Because such tests are labor intensive, they are normally made only to locate the cause of repeated trouble.

Local Test Desks

A *local test desk* (LTD) is a manually operated system that accesses a cable pair through the switching machine or over a trunk to the main distributing frame. The latter connection requires manually placing a device called a *test shoe* at the main distributing frame to access the cable pair directly. Many local loops are inaccessible through switching machines because they are dedicated to special services. These circuits may be tested from an LTD by use of an MDF shoe or by the use of switched access connectors and special service testing apparatus as described later. The principal faults that occur to local cable pairs are *shorts, crosses, grounds,* and *opens* as shown in Figure 23.3. Crosses and grounds cause a high level of noise on the cable pair and often result in unwanted voltage (called *foreign EMF*) from the central office battery of the interfering pair.

A diagram of a properly functioning circuit and its access from an LTD is shown in Figure 23.4. The LTD accesses the line through a circuit in the switching machine that connects the LTD to the line without ringing the telephone. The LTD evaluates the circuit by measuring capacitance between the two conductors of a cable pair and from each conductor to ground. Because the pair consists of two paralleling con-

FIGURE 23.3
Common Cable Faults

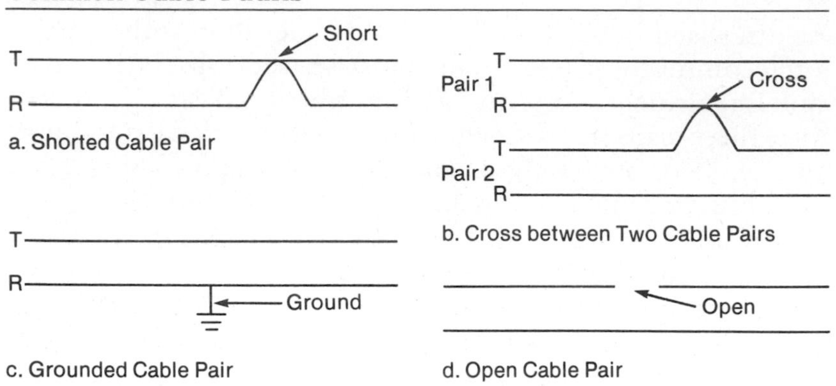

FIGURE 23.4
Cable Testing Circuit

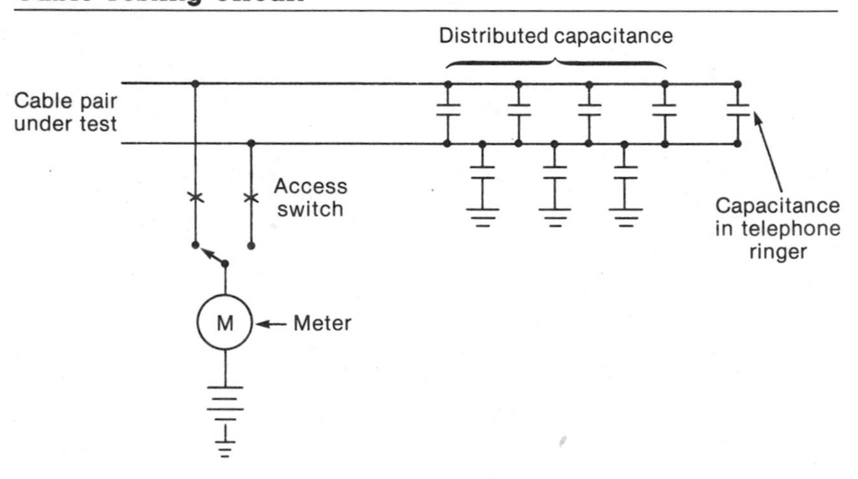

ductors separated by an insulator, the conductors form the two plates of a capacitor. When test voltage is applied, the meter on the LTD registers the amount of current flowing in the line as the capacitor charges. When the line reaches full charge, current ceases to flow and the meter returns to zero. An experienced test board operator can diagnose the condition of the line from the amount of "kick" to the meter. Shorted and open lines kick only a small amount. Properly functioning lines kick to a greater degree, with the amount of kick proportional to the length of the cable pair and the number of telephone sets installed on the line (the measurement charges the capacitor in the telephone ringer at the same time the cable pair is charged, so more telephones increase the amount of capacitance).

Foreign EMF is detected by using the LTD voltmeter to measure the battery from each side of the line to ground. When the line is connected to the testing circuit, the central office battery is removed. Any voltage observed is coming from an interfering source such as a cross with a line that is connected to central office battery. When water enters a cable, it often provides a conductive path permitting foreign EMF to reach many lines simultaneously.

The LTD is also able to make loss measurements on the line. Loss measurements require someone at the distant end of the circuit. Noise measurements can be made at the central office end of a circuit, but because the noise signal is attenuated by the loss of the cable pair, noise measurements made at only one end of a circuit do not fully express the degree to which noise interferes with the user.

Automatic Testing

Automatic test equipment accesses local loops through the switching machine to perform tests similar to LTD tests. The least complex type of automatic subscriber loop test system is the *line insulation test* (LIT) that has been used in central offices for many years. An LIT machine steps through each line in a central office, accessing the line through the switching machine and applying to each line tests similar to those applied by the LTD. LIT machines look for shorted and grounded cable pairs, and for foreign EMF. The results of the test are printed on a teletypewriter.

LIT machines are capable only of relatively crude qualitative tests and offer few clues to the nature and location of the trouble. Within the past decade, more sophisticated equipment has come into general use. This automatic test equipment also accesses the subscriber loop through the switching machine, but it is capable of testing the impedance of the subscriber loop and therefore determines more accurately what the trouble is and where it is located.

FIGURE 23.5
4TEL Central Office Line Testers

Courtesy Teradyne, Inc.

A mechanized line testing system functions by storing the electrical characteristics of normal lines and those with various faults in its data base. A remote test unit such as the one shown in Figure 23.5 is accessed manually through a console or it is driven automatically by a computer to perform routine measurements.

Manual Loop Tests

Local test desks and mechanized line testing equipment are limited in their ability to locate certain types of cable faults. Open and wet cable pairs are among the most difficult faults to locate and often require measuring interactively with a technician in the field. The *time domain reflectometer* (TDR) is a relatively new instrument that locates trouble with a high degree of accuracy in all types of cable including coaxial cable and fiber optics. It uses the same principle in cable with which radar operates in free space. A pulse is sent from the TDR. Any irregularity in the cable returns a reflected pulse that is displayed on a cathode ray tube calibrated in distance to the fault.

TDRs are excellent devices for locating all types of impedance irregularities, but their cost prevents their widespread use for less complex testing. Shorted and open cable pairs can be located by making precise resistance and capacitance measurements to a fault. Electronic instruments for these measurements are less expensive than TDRs and locate trouble with a satisfactory degree of accuracy.

Subscriber loop loss and noise measurements must be made from the user's premises to be accurate. Specialized test sets are available to measure the three variables that affect subscriber loop transmission: loss, noise, and loop current. Loss measurements are made by dialing a test signal supply in the central office. Noise measurements are made by dialing a termination or "quiet line" in the central office. Current measurements are made by measuring the off-hook current drawn by a telephone set.

Network Interface Devices

Test equipment located in the central office is often unable to distinguish between troubles located in outside cable plant and troubles in customer premise wiring and equipment. To aid in sectionalizing trouble, many types of *network interface device* (NIDs) are available and are sometimes installed at the interface between the customer's wiring and the telephone company's equipment. By applying voltage or an actuating tone from the LTD, the NID opens the line at the interface. Some types of NID also short the cable pair so tests can be made from the central office to the interface. With the NID actuated, if the trouble observed by the LTD remains, it is evidence that the fault is in the loop facilities. If the trouble disappears, the fault is in the customer premise wiring or equipment. NIDs are not universally applied because most ordinary telephone services do not experience trouble often enough to justify the cost of the NID. On data circuits, many data modems are equipped with internal test circuits that perform the same functions as a NID.

TRUNK TRANSMISSION MEASUREMENTS

The voice frequency measurements described earlier are of critical importance on all trunks. Transmission measurements are made either manually or automatically between TLPs, and adjustments are made by changing fixed loss pads or adjusting amplifier gains.

Manual Switched Circuit Test Systems

Circuit access for manual testing is obtained either through jacks wired to the circuit, by access through the switching machine, by removing jumper wires at a cross-connect point, or by switched access connectors. Either portable or fixed test equipment is used to measure transmission variables and supervision. Supervision tests, which are described in Chapter 8, are made on the signaling leads. These tests detect and register the status of supervisory leads under various conditions of circuit operation.

Manual tests are made either to a manual test board at the distant end or to responders. As described in Chapter 12, trunks may be equipped with a variety of test lines that are accessed by dialing special codes over the trunk. A typical test line is the 105, which permits two-way loss and noise measurements from an office equipped with a remote office test line (ROTL) and responder.

Automatic Switched Circuit Test Systems

In offices equipped with large numbers of circuits, the most economical way of testing is with automatic test equipment that conducts the tests under computer control. Figure 23.6 is a diagram of automatic circuit test equipment of the type discussed in Chapter 12. A typical system is AT&T Network Systems' Centralized Automatic Reporting on Trunks (CAROT). This computer-driven device actuates a ROTL, which communicates with a test line at the distant end. A full range of transmission and supervision tests can be made with this system. The results are stored in memory and printed out. Trunks exceeding design limits are automatically taken out of service.

Tests on Special Service Trunks

Tests on special service lines cannot be made with the test equipment described in this section because they are not accessible through a switching machine. Manual testing is particularly difficult in multipoint private lines because of the large number of technicians needed to sectionalize trouble. In locations with large concentrations of special services, switched access systems, illustrated in Figure 23.7, can be justified. Circuits are wired through electromechanical crosspoints that bridge the circuit on a monitoring basis or open it to sectionalize trouble. Testing terminations and signal sources can be connected to the circuit through switched access connectors. The test console, which is typically located centrally for a company or region, is connected to a near-end testing unit. Each remote location is equipped with a matching test unit, and the two units communicate over a dedicated or dialed access line.

Switched access connectors can provide the same types of access to circuits as jacks. Connectors are available to operate two, four, or six wires and allow the operator to monitor, split, test, and busy-out connections. Circuits can also be wired to enable patching hot standby spare equipment or circuits. Most systems also provide computer-supported operational functions such as logging trouble, recording circuit outage time, and maintaining trouble histories.

With a switched access testing system, a technician can obtain remote access to all points that are equipped. For example, if a multipoint circuit is singing, the technician can remotely terminate the legs of the circuit until the one caus-

FIGURE 23.6
Automatic Remote Trunk Testing System

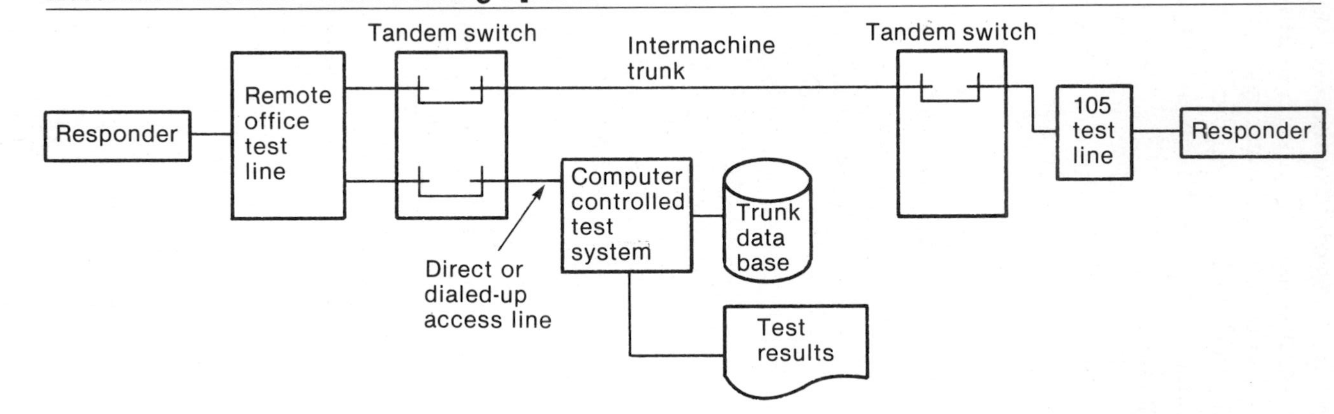

FIGURE 23.7
Switched Access Remote Testing

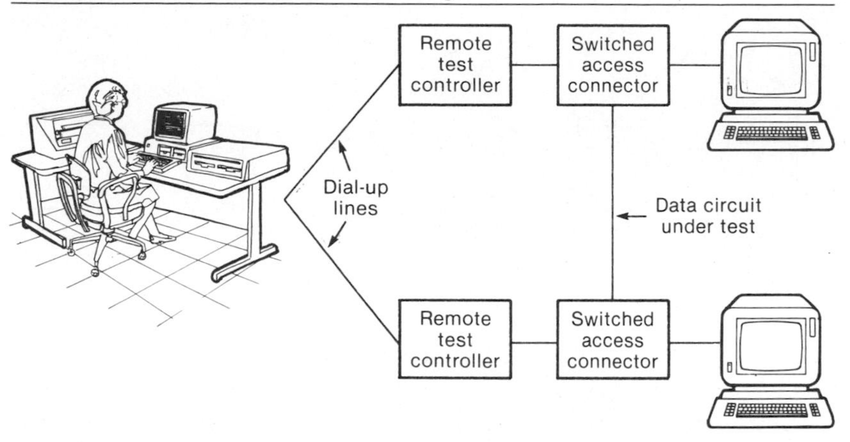

ing the trouble is found. Circuit lineup tests can be made by performing two-way transmission measurements to test equipment that is accessed through the switched access connector.

DATA CIRCUIT TESTING

Although nonconditioned data networks and voice networks share the same voice frequency facilities, data networks are more complex to test than voice networks. Voice networks are used by people who can give a qualitative analysis of what the trouble is. Data network troubles are apt to be reported as "a high error rate," with no clue as to whether the trouble is noise or momentary interruptions. Furthermore, data circuits are subject to incompatibilities such as protocol or addressing faults that can be caused by troubles in DCE, DTE, or software.

Tests are generally applied to data circuits for one of three purposes: to solve specific trouble reports, to monitor the circuit for proper operation, or to prevent trouble by detecting incipient faults with a preventive maintenance program. A large percentage of data network problems are found in the user's equipment: the software, the interface between

FIGURE 23.8
Modem Loopback Paths

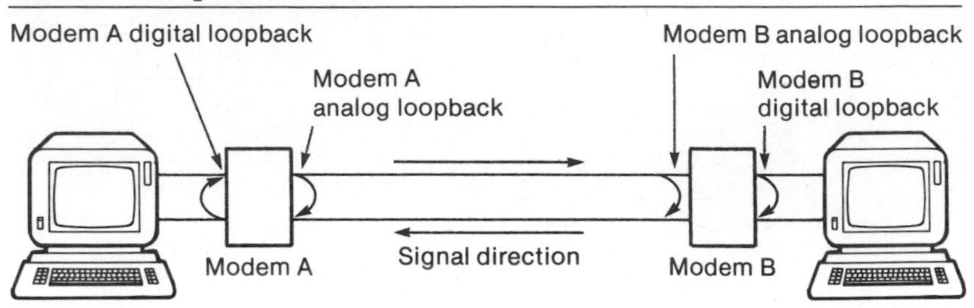

the DTE and DCE, or one of the hardware elements. Because each data network has a custom design, it is imperative that the user be prepared to perform the first elements of trouble analysis before the problem is referred to a vendor. Not only will such capability save money, it will also aid in rapid trouble restoral and may furthermore solve many problems before they are referred outside the organization.

Interface Tests

Numerous test sets are available to test the RS-232-C or equivalent interface. The test set plugs between the DCE and the DTE and provides access points to the principal signaling and communication leads. The status and polarity of the leads are often indicated with colored lamps. Switches are usually provided to reconfigure the interface. A breakout box is often used to provide access for connecting test probes and injecting test signals.

Loopback Tests

A loopback test on a full-duplex circuit is an effective way of locating faults and impairments in a data circuit. Figure 23.8 shows the different points at which a data circuit can be looped. Many modems contain integral loopback capability. If not, looping can be accomplished with an adapter plug. Tests are performed either by sending a phrase that uses all letters of the alphabet such as "fox" (the quick brown fox jumped over the lazy dogs back 1234567890), or by sending a standard test pattern and monitoring for errors with a bit error rate (BER) test set. Many modems also include built-in BER testing capability. By looping the circuit at progressively further points, the element causing the complaint will be identified.

Although loopback tests are useful in locating hard faults, some impairments such as data errors, phase jitter, and envelope delay are cumulative over the length of a circuit. Therefore the results of a loopback test must be tempered with the knowledge that the amount of impairment will be doubled because the length of the circuit is doubled. Such tests are more effectively made on an end-to-end basis.

End-to-End Tests

Loopback tests cannot be applied to half-duplex circuits because of the lack of return path. For half-duplex circuits and for some types of tests on full-duplex circuits, end-to-end tests are made with test sets operating in a master-slave relationship. The same kinds of test signals, fox or a fixed test pattern, are sent, and the error performance is observed.

Protocol Analyzers

Protocol analyzers are devices that are either inserted in or bridged on the digital portion of a data communications line to provide a character-by-character analysis of the data signal. Protocol analyzers such as the one shown in Figure 23.9 typically include several digital test functions in the same test set. Many analyzers measure bit error rate, block error rate, and percent error-free (or errored) seconds. These variables are useful for detecting the character of noise in a data circuit. For example, if a circuit has a high bit error rate but a low block error rate, it indicates that errors are coming in bursts with error-free intermediate periods. A low bit error rate with a high block error rate indicates that errors are more equally spaced.

Protocol analyzers operate in a monitor or simulation mode. In the monitor mode the test set is a passive observer of the bit stream between two devices. In the simulation mode the circuit is opened and the test set is used to simulate circuit elements. Test sets can simulate terminals, modems, and CPUs. Features commonly included are a terminal ex-

FIGURE 23.9
Protocol Analyzer

Courtesy Hewlett Packard.

erciser to facilitate diagnosing terminal troubles by sending data messages into a terminal. Some systems contain a polling generator to simulate the CPU in polling a multidrop line. Most units include some form of storage so an error sequence can be captured for off-line diagnosis. Programmable units allow the operator to set triggers or test sequences for trapping error conditions. Many such devices are capable of remembering test sequences so they can be recalled without reprogramming them.

PCM Test Sets

PCM test sets are used by both common carriers and end users to diagnose the condition of a PCM line. Test sets are available to monitor bipolar line violations (two adjoining pulses of the same polarity). Bipolar violations, however, are not available at the terminals of a high-speed data line (a DS-1 bit stream, for example) that transits any multiplexing equipment, so other indicators such as framing errors must be relied upon for performance measurement.

Technical Control

Technical control is a term applied to centralized network management and control systems that are used in large networks to monitor status and manage capacity. These systems include provisions for accessing data circuits by jacks or computer controlled switched access similar to that described in the automatic voice circuit testing section of this chapter. Technical control centers include alarm reporting, trouble history, and usually include a mechanized inventory of circuit equipment in addition to providing testing capability.

STANDARDS

Standard setting agencies issue standards on testing methods and procedures but generally do not produce standards on test equipment. Instead, manufactures design test equipment to their own specifications to conform to the standards of the circuits under test.

Bell Communications Research

Compatibility Specification 106 Guidelines relative to interface provisions and configurations to ensure compatibility between Bell Operating Company Automatic Transmission Measuring Systems (ATMS) and similar systems used by Independent Telephone Companies.

Technical Reference 55020 Subscriber Loop Transmission Test Set.

Technical Reference 60101 Test Lines—General Specifications.

PUB 53337 Transmission Test Requirement Tables—Description.

CCITT

G.222 Noise objectives for design of carrier-transmission systems of 2,500 km.

G.224 Maximum permissiblc value for the absolute power level (power referred to 1 milliwatt) of a signaling pulse.

G.225 Recommendations relating to the accuracy of carrier frequencies.

G.441 Permissible circuit noise on frequency-division multiplex radio-relay systems.

G.445 Noise objectives for communication-satellite system design.

J.62 Single value of signal-to-noise ratio for all television systems.

J.63 Insertion of test signals in the field-blanking interval of monochrome and color television signals.

J.64 Definitions of parameters for automatic measurement of television insertion test signals.

J.65 Standard test signal for conventional loading of a television channel.

J.74 Methods for measuring the transmission characteristics of translating equipments.

M.717 Testing point (transmission).

M.718 Testing point (line signaling).

M.731 Subjective testing.

M.732 Signaling and switching routine maintenance tests and measurements.

M.733 Transmission routine maintenance measurements on automatic and semiautomatic circuits.

O.21 CCITT automatic transmission measuring equipment ATME No. 1 (for telephone-type circuits).

O.22 Specification for the CCITT automatic transmission measuring and signaling testing equipment ATME No.2.

P.10 Vocabulary of terms on telephone transmission quality and telephone sets.

P.62 Measurements on subscribers' telephone equipment.

P.63 Methods for the evaluation of transmission quality on the basis of objective measurements.

P.64 Determination of sensitivity/frequency characteristics of local telephone systems to permit calculation of their loudness ratings.

P.71 Measurement of speech volume.

P.74 Methods for subjective determination of transmission quality.

Q.43 Transmission losses, relative levels.

Q.44 Attenuation distortion.

Q.330 Automatic transmission and signaling testing.

Q.331 Test equipment for checking equipment and signals.

T.20 Standardized test chart for facsimile transmissions.

T.21 Standardized test chart for document facsimile transmissions.

EIA

RS-189-A Encoded Color Bar signal.

RS-455 Standard Test Procedures for Fiber Optic Fibers, Cables, Transducers, Connections, and Terminating Devices (with 27 addenda covering specific tests).

APPLICATIONS

Anyone who has reviewed the advertisements for testing systems and equipment is aware that the market is flooded with alternatives. Any organization that owns or is planning to develop a telecommunications network must attend to testing strategy at the outset. After the overall strategy has been developed, individual test systems can be selected.

In general, the more sophisticated and expensive the test equipment, the higher the skill level or the more elaborate the software that is needed to obtain and interpret the results. Equipment should be acquired to match the level of communications expertise within the organization. It is a waste of money to purchase sophisticated equipment that is beyond the capabilities of the work force, and it is also a waste of money to buy simple equipment when more complex equipment can rapidly pay for itself in the hands of an experienced technician.

Evaluation Criteria

Accuracy

Accuracy and stability are of concern in all testing equipment and systems. This is of particular importance with analog test equipment, which must be held to precise level and frequency standards. Improperly calibrated test equipment and the failure to match the impedance of test equipment to the circuit are frequent causes of inaccuracy in measuring circuits and setting net loss. Every testing program must, therefore, include a procedure to ensure test equipment accuracy.

Portability

Test equipment is generally produced in three levels of portability: handheld, portable, and fixed. Portable test equipment is packaged in cases for mobility, but it is often heavy enough that it is not easily moved. Fixed test equipment is mounted in relay racks and is used in test centers where circuits are accessed through jacks or switched access connectors. The type of equipment selected for an application is based on the need for portability and the number of functions included in a single package. Handheld test equipment is generally capable of equally accurate results compared to portable or fixed equipment, but several units may be needed to perform the functions that are contained in one package with larger equipment.

Analog versus Digital Tests

On dedicated data circuits, tests can be made between either the analog or the digital side of modems. Digital tests are of little value in finding totally failed circuit conditions and are of no value in voice frequency circuit tests. Analog tests are useful only on dedicated circuits or in verifying the condition of a circuit up to the point of interface with a common carrier. There is little reason for a user to conduct analog tests over a dial-up circuit. When trouble is experienced on dial circuits, equipment can be tested to the interface and then referred to the common carrier.

Analog test sets are available to measure virtually all transmission variables in a single package. Measurements include attenuation distortion, intermodulation distortion, steady state and impulse noise, peak-to-average ratio, envelope delay, and phase jitter. Several test sets combine all these measurements into the same unit. These test sets are more flexible and more expensive than single-purpose units.

Centralized versus Distributed Testing

As networks become more complex, the usefulness of centralized testing becomes more apparent. From a central location where circuit status is continually monitored, reports of congestion and blockage can be most efficiently handled. The primary concern in determining whether to centralize is the availability of switched access connections to the circuits under test. If only jack access is provided, centralized testing is useful only on circuits terminating in the central location. In general, the larger and more complex the network, the greater the benefit from centralized testing. Also, the larger the network the greater the benefit from automated tests.

Protocol Flexibility

The cost of protocol test equipment is proportional to its flexibility in testing various data protocols. In selecting this type of test set, the present and future planned protocols should be examined and test equipment acquired that has the necessary capability without paying for unneeded features. Programmable test sets have the greatest flexibility in implementing new protocols. For example, CCITT is preparing new X.25 recommendations which, together with the number of options in X.25, make programmability a highly desirable feature in data test sets.

Security

A critical feature of any remote testing system is its provision for preventing unauthorized access. As these systems enable monitoring and testing of operating circuits, it is very important that unauthorized access be prevented. The most secure form of access is the use of dedicated circuits between the test console and the remote units, but over long distances, dedicated access is uneconomical. Access to long-haul facilities is usually provided over dial-up units. Where dial-up circuits are used, the telephone number of the remote units is difficult to keep secret, so means must be provided to prevent unauthorized access. If the "handshake" between the master and remote unit is complex, this may deter anyone who lacks knowledge of the protocol. Passwords are also useful but far from foolproof. The most effective method is to use a dialer at the remote to dial the master back from the remote. The problem with this method is that it allows testing from only one master location to each remote.

GLOSSARY

C notched noise: A measurement of C message-weighted noise in a circuit with a tone applied at the far end and filtered out at the near end.

Cross: A circuit impairment where two separate circuits are unintentionally interconnected.

Data line monitor: A data line impairment-measuring device that bridges the data line and observes the condition of data, addressing, and protocols.

Foreign EMF: Any unwanted voltage on a telecommunications circuit.

Ground: An impairment that exists when a circuit is unintentionally crossed with a grounded conductor.

Impulse noise: High-level and short-duration noise that is induced into a telecommunications circuit.

Line insulation test (LIT): A central office testing system that automatically measures each line for shorts, grounds, and foreign EMF.

Local test desk (LTD): A testing system that is used to access local loops and central office line equipment from a central location.

Loopback test: A test applied to a full-duplex circuit by connecting the receive leads to the transmit leads, applying a signal, and reading the returned test signal at the near end of the circuit.

Network interface device (NID): A device wired between a telephone station protector and the inside wiring to isolate the customer-provided equipment from the network.

Open: A circuit impairment that exists when one or both conductors are interrupted at an unintended point.

Oscillator: A device for generating an analog test signal.

Peak-to-average ratio (P/AR): An analog test that provides an index of data circuit quality by sending a pulse into one end of a circuit and measuring its envelope at the distant end of the circuit.

Short: A circuit impairment that exists when two conductors of the same pair are connected at an unintended point.

Switched access: A method of obtaining test access to telecommunications circuits by using electromechanical circuitry to switch test apparatus to the circuit.

Technical control center: A testing center for telecommunications circuits. The center provides test access and computer-assisted support functions to aid in circuit maintenance.

Test shoe: A device that is applied to a circuit at a distributing frame to gain test access to circuit conductors.

Time domain reflectometer (TDR): A testing device that acts on radar-like principles to determine the location of metallic circuit faults.

BIBLIOGRAPHY

American Telephone & Telegraph Co. *Notes on the Network.* 1980.

______. *Telecommunications Transmission Engineering.* Bell System Center for Technical Education, vol. 1, 1974; vol. 2, 1977; and vol. 3, 1975.

Freeman, Roger L. *Telecommunication System Engineering.* New York: John Wiley & Sons, 1980.

———. *Telecommunication Transmission Handbook.* New York: John Wiley & Sons, 1975.

Stuck, B. W. and E. Arthurs. *A Computer and Communications Network Performance Analysis Primer.* Englewood Cliffs, N.J.: Prentice-Hall, 1985.

MANUFACTURERS OF TESTING SYSTEMS AND EQUIPMENT

Centralized Trunk Testing Systems

ADC Telecommunications

AT&T Network Systems

Arus Corporation

Badger/TTI

Northern Telecom Inc., Northeast Electronics Division

Plantronics Wilcom

TTI Telecommunications Technologies, Inc.

Circuit Testing Equipment

ADC Telecommunications

AT&T Network Systems

Atlantic Research Corporation

Biddle Instruments

Calculagraph Co.

Communication Mfg. Co.

Design Development Inc.

Dynatel 3M

Fluke MFG. Co., Inc.

Halcyon Communications, Inc.

Hekimian Laboratories, Inc.

Hewlett-Packard Co.

Marconi Instruments

Northern Telecom Inc., Northeast Electronics Division

Plantronics Wilcom

Scientific-Atlanta, Inc.

Siemens Corporation, Telephone Division

Sierra Electronic Division, Lear Siegler, Inc.

Systron-Donner Corp.

Tau-Tron, Inc.

Tekno Industries, Inc.

Tektronix, Inc.

W & G Instruments Inc.

Wiltron Company

Data Test Equipment and Protocol Analyzers

ADC Telecommunications
AT&T Network Systems
Atlantic Research Corporation
Calculagraph Co.
Digitech Communications Fluke Mfg. Co., Inc.
Dynatech Data Systems
General Datacomm, Inc.
Halcyon Communications, Inc.
Hekimian Laboratories, Inc.
Hewlett-Packard Co.
Northern Telecom Inc., Northeast Electronics Division
Scientific-Atlanta, Inc.
Siemens Corporation, Telephone Division
Sierra Electronic Division, Lear Siegler, Inc.
Systron-Donner Corp.
Tau-Tron, Inc.
Tekno Industries, Inc.
Tektronix, Inc.
W & G Instruments Inc.

Centralized Line Testing Systems

AT&T Network Systems
Badger/TTI
Porta Systems Corp.
Teradyne Central, Inc.
Wiltron Company

CHAPTER **TWENTY–FOUR**

Network Management

A telecommunications system is an expensive resource that must be managed just like any other asset such as a computer, a building, or a motor-vehicle fleet. To do an effective job of controlling costs and to ensure that the system delivers the service it was acquired for requires knowledge that is gained only through training and experience. This chapter explains briefly the elements of a telecommunications management program. The principles apply whether the system is owned or leased. Telecommunications management applies to small systems as well as large, although in small networks some functions are more likely to be contracted than performed with in-house staff.

SYSTEM RECORDS

A complete record of all telecommunications resources is essential in any organization with a system that consists of more than single-line telephones. Records should consist of these elements:

- Equipment identification.

- Equipment documentation.
- Location records.
- Interconnection records.

These records are in addition to the property records that are kept for depreciation and accounting purposes. If the system is leased, the owner will probably retain the records, but they should be available to the user because they are essential for studies of future growth and rearrangement.

Equipment Identification

An inventory of equipment by type, model number, manufacture date or equipment list number, and wiring options is essential for managing a telecommunications system. For stored program controlled equipment, the program identification, serial number, and issue number should also be retained. Most equipment manufacturers have a systematic process for upgrading their equipment as improvements are made in both hardware and software. When these improvements are announced, the user must decide whether or not to upgrade. Equipment records are necessary for maintaining an upgrade program and for answering compatibility questions.

Equipment Documentation

Telecommunications equipment is typically documented with installation and maintenance manuals, schematic diagrams, wiring or interconnection diagrams, and circuit descriptions. All such documents should be acquired with the equipment and maintained, even if the vendor is maintaining the equipment under contract, because such documentation is the key to changing maintenance contractors if the need arises. Software documentation should also be acquired with any stored program controlled equipment. The documentation should be reviewed to ensure that it covers restart and error recovery procedures.

Location Records

The location of all fixed equipment should be documented, possibly on the equipment identification records. Location records are not needed for simple systems such as small PBX and key telephone systems, but they are vital for more complex networks that include transmission equipment. Location records are particularly essential for concealed equipment such as repeaters and junction boxes.

Interconnection Records

Telecommunications circuits are built up by interconnecting separate assemblies of equipment. Records of assignments and interconnection are essential for trouble shooting. Such records normally include both the drawings that the vendor assembles when equipment is installed and the records of the connections that are made when it is assigned. Interconnection drawings should show all options wired at the time of installation, designation of the wiring, and identification of the cross-connection points. Records should be kept of assignments of all services supplied by outside vendors. Station records should identify the user and the features of each station. Most modern key and PBX systems also require software translations. Records of these should be retained by the user.

SYSTEM USAGE MANAGEMENT

Although a telecommunications system is a high-cost resource, its costs are far outweighed in most organizations by personnel costs. The objective of an effective telecommunications system is to ensure that the productivity of the users is not hampered by non-availability of the service, poor response time, or cumbersome operating methods. At the same time, it essential to control telecommunications costs. Costs should be controlled through policies and supervision, not through underdesign of the network. The most essential part of a telecommunications management system is usage monitoring and translating the result into capacity changes. A second consideration in many organizations is allocating telecommunications costs to the unit incurring the expense.

These two elements are both measurable in switched networks equipped with a PBX or key system. Normally, cost allocation is based on line usage, while system utilization is measured by load indicators on the system components.

Line Usage Measurements

Line usage is determined by common carrier bills, user-dialed identification digits, or station message detail registers. In Centrex systems and PBXs with automatically identified outward dialing (AIOD), the toll bill is rendered directly to the end user's telephone number. Systems sharing a common group of trunks without AIOD need some other form of cost allocation.

The simplest form, but one that is the least convenient from a user's standpoint, is for the user to dial accounting digits with each call. Typically, the user dials the telephone number followed by an accounting code, and software in the switching machine or in an external computer tabulates costs by account code. The account code may be programmed into a "smart" telephone set that automatically pulses the code along with the dialed digits. However, this technique is not feasible for allocating the costs of telecommunication services from common facilities such as conference rooms.

PBXs and key telephone systems may be equipped with call management systems or station message detail recorders (SMDRs) to register calls in systems that are not equipped with call accounting software. An SMDR is a valuable tool for preventing abusive personal use of telephone services. The complete detail of each outgoing call is logged by the system and at the end of the accounting period a summary is produced. Authorized calls are screened automatically through a data base. Unauthorized or unknown calls are flagged for administrative action. Dedicated voice and data circuits are normally billed directly to the department incurring the expense and have no need for SMDRs and call management systems.

Traffic Usage Measurements

The second type of usage monitoring is system focused instead of user focused. Traffic usage monitoring supplies the information needed to determine when it is time to adjust circuit capacities. In systems with a processor-controlled switching machine, the machine will include software usage registers. Older electromechanical PBXs and KTs have no usage measuring provisions, so external recorders will be required. In systems based in the central office, such as Centrex, the telephone company monitors usage and recommends capacity changes. Many data networks include usage monitoring software in a control terminal or in a network monitoring system.

Three separate elements of a telecommunications system are monitored: line usage, trunk usage, and common equipment usage. As discussed in Chapter 13, usage measurements include the number of times a circuit is accessed, the total minutes of use for all circuits of the same type, and the amount of overflow.

Traffic usage readings must be distilled, interpreted, and translated into capacity changes. Any network management system requires the traffic usage monitoring function. Besides usage measurements, network service measurements should also be made. The average amount of dial tone delay on line circuits and the amount of blockage on trunk and common equipment should be monitored as service indicators. The objective of usage measurements is to detect bottlenecks so they can be corrected before service degenerates and to optimize line and equipment utilization.

TROUBLE HANDLING

Every telecommunications system, including those too small to justify a network management system, should have a procedure for handling trouble reports. In smaller offices the receptionist or attendant is usually the focal point for trouble reports. In systems with full-time network management, trouble report handling and analysis is a function of this center.

A trouble log is essential in either case. At a minimum, the log should include the following information:

- The identity of the person reporting trouble.
- Date and time of the trouble report.

- Nature of the trouble.
- Date, time, and identity of person to whom each trouble was referred.
- Date, time, and identity of person clearing trouble.
- Description of cause and work done to clear trouble.

Trouble reports should be summarized periodically—the frequency depends on the size of the system. Without regular analysis of trouble reports, unsatisfactory trends may escape notice. The trouble report summarization should reveal patterns of an excessive trouble report rate, excessive clearing time, or recurring failures of the same type. These patterns can be used to initiate corrective action with common carriers, equipment suppliers, and service organizations.

Unless cost is no objective, telecommunications system owners should be prepared to do a limited amount of self-maintenance. Simple tests can sectionalize trouble so the right service technician is called. Large organizations may have network testing centers. Small organizations can train one or two people on simple tests to make.

The network management center, or in smaller networks the PBX attendant, is also the focal point for system alarm reception. In an effective network, major trunk groups and critical equipment will be equipped with alarm points and will bring the alarm indications into a center for trouble diagnosis.

TELECOMMUNICATIONS COSTS

Every organization should maintain a detailed record of telecommunications costs segregated by major service categories such as:

- Local telephone service.
- Long-distance telephone service.
- Special services (data, private lines, foreign exchange, etc.).
- Maintenance and repair.

- Equipment lease costs.
- Administrative costs.

These costs are used to allocate overheads to the operating entities and as a basis for comparing alternative sources of telecommunications services. Long distance records arc essential for evaluating alternative carriers and for determining when it is economical to replace toll telephone service with some form of special service such as WATS, foreign exchange, and leased tie trunks. Organizations with large networks should be continually monitoring the use of common carrier message toll services to determine the economical point at which trunks should be added. Busy line or overflow studies from 800 numbers should be made to project a potential loss of revenue because of inability of customers to reach an incoming line.

Local service costs should be monitored to determine whether the economical quantity of local service lines has been provided into the system. Telephone companies can study the number of busy signals into a local number to project a potential loss of revenue because of inability of customers to access a local number. Where local service is measured, records of message charges can aid in determining when alternative types of service are economical. For example, the use of dial-up data on local lines may be more expensive than installing a private line to a public data network.

Trouble and cost records are useful in determining whether to maintain a system by contract, internal staff, or by maintenance contract. The cost of a maintenance contract is usually based on a worst-case assumption and is often considerably more expensive than paying for repairs on a case-by-case basis. However, maintenance contracts often include preventive maintenance services and equipment upgrades. A combination of trouble records and costs will aid in determining the best alternative.

Administrative costs, including the costs of internal maintenance staff, accounting for and billing the system, processing requests for additions and changes, handling trouble reports, and other such common functions should also be monitored.

SERVICE MONITORING

An effective network management system includes a process for measuring service levels. The key question is whether the system is supporting or hampering the productivity of its users. In a network with privately owned or leased facilities, the network management center should monitor call completions. Calls not completed because of blockage or equipment failures are indicative of lost time and are the responsibility of the network management center to correct. Busy signals are equally important because of the expense of setting up a circuit and consuming circuit time, only to find it blocked because of a busy signal at the far end. Call completions can often be improved by adding lines or forwarding busy calls to another number at the distant end. In data networks the error rate, response time, and throughput should be measured.

NETWORK RECONFIGURATION

A major goal of the network management center is to bypass failures when they occur and to reconfigure the network in response to changes in load. Ideally the network will be reconfigured dynamically as load patterns change and as failures occur. In networks composed of conventional private line services, network reconfiguration can take weeks or months because of the long lead times for ordering new circuits or equipment. As discussed in Chapter 4, the advent of digital cross-connect systems facilitates rearrangement under control of the user and aids in achieving the goal of rapid reconfiguration.

The network management system also provides status information on all parts of the network. If this status information is available in real time, the network management center can react quickly to changes.

PREVENTING FAILURES

The most important goal of network management is to prevent failures. The network management center should control preventive maintenance on all network elements including transmission testing on circuits as well as equipment adjustments if any are required.

The design of the system is the first line of defense against failures. Redundant equipment should be provided where failures cannot be tolerated. Diversity routes can be used to keep the system functioning with degraded service when one route fails. Backup equipment or facilities should be provided to restore a failed service rapidly. Dial-up facilities provide an excellent means of backing up dedicated voice and data services, and should be considered in most networks.

Stored program controlled equipment relies heavily on its software to prevent failures. All SPC software should devote a substantial portion of its resource to diagnosing the system and searching for potential weak spots.

STANDARDS

Few standards for network management are recommended by standards agencies. CCITT is primarily concerned with management of international networks, which are beyond the scope of this book. Networks within the United States are managed according to the policies of the network's owner.

CCITT

Q.256 Management signals.

Q.260 Management signals.

Q.506 Network management functions.

APPLICATIONS

Network management systems are not usually acquired as a product from a single vendor. Instead, they are assembled from equipment and software packages from a variety of sources and contain a considerable amount of custom software to accommodate the uniqueness of the network. Figure 24.1 is a photograph of an administrative position in the

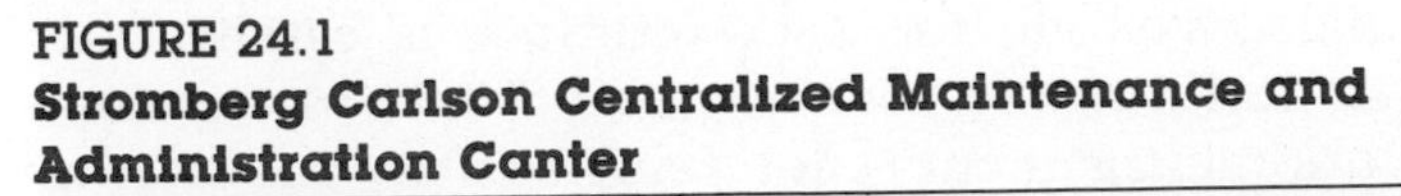
FIGURE 24.1
Stromberg Carlson Centralized Maintenance and Administration Canter

Courtesy Stromberg Carlson Corp.

Stromberg Carlson Centralized Maintenance and Administrative Center (CMAC). This center is supported by a computer that maintains records of the type discussed earlier. Lines and circuits can be tested from the center, and trouble history records retained in a data base.

Evaluation Criteria

As networks grow increasingly complex, network management systems grow in their complexity. Furthermore, as networks are continually changing, network management systems must possess the flexibility to follow the growth and rearrangement of the network without becoming so complicated that only experts can administer them. Network management can be simplified by using processors; and like any mechanized system, the user interface is essential to keeping it understandable. At the present state of development, network management is not a single entity but instead is a collection of techniques, most of which are computer supported. The primary issues in keeping a network management system manageable are:

- All aspects of management are covered in some way by a system, either manual or mechanical.
- The systems are interactive to drive each other to the maximum extent possible.
- The network layout records are accessible and easily understood.
- The system can infer the cause of the trouble from a diagnosis of a variety of inputs. For example, a hardware alarm could be diagnosed as the cause of a group of circuit failures.
- The system should aid the operators in determining appropriate action from the symptoms.
- The system should automatically reroute failed circuits and patch in standby equipment to guard against total failure.
- The system should provide output data summarized and sorted for analysis, or should provide a port to an external system to accomplish the same objective.
- The system should provide a trouble history log.

Method of Data Collection

A critical issue in network management systems is how data is collected from circuits and equipment and transported to a management center. The cost of equipment and facilities to connect data to the center can represent a significant portion of the network cost and must be justified by saving expense. Access switches, responders, and remote test equipment are most easily justified with large concentrations of circuits. In less dense locations it may be most economical to collect information manually and to access circuits through jacks. The network manager must examine each element of the network management system on the basis of value gained from the system compared to the costs of implementing it.

Centralized versus Decentralized Network Management

It is not possible to generalize about centralization or decentralization of a network management system because the benefits of centralization depend on the character of the network, the types of services being carried, the penalty for outage, and the ability of trained people to administer the system locally. If centralization can be justified, it is usually the most effective method of managing a network because diagnosing network troubles requires information that can often be analyzed more meaningfully from a central site. However, the people on the spot are usually the best equipped to take corrective action when troubles occur. Some of the most complex networks in the world, the telephone networks, are administered locally with central sites monitoring traffic flow and dealing with congestion. The issue of centralization is one that must be dealt with as part of the fundamental network design and reexamined as new services and equipment make it feasible to change the basic plan.

A parallel issue to that of centralization is whether testing will be performed manually or automatically. As networks gain more intelligence, automatic testing becomes more the rule than the exception. However, the network must be large enough to justify the cost of automatic testing before it can be considered. Equipment to perform automatic tests is advancing rapidly, and network management strategies should be reexamined periodically in the light of new developments.

Multiple Vendor Issues

With the divestiture of AT&T's operating telephone companies and the demise of end-to-end service responsibility on the part of the telephone companies, network users will increasingly obtain their equipment from multiple vendors. Although this offers opportunities for cost saving, it thrusts a much greater responsibility on the network manager to monitor service and to develop techniques for dealing with multiple vendors.

One problem is the lead time for obtaining additional capacity. The projects of multiple vendors must be coordinated or one vendor may install capacity that cannot be used because another vendor has not completed its work. Procurement policies must be closely administered to ensure that equipment is procured to the user's specifications. Compatibility problems become the user's responsibility, and this responsibility can be exercised best by preparing precise procurement requirements and specifications when acquiring equipment and services.

An allied issue is the degree of internal network management expertise that an organization develops. Although it is possible to turn the problem of network administration over to an outside contractor, to do so is to incur both a risk and an expense. The risk and the expense may be preferable to developing internal resources, but this is a decision that should be made only after an analysis of the alternative of developing internal staff.

Organizational Considerations

Many companies have not yet aligned their internal organizations with the realities of the new telecommunications networks. In most companies, data processing people administer the data network and a separate staff administers the voice network. The information resource that is supported by both may be managed by yet a third group. In the future, the integration of voice and data networks is inevitable for most large organizations. Even though the two may be separated at the source, bulk circuit procurement will be more economical and more flexible than obtaining individual circuits. When failures occur, a single organizational unit will be the most effective in dealing with the problems of restoring services and rerouting high-priority circuits over alternative facilities. In most organizations, the most effective structure will separate the information-generating entities from the information-transporting entities. As this is contrary to the way most companies are organized, this issue will need to be dealt with as the character of the network becomes more integrated.

Another issue that must be dealt with is flexibility of network planning. Changing tariffs and rates of long-haul carriers, access charges, and measured local service of local telephone companies make it essential that network plans be continually reexamined. Plans should be created that are flexible enough to enable the organization to react quickly as changes in tariffs and grades of service occur.

BIBLIOGRAPHY

American Telephone & Telegraph Co. Bell Laboratories. *Engineering and Operations in the Bell System.* Murray Hill, N.J.: AT&T Bell Laboratories, 1983.

Harper, William L., and Robert C. Pollard. *Data Communications Desk Book: A Systems Analysis Approach.* Englewood Cliffs, N.J.: Prentice-Hall, 1982.

Sharma, Roshan La.; Paulo T. deSousa; and Ashok D. Inglé. *Network Systems.* New York: Van Nostrand Reinhold, 1982.

Martin, James. *Design and Strategy for Distributed Data Processing.* Englewood Cliffs, N.J.: Prentice-Hall, 1981.

Meadow, Charles T., and Albert S. Teedesco. *Telecommunications for Management.* New York: McGraw-Hill, 1985.

MANUFACTURERS OF NETWORK MANAGEMENT SYSTEMS

Data Switch Corporation

Dynatech Data Systems

General Datacomm, Inc.

Infotron Systems Corporation

Northern Telecom Inc., Spectron Division

Paradyne Corporation

Stromberg Carlson Corp.

CHAPTER TWENTY-FIVE

Future Developments in Telecommunications

The bulk of this volume has been devoted to discussing how the present telecommunications network is configured. Previous chapters have not dealt with the future for a good reason—the only kind of services and technologies that can be applied are those that are available now. However, the twin forces of the AT&T divestiture and advancing technology arc rapidly driving the network toward different configurations. Managers must be aware of future trends because it is essential to design networks to take advantages of price reductions and improvements in technologies when they occur. We are living in a society that is increasingly becoming dependent on the flow of information. The demand for information transport is occurring at a time when the available bandwidth is expanding dramatically. Telecommunications managers must understand the impact of the following factors as they relate to the network of the future:

- Development of network intelligence.
- The integrated services digital network (ISDN).
- The information revolution.
- An explosion in the use of personal computers.

- Digital communications.
- Changes in local loops.
- Voice/data integration.
- Future deregulation.

DEVELOPMENT OF NETWORK INTELLIGENCE

Those networks composed of dedicated facilities tend to have static configurations because of the delays and high cost of facility rearrangements. Static networks, however, are contrary to the need for information flow in most modern organizations and contribute to low utilization of facilities. Network intelligence allows an organization to reconfigure the network based on instantaneous service demands. Products to accomplish this objective are based on digital circuit switches or on a digital cross-connect system that operates under some form of network management system. The critical issue is whether intelligence is located on the user's premises or at the common carrier's facility. In either case, the intelligent network provides users with a far greater degree of control than they now have.

Figure 25.1 shows the concept of an intelligent network. User services home on a service node that directs digital bandwidth where it is needed. Bit compression multiplex equipment compresses the bit stream to make the most effective use of digital facilities. The digital cross-connect system routes traffic to dedicated or switched services as needed. The network control system located on the user's premises dynamically monitors load and service and changes the network configuration in response to demand. For example, an airline reservation system extending across several time zones can be reconfigured to move calls to different answering centers as the load shifts during the day. Also, an intelligent network could provide the airline the capability of offering priority treatment to their best customers when all positions in the nearest reservation center are occupied. For example, a call from a customer identified as a frequent flier could be shifted over the intelligent network to the opposite end of

FIGURE 25.1
Intelligent Network Concept

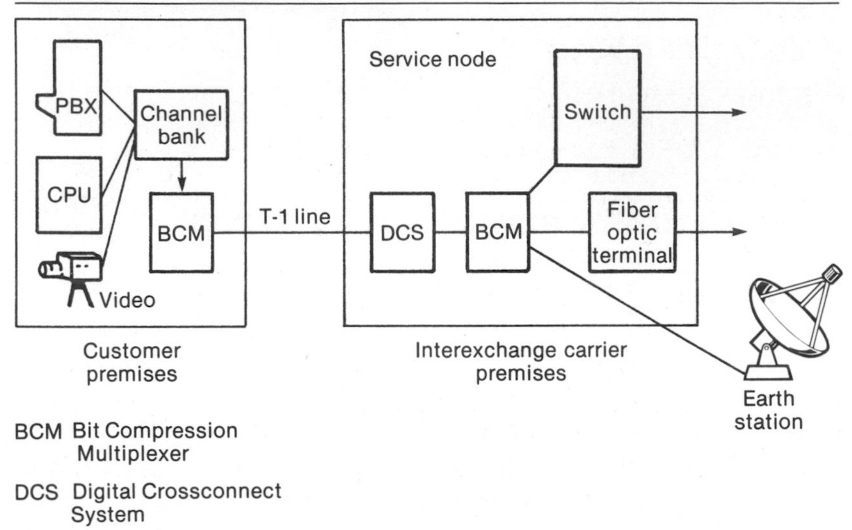

the country while less important customers are placed in queue at the local automatic call distributor.

Network intelligence is most effective with all-digital circuits where bandwidth can be reallocated according to demand. Digital circuits that are normally used for individual voice channels can be rerouted during off-peak hours to a computer center for high speed data transmission or can be reallocated to a video conference center. The advent of low-cost bulk digital facilities will have a significant impact on the demand among users for network intelligence.

The growth of network intelligence will greatly improve the utility of information resources and at the same time will demand a higher level of knowledge to use the network effectively. The users will have greater flexibility and control and will undoubtedly have to employ computer-based tools to make the maximum use of the network.

AT&T Communications has announced a new service called the Software Defined Network (SDN), which places intelligence in the network but allows the users to define their own services through access to the network control data base. The SDN consists of five major elements, as shown

FIGURE 25.2
Software Defined Network

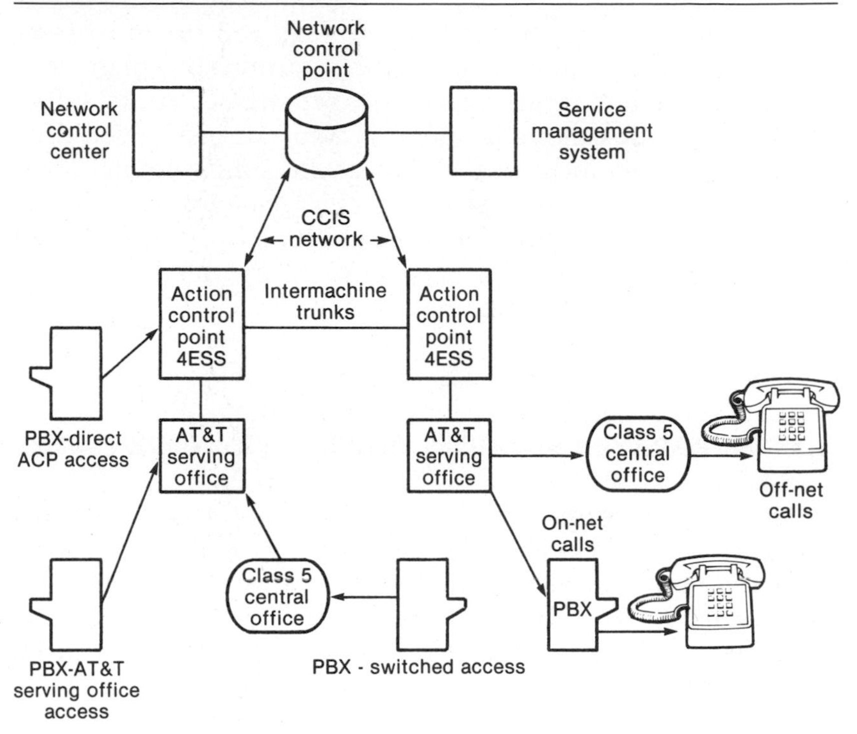

in Figure 25.2. SDN is a usage sensitive service that draws facilities from AT&T's switched voice network.

A data base, resident on an AT&T 3B20 computer in a Network Control Point (NCP), defines the user's network. The user accesses the network via dedicated lines to a local Class 5 switching machine, to an AT&T serving office, or directly to an Action Control Point (ACP). The ACP receives dialed digits from the user's location and obtains instructions over the CCIS network from the NCP regarding the user's dialing plan, features, and restrictions. The service provides calling to both on-net and off-net locations.

The users control their portion of the network data base through the Service Management System (SMS), a computer that accepts from the user's terminal, changes in such features as station restrictions. The fifth element, the Network Control Center, is AT&T's center for managing the network.

The NCC accepts trouble reports, tests the network, and resolves data base problems.

The SDN offers features that are required for most large private networks such as a uniform numbering plan, teleconferencing, and interface to other private networks. Other features include call management, remote access, seven-digit dialing to off-net locations, and location-level accounting detail in a single bill. The service offers large users an attractive long-distance billing rate plus the ability to administer their own network without the need to rearrange facilities to accommodate shifting traffic patterns.

THE INTEGRATED SERVICES DIGITAL NETWORK (ISDN)

Telecommunications networks to date have provided the user with little control over their options. With a few exceptions, if a call is not completed, about all a user can do is hang up and try again. Some benefits of network intelligence became available with the advent of stored program control central offices and PBXs. Such machines inform a user when a call is waiting on a busy line, allow users to transfer calls from one line to another, and enable the system to choose the least expensive of several long-distance routes to the terminating point. While the present state of communications offers many advantages to its users compared to the past, future networks will give users a quantum leap in their abilities to control how telecommunications is used and to customize its options.

Future network intelligence will allow the user to instruct the network to respond differently according to time of day, day of week, or who the calling party is. Calls arriving at a terminating location will be treated with the same kind of discrimination as if the caller arrived in person. For example, nuisance calls can be turned away, priority callers can be shunted to trained specialists, and callers can leave voice or data messages when it is unnecessary to talk to someone at the receiving end. Furthermore, dedicated services, as discussed in the previous section, will be rearranged by users without the need for intervention by the common carrier.

The primary problem with present telecommunications networks is that separate networks and separate interfaces are required for every application. Both voice and data circuits can have either a digital or an analog interface, but the interface and terminating equipment are far from uniform. Public data networks have different interfaces from the telephone network, and video networks such as CATV have yet another type of interface. Furthermore, the networks are incapable of handling information interchangeably. Analog voice networks are slow and inefficient at handling data, and low-speed data networks are ineffective for voice transmission. The Integrated Services Digital Network (ISDN) is proposed as a future network architecture that will provide connectivity over a single digital communications pipeline while opening the network to a range of services that are impractical with the amalgam of networks we have today.

The architecture of the ISDN was approved by a 1984 plenary assembly of CCITT, providing the foundation for cooperating nations to build upon. The objectives of the ISDN are:

- To provide end-to-end digital connectivity.
- To gain the economies of digital transmission, switching, and signaling.
- To provide users with direct control over their telecommunications services.
- To provide a universal network interface for voice and data.

The building blocks of the ISDN consist of four parts: bearer services, channels, interfaces, and message sets. Bearer network access (called "B" channel) is defined simply as a 64 kb/s digital signal. The basic ISDN service element consists of two bearer channels plus a 16 kb/s signaling channel (called a "D" channel) that enables the user to access services as shown in Figure 25.3. The availability of a separate signaling channel is one of the key features of the ISDN because it allows the user to send and receive signals while the channel is in use, a feature that is impossible today. By sending and receiving signals over this channel, users can tell who

FIGURE 25.3
Integrated Services Digital Network Concepts

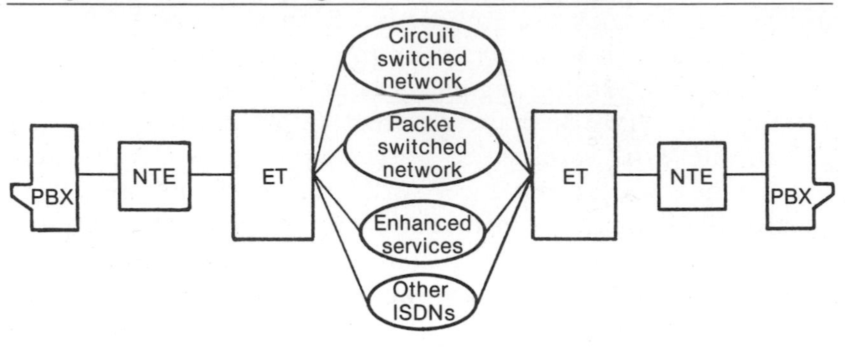

NTE—Network Terminating Equipment
ET—Exchange Terminator

is calling them, can find out what the call is about, and can decide to terminate the present call if they choose. This exchange of signals takes place without interrupting the call in progress.

The basic service element of two bearer and one signaling channel (called "2B+D" in ISDN vernacular) provides the user with the ability to transport information through the network. Through this service, users can access enhanced services that offer data processing capabilities. For example, an information provider might offer stock market quotations that are transported at 64 kb/s over the basic transport service.

Typical ISDN Services

The ISDN will make it possible to obtain many useful enhancements in addition to today's basic telephone service. Telephone instruments will have digital readouts that can display information transmitted over the signaling channel. For example, if a caller places a call to a busy line, a special message set will allow the caller to enter his or her identity and the subject of the call. This information can be transmitted over the signaling channel while a conversation is in progress, and the called party can decide whether to terminate the call or can even send a response over the signaling channel. The telephone network can be placed in a callback mode to automatically return calls in priority sequence.

A user's services are configured in the present telephone network by entering instructions into the switching system to indicate what optional services the user has chosen. Today, changes in service options require a communication between the user and the telephone company and often take several days to implement. The ISDN will offer flexibility so users can initiate and terminate services from their own terminals. Moreover, when users travel from one telephone to another such as between the office and home, their personal service profile can follow them so they can continue the method of operation they have become accustomed to.

The enormous expenditures needed to convert today's telephone systems into the ISDN will prevent the transition from happening overnight. The formative steps have been taken by international agreements on the nature of the services and interfaces, but the conversion will take several decades. Many observers question whether the ISDN is needed and doubt that it will ever be implemented in its presently conceived form. Whatever occurs as the ISDN emerges, telecommunications managers should stay attuned to its progress because it is important to ensure that equipment and network services procured in the meantime do not conflict with its ultimate direction.

THE INFORMATION REVOLUTION

The transformation of the U. S. economy from a manufacturing to an information-based economy has been well documented, but its implications are barely understood. The fact is, information is a valuable resource to companies in several ways. At one level, information is generated and sold outright. Publishing companies, legal firms, and accounting firms are examples of organizations whose mission is the production, distillation, and dissemination of information.

From another point of view, information can be conceived as part of the content of manufactured goods. Products assume a value far in excess of their physical size or of the resources required for their manufacture. High-technology products such as microprocessors and random access memory are constructed from resources that are only partially physical—the bulk of the resources are information-based.

From a third point of view, organizations that are not considered to be in the direct stream of the information economy, nevertheless possess an information resource that is valuable to their customers. For example, an airline is in the business of moving freight and passengers; yet without its information base of schedules, aircraft capacity, and seat vacancies, it would be unable to function. Virtually every organization has an information base of some sort that distinguishes it from its competitors and enables it to function efficiently. The companies that best learn how to use this information base to gain a strategic advantage over their competitors will be the ones that survive into the next century.

The role of telecommunications in this information society should be obvious: telecommunications is the movement of information. It is evident that the companies that learn how to use and move information and to make it accessible to their customers will be the strongest competitors in the future, which is to say that organizations must learn to manage their telecommunications resources well as an extension of their information resources.

AN EXPLOSION IN THE USE OF PERSONAL COMPUTERS

A few years ago the pundits of the office of the future were foreseeing the day when every desk would be equipped with an executive workstation. That day is rapidly approaching, although the workstation is taking a different form than most people foresaw a few years ago. The personal computer is becoming the workstation for a substantial number of workers, and this revolution shows no signs of abating. Many workers today have more processing power on their desks than was contained in the mainframe computers of less than a decade ago. Figure 25.4 shows how processing power is increasing at a geometric rate. Desktop computers of today are capable of executing as many instructions per second as mainframe computers of about six or seven years ago. (For

FIGURE 25.4
The Ratio between the Computing Power of Desktop and Mainframe Computers Is Narrowing

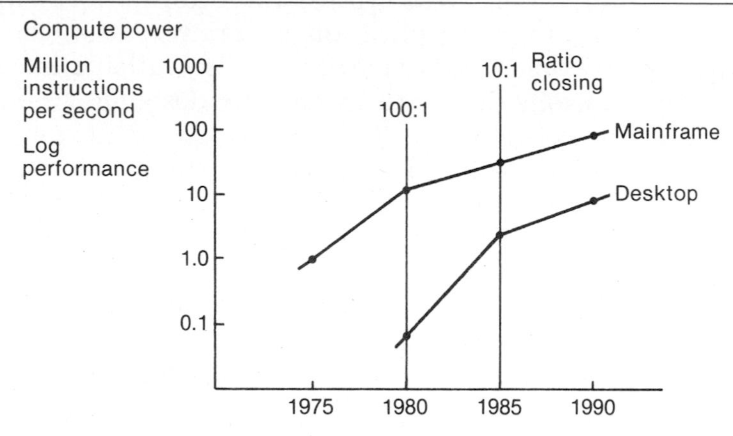

Courtesy Bell Northern Research, Inc.

the purposes of this discussion, mainframe and minicomputers are lumped together.)

Microcomputers have substantial advantages over mainframe computers. They are inexpensive, they are accessible, and they are far more flexible in their abilities to support personal applications through sophisticated software. This is not to say that microcomputers will replace mainframes. Mainframes are required to manage a data resource that must be centralized to ensure its validity. It would be unthinkable, for example, to have corporate financial results or personnel files distributed among thousands of personal computers, each with the ability to add, delete, and change data. It is clear that both types of computer have complementary roles and must be married through telecommunications. Specialized networks will grow to link microcomputers and data bases. This means that the data communications network of the future will take on an entirely different complexion than the network of the past.

In the past, data communications networks supported outlying terminals over either short loops or long, low-speed communications lines. The trend is now evolving toward high-speed networks supporting a large number of terminals that will do much of the processing outside the realm of the mainframe computer. To illustrate this point, an analyst will download information from the mainframe data base into a personal computer and will process the information with a spreadsheet or similar application program. Although the demand for this facility exists today, it is available in only a few organizations that have specifically designed a micro-to-mainframe connection.

Another important implication of the ubiquitous microprocessor is the fact that unlike dumb terminals that characteristically stayed within the boundaries of a single company, personal computers will be exploring multiple data bases over many networks. The personal computer will become the vehicle for extracting information from the data bases that organizations will increasingly offer to their customers or to the public.

DIGITAL COMMUNICATIONS

Communications networks are changing in another revolutionary way. Until now, most networks have been constructed for a predominance of voice communication. Because voice communication is analog, telecommunications networks have grown to support analog communications. Digital data has adapted to an analog world with numerous conversion devices. However, the trend is now toward digital facilities. With the exception of the local loop that connects the user's premises to the telephone central office, nearly all transmission facilities are now being constructed for digital communications. Instead of data signals being converted to resemble voice, voice signals are being converted to a digital format for transmission over these networks.

The trend is being accelerated by the development of fiber optics as a high-speed, high-capacity transmission medium. The competing long-distance common carriers are rushing to construct nationwide fiber optic transmission systems. According to a study by the Yankee Group, a Boston consulting firm, if all the planned fiber optic capacity is put into service, it will provide about seven times the capacity of AT&T's present network.[1] Whereas today, digital facilities

[1]Reported in *The Wall Street Journal,* July 26, 1985.

are expensive and unavailable in many places, tomorrow's networks will be composed almost entirely of digital facilities. This change, in turn, brings about several beneficial results. Two of the most important are a significant increase in circuit quality and a decrease in the cost of bulk facilities delivered directly to the end user. Organizations that develop the ability to apply bulk digital communications facilities will obtain the most cost effective telecommunications service in the future.

CHANGES IN LOCAL LOOPS

We have described in this book how the local loop is the link in the telecommunications network most vulnerable to damage and interference and is the most restrictive part of the network in bandwidth. Many articles have been written and discussions held about the problems of the "last mile," the link between the interexchangc carrier and the end user. Alternatives are being developed to bypass this part of the telephone network, and for good reason; besides problems of noise and bandwidth, it has by far the highest per mile cost of the entire network. Installing local loops has required digging countless miles of trench through some of the most expensive real estate in the world to serve virtually every location with metallic facilities.

In the past there were excellent reasons for using metallic facilities to the user's premises. First, the alternatives to cable were prohibitively expensive unless the served location was several miles from the central office. Second, the telephone company owned and installed the station equipment and was responsible for providing inexpensive and reliable telephone service. The conventional telephone set is not only cheap, but it is independent of local electric power. Furthermore, its signaling method requires sufficient line current to drive an electromechanical ringer. The technology to perform these functions electronically has existed for years and would have been practical if the user furnished the electric current to operate them during power outages. Now, through the evolutionary forces of regulation, legislation, and judicial action, it is feasible for telephone companies to deliver an interface link to the users' premises and let them provide their own talking and signaling power. How soon this will occur is open to conjecture, but it is clear that telephone companies will increasingly use fiber optics to serve local loops, and power cannot be transmitted over fiber optics. Several forces are bringing about the evolution to fiber optics in the loop:

- As services grow, conduits fill, and construction costs increase, it becomes economically feasible to remove existing cables from conduits and replace them with compact fiber optic cables. This avoids the need to augment existing conduit routes.
- Multiplexing technologies continue to improve, bringing smaller, less expensive, and more reliable multiplex equipment that can concentrate quantities of loops and connect them to the central office over fiber optics.
- The latest generation of digital central offices and PBXs provide line modules that can economically be moved closer to the users' premises, or even dedicated to large users.
- Present analog trunk facilities are gradually being replaced with digital facilities, and the ISDN requires digital facilities to the end user. While metallic local loops are capable of transmitting digital signals, they are limited in bandwidth.

The death knell of the metallic local loop should not be sounded yet because this process will be evolutionary over several decades. The most cost effective areas to replace are large business concentrations, high-rise apartments, and other such locations where the demand for service can support the cost of nonmetallic loops and where real estate can be acquired to house the terminating equipment. It is by no means certain that nonmetallic loops will inevitably transfer the problem of powering station equipment to the user, but as the switching and terminating equipment is distributed, a major part of the space requirements will be for battery plants and emergency power equipment, and this may lead to user-powered station equipment.

VOICE/DATA INTEGRATION

Voice/data integration today is perceived by many people as an ideal, but it is one that is realized by only a handful of PBXs at a much higher cost than the analog voice circuits used for telephone service. On PBX user's premises, one advantage of voice/data integration is the saving in wiring that is achieved by digitizing the voice at the station set and integrating the voice stream with a data stream from a colocated terminal. Although a few PBXs today furnish this kind of service, the majority of data applications still involve separate voice and data circuits with the data circuit terminated on a separate PBX port, or bypassing the PBX altogether. As combined voice and data switching machines become more prevalent, this separation will diminish, but as with the nonmetallic local loop, this evolution will require many years.

Voice/Data Integration in the Switching System

When many people consider voice/data integration, they assume that a switching machine will handle digitized voice and data bit streams interchangeably. However, this viewpoint does not consider some of the practical barriers to full integration. Data and voice signals differ considerably in their requirements for the following reasons:

- The bursty nature of data traffic makes packet switching appropriate for data. Packet switching is not yet practical for voice.
- A large portion of data communication today is conducted over SNA networks. These networks are not suitable for voice traffic.
- Voice switches are inefficient for handling the multiple paths required by many data applications.

Full voice/data integration is probably years in the future, but a limited number of data applications can now be integrated with voice. These applications are the ones that are handled over dial-up circuits today, often using asynchronous terminals. However, in most large organizations these are not the majority of applications of data transmission. The host-to-terminal applications that connect mainframe computers to multiple terminals cannot practically be switched through a voice switch, and voice traffic cannot be switched through a front-end processor. As a result, these services will remain separate for the foreseeable future and will make the full integration of data and voice unnecessary.

Voice/Data Integration in Transmission Facilities

Although full integration is impractical in the switch, a change that can be expected to occur rapidly for most large users is voice/data integration in transmission facilities. Long-haul fiber optic routes are being installed in the United States at breakneck pace, to the point that some observers believe this country will have an oversupply of digital facilities by the end of the 1980s. However, this viewpoint reckons without knowledge of what will happen when declining circuit costs fuel the demand for additional telecommunications facilities. It is safe to assume that we have hardly glimpsed the limits of demand for facilities. It is also safe to predict that if an oversupply occurs, analog facilities will be the ones to disappear and prices will decline because of the oversupply.

The trend in the mid-1980s is toward bulk pricing of digital transmission facilities. Even before intercontinental facilities are completed, carriers are offering T-1 facilities at such attractive prices that one T-1 facility is cost competitive with less than a dozen individual analog circuits. At the same time, many local telephone companies have been freed from tariff restrictions on local T-1 facilities and are basing their rates on the cost of developing the facility.

These changes will cause most large users to reexamine their networks. Not only will T-1 facilities reduce circuit costs, they will eliminate many modems and will also gain the full 64 kb/s bandwidth capability of a voice grade circuit. As these facilities are redesigned it will become evident that voice, data, and video services must share the T-1 facilities to gain the maximum economy. To do so, however, will not require voice/data integration in the switch.

FURTHER DEREGULATION

Once the forces of deregulation were set in motion, the demand to free additional services continues to swell. Companies that once existed comfortably under the regulatory blanket are now petitioning to escape its clutches. The question is no longer whether regulation is necessary to protect consumers. That decision was made when competition was introduced into telecommunications network. Now the issue is the ability of small companies to survive if AT&T's market power is unleashed.

Another allied issue of utmost concern to all users is what will happen to local telephone rates. Just as local rates in total were once subsidized by long-distance services, now users in high-cost rural areas are being subsidized by access charges and by rates from low-cost areas. While it would be foolhardy to predict the outcome of the regulatory melee that is visiting the telephone industry in the mid-1980s, some changes are inevitable, and astute managers will plan accordingly.

The Advent of Equal Access

Equal access to local telephone exchanges was mandated by the 1982 settlement between the Department of Justice and AT&T. Equal access is programmed for most central offices by the end of 1986. As this occurs, exchange access charges for AT&T will drop and those for its competitors will increase. Prior to the vigorous construction programs most large carriers are undertaking, AT&T had lower facility costs than its competitors because of real estate and equipment that had been acquired over the past hundred years. What will happen as a result of the expansion of most carriers into fiber optic facilities is difficult to predict, but it is safe to predict that changes in long-distance rates will occur, that prices will probably drop, and that new service offerings will appear on the market.

Effect on Resellers

When the FCC opened long-distance service to competition, a large number of "resellers" entered the long-distance market. Equipped with low-cost switching machines, these carriers obtain bulk switched services such as WATS from the interexchange carriers and rebill these services to their customers. As deregulation and competition drive down the rates of the major interexchange carriers, resellers will be faced with the alternative of building their own networks from bulk T-1 facilities, merging with other carriers, or going out of business. A substantial shakeout in the reseller business is likely to occur in the next few years. Many observers predict that the long-distance market will be reduced to a handful of the largest and best financed of today's interexchange carriers.

Coin Telephone Deregulation

An aspect of deregulation that has not been widely publicized is that of coin telephones. Once the exclusive province of operating telephone companies, coin telephone service has been opened to competition by some state regulatory agencies, and it is likely that most states will follow suit in the next few years. Although coin service has been a lucrative business for telephone companies, they are now gaining high rates for the local access lines and are freed from the troublesome aspects of coin service such as collecting coins and coping with theft and vandalism. The effect on the user will be a significant increase in the number of telephones and confusion over rates.

Telephones that accept credit cards are appearing on the market. Within the next few years most public locations will be equipped with telephones that provide coinless access to long-distance carriers.

Competition in the Local Network

The theory of the AT&T divestiture was that the competitive elements of telecommunications service—long-distance and station equipment—would be assigned to AT&T and that local operating companies would be assigned the "noncompetitive" portion of the network, local service. However, this theory overlooks the reality of competition in the local network. Local competition has two primary sources: direct competition for point-to-point services from cable television companies and bypass competition from the users and interexchange carriers.

In both cases, the unfortunate result for residential telephone users is that local network competition skims the lucrative accounts and leaves the high-cost accounts for telephone companies that are required by franchise to provide service. This leaves the state regulatory commissions with several equally unpalatable alternatives. One is to allow the telephone companies to lower their rates selectively to compete for the more lucrative accounts. The second is to require the telephone companies to keep their rates based on statewide average cost, in which case they will lose business to the competition. The third alternative is to regulate the competitors, a move that is both resisted by the competitors and is contrary to current national policy. The fourth alternative is to discard the theory of statewide average pricing and allow the telephone companies to price in more narrowly defined jurisdictions, which will hold down rates for urban users and drive them out of reach for many rural users.

None of these alternatives bodes well for local telephone rates. Legislative solutions have been debated for the past decade, but no one has yet developed a satisfactory compromise between the fully regulated environment of the past and the partially deregulated environment we are now in. The outcome of this dilemma is unpredictable because so many forces are converging on the problem from divergent points of view.

SUMMARY

This chapter should not be seen as a set of predictions of what will inevitably occur in the telecommunications industry. No one attempting to predict the shape of the industry in 1975 at the outset of the Department of Justice's suit against AT&T could possibly have predicted the outcome. A few people at the forefront of technology might have predicted which way technology would drive the industry, but it is doubtful that anyone a decade ago could have predicted the technical shape of the industry in the mid-1980s.

Instcad of a collection of predictions, this chapter should be considered as a brief explanation of some of the forces that inevitably will alter the shape of the telecommunications system between now and the beginning of the 21st century. The general direction of telecommunications services can be predicted, but to forecast the direction accurately would require knowledge of the outcome of several political decisions, and those are far less predictable than the technical forces. Whatever the outcome, no field in the world today offers more promise than the twin technologies of computers and telecommunications.

APPENDIX A

Principles of Electricity Applied to Telecommunications

Although it is possible to gain an appreciation of telecommunications with little or no technical background, the technology is based on the applied science of electronics. Without some understanding of the principles, the reader is apt to be confused by some concepts. A complete understanding of communications technology cannot be developed without understanding the mathematics involved, some of it quite complex. However, a working knowledge of the concepts presented in this book can be gained with no more than elementary mathematics and some fundamental principles of electrical theory. For those readers who lack understanding of electricity, this appendix is an overview of the basic principles of electricity as used in telephony. This appendix also explains most of the terminology used in basic electricity.

An understanding of the terminology begins with the units of measure. These are multiplied or divided by factors ranging from one to one billion. For those who are unfamiliar with electrical units, Table A.1 lists the electrical units most frequently encountered.

TABLE A.1
Common Units of Electric Measurements

	Prefix	*Abbreviation*	*Frequency*	*Power*	*Current*	*Resistance*	*Voltage*	*Other*	*Data*
1,000,000,000	Giga	g	Gigahertz (GHz)						Gigabit (gb)
1,000,000	Mega	Meg or M	Megahertz (MHz)	Megawatt (Mw)		Megohm	Megavolt		Megabit (mb)
1,000	Kilo	K	Kilohertz (KHz)	Kilowatt (Kw)		Kilohm	Kilovolt		Kilobit (kb)
1			Hertz (Hz)	Watt (w)	Ampere (amp or A)	Ohm (Ω)	Volt (V)		Bit
1/10	Deci	d						Decibel (dB)	
1/1,000	Milli	m		Milliwatt (mw)	Milliamp (ma)		Millivolt (mv)		
1/1,000,000	Micro	μ		Microwatt (μw)	Microamp (μa)		Microvolt (μv)	Microfarad (mf) Microhenry (μh)	
1/1,000,000,000	Pico	p						Picofarad (pf)	

FIGURE A.1
Schematic Diagram of a Relay

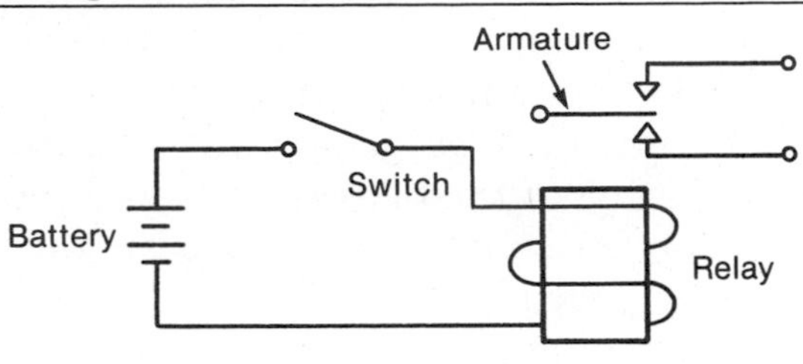

ELECTRICITY AND MAGNETISM

The characteristics of permanent magnets are familiar to all who have experimented with magnetized iron bars. In iron's natural state, electrons array themselves in disorderly patterns and have no attracting force. When the material is magnetized, the electrons align themselves neatly, creating the power to attract and repel other elements. The opposite ends of magnetized material are called its poles. Anyone who has experimented with magnets knows that like poles repel and opposite poles attract, and that either pole attracts some metals but not others. Metals such as iron and its alloys exhibit strong magnetic properties while others such as copper and aluminum cannot be magnetized.

Relays and Solenoids

Permanently magnetized materials have application in telecommunications; for example, loudspeakers and the ear piece in an ordinary telephone set contain permanent magnets. A more important application of magnetism in telecommunications is the *electromagnet,* a device that becomes magnetized when external current is applied and nonmagnetized when it is removed. This principle is used in the millions of relays and electromechanical switches in telephone central offices. Although relays are being replaced by solid-state electronic circuits, vast quantities of them remain in service and are important to understanding telecommunications.

A relay is shown schematically in Figure A.1. When the coil surrounding the relay's core is energized by closing a switch to the battery, the core attracts the armature and the movement closes or opens contacts. Relays are used to control large electric currents with a small current flowing through the winding. For example, the starter solenoid in an automobile allows the ignition switch, which is a low-current device, to control the large amount of current needed to connect the starter to the battery. Relays are often made with multiple sets of contacts so that a single source can control current to multiple paths. A relay is a binary device; that is, its coil and contacts are either open or closed, corresponding to the zeros and ones of logic circuitry. These principles are used in chaining relays to form logic circuits in electromechanical central offices.

Electric Current and Voltage

When an electric source is applied to a circuit by closing a switch, *current* begins to flow. A circuit is composed of three variables: current, *voltage*, and *resistance*.

- **Current** is the quantity of electricity flowing in a circuit. The unit of current is the ampere or "amp." Its symbol in circuits and formulas is "**I**."
- **Voltage** (sometimes called **electromagnetic force** or **EMF**) is the pressure forcing current to flow. The unit of voltage is the volt. Its symbol is "**E**."
- **Resistance** is the opposition to flow of current. The unit of resistance is the ohm (Ω). Its symbol is "**R**."

These three variables are interrelated by Ohm's law, which states that the amount of current flowing in an electric circuit is directly proportional to the voltage and inversely proportional to the resistance. Stated as a formula, Ohm's law is:

$$I = \frac{E}{R}$$

This simple formula is the basis for understanding the behavior of electricity. To illustrate with a simple example, assume the circuit in Figure A.2. A 12-volt battery such as

FIGURE A.2
Current Flow in a Resistive Circuit

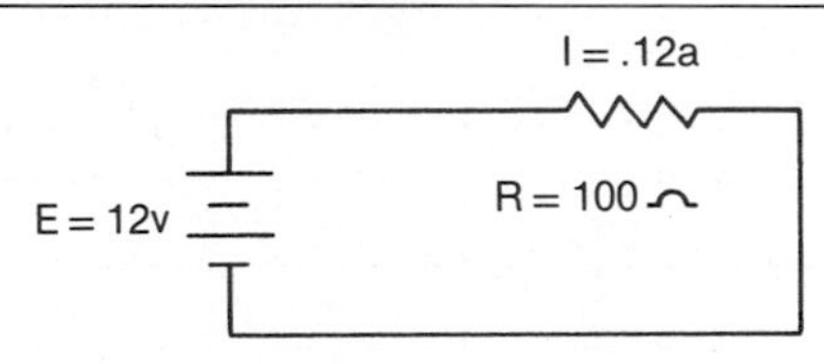

that used in automobiles is feeding current to a 100-ohm resistance such as a light bulb. In this circuit, 12 volts/100 ohms, or 120 milliamps, of current flows.

Power

Because current and voltage are inversely related and may vary with changing circuit conditions, neither is an adequate measure of the power consumed. The power in a circuit is defined as the product of the current and voltage, and it is measured in *watts*. For example, the circuit in Figure A.2 consumes 12 volts × .12 amps = 1.44 watts of power. The measurement of power is usually combined with the length of time the current flows and is normally expressed in kilowatt hours.

The Decibel

The power in telecommunications circuits is so low that it is normally measured in milliwatts. However, the milliwatt is not a convenient way to express differences in power level between circuits. Voice frequency circuits are designed around the human ear, which has a logarithmic response to changes in power. Therefore, in telephony the *decibel* (dB), a logarithmic rather than a linear measurement, is used as a measure of relative power between circuits or transmission level points. A change in level of 1 dB is barely perceptible under ideal conditions. The number of dB corresponding to a power ratio is expressed as:

$$dB = 10 \log P2/P1$$

The dB can also be used to express voltage and current

FIGURE A.3
Chart of Power and Voltage Ratios versus Decibels

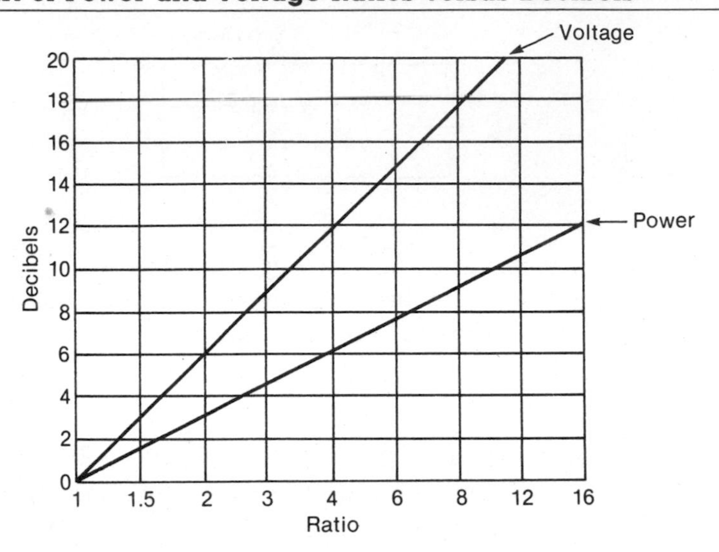

ratios if the impedance, which will be discussed in a later section, is the same for both values of voltage or current. The dB ratio between voltage and current is:

$$dB = 20 \log V2/V1$$

$$dB = 20 \log I2/I1$$

Figure A.3 is a chart showing the ratios between power, voltage, and current plotted as a function of decibels. Note that the horizontal scale is logarithmic. Also note that increases or reductions of 3 dB results in doubling or halving the power in a circuit. This ratio is handy to remember when evaluating power differences. The corresponding figure for doubling or halving voltage and current is 6 dB.

Series and Parallel Resistance

Figure A.4 shows resistors connected in series and parallel. In a series circuit, the total resistance is the sum of the individual resistors. In a parallel circuit, current divides between the resistors. If the resistors are of equal value, the

FIGURE A.4
Series and Parallel Resistance

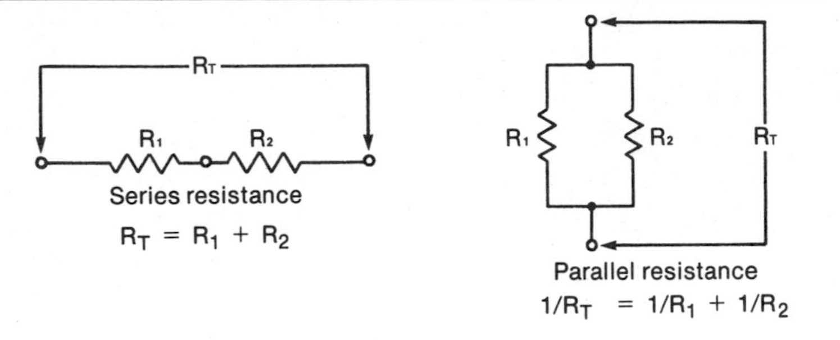

current divides equally; if they are unequal, a greater current flows through the smaller resistance.

DIRECT AND ALTERNATING CURRENT

So far we have been discussing only the flow of direct current (DC) in a circuit. In a DC circuit the source or battery supplies voltage with fixed *polarity*. The polarity in a battery corresponds to the poles in an electromagnet and is designated as positive (+) or negative (−). All telecommunications apparatus is powered by a DC source. However, the signals carried by the telecommunications apparatus and the power source that charges the batteries come from commercial alternating current (AC).

The polarity from an AC source is constantly reversing. If voltage is measured against time, the result is a *sine wave* as shown in Figure A.5. The wave is so named because of its relationship to a trigonometric function, the sine, in a manner that is beyond the scope of this discussion.

In AC circuit analysis, we are not only concerned with the magnitude or voltage of the source but also with *frequency* of its reversals. A complete cycle carries the voltage from its zero starting point, to peak positive, back through zero to peak negative, and back to zero again. The unit of measurement of frequency is the hertz (Hz). The distance between corresponding points on a cycle is called its *wavelength*, which is inversely proportional to frequency. Table

FIGURE A.5
Alternating Current Sine Wave

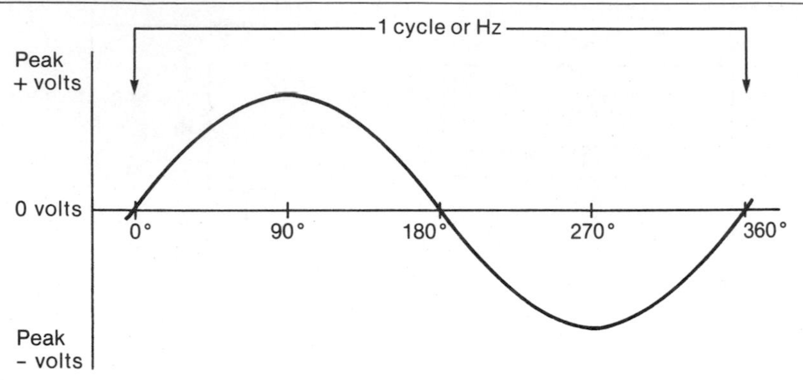

TABLE A.2
Frequencies and Wavelengths of Radio Frequencies

Frequency	*Wavelength*	*Class*	*Abbreviation*
30 KHz	10,000 meters	Very low frequency	VLF
300 KHz	1,000 meters	Low frequency	LF
3,000 KHz	100 meters	Medium frequency	MF
30 MHz	10 meters	High frequency	HF
300 MHz	1 meter	Very high frequency	VHF
3,000 MHz	.1 meter	Ultra high frequency	UHF
30,000 MHz	.01 meter	Superhigh frequency	SHF

A.2 shows the frequency and corresponding wavelengths of the radio frequency spectrum.

Another property of AC is *phase*, which describes the relationship between the zero-crossing points of signals. Figure A.6 shows two voltages that are 90 degrees out of phase with each other. It is convenient to measure phase in degrees as illustrated in the figure. A full cycle describes a 360-degree arc. The two signals shown are out of phase, with Curve B lagging Curve A by 90 degrees.

AC follows Ohm's law in a circuit composed only of resistance. Because the voltage in an AC circuit is varying continually, the current flow in such a circuit is proportional to the *root mean square* (RMS) voltage, which is .707 of the peak voltage. This voltage is normally used in describing AC; for example, the 120 volts of electric house current actually has a peak value of 170 volts.

FIGURE A.6
Phase Relationship between Two Sine Waves

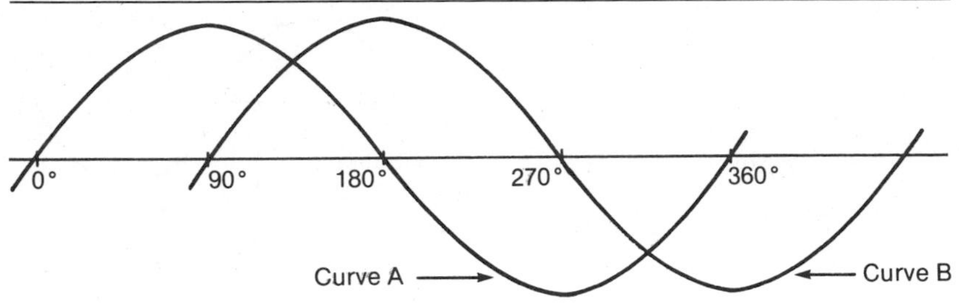

A sine wave consists of a single fundamental frequency. Voice signals are a complex composite of several fundamental tones rapidly varying in both frequency and amplitude. Because of the limited passband of a voice frequency telecommunications channel, higher and lower frequencies are cut off, but enough frequencies remain to ensure intelligibility. A digital signal, however, is not a sine wave. Instead, it takes the shape of a square wave as shown in Figure A.7a.

If a square wave is examined, it is found to consist of a fundamental frequency and numerous *harmonics*, which are multiples of the fundamental frequency. Figure A.7b shows the derivation of a square wave from the fundamental frequency and its harmonics. The harmonics consist of high-frequency components, many of which are filtered out by a telephone channel.

As shown in Figure A.8, the attenuation in a cable pair is generally limited, but as loading is applied, as described in Chapter 2, the cutoff frequency becomes sharper. The filters in amplifiers and multiplex equipment are sharper yet, approximating the curve shown in Figure A.9.

When a square wave signal is passed through a voice channel, the high-frequency components are filtered out. The signal is reduced to its fundamental frequency and assumes the shape of a sine wave. The degree of filtering in a voice channel is in direct proportion to the frequency of the square wave (or speed of a data signal) and the distance the signal travels. It is this filtering effect that makes it necessary to use modems to pass data signals over voice frequency channels.

FIGURE A.7
Derivation of Square Waves

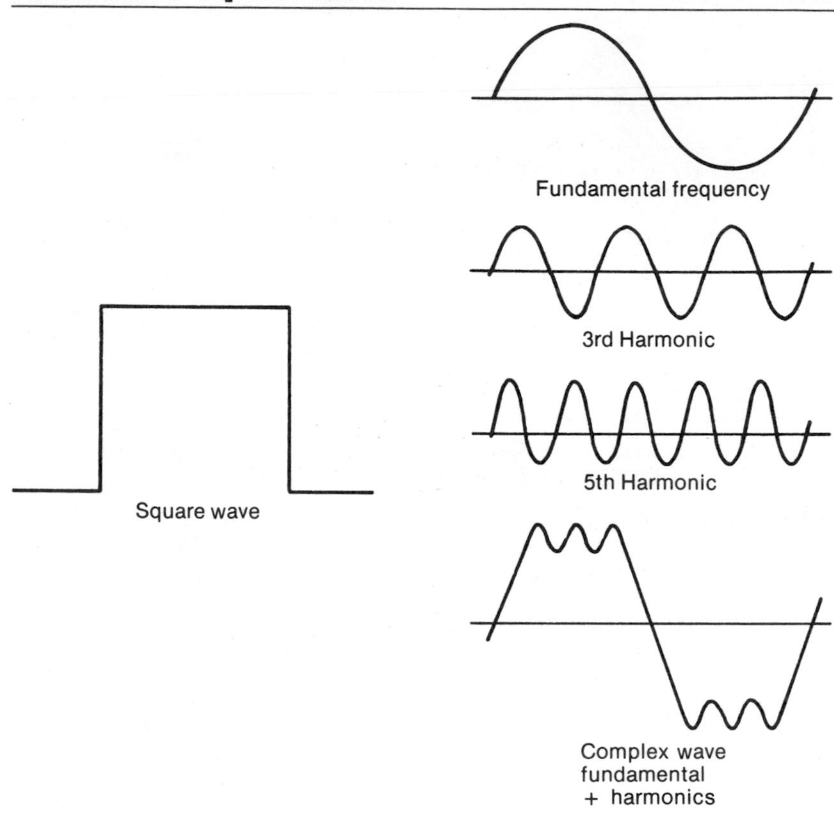

Inductance and Capacitance

All circuits also contain two more variables—capacitance, which is the property of storage of an electric charge, and inductance, which is the property of an electric force field built up around a conductor. Although inductance and capacitance are present in all circuits, their effects are slight in many circuits and can be ignored. However, in other circuits they are deliberately introduced to produce an intended effect.

Capacitance

Capacitance occurs when two plates of conducting material are separated by an insulator called a *dielectric*. Com-

FIGURE A.8
Frequency Response of Loaded and Nonloaded Voice Cable

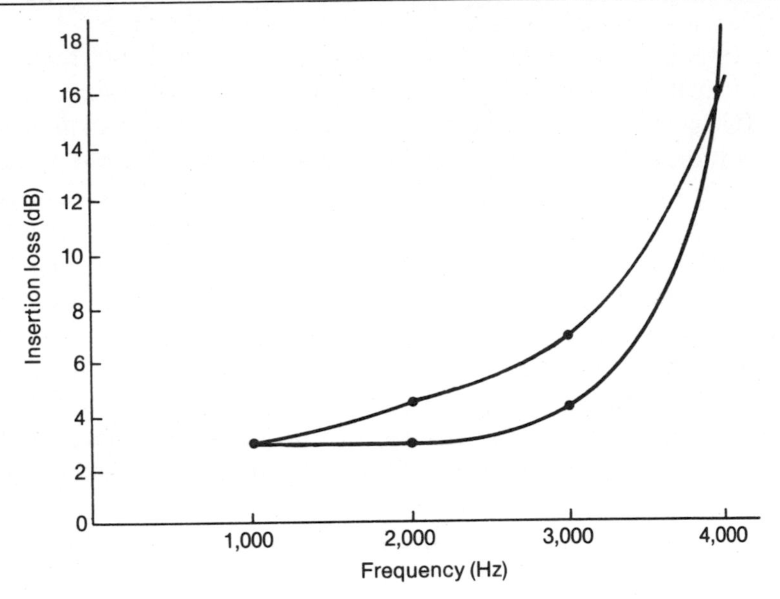

FIGURE A.9
Frequency Response of a Typical Telephone Channel

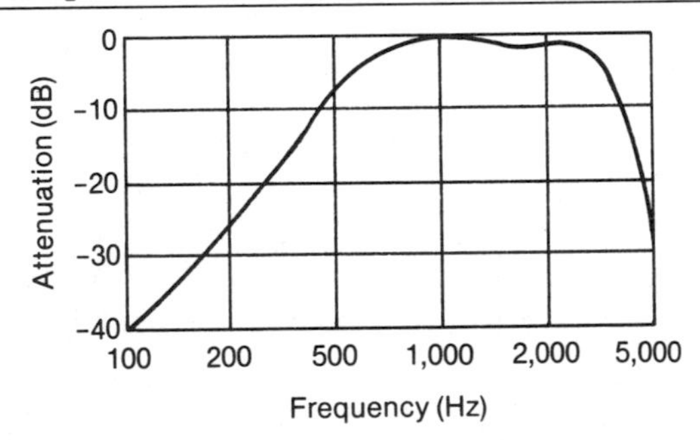

mon dielectrics are air, mica, ceramics, and plastics. Capacitance has the property of blocking the flow of DC but permitting some AC to flow. Capacitors are constructed with fixed values or with movable plates so the amount of capacitance is variable.

The unit of capacitance is the farad, although the farad is so large that capacitors are measured in microfarads or picofarads. The amount of capacitance is proportional to the amount of area of the conducting plates and the insulating properties or *dielectric constant* of the insulator.

A capacitor permits alternating current to flow through it, but not without some opposition. The opposition to flow of AC is called the *reactance* of a capacitor. For a capacitor of a given size, its reactance is inversely proportional to the AC frequency of the current flowing in the circuit. Capacitors are used in any electronic circuit where it is desirable to block the flow of DC while permitting the flow of AC, or to permit higher frequencies to pass while attenuating lower frequencies.

Of great pertinence to telecommunications is the fact that capacitance occurs naturally whenever two conductors parallel one another. Thus, the two wires of a cable pair form the plates of a capacitor and the wire insulation forms the dielectric. In a cable circuit, some current flows between the two wires, attenuating the higher frequencies more than the low.

When capacitors are connected in parallel, the effect is directly additive. For example, if two 1.0 mf capacitors are connected in parallel, the resulting capacitance is 2.0 mf. When they are connected in series, the total capacitance is reduced according to the same principle as when resistors are connected in parallel.

Inductance

An inductor is formed by winding a conductor into a coil. The amount of inductance is a function of the number of turns, the length and diameter of the coil, and the material used in the core of the coil. Many coils are wound with air cores, but the inductance of an air core coil can be increased by using magnetic material such as iron as its core. Inductors can be made variable by moving the core inside the winding.

When current flows through a wire, lines of force are built up around the wire. The field created by DC current is steady and unvarying, but when AC flows through a wire, the lines of force are constantly building and collapsing. It is these lines of force that impede the flow of AC in a coil. The effect of inductance on current flow is the opposite of the effects of capacitance. The flow of AC is impeded by an inductor while DC is passed. The unit of inductance is the henry, or in smaller coils, the millihenry or microhenry. The higher the frequency of an AC signal, the higher the *inductive reactance* of the coil.

The paralleling wires of a cable pair present some inductive reactance, but the effect is outweighed by the much greater effects of capacitive reactance. To counter capacitive reactance, inductance is often deliberately introduced into telephone circuits. For example, inductors known as *load coils* are connected in series with the two wires of long subscriber loops to counteract the effects of capacitive reactance and thus reduce the loss of the loop as shown in Figure A.8. Inductors behave in the same way as resistors when connected in series and parallel; that is, paralleling inductors reduces inductance, and adding them in series increases the inductance.

Impedance

The algebraic sum of inductive and capacitive reactance effects is known as the *impedance* of a circuit. Impedance in a circuit describes the opposition to the flow of current and varies with the frequency of an AC signal. When capacitive reactance and inductive reactance are equal to each other, the circuit is said to be in *resonance*. At its resonant frequency, a circuit offers minimum opposition to current flow.

The principles of resonance are used in electronics to tune circuits to a desired frequency. When a small amount of energy is added to a resonant circuit and some of the output is fed back into the input, the circuit *oscillates* at its resonant frequency. Figure A.10 illustrates the concept of oscillation in a parallel resonant circuit. Oscillators are widely used in telecommunications to generate audible tones and radio frequencies.

Transformers

When a coil is moved into the electrical field caused by the rising and collapsing lines of force of another coil fed by an AC source, a portion of the energy is coupled from the source into the second coil. This effect, shown in Figure A.11, is known as *mutual inductance* or the *transformer*

FIGURE A.10
Oscillator Circuit

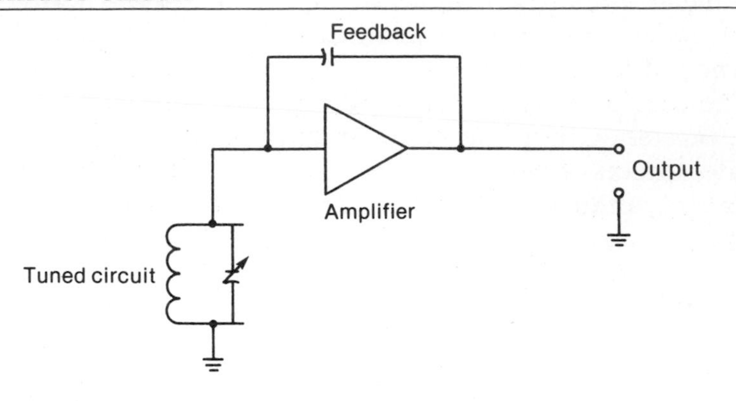

FIGURE A.11
Transformer

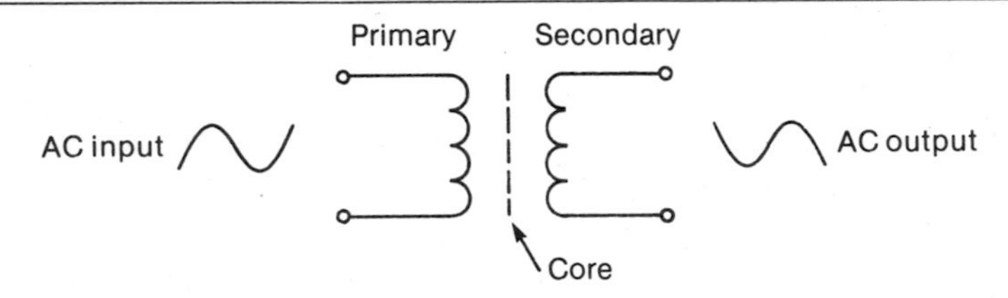

effect. The transformer windings connected to the source are known as the primary, and those connected to the load, the secondary. The amount of voltage induced into the secondary is a function of how closely coupled the two windings are, the frequency of the AC source, and the ratio of turns between the primary and secondary windings.

If the secondary consists of more turns than the primary, the voltage is stepped up; with fewer turns in the secondary the voltage is stepped down. Many transformers have multiple windings in the secondary to create several voltages from a single source, but of course, the total power in the secondary windings cannot exceed the power of the primary. For example, if a transformer that draws one amp of current at 120 volts in the primary has a 1:2 turns ratio, the voltage in the secondary is 240 volts. However, no more than 0.5 amps of current can be supplied by the secondary winding.

Actually, the power supplied in the secondary is somewhat less because the transformer is not 100 percent efficient.

Transformers are widely used in telecommunications, where they are often called *repeat coils*. Not only do transformers convert AC power to the voltages used to charge batteries and to power apparatus, they are also used as impedance matching devices.

With two circuits interconnected, the maximum transfer of power occurs when their impedances are exactly matched. Telephone circuits are designed to present impedances of 600 ohms in the case of toll equipment or 900 ohms for local equipment. These impedances were chosen because of the characteristic impedances of wire transmission lines. The characteristic impedance of equipment used in amplifiers and multiplex is much higher than the 600 or 900 ohms of telephone circuits. Therefore, wherever equipment connects to external telephone circuits, the impedances are matched by repeat coils or other matching circuits that are beyond the scope of this presentation.

All users of telecommunications services must be alert to the hazards of mismatched impedances. Wherever a mismatch occurs, part of the signal is reflected to the source and a reduced transfer of energy results. Impedance mismatches are the source of echo, unwanted oscillations, and excessive loss. It is important in telecommunications not to connect circuits without the use of impedance matching devices.

Filters

A filter is any device that rejects a frequency or band of frequencies while allowing other frequencies to pass. Capacitors and inductors, as we have seen, make very effective filters with opposite effects. A capacitor presents high impedance to low-frequency signals, blocks DC entirely, and allows high-frequency signals to pass with little or no attenuation. Inductors, on the other hand, offer no resistance to the flow of DC but high resistance to AC. As a practical matter, an instantaneous impedance is offered to AC by a capacitor until an initial charge is established. Similarly, an initial impedance is offered to the flow of DC by an inductor until the lines of force are built up. These effects are used

FIGURE A.12
Band Pass Filter

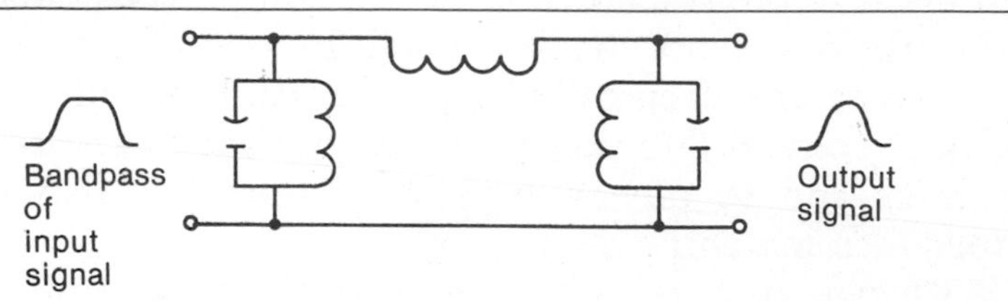

in pulse generating circuits, but in filters where the flow of current is fairly steady, this effect can be ignored.

Capacitors and inductors are used in filter networks such that shown in Figure A.12. As shown in the figure, if a broad band of frequencies is applied to the input of the circuit, the upper and lower frequencies are attenuated while the mid-band is passed. Although such a passband is adequate where it is unnecessary to reject the high and low frequencies entirely, in many telecommunications applications this "leak through" of signal power is unacceptable. For example, if two channels share the same medium and both have response curves similar to those shown in the output of Figure A.12, the overlapping passbands would result in audible crosstalk, which is clearly undesirable. For most telecommunications circuits, filters with steep-skirted response curves and a passband with little amplitude distortion are required. This is accomplished by inserting piezoelectric crystals in the filter, resulting in a passband similar to that in Figure A.9.

RADIO

The principles of radio transmission are fundamental to understanding telecommunications. Not only is radio widely used for carrying voice channels and for mobile telephones, the same principles are also used in analog multiplex. An analog carrier system resembles a multichannel radio except the signal is low powered and is not radiated into free space.

FIGURE A.13
Block Diagram of a Radio Transmitter

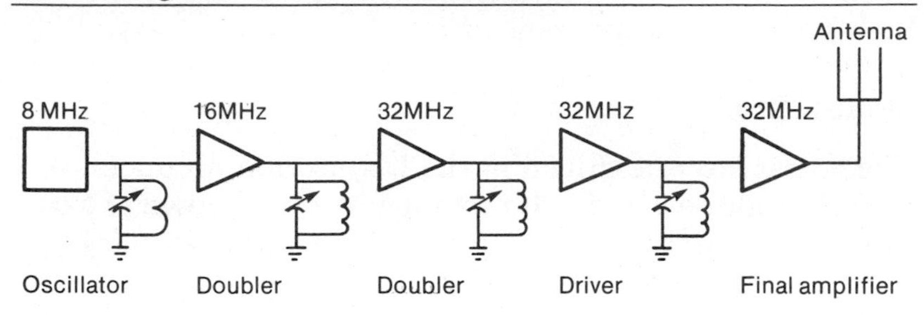

Transmitters

Tuned or resonant circuits are at the heart of every radio. A tuned circuit is a type of filter in which the capacitance and inductance are tuned to a resonant frequency and pass only a very narrow band of frequencies. As we discussed earlier, when amplification is added to a tuned circuit and a portion of the output is fed back into the input of the amplifier, the amplifier oscillates. Oscillators are used to generate radio frequencies (**rf**). Piezoelectric crystals are often employed in oscillator circuits to ensure frequency stability and accuracy. The output of the oscillator is a sine wave of a fundamental frequency and is often rich in harmonic frequencies as well.

Harmonic frequencies are used in circuits called *frequency multipliers*. In the block diagram of Figure A.13, an oscillator and a chain of frequency multipliers are used to raise the fundamental frequency to the desired rf output frequency. Low-frequency oscillators are used in radio because it is easier to control the stability at low frequencies than at high.

Frequency multiplier stages employ tuned circuits to select the desired harmonic. For example, in Figure A.13 the first stage frequency doubler is tuned to the second harmonic of the 8 MHz oscillator to select an output frequency of 16 MHz.

The output of the multipliers is connected to a *driver* that amplifies the signal to the higher level required by the final rf amplifier. The final amplifier boosts the signal to the desired output level. The output may range from a watt or less in some microwave transmitters to a megawatt or more in high-power broadcast transmitters.

Modulation

The transmitter described in the last section produces only a single high-powered radio frequency wave known as a *carrier*. The carrier contains no intelligence. Radio telegraph transmitters carry information by interrupting the carrier with a coded signal of dots and dashes. This type of modulation is known as *continuous wave* (cw).

Amplitude Modulation

Voice communication is far more essential to telecommunications than cw, which is disappearing as a means of communication. A voice signal is impressed on an rf carrier by a process called modulation. The simplest form is called *amplitude modulation* (AM). An AM modulator consists of several stages of audio amplification that are coupled to an rf amplifier through a transformer. The resultant signal is shown in Figure A.14.

Single Sideband

When amplitude modulation occurs, four frequencies result: the original voice frequency, the carrier frequency, the sum of the two frequencies, and the difference between the two frequencies. The sum and difference frequencies are called *sidebands*. In an AM signal all of the information is contained in the two sidebands. The carrier contributes nothing to communication, and the two sidebands are redundant. Moreover, 75 percent of the signal power is contained in the carrier and the redundant sideband. A large improvement in efficiency is obtained by using *single sideband* (**SSB**) telephony.

In an SSB transmitter the carrier is suppressed in a circuit called a *balanced modulator*. The output of this modulator consists of the upper and lower sidebands, only one of which is needed. The unwanted sideband is eliminated with a filter. The carrier suppression and filtering are done in low-level stages of the transmitter where little power handling capac-

FIGURE A.14
Amplitude Modulation

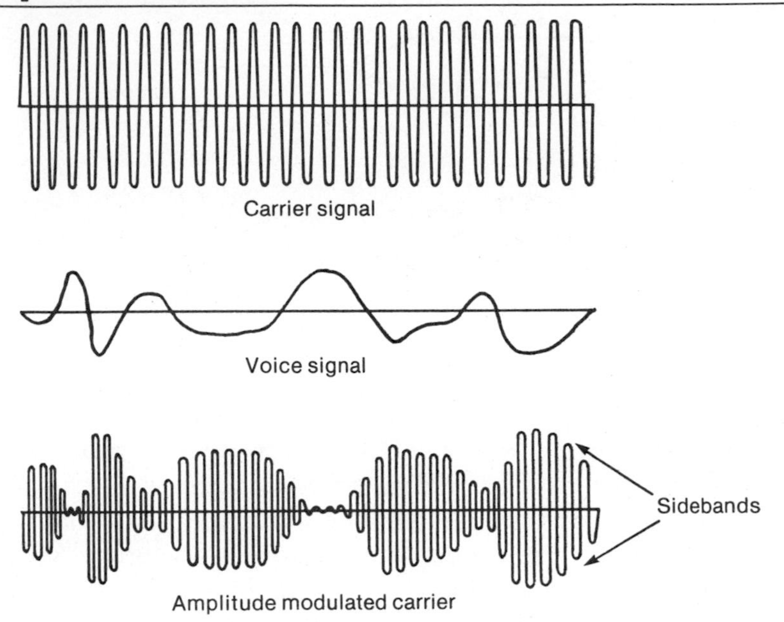

ity is needed. After filtering, the SSB signal is boosted to the desired output power by the rf amplifier.

Frequency Modulation

Telecommunications apparatus also makes wide use of *frequency modulation* (FM) in radio and, to a limited degree, in multiplex systems. In an FM transmitter an rf signal is generated by an oscillator. With no modulating voice signal, the carrier rests at its center frequency. The modulating signal shifts the carrier above and below its center frequency in proportion to the frequency and amplitude of the modulating signal. The result is a broad band signal of varying frequencies.

FM passes a wide band signal with a great deal of linearity. Noise, which tends to affect the amplitude of a signal, can be eliminated in the receiver by filtering the received signal through limiting amplifiers that chop off the noise peaks. The result is a low noise output with wide bandwidth. FM is widely used in mobile and microwave radio, but it is rarely used in analog multiplex systems.

FIGURE A.15
Mixer

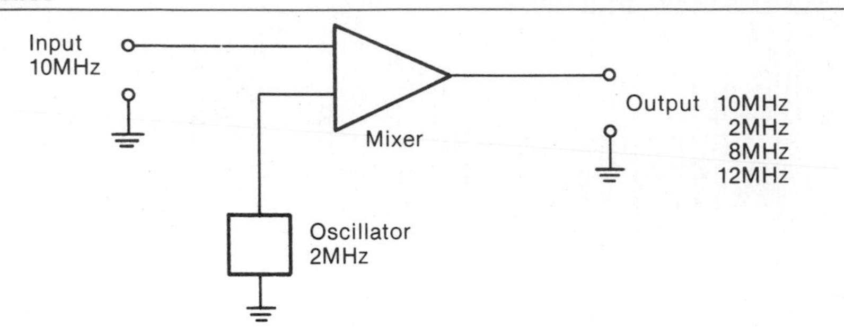

Heterodyning

The transmitter in Figure A.13 uses frequency multipliers to generate the desired transmitting frequency. Both transmitters and receivers also use an effect called *heterodyning* to raise or lower the fundamental frequency. Figure A.15 is a block diagram that illustrates the principle. As discussed in the modulation section, when two frequencies are applied to a mixer, the output consists of the two original frequencies, the sum of the two frequencies, and the difference between the frequencies. Tuned circuits are used to select the desired frequency.

Heterodyning is used in many microwave repeaters to change the frequency of the received signal to a different transmit frequency. Most receivers also use heterodyning to change an incoming signal to a fixed intermediate frequency as we discuss in the next section.

Receivers

A block diagram of a typical radio receiver using *superheterodyne* technology is shown in Figure A.16. The signal from the antenna is boosted by an rf amplifier and coupled to a mixer. A signal from an oscillator is coupled into the other mixer port. The selected output is a fixed *intermediate frequency* (**if**) that is coupled to succeeding if amplifier stages. The superheterodyne technique is used to improve selectivity. Intermediate frequency amplifier stages are designed with steep-skirted filters to reject unwanted frequencies.

FIGURE A.16
Block Diagram of a Superheterodyne Radio Receiver

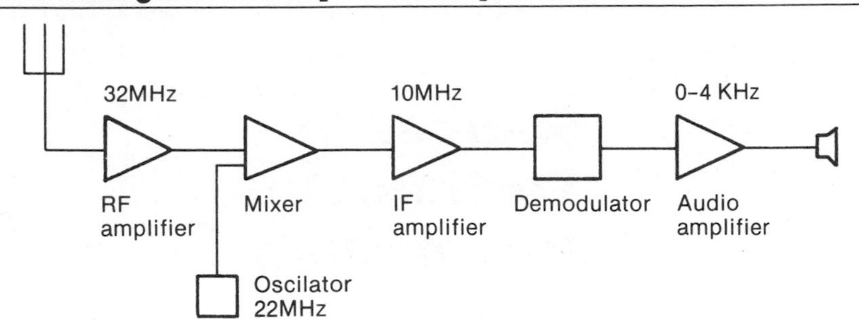

The superheterodyne technique is used in virtually all receivers ranging from simple broadcast receivers to microwave radios. The intermediate frequency is chosen to be high enough to pass the entire range of desired frequencies. Broadcast band radios, which are tuned to only one voice channel, typically use a 455 kHz if, FM radios use 10.7 MHz, and most microwave systems use 70 MHz as the if. An intermediate frequency this high is required to handle the bandwidth of a microwave system. Bandwidths of 10 or 20 MHz are common.

The demodulator stage in a receiver is designed to select the audio output from the if stages. In a broadcast receiver only a simple device such as a diode is needed to detect the audio envelope from the AM signal. In single sideband receivers a frequency identical to the original suppressed carrier frequency must be injected into the detector by means of an oscillator. In FM receivers the detector is a linear circuit known as a *discriminator* that translates the frequency excursions into a signal identical to the original modulating frequencies.

This overview of electrical theory has only brushed the surface of the technology. Readers who require more information are advised to consult one of the many manuals on electric theory that are available in most bookstores.

APPENDIX B

Sources of Technical Information

The following are the addresses of organizations that prepare standards and technical information pertaining to telecommunications:

AMERICAN NATIONAL STANDARDS INSTITUTE

ANSI standards and ANSI's *Catalog of American National Standards* can be ordered from:

American National Standards Institute
1430 Broadway
New York, NY 10018
(212) 354-3471

AT&T TECHNOLOGIES

AT&T publications and their *Commercial Sales Documentation Catalog* are available from:

AT&T Customer Information Center
Commercial Sales Representative
P.O. Box 19901
Indianapolis, IN 46219
(317) 352-8557 or (800) 432-6600

BELL COMMUNICATIONS RESEARCH

Bell Communications Research publications and its *Catalog of Technical Information* listing publications of Bellcore and the seven operating telephone regions are available from:

Bell Communications Research, Inc.
Information Operations Center
60 New England Ave.
Piscataway, NJ 08854
(201) 699-3047

ELECTRONIC INDUSTRIES ASSOCIATION

EIA standards and their publication *Catalog of EIA and JEDC Standards and Engineering Publications* are available from:

Electronic Industries Association
2001 Eye St., N.W.
Washington, DC 20006
(202) 457-4966

INSTITUTE OF ELECTRICAL AND ELECTRONIC ENGINEERS

IEEE standards are available from:

IEEE Computer Society
Suite 608
111 19th St, N. W.
Washington, DC 20036
(202) 785-0017

INTERNATIONAL ORGANIZATION FOR STANDARDIZATION

ISO standards are available from:

Central Secretariat
1, rue de Varembe
CH–1211 Geneva, Switzerland
41 22 34–12–40

or

American National Standards Institute
1430 Broadway
New York, NY 10018
(212) 354–3471

INTERNATIONAL RADIO CONSULTATIVE COMMITTEE

CCIR standards are available from:

General Secretariat
International Telecommunications Union
Place de Nations
1211 Geneva 20 Switzerland

or

United Nations Bookstore
Room 32B
U.N. General Assembly Building
New York, NY 10017

INTERNATIONAL TELEPHONE AND TELEGRAPH CONSULTATIVE COMMITTEE

CCITT standards are available from:

General Secretariat
International Telecommunications Union
Place de Nations
1211 Geneva 20 Switzerland

or

United Nations Bookstore
Room 32B
U.N. General Assembly Building
New York, NY 10017

APPENDIX C

List of Selected Manufacturers and Vendors of Telecommunications Products and Services

This appendix lists manufacturers and vendors whose products are included in the lists at the end of selected chapters. This list is not a complete buyers guide. This list and the lists at the ends of the chapters are a representative sample of the vendors on the market. Because addresses and telephone numbers are changing frequently, it is advisable to check these against sources that are kept updated. Readers are advised to consult one of the following sources for current and more complete manufacturer information:

Data Communications. Buyer's Guide Issue Data Communications Magazine, McGraw-Hill Building, 1221 Avenue of the Americas, New York, NY 10020, (212) 512-2000.

Data Decisions Communications Systems. Data Decisions, 20 Brace Rd., Cherry Hill, NJ 08034, (609) 429-7100.

Datapro Reports on Telecommunications and *Datapro Reports on Data Communications.* Data Pro Research Corporation, 1805 Underwood Blvd., Delran, NJ 08075, (609) 764-0100.

Telecommunications Reference Data and Buyers Guide. Telecommunications, 610 Washington St., Dedham, MA 02026, (617) 326-8220.

Telephone Engineer and Management Directory. Telephone Engineer and Management Magazine, 124 South First St., Geneva, IL 60134, (312) 232-1400.

Telephony Directory and Buyers Guide. Telephony Publishing Co., 55 E. Jackson Blvd., Chicago, IL 60604.

MANUFACTURERS AND VENDORS OF TELECOMMUNICATIONS PRODUCTS AND SERVICES

Account-A-Call Corporation, 4450 Lakeside Dr., Burbank, CA 91505, (818) 846-3340.

Action Communication, Division of Honeywell, Inc., 4401 Beltwood Pkwy. S., Dallas, TX 75234, (214) 386-3500.

ADC Telecommunications, 4900 W. 78th St., Minneapolis, MN 55435, (612) 835-6800.

ADP Autonet, 175 Jackson Plaza, Ann Arbor, MI 48106, (313) 769-6800.

Alpha Wire, Box 711, Elizabeth, NJ 07207, (201) 925-8000.

Alston Division, Conrac Corporation, 1724 S. Mountain Ave., Duarte, CA 91010, (818) 357-2121.

Amdahl Communications Systems Division, 2500 Walnut Ave., Marina del Rey, CA 90291, (213) 822-3202.

American Laser Systems, Inc., 106 James Fowler Rd., Goleta, CA 93117, (805) 967-0423.

American Satellite Corporation, 20301 Century Blvd., Germantown, MD 20767, (301) 428-6040.

AMP Telecom Division, P. O. Box 1776, Southeastern, PA 19399, (215) 647-1000.

Amphenol, 2122 York Rd., Oak Brook, IL 60521, (312) 986-2300.

Anaconda-Ericsson Inc., Wire & Cable Division-Telcom Cable, P.O. Box 4405, Overland Park, KS 66204, (913) 677-7500.

Andrew Corporation, 10500 W. 153rd St., Orland Park, IL 60462, (312) 349-3300.

Anritsu America, Inc., 128 Bauer Dr., Oakland, NJ 07436, (201) 337-1111.

Arus Corporation, 1305 Conkling Ave., Utica, NY 13501, (800) 824-4357.

Atlantic Research Corporation, 5390 Cherokee Ave., Alexandria, VA 22312, (703) 642-4000.

AT&T Communications, Bedminister, NJ 07921, (201) 234-4000.

AT&T Information Systems, 1 Speedwell Ave., Morristown, NJ 07960, (800) 247-1212.

AT&T Network Systems, P.O. Box 1278R, Morristown, NJ 07960.

AT&T Technologies Consumer Products, 5 Woodhollow Rd., Parsippany, NJ 07054, (201) 581-3000.

Audichron Company, 3620 Clearview Pkwy., Atlanta, GA 30340, (404) 455-4890.

Avantek, Inc., 481 Cottonwood Dr., Bldg. 5, Milpitas, CA 95035, (408) 943-4412.

Aydin Microwave Division, 75 E. Trimble Rd., San Jose, CA 95131, (408) 946-5600.

Aydin Monitor System Digital Communications Group, 502 Office Center Dr., Ft Washington, PA 19034, (215) 646-8100.

Badger/TTI, 150 E. Standard Ave., Richmond, CA 94804, (415) 233-8220.

BBL Industries, 2935 Northeast Pkwy., Atlanta, GA 30360, (404) 449-7740.

BBN Communications Corporation, 70 Fawcett St., Cambridge, MA 02238, (617) 497-2800.

Belden Corp., 2000 S. Batavia Ave., Geneva, IL 60134, (312) 232-8900.

Benner-Nawman, Inc., 3070 Bay Vista Ct., Benicia, CA 94510, (707) 746-0500.

Berry Electronics, Inc., 719 Swift St., #B12, Santa Cruz, CA 95060, (408) 429-1857.

BFI Communications Systems Inc., 109 N. Genesee St., Utica, NY 13502, (315) 735-8867.

Biddle Instruments, 510 Township Line Rd., Blue Bell, PA 19422, (215) 646-9200.

Bitek International, Inc., 6015 Obispo Ave., Long Beach, CA 90805, (213) 634-8950.

Bogen Division/Lear Siegler, Inc., P.O. Box 500, Paramus, NJ 07652, (201) 343-5700.

Bowmar/ALI, Inc., 531 Main St., Acton, MA 01720, (617) 263-8365.

Brand-Rex Company, Telecommunications Division, P.O. Box 498, Willimantic, CT 06226, (203) 423-7783.

Bridge Communications, 1345 Shoebird Way, Mountain View, CA 94043, (415) 969-4400.

Burroughs Corporation Imaging Systems Division, Corporate Drive, Commerce Park, Danbury, CT 06810, (800) 243-7046.

Calculagraph Co., 272 Ridgedale Ave., East Hanover, NJ 07936, (201) 887-5000.

California Microwave, 990 Almanor Ave., Sunnyvale, CA 94086, (408) 732-4000.

Candela Electronics, 550 Del Rey Ave., P.O. Box 461, Sunnyvale, CA 94088, (408) 738-3800.

Canon USA, One Cannon Plaza, Lake Success, NY 11042, (516) 488-6700.

Case-Rixon Communications, Inc., 2120 Industrial Parkway, Silver Spring, MD 20904-1999.

C-COR Electronics, 8285 S.W. Nimbus Ave, Beaverton, OR 97005, (503) 644–0551.

Cermetek Microelectronics, Inc., 1308 Borregas Ave., Sunnyvale, CA 94088-3565, (408) 734–8150.

Chatlos Systems, Inc., 125 Algonquin Pkwy., Whippany, NJ 07981, (201) 887–1456.

CIT-Alcatel, Inc., 10800 Parkridge Blvd., Reston, VA 22091, (703) 476–7300.

Coastcom, 2312 Stanwell Dr., Concord, CA 94520, (415) 825–7500.

Cobra/Dynascan Corp., 6460 West Cortland St., Chicago, IL 60635, (312) 889–8870.

Code-A-Phone Corporation, P.O. Box 5656, Portland, OR 97228, (503) 655–8940.

Codex Corp., 20 Cabot Blvd., Mansfield, MA 02408, (617) 364–2000.

Cointel Communications, 23801 Calabasas Rd., Calabasas, CA 91302, (818) 716–5889.

Collins Transmission Systems Division, Rockwell International Corp., P.O. Box 10462, Dallas, TX 75207, (214) 996–5000.

Colorado Video, Box 928, Boulder, CO 80306, (303) 444–3972.

Comdial Corporation, P.O. Box 7266, Charlottesville, VA 22906, (804) 978–2458.

Communication Mfg. Co., P.O. Box 2708, Long Beach, CA 90801, (213) 426–8345.

Complexx Systems, Inc., 4930 Research Dr., Huntsville, AL 35805, (205) 830–4310.

Compression Labs, Inc., 2305 Bering Dr., San Jose, CA 95131, (408) 946–3060.

CompuServe, Inc., 5000 Arlington Centre Blvd., Columbus, OH 43220, (614) 457–8600.

Computer Sciences Corporation, 2100 East Grand Ave., El Segundo, CA 90245, (213) 615–0311.

Concord Data Systems, 303 Bear Hill Rd., Waltham, MA 02154, (617) 890–1394.

Control Data Corporation, P.O. Box 0, Minneapolis, MN 55440, (612) 853–8100.

Contel Information Systems, 130 Steamboat Rd., Great Neck, NY 11024, (516) 829–5900.

Corning Glass Works, Telecommunications Products Division, Corning, NY 14831, (607) 974–4411.

Corvus Systems, Inc., 2100 Corvus Dr., San Jose, CA 95124, (408) 559–7000.

Crest Industries, Inc., 6922 N. Meridian, Puyallup, WA 98371, (206) 927–6922.

Cushman Electronics, 2450 N. First St., San Jose, CA 95131, (408) 263-8100.

Data General Corp., 4400 Computer Dr., Westboro, MA 01580, (617) 366-8911.

Data Switch Corporation, 444 Westport Ave., Norwalk, CT 06851, (203) 847-9800.

Databit Inc., 110 Ricefield Lane, Hauppauge, NY 11788, (800) 328-2248.

Datapoint Corporation, 9725 Datapoint Dr., San Antonio, TX 78284, (512) 699-7000.

Datatel Inc., Cherry Hill Industrial Center, Cherry Hill, NJ 08003, (609) 424-4451.

Datec Inc., 200 Eastowne Dr., Chapel Hill, NC 27514, (919) 929-2135.

Design Development Inc., 1440 Atteberry La., San Jose, CA 95131, (408) 946-6770.

Dictaphone Corporation, 120 Old Post Rd., Rye, NY 10580, (914) 967-7300.

Digital Communications Associates, Inc., 303 Research Dr., Norcross, GA 30092, (800) 241-5793.

Digital Equipment Corporation, Box 430, Merrimack, NH 03054, (603) 884-5039.

Digital Microwave Co., 2363 Calle del Mundo, Santa Clara, CA 95054, (800) 362-9283.

Digitech Communications, 914 Bob Wallace Ave., Huntsville, AL 35801, (205) 533-5941.

DSC Communications Corporation, 707 E. Arapaho, Richardson, TX 75081, (214) 238-4000.

Dynatech Data Systems, Division of Dynatech Corporation, 7644 Dynatech Ct., Springfield, VA 22153, (703) 569-9000.

Dynatel 3M, 380 N. Pastoria Ave., Sunnyvale, CA 95086, (408) 733-4300.

Eagle Telephonics, Inc., 375 Oser Ave., Hauppauge, NY 11788, (516) 273-6700.

Equatorial Communications Co., 300 Ferguson Dr., Mountain View, CA 04043, (415) 969-9500.

Ericsson Inc., Communications Division, 7465 Lampson Ave., Garden Grove, CA 92641, (714) 895-3962.

Executone, Inc., 29-10 Thomson Ave., Long Island City, NY 11101, (212) 392-4800.

Exide Corporation, 101 Gibralter Rd., Horsham, PA 19044, (215) 674-9500.

Farinon Division, Harris Corporation, 1691 Bayport Ave., San Carlos, CA 94070, (415) 592-4120.

Fluke MFG. Co., Inc., P.O. Box C9090, Everett, WA 98206, (206) 365-5400.

Frederick Electronics Corp., Hayward Rd., P.O. Box 502, Frederick, MD 21701, (301) 662-5901.

Fuijitsu America, Inc., Data Communications Division, 3055 Orchard Dr. San Jose, CA 95134, (408) 946-8777.

Fuijitsu Business Communications, 3190 MiraLoma Ave., Anaheim, CA 92806, (714) 630-7721.

Gabriel Electronics, Inc., P.O. Box 70, Scarborough, ME 04704, (207) 883-5161.

Gandalf Technologies, Inc. 350 E. Dundee Rd., Wheeling, IL 60090, (312) 541-6060.

General Cable Company, 500 W. Putnam Ave., Greenwich, CT 06830, (203) 661-0100.

General Cable Co., Fiber Optics Division, 160 Fieldcrest Ave., Raritan Center, Edison, NJ 08810, (201) 225-4780.

General Datacomm, Inc. One Kennedy Ave., Danbury, CT 06762-1299, (203) 797-0711.

General Electric, Instrument & Communications Equpt. Services Centers, Bldg. 4, Room 210, 1 River Rd., Schenectady, NY 12345, (518) 385-9912.

General Electric Mobile Communications Division, P.O. Box 4197, Lynchburg, VA 24502, (804) 528-7721.

Glenayre Electronics, 12 Pacific Highway, Blaine, WA 98230, (206) 676-1980.

Grafnet, Inc., 329 Alfred Ave., Teaneck, NJ 07666, (201) 837-5100.

Granger Associates, 3101 Scott Blvd., Santa Clara, CA 95051, (408) 727-3101.

GTE Communication Systems, 2500 W. Utopia Rd., Phoenix, AZ 85027, (602) 582-7000.

GTE Corporation Business Communications Systems, Inc., 12502 Sunrise Valley Dr., Reston, VA 22090, (703) 435-7643.

GTE Satellite Corporation, One Stamford Forum, Stamford, CT 16904, (203) 965-3400.

GTE Telenet Communications Corporation, 8229 Boone Blvd., Vienna, VA 22180, (703) 442-1000.

Halcyon Communications, Inc., 2121 Zanker Rd., San Jose, CA 95131, (408) 293-9970.

Harris Corporation, Digital Telephone Systems Division, 1 Digital Dr., P.O. Box 1188, Novato, CA 94947, (415) 472-2500.

Harris Corporation, RF Communications Group, 1680 University Ave., Rochester, NY 14610, (716) 244-5830.

Harris Corporation, Fiber Optic Systems, P.O. Box 37, Melbourne, FL 32901, (305) 724-3600.

Hayes Microcomputer Products, Inc., 5923 Peachtree Industrial Blvd., Norcross, GA 30092, (404) 449–8791.

Hekimian Laboratories, Inc., 9298 Gaither Rd., Gaithersburg, MD 20877, (301) 840–1217.

Hewlett-Packard Co., 1501 Page Mill Rd., Palo Alto, CA 94304, (415) 493–1501.

Hitachi America, LTD. Telecommunications Research and Sales Division, 2990 Gateway Dr., Suite 1000, Norcross, GA 30071, (404) 446–8820.

IBM Corporation, Old Orchard Rd., Armonk, NY 10504, (914) 765–1900.

IBM Information Network, P.O. Box 30021, Tampa, FL 33630-9948, (800) 426–2468.

IBM Telecommunications Carrier Products, P.O. Box 10, Dept. T30-02, Princeton, NJ 08540, (609) 734–8748.

Infotron Systems Corporation, Cherry Hill Industrial Cntr., Cherry Hill, NJ 08003, (800) 257–8352.

Intecom, Inc., 601 Intecom Dr., Allen, TX 75002, (214) 727–9141.

Interlan, Lyberty Way, Westford, MA 01886, (617) 692–3900.

IPC Technologies, Inc., Winthrop Rd., Chester, CT 06412, (203) 526–9574.

ITEC Inc., P.O. Box 4147, Huntsville, AL 35815, (205) 881–7944.

ITT PowerSystems Corporation, P.O. Box 688, Galion, OH 44833, (419) 468–8100.

ITT Telecom, Network Systems Division, 3100 Highwoods Blvd., Raleigh, NC 27604, (919) 872–3359.

ITT Telecommunications Business and Consumer Division, 133 Terminal Ave., Clark, NJ 07066, (201) 348–7000.

ITT Worldcom, 100 Plaza Dr., Secaucus, NJ 07906, (201) 330–5000.

Jerrold Division of General Instrument Corp., 2200 Byberry Rd., Hatboro, PA 19040, (215) 674–4800.

E. F. Johnson Company, 299 Johnson Ave., Waseca, MN 56093, (507) 835–6222.

Joslyn Electronic Systems, Box 817, Goleta, CA 93116, (805) 968–3551.

Kentrox Industries, Inc., P.O. Box 10704, Portland, OR 97201, (503) 643–1681.

Kohler Co., Kohler, WI 53044, (414) 565–3381.

Lanier Business Products Inc., A Harris Company, 1700 Chantilly Dr., N.E., Atlanta, GA 30324, (404) 329–8000.

Lear Siegler, Inc., Electronic Instrumentation Division, 714 N. Brookhurst St., Anaheim, CA 92803, (714) 774–1010.

Light Communications Corporation, 25 Van Zant St., Norwalk, CT 06855, (203) 866–6858.

LorTec Power Systems, Inc., 5214 Mills Industrial Parkway, North Ridgeville, OH 44039, (216) 327-5050.

Lynch Communications Systems Inc., 204 Edison Way, Reno, NV 89520, (702) 786-4020.

3M Company, Telcomm Products Division, 3M Center, Bldg. 225-4S-06, St. Paul, MN 55144, (612) 733-9646.

3M Company, Business Communication Products Division, 3M Center, St. Paul, MN 55101, (612) 733-1110.

M/A-COM DCC Inc., 1171 Exploration La., Germantown, MD 20874, (603) 424-3400.

M/A-COM Land Mobile Communications, Inc., 21 Continental Blvd., Merrimack, NH 03054, (603) 424-3400.

M/A-COM MAC, Inc. 15 Fortune Dr., Billerica, MA 01821, (800) 343-0864.

Marconi Instruments, Division of Marconi Electronics Inc., 100 Stonehurst Ct., Northvale, NJ 07647, (201) 767-7250.

MCI International, International Dr., Rye Brook, NY 10573, (914) 937-3444.

MCI Telecommunications Corporation, 1133 19th St. N.W., Washington, DC 20036, (202) 872-1600.

Melco Labs, 14408 N.E. 20th St., Bellevue, WA 98007, (206) 643-3400.

Micom Systems, Inc., 20151 Nordhoff St., Chatsworth, CA 91311, (213) 998-8844.

Mitel, Inc., 5400 Broken Sound Blvd., N.W. Boca Raton, FL 33431, (305) 994-8500.

Motorola Communications & Electronics, Inc., 1301 E. Algonquin Rd., Schaumburg, IL 60196, (312) 397-1000.

NEC America Inc. Switching Systems Division, 1525 Walnut Hill La., Irving, TX 75062, (214) 257-9100.

NEC America Inc., Radio & Transmission Division, 2740 Prosperity Ave., Fairfax, VA 22031, (703) 560-2010.

NEC Telephones, Inc., 532 Broad Hollow Rd., Melville, NY 11747, (800) 626-4952.

Newton Instrument Co., Inc., 111 East A St., Butner, NC 27509, (919) 575-6426.

Northcom, P.O. 600, Industrial Parkway, Industrial Airport, KS 66031, (913) 791-7000.

Northern Telecom Inc., Digital Switching Systems, 4001 E. Chapel Hill-Nelson Hwy., Research Triangle Pk., NC 27709, (919) 549-5000.

Northern Telecom Inc., Integrated Office Systems, 1001 E. Arapaho Rd., Richardson, TX 75081, (214) 234-5300.

Northern Telecom Inc., Network Systems, 1201 E. Arapaho Rd., Richardson, TX 75081, (214) 234-7500.

Northern Telecom Inc., Northeast Electronics Division, Airport Rd., P.O. Box 649, Concord, NH 03301, (603) 224-6511.

Northern Telecom Inc., Optical Systems Division, 1555 Roadhaven Dr., Stone Mountain, GA 30083, (404) 491-7717.

Northern Telecom Inc., Spectron Division, 8000 Lincoln Dr. East, Marlton, NJ 08053, (609) 596-2500.

Northern Telecom/General Electric Cellular System, 1201 Arapaho Rd., Richardson, TX 75081, (214) 234-7500.

Novell, Inc. 1170 N. Industrial Park Dr., Orem, UT 84057, (801) 226-8202.

Oki Telecom, 5901 Peachtree-Dunwoody Rd., Suite 100, Atlanta, GA 30328, (800) 554-3112.

Panafax Corporation, 10 Melville Park Rd., Melville, NY 11747, (516) 420-0055.

Panasonic Co., Telephone Products Division, 1 Panasonic Way, Secaucus, NJ 07094, (201) 348-7000.

Paradyne Corporation, 8550 Ulmerton Rd., Largo, FL 33540, (813) 536-4771.

Penril DataComm, 207 Perry Parkway, Gaithersburg, MD 20877, (301) 921-8658.

Phone-Mate, Inc. 325 Maple Ave., Torrance, CA 90503, (213) 618-9910.

Pitney Bowes Facsimile Systems, 1515 Summer St., 5th Floor, Stamford, CT 06926, (203) 356-7178.

Plantronics Wilcom, P.O. Box 508, Laconia, NH 03246, (603) 524-2622.

Porta Systems Corp., 575 Underhill Blvd., Syosset, NY 11791, (516) 364-9300.

Power Conversion Products, Inc., 42 East St., Crystal Lake, IL 60014, (815) 459-9100.

Preformed Line Products Company, P.O. Box 91129, Cleveland, OH 44101, (216) 461-5200.

Prentice Corp., 266 Caspian Dr., Sunnyvale, CA 94088, (408) 734-9810.

Proctor & Associates Co., Inc., 15050 N.E. 36th, Redmond, WA 98052, (206) 881-7000.

Pulsecom Division, Harvey Hubbell, Inc., 2900 Towerview Rd., Herndon, VA 22071, (703) 471-2900.

Racal-Milgo, Inc., 8600 N.W. 41st St., Miami, FL 33166, (305) 592-8600.

Rapicom, Inc., 7 Kingsbridge Rd., Fairfield, NJ 07006, (201) 575-6010.

RCA Americom, 400 College Rd. East, Princeton, NJ 08540, (609) 734-4000.

RCA Cylix Communications Network, 800 Ridge Lake Blvd., Memphis, TN 38119, (901) 761-1177.

RCA Global Communications, 60 Broad St., New York, NY 10004, (212) 806-7000.

Reliance Comm/Tec, Lorain Products, 1122 F. St., Lorain, OH 44052, (216) 226-1122.

Reliance Comm/Tec, R-Tec Systems, 2100 Reliance Pkwy., P.O. Box 919, Bedford, TX 76021, (817) 267-3141.

Reliance Comm/Tec, Reliable Electric/Utility Products, 11333 Addison Dr., Franklin Park, IL 60131, (312) 455-8010.

Ricoh Corp., 7 Kingsbridge Rd., Fairfield, NJ 07006, (201) 575-6010.

Rockwell International Corp., Collins Transmission Systems Division, 1200 N. Alma Rd., Richardson, TX 75081, (214) 996-5000.

Rockwell International Corp., Switching Systems Division, 1431 Opus Pl, Downers Grove, IL 60515, (312) 852-5700.

Rolm Corporation, 4900 Old Ironsides Dr., Santa Clara, CA 95050, (408) 986-1000.

San/Bar Corporation, 9999 Muirlands Blvd., Irvine, CA 92714, (714) 855-9911.

Satellite Business Systems, 8283 Greensboro Dr., McLean, VA 22102, (703) 442-5000.

Satellite Transmission Systems, Inc., 125 Kennedy Dr., Hauppauge, NY 11788, (516) 231-1919.

Scientific-Atlanta, Inc., Box 105600, Atlanta, GA 30348, (404) 411-4000.

Seiscor Inc., P.O. Box 470580, Tulsa, OK 74147, (918) 252-1578.

Siecor Corporation, 489 Siecor Pk., Hickory, NC 28603, (704) 328-2171.

Siecor Fiberlan, Box 12726, Research Triangle Park, NC 27709, (919) 544-3791.

Siemens Corporation, Telephone Division, 1001 N.W. 58th St., Boca Raton, FL 33431, (201) 494-1000.

Sierra Electronic Division, Lear Siegler, Inc., 3885 Bohannon Dr., Menlo Park, CA 94025, (415) 321-5374.

Simplex Wire and Cable Company, P.O. Box 479, Portsmouth, NH 03801, (603) 436-6100.

SNC MFG. Co., Inc., 101 Waukau Ave., Oshkosh, WI 54901, (414) 231-7370.

Standard Wire and Cable Co., 2345 Alaska Ave., El Segundo, CA 90245, (213) 536-0006.

Stromberg Carlson Corp., 400 Rinehart Rd., Lake Mary, FL 32746, (305) 849-3000.

Sumitomo Electric U.S.A., Inc. Fiber Optics Division, 551 Madison Ave., New York, NY 10022, (212) 308-6444.

Sykes Datatronics, Inc., 159 Main St. E., Rochester, NY 14604, (716) 325-9000.

Systron-Donner Corp., 2727 Systron Dr., Concord, CA 94518, (415) 671-6589.

Sytek, Inc., 1225 Charleston Rd., Mountain View, CA 94039, (415) 966-7333.

Tau-Tron, Inc., 27 Industrial Ave., Chelmsford, MA 08124, (617) 256-9013.

Telecommunications, Inc., 5 Research Dr., Shelton, CT 06484, (203) 926-2000.

Tekno Industries, Inc., 795 Eagle Dr., Bensenville, IL 60106, (312) 766-6960.

Tektronix, Inc., P.O. Box 500, Beaverton, OR 97077, (503) 644-0161.

Telco Systems Fiber Optics Corp., 33 Boston-Providence Highway, Norwood, MA 02062, (617) 769-7510.

Telco Systems, Inc., 1040 Marsh Rd., Suite 100, Menlo Park, CA 94025, (415) 324-4300.

TeleBit, Inc., 328 Eisenhower Lane, North, Lombard, IL 60148, (312) 932-9180.

Telesciences, Inc., 351 New Albany Rd., Moorestown, NJ 08057, (609) 235-6227.

Teletype Corporation, 5555 Touhy Ave., Skokie, IL 60076, (312) 982-2000.

Tellabs Inc., 4951 Indiana Ave., Lisle, IL 60532, (312) 969-8800.

Tel-Tone Corporation, 10801 120th N.E., Kirkland, WA 98033, (800) 426-5918.

Teradyne Central, Inc., 1405 Lake Cook Rd., Deerfield, IL 60015, (312) 940-9000.

Terracom, Division of Loral Corp., 9020 Balboa Ave., San Diego, CA 92123, (714) 278-4100.

TIE/communications, Inc., 5 Research Dr., Shelton, CT 06484, (203) 929-7373.

Timeplex Inc., 400 Chestnut Ridge Rd., Woodcliff Lake, NJ 07675, (201) 930-4600.

Times Fiber Communications, Inc., P.O. Box 384, Wallingford, CT 06492, (203) 265-8500.

Timeplex, Inc., 400 Chestnut Ridge Rd., Woodcliff Lake, NJ 07675, (201) 930-4600.

Tone Commander Systems, Inc., 4320 150th N.E., Redmond, WA 98052, (206) 883-3600.

Toshiba Telecom, 2441 Michelle Dr., Tustin, CA 92680, (714) 730-5000.

Triplett Corporation, One Triplett Dr., Buffton, OH 45817, (419) 358-5015.

TTI Telecommunications Technology, Inc., P.O. Box 3527, Sunnyvale, CA 94088-3527, (408) 735-8080.

Tymnet, Inc., 2710 Orchard Parkway, San Jose, CA 95134, (408) 946-4900.

Ungermann Bass Inc., 2560 Mission College Blvd., Santa Clara, CA 95050, (408) 496-0111.

Uniden, 15161 Triton Lane, Huntington Beach, CA 92649, (714) 898-0576.

Uninet, Inc., 10951 Lakeview Ave., Lenexa, KS 66219, (913) 541-4400.

Universal Data Systems Inc. 5000 Bradford Dr., Huntsville, AL 35805, (205) 837-8100.

U S Telecommunications, 3118 62nd Ave. N, St. Petersburg, FL 33702, (813) 527-1107.

Valtec, 99 Hartwell St., West Boylston, MA 01583, (617) 835-6082.

W & G Instruments Inc., 119 Naylon Ave., Livingston, NJ 07039, (201) 994-0854.

Wang Laboratories, Inc., One Industrial Ave, Lowell, MA 01851, (617) 459-5000.

Wescom Telephone Products Division, Rockwell Telecommunications, Inc., 8245 S. Lemont Rd., Downers Grove, IL 60515, (312) 985-9000.

Western Union Telegraph Co., 1 Lake St., Upper Saddle River, NJ 07458, (201) 825-5000.

Wiltron Company, 805 E. Middlefield Rd., P.O. Box 7290, Mountain View, CA 94042, (415) 969-6500.

Xerox Corporation Information Products Division, 1301 Ridgeview Dr., Lewisville, TX 75067, (214) 412-7200.

Ztel, Inc., Andover Industrial Center, York St., Andover, MA 01810, (617) 470-290

INDEX

A

Absorption, 422, 437
Access, 231, 531–32, 603–4
  to local networks, 255–59
Accounting information, 332
Action Control Point (ACP), 647
Adaptive differential pulse code modulated transcoder (ADPCM), 128, 135
Adaptive equalizer, 70, 99, 401, 413
Adaptive predictor, 127
Add-on conferencing, 50
Addressing, 213, 227, 474–75
  signals, 215, 220–21, 227
Aerial cable, 191
Aerial splice cases, 194
Alarms, 277–78, 411–12
  for CATV, 550
  equipment, 383–84
Alerting, 213, 227
Algorithms, 370, 372
Alternate routing, 12, 262, 366–67
Alternating current, 669–78
American National Standards Institute (ANSI), 19
American Standard Code for Information Interchange (ASCII), 61–62, 99
Amplitude distortion, 52
Amplitude modulation (AM), 680
Analog, 9–10
  carrier technology, 141–47
  circuit, 604–8
  to digital connectors, 147–49
  versus digital tests, 625
  facilities, 77–78, 99
  modulation, 395
  multiplex equipment manufacturers, 155
Angle of acceptance, 421, 437
Answering sets, 182
Antennas, 393–94, 404–5
Aspect ratio, 558
Asynchronous data transmission, 66, 99
AT&T divestiture, 15–17, 255–59, 641, 659–61
AT&T Information Systems Net 1000, 486
Audible ring, 216, 227
Automatic call distributors (ACD), 304

Automatic message accounting (AMA), 227, 262, 278
Automatic number identification (ANI), 278, 287
Automatically identified outward dialing (AIOD), 633
Availability, 78–79
Avalanche photo diode (APD), 418, 437

B
Backoff algorithm, 507, 532
Back-to-back channels, 120, 135
Balance, 36–37, 52
Balanced modulator, 153
Balancing network, 52
Bandpass, 52
Bandwidth, 99, 392, 428
  requirements, 316
Baseband, 402–3, 509–11, 532
Baud, 65, 99
Bell, Alexander Graham, 156, 417
Binary digit (bit), 60, 99
Bipolar coding, 136
Bipolar violation, 119, 136
Bit, 99
  error rate (BER), 78, 99, 396–97
  error rate test, 619
  overhead, 67
  robbing, 136
  stuffing, 136
Blockage, 362–63
Blocked calls cleared (BCC), 355
Blocked calls delayed (BCD), 356, 375
Blocked calls held (BCH), 355, 376
Blocked calls released (BCR), 376
Blocking switching network, 183, 237
Branch feeder, 209
Branching filters, 405, 413
Bridge, 99, 517, 532
Bridged tap, 196, 209
Bridgers, 544, 558
Broadband, 509–11, 532
Busy hour, 354, 361–62, 376
Byte, 60, 99

C
Cable, 191–92, 388–89
  characteristics, 192–94, 436
Cable—*Cont.*
  common faults, 610
  distribution, 196–99, 209
  feeder, 196–99
  pressurization, 203–208
  racking, 378–79, 389
  structural quality, 208
  testing circuit, 610
Cable television; *see* CATV
Call forwarding, 283
Call hold, 167
Call pickup, 166–67
Call-processing features, 282
Call progress, 215, 227
  supplies, 254, 262
Call store, 269, 287
Call transfer, 282
Call waiting, 282
Call warning tone, 328, 351
Capacitance, 672–74
Capacity, 231, 287
Carbon block, 204–5, 209
Carrier, 680
Carrier sense multiple access with collision detection (CSMA/CD), 507, 532
Carrier to noise ratio, 454, 467
CATV, 535–36, 541–46
  feeder and drop system, 544
  head-end equipment, 541–43, 558
  pay television equipment, 544–45
  trunk-cable systems, 543–44
  two way systems, 545–46
CCIS network, 221–22
CCITT X.25 Protocol, 483–84
Cells, 583–84, 598
Cell-site, 585–89, 599
  controller, 585, 599
Cellular geographic serving area (CGSA), 599
Cellular radio switching office (CRSO), 585, 599
"Cent call seconds" (CCS), 356
Centralized algorithm, 372
Centralized attendant service (CAS), 304
Central office; *see also* Digital central offices (DCOs)
  common equipment, 280

Central office—*Cont.*
power plant, 385
service features, 282
switching machines, 266
Central processing unit (CPU), 99
Centrex, 283, 287
Channel banks, 113–17, 136, 120
Channel service unit (CSU), 123, 136, 477
Channel signaling, 217
Character parity, 80
Chrominance signal, 539, 558
Circuit switching, 58–59, 72, 99, 231
Circuit switched network systems, 230–64; *see also* Network
comparison of digital and analog, 251–52
direct control switching system, 239–41
standards, 259–62
Cladding, 420, 437
Clear channel, 110, 136
Clear to send (CTS), 373
Closed circuit television (CCTV), 558
C message weighting, 52
C notched noise, 626
Coaxial cable, 513–14
Code, 301
Code blocking, 350–51
Codec, 287, 558
Code conversion, 288, 331
Coding, 61–62
Coin telephone interface, 278–79
Collimate, 437
Collision window, 507, 532
Common bell, 167
Common carrier, 71–72
Common channel signaling, 221–22, 227, 321, 351
Common control, 236, 262
central offices, 241–46
Common equipment, 378–90
applications, 387–89
evaluating, 387–89
manufacturers of, 390
standards, 386
Companding, 109, 136
Competition in toll networks, 256–58
Composite (CX) signaling, 227
Communicating word processor, 576
Communications Network Service (CNS), 455
Communications Satellite Corporation (CONSAT), 441
Community antenna television; *see* CATV
Community dial office (CDO), 288
Computer branch exchange (CBX), 316
Concentration, 73–76, 200, 209, 286, 288
ratio, 209, 273, 288
Conditioning, 53, 499
Contention, 506–7, 532
Continuous wave (CW), 680
Control equipment, 383–84, 389
Control systems, CATV, 550–51
Conventional mobile telephone technology, 580–83
Converters, 386, 389
Core, 437
Cross, 626
Crossbar analog, 236
Crossbar switching machines, 242–46, 262
Cross-connect, 136
Cross-polarization, 413
discrimination (XPD), 413
Crosstalk, 53
Current, 666
Customer premises equipment, 2–4, 156–57, 185
Cutoff wavelength, 429
Cyclical redundancy checking (CRC), 82, 99

D

Data base integrity, 273
Data base updates, 315
Data circuit-terminating equipment (DCE), 60, 100
Data circuit testing, 617–21
Data communications, CATV, 550
Data communications networks, 471–501
access methods, 482–83
applications, 494–99

Data communications networks—*Cont.*
availability, 499
costs, 495
evaluations, 494–99
features, 498–99
interface, 497
line conditioning, 499
standards, 491–94
vendors, 501
Data communications systems, 56–104
applications, 94–98
data network, 57–60
data network facilities, 71–78
evaluating, 68–88
fundamentals, 60–88
manufacturers, 104–5
modulation methods, 65–66
speeds, 62–65
standards, 88–94
Data compression, 70, 100
equipment, 126–27
Datagram, 100, 481, 499
Data line monitor, 626
Data multiplexing, 9
Data network, 57–60
design, 370–73
Data Network Identification Code (DNIC), 499
Data over voice (DOV), 71, 100, 295–96, 316
Dataphone Digital Service (DDS), 397, 477
Dataport service units (DSU), 115, 136
Data terminal equipment (DTE), 60, 100
Data traffic, 473–74
Data under voice (DUV), 320, 397, 477
Decibel, 667–68
Dedicated circuit, 4, 10
Dedicated facilities, 72–73
DeForest, Lee, 139
Delay systems, 362–63
Delta modulation, 128–29, 136
Demand assigned multiple access (DAMA), 450, 468
Deregulation, 659–61; *see also* AT&T divestiture
Dialing frequency combinations, 159
Dials, 160
Dielectric constant, 674
Dielectric printing, 570
Digital carrier, 106–38; *see also* T carrier
applications, 131–35
digital transmission facilities, 111–12
disadvantages, 107
manufacturers, 138
standards, 129–31
technology, 107–12
T-1 carrier system, 112–20
Digital central offices (DCOs); 265; *see also* Central office
line circuit architecture, 274, 288
mainstream and administrative features, 271–73
memory, 267–69, 288
technology, 266–69
Digital communications, 654–55
Digital cross-connect panel, 117, 136
Digital cross-connect systems (DCS), 120–21
Digital facilities, 77–78, 100
Digital modulation, 395–96
Digital signal cross-connect (DSX), 136
Digital signal hierarchy, 121–22
Digital signal timing, 112
Digital speech interpolation (DSI), 150
Digital termination system (DTS), 406
Digital transmission concept, 9–10
Digital transmission systems
availability of special service features, 132–33
backup power, 134
compatibility, 134–35
density, 134
maintenance features, 133
operating temperature range, 134
power consumption, 133
reliability, 132
Diplexer, 591, 599
Direct broadcast satellite (DBS), 457–58, 468
Direct burial, 191
Direct control, 236
Direct control switching systems, 239–41, 262
Direct current, 669–78

Direct inward dialing, 283
Directional couplers, 405, 413, 544, 558
Direct outward dialing (DOD), 283
Direct station selection, (DSS), 303–4
Direct-to-line multiplex unit, 145, 154
Direct trunks, 233, 262
Discriminator, 683
Disgroup, 113, 137
Dispersion, 421–22, 437
Dispersive fade margin, 401, 413
Distortion, 401
Distributed algorithm, 372
Distributed processing, 288
Distributed switching, 275, 288
Distributing frame, 117–18, 388–89
Diversity, 413; *see also specific types*
Divestiture; *see* AT&T divestiture
Downlink, 442, 468
Downstream channel, 558
Driver, 679–80
Dropwire, 209
DTMF, 220–21
Dual cable system, 511
Dual tone multifrequency (DTMF), 159, 185
Duplex, 218, 227
Dynamic overload control (DOC), 282, 288

E

Early Bird, 445
Earth station technology, 448–54, 468
Ease of addressing, 231
Echo, 26–29, 53
  cancelers, 28–29, 53
  checking, 80–81, 100
  control, 41
  four-wire terminating set, 27–28
  return loss, 28, 34, 53
  sidetone, 27
  suppressors, 28, 53
  via net loss, 29
Effective isotropic radiated power (EIRP), 454, 468
Electricity, 663–83
  and magnetism, 665–69
Electrolytic printing, 570
Electromagnet, 665
Electronic filing, 305–6
Electronic Industries Association (EIA), 19
Electronic mail, 305, 316, 562
Electronic switching networks, 248–50
Electrosensitive printing, 569
Electrostatic printing, 570
Emergency reporting, 283
E&M signaling, 217, 227
Encoding, 568
End-to-end signaling, 226–27
End-to-end tests, 619
Entrance links, 406, 413
Envelope delay, 53, 607
Equal access, 258–62, 659
Erlang, A.K., 356
Erlangs, 356, 359, 376
Errors, 79–82, 100; *see also specific types*
Essential service and overload control, 271–72
Ethernet, 532
Exchange Carrier Standards Association (ECSA), 19
Expanded Binary Coded Decimal Interchange Code (EBCDIC), 61–63, 100
Exponential, 376
Express office repeater, 120
Extended super frame (ESF), 100, 137
External interfaces, 299–300

F

Facility, 53
Facsimile transmission, 561–77
  advantages and disadvantages, 562
  applications, 574–76
  evaluation, 574–76
  group, 4, 571–72
  machine characteristics, 563–67
  manufacturers, 577
  standards, 572–74
  technology, 563–72
  telecommunications features, 570–71
Fade margin, 401, 410–11, 414
Fading, 394, 414
"Fast busy" tone, 356
Fast select, 481, 499
Fault detection and correction, 271; *see also* Alarms

Fiber optics, 514–15, 656; *see also* Lightwave communications
diagram of system, 419
terminal equipment, 424–26
Filters, 677–78
Fixed frequency modems, 512
Flooding algorithm, 372
Flow control, 360, 376
Footprint, 441, 468
Foreign EMF, 626
Foreign exchange (FX), 115, 137
Forward error correction (FEC), 70, 100
45BN carrier, 141
Four-wire circuit, 53
Four wire terminating set, 53
Frame, 110, 537, 558
Free space, 417, 429–30
Frequency, 221, 669
agile modems, 512, 584, 599
allocations for CATV, 511
band, 412
diversity, 398, 414
Frequency division multiple access (FDMA), 450, 468
Frequency division multiplexing, 7, 139–55
applications, 152–53
standards, 151–52
Frequency modulated receiver (FMR), 404
Frequency modulated transmitter (FMT), 404
Frequency modulation (FM), 681
Frequency multipliers, 679
Frequency synthesizer, 585
Frogging, 147, 154
Front-end processor, 100
Full duplex mode, 66, 100

G

Gas tubes, 205, 209
Gateway, 100, 299–300, 517, 533
Gauge, 209
Gel cells, 385
Generic program, 247, 262, 267–68
Geosynchronous, 468
Glare, 228
Glass fiber, 418
Graded index, 421, 438
Grade of service, 53, 362–63, 376
Ground, 626
Ground start, 218, 228
Group, 154
Group 4 facsimile, 571
GTE Telenet, 485–86

H

Half-duplex mode, 66, 100
Hand off, 585, 599
Harmonics, 671
distortion, 608
ringing, 162, 185
Head-end equipment, 511
Heat coils, 205, 209
Hercules Corporation, 460–66
Heterodyning, 402–3, 414, 682
High definition television (HDTV), 548–49, 558
Higher order multiplexing, 9
High-level data link control (HDLC), 67
High-loss PBX switching networks, 50–51
High-speed switching capability, 351
High-usage groups, 263
Holding time, 368
Hot standby diversity, 398, 414
Hot standby redundancy, 270
Hub head ends, 543, 558
Hub polling, 499
Hundred call seconds (CCS), 376
Hybrid, 53, 170
Hz, 53

I

IEEE 802 committee, 518–23
Impedance, 675
"Improved Mobile Telephone Service" (IMTS), 582–599
Impulse noise, 607, 626
Incoming matching loss, 281
Independent company (IC), 263
Inductance, 674–75
Inductive reactance, 675
Information revolution, 651–52
Information transfer rate, 428

Insertion loss, 54
Inside wiring, 162, 185
Institute of Electrical and Electronic Engineers (IEEE), 19
  802 committee, 518–23
Integrated Services Digital Network (ISDN), 18, 316, 472, 648–51
  services, 650–51
  Integrated voice/data, 316
  Intelligent channel banks, 116–17
  INTELSAT VI, 440
  Interbay wiring, 378–79
  Interchange carriers (ICs), 15–17
  Interconnectibility, 231
  Interconnection records, 632
  Interexchange trunks, 5, 263
  Interfaces, 296–301, 306, 312–13; *see also special types*
  microwave radios, 412
  tests, 618
Interference, 402
Interframe encoding, 547, 558
Interlaced scanning, 538
Intermediate frequency (if), 682
Intermodulation distortion, 54
International Maritime Satellite Service (INMARSAT), 456–57, 468
International Radio Consultative Committee, 18
International Standards Organization, (ISO), 19
International Telecommunications Satellite Organization (INTELSAT), 441
International Telecommunications Union (ITU), 18
International Telephone and Telegraph Consultative Committee (CCITT), 18
Interoffice connection, 214
Interoffice trunks, 4–5
Intertoll trunks, 38–39, 263
Intracalling, 200–201, 209
Intraframe encoding, 547, 558
Inverter, 384, 389
Isolated algorithm, 372

J

J carrier system, 139
Jitter, 119, 137
Jumbo group, 154
Jumper, 389

K

K carrier, 140
Key telephone system, 166–77, 185
  cost, 183–84
  equipment, 2–4
  versus PBX, 183
  power failure, 184
  size, 185
  typical features, 171
  wired versus stored program logic, 184–85
KHz, 54
Korzybski, Alfred, 1

L

Laser, 418
L carrier, 140
  microwave radio, 146
L Coaxial transmission lines, 145–46
Least cost routing (LCR), 301, 316
Level, 54
Light emitting diode (LED), 425
Lightguide cables, 420–24
Lightwave communications, 416–39
  applications, 431–37
  design criteria, 427–29
  evaluation, 435–37
  manufacturers, 439
  standards, 429–31
  technology, 417–27
Lines, 232
  circuit functions, 252–53
  circuits, 266
  concentrator, 200
  conditioning, 88, 100
  equipment features, 273–75
  insulation tests (LIT), 272–73, 288, 611–12, 627
  interfaces, 297
  maintenance features, 272–73
  status, 228
Link-by-link signaling, 226, 228
L multiplex (LMX), 154
Load coils, 675

Loading, 35, 54
Load/service curve, 361
Local access transport areas (LATAs), 17, 259, 263
Local area data transport (LADT), 487–90, 500
Local area networks (LANs), 2, 373, 502–34
  access method, 505–6, 532
  applications, 524–32
  compatibility, 530
  contention access, 506–7, 532
  costs, 527–28
  evaluating, 526–32
  functions of, 503
  interconnecting, 517–18
  manufacturers, 534
  modulation methods, 509–10
  network topology, 504–5
  noncontention access, 507–8
  off-net communications, 529–30
  reliability, 529
  security, 530–31
  size of network, 516–17
  standards, 518–24
  traffic characteristics, 528–29
  vendor support, 527
Local central offices, contrasted with PBX, 292–95
Local interoffice trunks, 41
Local loop, 4
  changes in, 655–57
Local measured service (LMS), 280–81
Local switching systems, 4, 265–89; *see also individual systems*
  applications, 285–86
  digital remote line equipment, 276
  distributed switching, 275
  local central office equipment features, 277–84
  manufacturers, 289
  standards, 284–85
Local test desk (LTD), 609–11, 627
L multiplex system, 142, 144–45
Long-haul carrier, 141
Longitudinal redundancy checking, 81
Loopback test, 618–19, 627
Loop loss, 35
Loop start, 228
Loop timing, 137
Loss, 54
  budget, 428–29, 438
  measurement, 604–5
  systems, 362–63
L-to-T connectors, 147–49, 154
L-to-T transmultiplexers, 147–49
Luminance signal, 539, 559

M

Magnetism, 665–69
Main feeder, 209
Maintenance Control Center, (MCC), 288
Manufacturers, 689–99; *see also specific products*
  current information, 688–89
Marker, 263
Martin Marietta, fiber optic network, 431–35
Master group, 142, 144, 154
Material dispersion, 422, 438
Meantime between failures, (MTBF), 78, 101, 132
Meantime to repair (MTTR), 78, 101
Media access controller, 533
Memory, 267–69
Message switching, 58–59, 72, 101, 231
Message weighting response curve, 33
Messenger, 209
Microwave radio, 391–415, 515
  applications, 409–13
  characteristics, 391–94
  environmental factors, 412–13
  evaluation, 409–13
  impairments, 399–400
  manufacturers, 415
  standards, 407–9
  technology, 394–407
  test equipment, 413
Milliwatt, 54
Mixed media, 577
Mobile radio, 578–600
  applications, 597–98
  cellular technology, 583–85
  manufacturers, 600
  mobile data transmission, 594–95

Mobile radio—*Cont.*
  radio paging, 594
  standards, 595–97
Mobile telephones, 578–79, 589–93; *see also* Mobile radio
Modes, 420, 438
Model dispersion, 422, 438
Modeling, 376
Modems, 60, 101
  compatibility, 68–69
  features, 69–70
  pooling, 300, 316
  selection, 68–71
  special types, 70–71
  turnaround time, 500
Modified chemical vapor deposition (MCVD), 422–23, 438
Modulation, 395–96, 680–81
  methods, 65–66
M1-3 multiplexer, 123
Morse, F.B., 56
Multidrop polled network, 371–72
Multifrequency pulses, 215, 228
Multiline hunt, 283–84, 288
Multimode fiber, 420–21
Multipair cable, 193
Multipath distortion, 401–2
Multipath fading, 401–2, 414
Multiple access, 533
Multiple channel analog, 200
Multiple vendors, 641–42
Multiplexers, 6–9, 74–76, 101, 122–25
  earth station, 449
  interface, 403–4
Mutual inductance, 675

N

National Television Systems Committee (NTSC), 537–39
N carrier lines, 147
Network
  as an ambiguous term, 230–31
  modeling, 350
  setup time, 495
  statistical information, 331
  terminology, 232
Network Access Center (NAC), 455
Network Access Unit (NAU), 503–533
Network architecture, 232–36
  predivestiture, 255–56
Network channel terminating equipment (NCTE), 172–73, 185
Network configuration, 76
Network control centers (NCC), 491
Network control point (NCP), 647
Network design concepts, 353–77
  administration, 373–76
  automatic network controls, 375
  data network design, 370–73
  design problem, 354–70
  grade of service, 362–63, 376
  hourly variation in calls, 355
  standards, 375
  topology, 368–69
Network intelligence, 645–48
Network interface device (NID), 613–14, 627
Network management, 282, 630–43
  applications, 638–43
  evaluation, 639–43
  line usage measurements, 633
  manufacturers, 643
  network reconfiguration, 637
  preventing failures, 637–38
  service monitoring, 637
  standards, 638
  system records, 630–32
  system usage management, 632–33
  telecommunications costs, 635–36
  traffic usage measurements, 633–34
  trouble handling, 634–35
Network management control, 349–50
Network Management Control Center (NMCC), 332–33, 352
Network terminal number (NTN), 500
Node, 232, 263
Nonblocking switching network, 183
No signaling, 222–23
Noise
  definition, 54
  measurement, 605–7
  notched noise, 607
  sources, 39–41
N-1 carrier, 141, 148
Numbering systems, 13–15

O

Octet, 60, 101
Office location, 52
Ohm's law, 666
On hook/off hook, 159, 186
Open, 627
Open Systems Interconnection (OSI), 19
Opinion polling, 551
Oscillator, 604, 627, 675–76
Out-of-band signaling, 219, 228
Outside plant, 189–211
  applications, 207–8
  manufacturers of products, 211
  standards, 206–7

P

Packet, 101
Packet assembler/disassembler (PAD), 54, 101, 478–79, 500
Packet network technology, 478–84
Packet switching, 59–60, 72–73, 101, 231, 485
Pair gain, 202, 209
Parallel resistance, 668–69
Parameters, 267, 288
Parity checking, 79–80, 101
Patch, 137
Path engineering, 400, 412
Partial dial, 358–376
PBX (private branch exchange), 2, 290–318
  applications, 307–16
  contrasted with Local Central Offices, 292–95
  diagnostic capability, 315
  evaluation considerations, 311–16
  fourth generation, 293
  interfaces, 296–301
  versus KTS, 183
  to LAN interface, 300
  manufacturers, 318
  and office automation, 311–12
  principle features, 301–6
  program storage, 314
  standards, 306–7
  technology, 291–92
  trunk interfaces, 297–98
  voice-data integration, 295
PBX (private branch exchange)—*Cont.*
  voice features, 302–3
PCM test sets, 620
Peak-to-average ratio, 608, 627
Percussive printing, 570
Permanent signal tones, 254, 263
Permanent virtual circuit, 500
Personal computers, 652–54
  network (PCN), 516–17
Personal identification number (PIN), 16, 262, 327–28
  diode, 418, 438
Phase, 670
  jitter, 608
  shift keying (PSK), 396, 414
Phasing, 565, 577
Photographic printing, 570
Photophone, 417
Picture element (Pixel), 539–559
Pilot, 154
Ping pong, 297, 316
Point-to-point circuit, 10, 57
Point-to-point network technology, 474–77
Poisson distribution, 358–59, 376
Poisson, S.D., 358
Polarity, 669
Polled multidrop network, 58, 101
Polling, 475–76, 500
Positive action digit, 328, 352
Power, 667
  equipment, 384–85, 388
  fail transfer, 186
Preform, 423, 438
Pressurization, 210
Printing; *see also specific methods*
  facsimile transmission, 569, 577
Private automatic branch exchange (PABX), 317
Private branch exchange; *see* PBX
Private line circuit, 4
Private line signaling, 222–23
Private telephone systems, 17
Program (PG) channel units, 116
Program store, 288
Propagation delay, 101
Protection systems, 398–99, 411
Protector frames, 381, 388

Protectors, 162, 186, 203–6, 210
Protocol, 82–86, 102, 301
  analyzers, 619–20
  X.25, 86–87
Pulse amplitude modulated (PAM), 107, 137
  analog, 236
  digitals, 237
  networks, 249–50
Pulse amplitude switching, 248–49
Pulse code modulation (PCM), 10, 106–38

Q

Quadrature amplitude modulation (QAM), 396, 414
Quantizing noise, 108, 137
Queuing, 330, 348–49, 352, 355–56

R

Radio, 678–83
Radio frequency (rf), 391–92
  modem, 510
Radio relay equipment, 448
Rain absorption, 400–401, 453
Range extenders, 202–3, 210
Reactance, 674
Read access memory (RAM), 314
Read only memory (ROM), 314
Received signal level (RSL), 401, 414
Receivers, 682–83
Receiver sensitivity, 414
Recent change, 269, 288
Recorded announcements, 254
Redundancy, 269–71, 289
Reed relay analog, 236
Reeves, Alec, 106
Reference noise (rn), 54
Refractive index, 420, 438
Registration, 186
Relay rack, 389
Relays, 665–66
Reliability, 78–79
Remote access, 328–30, 351–52
Remote line capability, 287
Remote line concentrator, 201, 210
Remote line switch, 289
Remote line unit, 289
Remote office test line (ROTL), 323, 352
Reorder, 215, 228
Repeat coils, 677
Repeater, 54
Resistance, 666
  design, 194–96, 210
Resolution, 577
Resonance, 675
Responder, 323, 352
Response time, 495–97, 500
Restriction, 317
Return loss, 54, 607–8
Ribbon cable, 513
Ring, 159, 186
Ringdown circuits, 223
Ringer isolators, 162, 186
Ringing supplies, 254, 383
Ring topology, 505
Roamers, 583, 599
Robustness, 231
Rockwell International Telecommunications Network, 333–48
Rootmean square (RMS), 670
Rotary line group, 283–84
Routing, 330–31, 350–51
Run-length encoding, 566, 577

S

San Diego State University, network, 525–26
Satellite circuits, 73
Satellite common carriers applications, 131–35
Satellite communications, 440–70
  advantages, 442–43
  applications, 460–67
  attitude control apparatus, 446–47
  business service, 455–56
  controller (SCC), 449, 468
  evaluations, 466–67
  frequency bands, 441
  interference, 454
  limitations, 443–44
  maritime service, 456–57
  physical structure, 445
  power supply, 447
  satellite circuit, 445

Satellite communications—*Cont.*
satellite transmission, 451–53
standards, 458–60
technology, 444–45
vendors, 470
and video systems, 536
Satellite delay compensator, 468
Satellite services, 51–52
Scanners, 567–69, 577
Scattering, 422, 438
Sectoring, 599
Security blanking, 328, 352
Selective signaling, 222–23, 228
Sender, 228
Serial interface, 61, 102
Series circuit, 668–69
Service circuits, 253
Service Management System (SMS), 647
Shared-load redundancy, 270
Sheath, 210
Shielding, 35–36
Short, 627
Short haul carrier, 141, 146–47
Short haul digital microwave, 407
Sidebands, 680–81
Sidetone, 54
Singing, 54
Singing return loss, 54
Single cable system, 511
Single-frequency signaling, 219–20, 228
Single mode fiber, 420–21, 438
Single sideband suppressed carrier (SSBSC), 141–43
Signaling systems, 212–29, 323–24
applications, 224–27
common channel signaling, 221–22, 227
direct current, 218–19
E&M, 217, 227
evaluating equipment, 224–27
interoffice connection, 214
irregularities, 217–18
manufacturers of equipment, 229
overviews, 213–17
private line, 222–23
standards, 223–24
technology, 213–17
trunk signaling systems, 219–23
Signal processing, 450–51
Signal-to-noise ratio, 54
Simulation, 367–68, 376
Software defined network, 646
Solenoids, 665–66
Spacecraft Switched Time Division Multiple Access (SSTDMA), 450, 468
Space diversity, 397–98, 414
Space division, 249–50, 263
Special feature telephones, 180–81
Special-purpose trunks, 41–42
Spectral efficiency, 410, 414
Speed, 301
Speed calling, 282
Spin stabilized satellite, 446–48
Splicing, 193–94
Split channel modem, 66, 102
Split pair, 210
Splitters, 544, 559
Spot beam antenna, 468
Standard equipment
applications, 179–85
evaluation, 179–80
standards, 176–79
Standard Metropolitan Statistical Area (SMSA), 599
Star couplers, 427, 438
Static algorithm, 372
Station equipment, 2–4, 156–88
manufacturers, 187–88
Station keeping, 447–48, 469
Station message detail recording (SMDR), 302, 317, 633
Stations, 232
Station wiring limits, 315–16
Station wiring plans, 173–76
Statistical multiplexing, 102; *see also* Multiplexers
Step-by-step central office, 240, 263
Step-by-step system, 11
Step index, 421, 438
Store and forward, 58–59
switching, 72, 102
Stored program control (SPC), 236, 246–52, 263
Strowger, Almon B., 11
Strowger system, 11

Subscriber carrier, 199–202
Subscriber line usage, 281
Subscriber loop, 4, 189, 210, 609–14
  evaluating equipment, 208
  protection, 208
  supporting structure, 191–92
  transmission, 34–37
Sun transit outage, 454, 469
Supergroup, 154
Superheterodyne technology, 682
Superimposed ringing, 186
Super trunk, 543, 559
Supervising, 212–13, 228
Supervisory audio tones (SAT), 589
Supervisory signals, 167
Suppression, 360–61, 377
Switched access, 627
Switched facilities, 72–73
Switched 56 kb/s Service, 490–91
Switched virtual circuit, 500
Switching hierarchy, 13–14
Switching networks, 236–38, 313–14
Switching systems, 10–13, 231, 234–36
  common control, 11–12
  computer controlled, 12
  crossbar, 12
  digital central offices, 12–13
  early systems, 11
  panel system, 12
Synchronizing methods, 66–68, 102
Synchronous redundancy, 270
System gain, 404, 414
System records, 630–32
Systems Network Architecture (SNA), 18, 94–99, 121

T

Talk-off, 220, 228
Tandem switching, 5, 233, 263, 319–52
  applications, 333–51
  evaluation, 348–51
  features, 322–32
  manufacturers, 352
  private, 327–32
  public, 322–27
  standards, 333
  tandem trunking facilities, 320
  technology, 320–21
Tandem switching—*Cont.*
  wideband switching, 321
Tandem trunks, 263
Taps, 544, 559
T carrier; *see* Digital carrier
T carrier data compression systems, 125–29
  line coding, 126
T carrier lines, 118–20
Technical control, 621, 627
Technical information, sources of, 684–87
Telecommunications
  and the computer, 56–57
  future development in, 644–62
  introduction to, 1–21
  major systems, 2–6
  manufacturers and vendors, 688–99
  and principles of electricity, 663–83
Telemetry equipment, 447
Telephones
  evaluation, 180
  special features, 180–81
  transmission quality, 181–82
Telephone set technology, 157–73
  coin telephones, 163–64
  cordless telephones, 164–65
  elements, 157–62
  multiple line equipment, 166–77
  party lines, 160–62
  protection, 162–63
  voice recording equipment, 165–66
Teletext, 549–50, 559, 577
Telstar 1, 440
Terminal equipment, 2–4
Terminals, 198, 210
Terrestrial circuits, 73
Testing circuits, 254–55
Testing principles, 601–29
  analog circuit, 604–8
  applications, 623–26
  automatic testing, 611–12
  data circuit testing, 617–21
  evaluation, 624–26
  manual loop tests, 613
  manufacturers, 628–29
  standards, 621–23
  test access methods, 603–4

Testing principles—*Cont.*
  trunk transmission measurement, 614–17
Testshoe, 627
Thermal noise, 54
Three-axis stabilization, 447, 469
Three-way calling, 282
Throughput, 87–88, 102, 515–16
Tie trunk, 317
Time assignment speech interpolation (TASI), 149–51
Time division multiple access (TDMA), 450, 469
Time division multiplexing, 7–9, 102; *see also* Multiplexers
Time division switching, 263
Time domain reflectometer (TDR), 613, 627
Time slot interchange (TSI), 250
Tip, 159, 186
Token, 533
Token passing networks, 373, 508, 533
Toll connecting trunks, 37–38
T-1 carrier system, 112–20
  special transmission functions, 115–16
T-1 data multiplexers, 122–25
T-1 multiplex, evaluating, 135
Tone supplies, 383
Topology, 76, 102, 368–70
Toroidal load coil, 199
Towers, 404–5
Traffic engineering, 353–77; *see also* Network design concepts
Traffic load, 356–61
Traffic measurement, 363–64
  equipment, 281
Traffic tables, 364
Traffic usage recording (TUR), 363–64
Transceiver, 510
Transcoder, 137
Transducers, 157, 186
Transformers, 675–77
Translation, 263, 269, 350–51
Transmission, 54, 315
  equipment, 5–6
  only to channel units, 116
Transmission concepts, 22–55
  applications, 48–52
  traps, 50–52
Transmission design, 30–43
  echo return loss, 34
  insertion loss, 33
  level points, 32–33
  loss and noise grade of service, 42–43
  quality, 31–32
  reference noise, 33–34
  subscriber loop transmission, 34–37
Transmission impairments, 23–30
  amplitude distortion, 29–30
  bandwidth, 25–26
  echo, 26–29
  envelope delay, 30
  noise, 24–25
  volume, 23–24
Transmission level point (TLP), 54
Transmission measurements, 43–45
  envelope delay, 45
  loss measurement, 43–44
  noise measurements, 44
  return loss measurements, 44–45
Transmission media, 512–13
Transmission performance, 275
Transmission units (TUs), 455
Transmitters, 679–80
Transmultiplexer, 154
Transponder, 445–46, 469
Trunk circuits, 253, 266
Trunk equipment
  features of, 277
  maintenance features, 272
Trunk maintenance test lines, 324
Trunks, 4–6, 37–39, 41–42, 232, 297–99; *see also specific types*
  final, 13
  high usage, 13
Trunk testing, 614–17
Twisted pair wire, 513
Tymnet, 486

U

Uplink, 442, 469
Upstream channel, 559
User-dialed billing arrangement, 326–27

V

Vacuum tube, 139
Value added carrier, 71–72, 102
Value added networks, 484–87, 500
Vendors, 689–99; *see also specific products*
Vertical blanking interval, 539, 559
Vertical redundancy checking (VRC), 79–80
Via net loss (VNL), 39, 54
Video compression techniques, 536–37, 559
Video conferencing, 551–53
Video meeting services, 537
Video systems, 535–60
  applications, 555–57
  compression, 546–48, 556–57
  evaluation, 555–57
  freeze-frame video, 548, 558
  manufacturers, 560
  services and applications, 549–53
  standards, 553–55
  technology, 537–41
Videotex services, 472, 500
Virtual circuits, 59–60, 102, 481–82, 500
Voice channel attenuation, 30
Voice/data integration
  in the switching system, 657–58
  in transmission facilities, 658–59
Voice filters, 220
Voice frequency terminating and signaling units, 225
Voice mail, 166, 186, 302, 317
Voice sampling, 109
Voltage, 666
Volume unit, 55
Voting receivers, 580, 599

W

Waiting time, 357
Watts, 667
Waveguides, 404–5, 414
Wavelength, 436–37, 669
Wavelength division multiplexing (WDM), 425–27, 437–38
White noise, 55
Wink, 215, 228
Wire center, 4
Wireline, 599